T0189718

Studies in Systems, Decision and Control

Volume 3

Series editor

Janusz Kacprzyk, Polish Academy of Sciences, Warsaw, Poland
e-mail: kacprzyk@ibspan.waw.pl

For further volumes:
http://www.springer.com/series/13304

Maciej Ławryńczuk

Computationally Efficient Model Predictive Control Algorithms

A Neural Network Approach

Maciej Ławryńczuk
Institute of Control and Computation
Engineering
Faculty of Electronics and Information
Technology
Warsaw University of Technology
Warsaw
Poland
e.mail: M.Lawrynczuk@ia.pw.edu.pl
www.ia.pw.edu.pl/~maciek

ISSN 2198-4182 ISSN 2198-4190 (electronic)
ISBN 978-3-319-35021-9 ISBN 978-3-319-04229-9 (eBook)
DOI 10.1007/978-3-319-04229-9
Springer Cham Heidelberg New York Dordrecht London

© Springer International Publishing Switzerland 2014
Softcover reprint of the hardcover 1st edition 2014
This work is subject to copyright. All rights are reserved by the Publisher, whether the whole or part of the material is concerned, specifically the rights of translation, reprinting, reuse of illustrations, recitation, broadcasting, reproduction on microfilms or in any other physical way, and transmission or information storage and retrieval, electronic adaptation, computer software, or by similar or dissimilar methodology now known or hereafter developed. Exempted from this legal reservation are brief excerpts in connection with reviews or scholarly analysis or material supplied specifically for the purpose of being entered and executed on a computer system, for exclusive use by the purchaser of the work. Duplication of this publication or parts thereof is permitted only under the provisions of the Copyright Law of the Publisher's location, in its current version, and permission for use must always be obtained from Springer. Permissions for use may be obtained through RightsLink at the Copyright Clearance Center. Violations are liable to prosecution under the respective Copyright Law.
The use of general descriptive names, registered names, trademarks, service marks, etc. in this publication does not imply, even in the absence of a specific statement, that such names are exempt from the relevant protective laws and regulations and therefore free for general use.
While the advice and information in this book are believed to be true and accurate at the date of publication, neither the authors nor the editors nor the publisher can accept any legal responsibility for any errors or omissions that may be made. The publisher makes no warranty, express or implied, with respect to the material contained herein.

Printed on acid-free paper

Springer is part of Springer Science+Business Media (www.springer.com)

To the memory of my parents

For my wife and children

Preface

In the Proportional-Integral-Derivative (PID) controllers the control signal is a linear function of: the current control error (the proportional part), the past errors (the integral part) and the rate of change of the error (the derivative part). The PID controllers, in particular when used for processes with one input and one output, are very successful provided that properties of the process are (approximately) linear and the time-delay is not significant.

In Model Predictive Control (MPC) algorithms [41, 186, 268, 278, 316] a future control policy is successively calculated on-line from an optimisation problem which takes into account some predicted future control errors. An explicit dynamic model of the process is used for prediction. In comparison with existing control techniques, particularly with the PID controllers, the MPC algorithms have a few advantages, the most important of which are:

a) constraints imposed on process variables can be easily taken into account in a systematic way (i.e. constraints of the input variables, of the predicted output variables and (or) of the predicted state variables),
b) the MPC algorithms can be used for multi-input multi-output processes and for processes which are difficult to control (i.e. processes with significant time-delays or with the inverse step-response).

It is an interesting fact that the first versions of the MPC algorithms were developed by practitioners who worked in industry. The MPC algorithms turned out to be very successful. After some time the new control approach attracted scientists who recognised an interesting new research field. Nowadays, the MPC algorithms are used in numerous applications, in particular in the chemical industry, in the petrochemical industry, in the paper industry and in food processing [262]. M. Morari and J. Lee state it clearly: "*DMC had a tremendous impact on industry. There is probably not a single major oil company in the world, where DMC (or a functionally similar product with a different trade name) is not employed in most new installations or revamps.* [215]." (DMC stands for Dynamic Matrix Control algorithm).

In consequence, applications influence research directions and development of new MPC algorithms.

In the simplest case linear dynamic models are used for prediction (in linear MPC algorithms). In such a case the future control sequence can be relatively easily calculated from a quadratic optimisation task. Unfortunately, properties of many technological processes are usually nonlinear. Because of significant differences between processes and their linear models, the quality of control may be not satisfactory, in particular when the operating point changes fast and significantly. That is why for nonlinear processes nonlinear MPC algorithms should be used, in which the prediction is calculated from nonlinear models. Two main problems which must be addressed during development of any nonlinear MPC algorithm are:

a) the choice of the nonlinear model,
b) the way the model is used on-line for prediction.

Modelling is a very important part of development of the MPC algorithms. Although the idea of MPC is very general and it does not impose any restrictions on the model type, its structure significantly influences complexity of the MPC algorithm and determines the possible applications. Two general model classes are: fundamental (first-principle models) and empirical models. Theoretically, it is best to formulate a complete fundamental model comprised of differential and algebraic equations which describe all parts of the process [141, 193]. Unfortunately, development and validation of such a model are likely to be very difficult, in particular for non-technologists, e.g. control engineers. Furthermore, fundamental models may be too complicated to be directly used for prediction in MPC. That is why this monograph is concerned with empirical models, the structure of which is chosen arbitrarily and parameters are determined by means of an identification algorithm. During identification some data sets previously recorded are only used, no technological knowledge is necessary. Among many different structures of empirical models, neural models based on the perceptron network with one hidden layer are chosen [66, 94, 119, 235, 274, 283, 309]. It is because of their advantages: good approximation accuracy, a reasonably low number of parameters (in comparison with alternatives) and a simple structure. Moreover, a great number of training and structure optimisation algorithms are available for neural models, which makes the identification process relatively simple. Unlike the fundamental models, the neural models consist of neither differential nor algebraic equations which must be successively solved on-line in MPC.

When a nonlinear model (e.g. a neural one) is directly used for prediction without any simplifications, one obtains a nonlinear MPC optimisation problem which must be solved on-line at each sampling instant. Nonlinear optimisation is not only computationally demanding (the calculation time may be longer than the sampling time), but it is also impossible to guarantee that the global solution is always found. A very practical approach is MPC with successive on-line linearisation [316], also called MPC with instantaneous linearisation [227]. A linear approximation of the nonlinear model is repeatedly found on-line for the current operating point of the process and used for prediction. Thanks to linearisation, the MPC optimisation problem is of quadratic optimisation type. Such MPC algorithms are inherently suboptimal, because the predictions calculated by means of the linearised model are likely to be different from those obtained by the original nonlinear one. On the one hand, it is a well-known fact that suboptimal MPC algorithms are appreciated by practitioners and used in many applications: *"Linearization is the only method which has found any wider use in industry beyond demonstration projects. For industry there has to be clear justification for solving non-linear programs on-line in a dynamic setting and there are no examples to bear that out in a convincing manner."* [215]. On the other hand, there are relatively few publications concerned with new versions or applications of suboptimal MPC algorithms, simulation results or applications of the "ideal" MPC approach with on-line nonlinear optimisation can be found much more frequently. Unfortunately, the problem of apparent computational inefficiency of such an approach is usually simply overlooked. It is also necessary to emphasise the fact that there are numerous theoretical investigations of the MPC approach with nonlinear optimisation whereas properties of the MPC algorithms with on-line linearisation are not discussed frequently in the literature ("*the theoretical purists tend to stay away from linearization approaches*" [215]). It is because stability and robustness analysis of the suboptimal MPC approaches with successive on-line linearisation is much more difficult than in the case of MPC with nonlinear optimisation.

The objective of this monograph is to thoroughly discuss the possibilities of using neural networks in nonlinear MPC algorithms, in particular in computationally efficient suboptimal MPC structures. The book results mainly from the author's research conducted in years 2004–2011 in the Institute of Control and Computation Engineering at the Faculty of Electronics and Information Technology, Warsaw University of Technology. In short:

a) A few types of suboptimal MPC algorithms in which a linear approximation of the model or of the predicted trajectory is successively calculated on-line and used for prediction are discussed.
b) Implementation details of the MPC algorithms for the feedforward perceptron neural models, the neural Hammerstein models, the neural Wiener models and the state-space neural models are given.
c) The MPC algorithms based on neural multi-models (inspired by the idea of predictive control) are presented.
d) The MPC algorithms with neural approximation with no on-line linearisation are discussed.
e) The most important methods for assuring stability and robustness of the MPC algorithms are reviewed. A suboptimal MPC algorithm with guaranteed stability and a suboptimal MPC algorithm with guaranteed robustness are detailed.

f) Cooperation between the MPC algorithms and set-point optimisation is discussed.

Thanks to linearisation (or neural approximation), the presented suboptimal algorithms do not require demanding on-line nonlinear optimisation. The presented simulation results demonstrate high accuracy and computational efficiency of the algorithms. For a few representative nonlinear benchmark processes, such as chemical reactors and a distillation column, for which the classical MPC algorithms based on linear models do not work properly, the trajectories obtained in the suboptimal MPC algorithms are very similar to those given by the "ideal" MPC algorithm with on-line nonlinear optimisation repeated at each sampling instant. At the same time, the suboptimal MPC algorithms are significantly less computationally demanding. Due to the fact that neural models have a simple, regular structure, they can be used in the discussed algorithms relatively easily in practice.

The monograph consists of 9 chapters. It also includes the list of symbols and acronyms used, the list of references and the index.

The first chapter introduces the reader into the field of MPC. The MPC optimisation problem is defined, linear and nonlinear MPC approaches are characterised, some methods aiming at reducing computational complexity are reviewed. A short history of MPC is also given. Next, a few classes of nonlinear models and possibilities of using them in MPC are discussed. In particular, interesting advantages of neural models are emphasised.

The second chapter details a few rudimentary types of nonlinear MPC algorithms (the prototypes). In addition to the "ideal" MPC algorithm with nonlinear optimisation, the suboptimal algorithms are discussed (the differences are associated with the linearisation method, e.g. model linearisation at the current operating point or predicted output trajectory linearisation along some future input trajectory). Next, the implementation details for the classical feedforward perceptron neural networks with one hidden layer are given. Explicit (analytical) versions of the suboptimal MPC algorithms are also discussed in which the current values of the control signals can be calculated explicitly, without any optimisation. It is only necessary to carry out a matrix decomposition task and solve linear equations.

The third chapter is devoted to nonlinear MPC algorithms based on cascade neural Hammerstein and Wiener models. The cascade models consist of linear dynamic parts and nonlinear steady-state ones. In the simplest case polynomials can be used in the steady-state parts of the cascade models, but such an approach has some important disadvantages (among others: low accuracy and a high number of parameters, in particular when the process has many inputs and many outputs). Although the cascade models are typically used if the steady-state nonlinearity is associated with sensors or actuators, in practice they may turn out to be more precise and have a lower number of parameters than the classical black-box models. Using the prototype MPC algorithms presented in the previous chapter, algorithms for neural Hammerstein and Wiener models (their steady-state parts are neural networks) are detailed. Furthermore, some MPC algorithms with simplified on-line linearisation are also discussed. It is possible because of the cascade structure of Hammerstein and Wiener models. Unlike many control algorithms published elsewhere, the discussed MPC algorithms do not need the inverse steady-state model (apart from one algorithm, only for the Wiener model).

The fourth chapter is concerned with MPC algorithms based on state-space neural models. Implementation details of two approaches are detailed: the output set-point trajectory or the state set-point trajectory may be considered. In spite of the fact that in the majority of MPC algorithms applied in practice the input-output models are usually used, it is necessary to emphasise the fact that the state-space description is much more general, very convenient for theoretical investigations. The state-space approach is natural for stability and robustness analysis.

The objective of the next two chapters is to discuss MPC algorithms based on some models inspired by the idea of MPC. The fifth chapter deals with MPC algorithms based on neural multi-models. The model is considered as good for MPC provided that it is able to precisely find the predictions of the outputs (or of the states) over the whole prediction horizon. In practice the order of dynamics of the model used in MPC is usually lower than the order of the process (quite frequently the real order of the process is unknown). Such a model may give good predictions when the horizon is short but for long horizons prediction accuracy is likely to be low, because the prediction error is propagated. A practical method which solves the problem is to use the multi-models: for each sampling instant within the prediction horizon one independent submodel is used. The consecutive submodels may be significantly less complicated than the classical model, the role of which is to calculate predictions over the whole prediction horizon. Furthermore, the submodels are trained easily as one-step ahead predictors.

The sixth chapter describes MPC algorithms with neural approximation in which no on-line linearisation is necessary. A direct consequence of this fact is low computational complexity. In the numerical versions of such MPC algorithms the neural approximator calculates on-line, for the current operating point of the process, the step-response coefficients of the linearised model or the derivatives of the predicted output trajectory with respect to the future control sequence. The explicit versions of MPC algorithms with neural approximation are particularly interesting, because the neural approximator finds on-line coefficients of the control law, successive on-line linearisation and calculations typical of the classical explicit MPC algorithms are not necessary. The MPC techniques with neural approximation are recommended when low computational burden is a priority. Inexpensive hardware can be used for implementation.

The seventh chapter is concerned with stability and robustness of nonlinear MPC algorithms. The most important general approaches for assuring stability are reviewed, in particular the dual-mode MPC algorithm with an additional stabilising controller. Its great advantage is the fact that it is not necessary to find on-line the global solution of the nonlinear MPC optimisation problem at each sampling instant. A feasible solution (which satisfies all constraints) is only necessary for stability. Thanks to such a property, it is possible to develop the dual MPC algorithm with on-line linearisation, which is an extension of the prototype MPC strategies discussed in the previous chapters. Next, the most important general approaches for assuring robustness are reviewed, in particular the MPC algorithm with additional constraints. Similarly to the case of the dual MPC algorithm, a feasible solution is only necessary for guaranteeing robustness. Finally, the MPC algorithm with additional constraints and on-line linearisation is discussed.

The eighth chapter describes cooperation between the suboptimal MPC algorithms and the set-point optimisation algorithms. The classical multilayer control system structure is discussed and its main disadvantage, i.e. the necessity of on-line nonlinear optimisation, is pointed out. Three alternative control structures with on-line linearisation used for set-point optimisation are investigated. The implementation details for three classes of neural models are given. In the first case two independent models are necessary: a steady-state model for set-point optimisation and a dynamic one for MPC. Alternatively, it is also possible to use only one neural dynamic model: the cascade structure of Hammerstein or Wiener type. The model is not only used for prediction in MPC, but also the corresponding steady-state description is derived on-line for the current operating point of the process. In order to reduce computational complexity, dynamic and steady-state models are successively linearised on-line, which makes it possible to eliminate the necessity of nonlinear optimisation.

The ninth chapter summarises the whole book, some possible directions of future research are also given.

The chapters from the second to the sixth and the eighth one include the literature reviews. Because those chapters mainly describe the algorithms developed by the author of this book, the literature reviews emphasise the fact that there are plenty of works concerned with the "ideal" MPC algorithm with nonlinear optimisation whereas the publications devoted to suboptimal MPC solutions with on-line linearisation are rare. The literature references are given throughout the first chapter because it introduces the general idea of MPC and considers modelling issues. The seventh chapter, which discusses the existing methods used for assuring stability and robustness of MPC algorithms as well as the stable and robust algorithms developed by the author, is organised in the same way. Finally, the most important differences between the author's concepts and other MPC approaches described elsewhere are shortly pointed out in the ninth chapter.

Warsaw, Maciej Ławryńczuk
October 2013

Contents

Notation

General Notation

a, b, ...	variables or constants, scalar or vectors
$\boldsymbol{A}$, $\boldsymbol{B}$, ...	real matrices
a^{T}, $\boldsymbol{A}^{\mathrm{T}}$	transpose of vector a and of matrix $\boldsymbol{A}$
$\mathcal{A}$, $\mathcal{B}$, ...	sets
$\mathrm{diag}(a_1, \ldots, a_n)$	diagonal matrix with $a_1, \ldots, a_n$ on the diagonal
$\left.\frac{\mathrm{d}y(x)}{\mathrm{d}x}\right\|_{x=\bar{x}}$	derivative of function $y(x)$ at point $\bar{x}$ (scalar) or derivative of function $y(x_1, \ldots, x_{n_{\mathrm{x}}})$ with respect to vector $x = [x_1 \ldots x_{n_{\mathrm{x}}}]^{\mathrm{T}}$ at point $\bar{x} = [\bar{x}_1 \ldots \bar{x}_{n_{\mathrm{x}}}]^{\mathrm{T}}$ (vector of length n_{x}) or derivative of function $y(x) = \left[y_1(x_1, \ldots, x_{n_{\mathrm{x}}}), \ldots, y_{n_{\mathrm{y}}}(x_1, \ldots, x_{n_{\mathrm{x}}})\right]^{\mathrm{T}}$ with respect to vector $x = [x_1 \ldots x_{n_{\mathrm{x}}}]^{\mathrm{T}}$ at point $\bar{x} = [\bar{x}_1 \ldots \bar{x}_{n_{\mathrm{x}}}]^{\mathrm{T}}$ (matrix of dimensionality $n_{\mathrm{y}} \times n_{\mathrm{x}}$)
$f(\cdot)$, $g(\cdot)$, ...	scalar or vector functions
0_n, 1_n	zeros and ones vectors of length n
$\mathbf{0}_{m \times n}$, $\boldsymbol{I}_{m \times n}$	zeros and ones matrices of dimensionality $m \times n$
n_{a}, n_{b}, ...	vector lengths, $n_{\mathrm{a}} = \dim(a)$, $n_{\mathrm{b}} = \dim(b)$
q^{-1}	discrete unit time-delay
$\left.\frac{\partial y(x)}{\partial x_i}\right\|_{x=\bar{x}}$	fractional derivative of function $y(x) = y(x_1, \ldots, x_{n_{\mathrm{x}}})$ with respect to scalar x_i $(i = 1, \ldots, n_{\mathrm{x}})$ at point $\bar{x} = [\bar{x}_1 \ldots \bar{x}_{n_{\mathrm{x}}}]^{\mathrm{T}}$ (scalar) or fractional derivative of function $y(x) = \left[y_1(x_1, \ldots, x_{n_{\mathrm{x}}}), \ldots, y_{n_{\mathrm{y}}}(x_1, \ldots, x_{n_{\mathrm{x}}})\right]^{\mathrm{T}}$ with respect to vector $x_i = \left[x_1 \ldots x_{n_{\mathrm{x_i}}}\right]^{\mathrm{T}}$ $(n_{\mathrm{x_i}} < n_{\mathrm{x}})$ at point $\bar{x}_i = \left[\bar{x}_1 \ldots \bar{x}_{n_{\mathrm{x_i}}}\right]^{\mathrm{T}}$ (matrix of dimensionality $n_{\mathrm{y}} \times n_{\mathrm{x_i}}$)

$\lvert x \rvert$	absolute value of x
$\lVert x \rVert^2$	$x^{\mathrm{T}} x$
$\lVert x \rVert_{\boldsymbol{A}}^2$	$x^{\mathrm{T}} \boldsymbol{A} x$
$\lVert x \rVert_1, \lVert x \rVert_2, \ldots$	vector norms
$\lVert \boldsymbol{A} \rVert_1, \lVert \boldsymbol{A} \rVert_2, \ldots$	matrix norms

Specific Notation: Processes and Models

a_i^m, $b_i^{m,n}$	parameters of linear model or parameters of linear dynamic part of Hammerstein or Wiener model
$a_i^m(k)$, $b_i^{m,n}(k)$	parameters of linearised model at current operating point for sampling instant k
$\boldsymbol{A}(q^{-1})$, $\boldsymbol{B}(q^{-1})$	polynomials of unit time-delay operator q^{-1}
$f(\cdot)$	general function describing dynamic model
$f^{\mathrm{ss}}(\cdot)$	general function describing steady-state model
$g(\cdot)$	general function describing nonlinear steady-state part of Hammerstein or Wiener models
$h(k)$	measured disturbances vector at sampling instant k
k	discrete time (sampling instant, algorithm iteration)
K^m	number of hidden nodes
n_{A}^m, $n_{\mathrm{B}}^{m,n}$, $n_{\mathrm{C}}^{m,n}$	constants defining order of dynamics of dynamic model
n_{h}	number of measured disturbances
n_{u}	number of inputs
n_{v}	number of auxiliary variables in Hammerstein or Wiener model
n_{x}	number of state variables
n_{y}	number of outputs
$u(k)$	input vector (manipulated variables vector) at sampling instant k
$v(k)$	auxiliary variables vector in Hammerstein or Wiener model at sampling instant k
$w_{i,j}^{1,m}$, $w_i^{2,m}$	weights of first and second layer of neural network
$x(k)$	state vector at sampling instant k
$\boldsymbol{x}_m(k)$	vector of model arguments
$y(k)$	output vector (controlled variables vector) at sampling instant k
$z_i^m(k)$	sum of input signals connected to i^{th} hidden node
$\triangle u(k+p\mid k)$	$u(k+p\mid k) - u(k+p-1\mid k)$
φ	transfer function
$\tau^{m,n}$	time-delay of input-output channel

$\tau_{\mathrm{h}}^{m,n}$	time-delay of measured disturbance-output channel

Specific Notation: MPC Algorithms

$d(k)$	unmeasured disturbances vector at process output at sampling instant k
D	horizon of dynamics
$\boldsymbol{G}(k)$	dynamic matrix (of step-response coefficients) of model linearised at sampling instant k
$\boldsymbol{H}(k)$	matrix of derivatives of predicted output or state trajectory with respect to future control sequence calculated at sampling instant k
$J(k)$	cost-function minimised in MPC
$\boldsymbol{K}(k)$, $\boldsymbol{K}^{n_{\mathrm{u}}}(k)$	control law matrices of explicit MPC algorithm
N	prediction horizon
N_{u}	control horizon
$s_p^{m,n}$, $\boldsymbol{S}_p(k)$	step-response coefficient and step-response matrix of model linearised at sampling instant k
$u(k+p\|k)$	process input vector predicted for sampling instant $k+p$ at instant k
$\boldsymbol{u}(k)$	process input vector predicted at sampling instant k over control horizon
$u^{\min}$, $u^{\max}$	vectors of magnitude constraints imposed on process inputs
$\boldsymbol{u}^{\min}$, $\boldsymbol{u}^{\max}$	vectors of magnitude constraints imposed on process inputs over control horizon
$\mathcal{U}$	set of constraints imposed on process inputs
$x^0(k+p\|k)$	state free trajectory vector predicted for sampling instant $k+p$ at instant k
$\boldsymbol{x}^0(k)$	state free trajectory vector predicted at sampling instant k over prediction horizon
$\hat{x}(k+p\|k)$	state trajectory vector predicted for sampling instant $k+p$ at instant k
$\hat{\boldsymbol{x}}(k)$	state trajectory vector predicted at sampling instant k over prediction horizon
$\bar{\boldsymbol{x}}_m(k)$	vector defining current operating point in MPC algorithms with on-line model linearisation
$\widetilde{\boldsymbol{x}}_m(k)$	vector defining current operating point in MPC algorithms with neural approximation
$x^{\min}$, $x^{\max}$	vectors of magnitude constraints imposed on predicted state variables
$\boldsymbol{x}^{\min}$, $\boldsymbol{x}^{\max}$	vectors of magnitude constraints imposed on predicted state variables over prediction horizon
$x^{\mathrm{sp}}(k+p\|k)$	state set-point trajectory vector for sampling instant $k+p$ known at instant k

$\boldsymbol{x}^{\mathrm{sp}}(k)$	state set-point trajectory vector at sampling instant k over prediction horizon
$\mathcal{X}$	set of constraints imposed on state variables
$y^0(k+p\|k)$	output free trajectory vector predicted for sampling instant $k+p$ at instant k
$\boldsymbol{y}^0(k)$	output free trajectory vector predicted at sampling instant k over prediction horizon
$\hat{y}(k+p\|k)$	output trajectory vector predicted for sampling instant $k+p$ at instant k
$\hat{\boldsymbol{y}}(k)$	output trajectory vector predicted at sampling instant k over prediction horizon
$y^{\min}$, $y^{\max}$	vectors of magnitude constraints imposed on predicted output variables
$\boldsymbol{y}^{\min}$, $\boldsymbol{y}^{\max}$	vectors of magnitude constraints imposed on predicted output variables over prediction horizon
$y^{\mathrm{sp}}(k+p\|k)$	output set-point trajectory vector for sampling instant $k+p$ known at instant k
$\boldsymbol{y}^{\mathrm{sp}}(k)$	output set-point trajectory vector at sampling instant k over prediction horizon
$\triangle u(k+p\|k)$	input increments vector for sampling instant $k+p$ at instant k
$\triangle \boldsymbol{u}(k)$	input increments vector for sampling instant k over control horizon
$\triangle u^{\max}$	vectors of increment constraints imposed on inputs
$\triangle \boldsymbol{u}^{\max}$	vectors of increment constraints imposed on inputs over control horizon
$\varepsilon(k)$, $\boldsymbol{\varepsilon}(k)$	additional decision variables of MPC optimisation problem defining degree of output or state constraint violation
$\lambda_{n,p}$, $\boldsymbol{\Lambda}$, $\boldsymbol{\Lambda}_p$	weighting coefficient and weighting matrices of control increments in MPC cost-function
$\mu_{m,p}$, $\boldsymbol{M}$, $\boldsymbol{M}_p$	weighting coefficient and weighting matrices of predicted output control errors in MPC cost-function
$\nu(k)$	unmeasured state disturbances vector at sampling instant k
$\phi_{m,p}$, $\boldsymbol{\Phi}$, $\boldsymbol{\Phi}_p$	weighting coefficient and weighting matrices of predicted state control errors in MPC cost-function
Ω	terminal set

Specific Notation: Set-Point Optimisation

c_{u}, c_{y}	weighting coefficients used in set-point optimisation cost-function (costs)
$\boldsymbol{G}^{\mathrm{ss}}$	steady-state gain matrix derived from linear dynamic model used in MPC
$\boldsymbol{H}^{\mathrm{ss}}(k)$	matrix defining linear approximation of steady-state nonlinear model at sampling instant k
J_{E}	cost-function minimised in set-point optimisation

Acronyms: MPC Algorithms

CRHPC	Constrained Receding Horizon Predictive Control
DMC	Dynamic Matrix Control
DMC-NA	DMC algorithm with Neural Approximation
GPC	Generalized Predictive Control
GPC^{∞}	GPC algorithm with infinite horizon
MAC	Model Algorithmic Control
MHPC	Model Heuristic Predictive Control
MPC	Model Predictive Control
MPC-NNPAPT	suboptimal MPC Newton-like algorithm with Nonlinear Prediction and Approximation along the Predicted Trajectory
MPC-NO	MPC algorithm with Nonlinear Optimisation
MPC-NPL	suboptimal MPC algorithm with Nonlinear Prediction and Linearisation for the current operating point
MPC-NPL-NA	suboptimal MPC-NPL algorithm with Neural Approximation
MPC-NPLT	suboptimal MPC algorithm with Nonlinear Prediction and Linearisation along the Trajectory
MPC-NPLT-NA	suboptimal MPC-NPLT algorithm with Neural Approximation
$\mathrm{MPC\text{-}NPLT}_{\boldsymbol{u}(k-1)}$	suboptimal MPC-NPLT algorithm with linearisation along the trajectory defined by input signals applied at previous sampling instant
$\mathrm{MPC\text{-}NPLT}_{\boldsymbol{u}(k-1)}\text{-NA}$	suboptimal $\mathrm{MPC\text{-}NPLT}_{\boldsymbol{u}(k-1)}$ algorithm with neural approximation
$\mathrm{MPC\text{-}NPLT}_{\boldsymbol{u}(k\vert k-1)}$	suboptimal MPC-NPLT algorithm with linearisation along the trajectory defined by optimal input signals calculated at previous sampling instant
MPC-NPLPT	suboptimal MPC algorithm with Nonlinear Prediction and Linearisation along the Predicted Trajectory

MPC-NPSL	suboptimal MPC algorithm with Nonlinear Prediction and Simplified Linearisation for the current operating point
MPC-SL	suboptimal MPC algorithm with Successive Linearisation for the current operating point
MPC-SSL	suboptimal MPC algorithm with Simplified Successive Linearisation for the current operating point
QDMC	Quadratic Dynamic Matrix Control

Acronyms: Set-Point Optimisation

ASSTO	Adaptive Steady-State Target Optimisation layer
LSSO	Local Steady-State Optimisation layer
SSTO	Steady-State Target Optimisation layer

Acronyms: Miscellaneous

BFGS	Broyden-Fletcher-Goldfarb-Shanno variable metrics nonlinear optimisation algorithm
DFP	Davidon-Fletcher-Powell variable metrics nonlinear optimisation algorithm
DLP	Double-Layer Perceptron feedforward neural network
FSQP	Feasible Sequential Quadratic Programming
IMC	Internal Model Control
LRGF	Locally Recurrent Globally Feedforward neural network
LS-SVM	Least Squares Support Vector Machine
MFLOPS	Million of FLOating Point operationS
MLP	Multi-Layer Perceptron feedforward neural network
MSE	Mean of Squared Errors
OBD	Optimal Brain Damage
OBS	Optimal Brain Surgeon
PID	Proportional Integral Derivative
RBF	Radial Basis Function feedforward neural network
SQP	Sequential Quadratic Programming
ss	steady state
SSE	Sum of Squared Errors
SVM	Support Vector Machine

Symbols and Indices, Vectors and Scalars

All scalars, vectors, matrices and indices are typeset in italics whereas non-mathematical elements are typeset in roman. E.g. in the case of the symbol $y_m^{\mathrm{sp}}(k+p|k)$, y^{sp} denotes the variable, k, m and p are the indices. Both scalars and vectors are used, e.g. scalars $y_1^{\mathrm{sp}}(k+p|k), \ldots, y_{n_{\mathrm{y}}}^{\mathrm{sp}}(k+p|k)$ comprise the vector $y^{\mathrm{sp}}(k+p|k) = \left[y_1^{\mathrm{sp}}(k+p|k) \ldots y_{n_{\mathrm{y}}}^{\mathrm{sp}}(k+p|k)\right]^{\mathrm{T}}$.

1

MPC Algorithms

This chapter introduces the reader into the field of MPC. The basic MPC optimisation problem is defined, the fundamental role of the model is emphasised. The general classification of MPC algorithms is given, i.e. linear and nonlinear approaches are characterised. Next, some methods which make it possible to reduce computational burden of nonlinear MPC algorithms are shortly described, including the on-line linearisation approach. A history of MPC algorithms is given. Finally, a short review of nonlinear model structures is included, their advantages and disadvantages as well as possibilities of using them in MPC are pointed out.

1.1 Principle of Predictive Control

In spite of the fact that numerous MPC algorithms have been developed [41, 186, 268, 278, 316], the principal idea is always the same. At each consecutive iteration of the algorithm k (sampling instant), $k = 1, 2, 3, \ldots$, the whole vector of the future values of the manipulated variable

$$\boldsymbol{u}(k) = \begin{bmatrix} u(k|k) \\ \vdots \\ u(k+N_{\mathrm{u}}-1|k) \end{bmatrix} \tag{1.1}$$

is calculated on-line. The symbol $u(k+p|k)$ denotes the value of the manipulated variable for the sampling instant $k+p$ calculated at the current iteration, N_{u} is the control horizon. Alternatively, the vector of increments of the future values of the manipulated variable

$$\triangle\boldsymbol{u}(k) = \begin{bmatrix} \triangle u(k|k) \\ \vdots \\ \triangle u(k+N_{\mathrm{u}}-1|k) \end{bmatrix} \tag{1.2}$$

M. Ławryńczuk, *Computationally Efficient Model Predictive Control Algorithms*,
Studies in Systems, Decision and Control 3,
DOI: 10.1007/978-3-319-04229-9_1, © Springer International Publishing Switzerland 2014

where

$$\Delta u(k+p|k) = \begin{cases} u(k|k) - u(k-1) & \text{if } p = 0 \\ u(k+p|k) - u(k+p-1|k) & \text{if } p \geq 1 \end{cases}$$

can be determined. The vector of decision variables (1.1) or (1.2) is successively found on-line as a result of solving an optimisation problem. The minimised objective function (the cost-function) usually consists of two parts. The first one takes into account the differences between the predicted trajectory of the output variable and the set-point trajectory (i.e. the predicted control errors) over the prediction horizon N. The role of the second part of the objective function (the penalty term) is to reduce excessive (and hence disadvantageous) changes of the manipulated variable. Typically, the following quadratic cost-function is used

$$J(k) = \sum_{p=1}^{N} (y^{\mathrm{sp}}(k+p|k) - \hat{y}(k+p|k))^2 + \lambda \sum_{p=0}^{N_{\mathrm{u}}-1} (\Delta u(k+p|k))^2 \quad (1.3)$$

where $\lambda > 0$ is a weighting coefficient (the greater its value, the lower the increments of the manipulated variable and, hence, the slower control). The set-point value for the sampling instant $k+p$ known at the current iteration (at the instant k) is denoted by $y^{\mathrm{sp}}(k+p|k)$, the predicted value of the output variable for the sampling instant $k+p$ calculated at the current iteration is denoted by $\hat{y}(k+p|k)$. Consecutive output predictions, for the whole prediction horizon, i.e. for $p = 1, \ldots, N$, are calculated by means of a dynamic model of the process. It is assumed that $u(k+p|k) = u(k+N_{\mathrm{u}}-1|k)$ for $p = N_{\mathrm{u}}, \ldots, N$ (it means that $\Delta u(k+N_{\mathrm{u}}|k) = \ldots = \Delta u(k+N|k) = 0$). Fig. 1.1 shows the general structure of the MPC algorithm, whereas Fig. 1.2 depicts three important trajectories: the predicted output trajectory, the set-point trajectory and the calculated future control trajectory.

In spite of the fact that the whole future control sequence (1.1) or the increments (1.2) are calculated at each iteration of the MPC algorithm, only its first element is actually applied to the process at the current sampling instant k

$$u(k) = u(k|k) \text{ or } u(k) = \Delta u(k|k) + u(k-1)$$

In the next algorithm iteration $(k+1)$ the output measurement is updated (i.e. the value of $y(k)$ recorded), the prediction horizon is shifted one step forward and the whole procedure described above is repeated.

In short, there are three common important features of all MPC algorithms: the receding horizon, successive on-line optimisation of the cost-function and the direct use of a dynamic model of the process for prediction calculation. Unlike the classical control algorithms, e.g. the PID controller, not only the current value of the manipulated variable is calculated, but the whole future control policy.

Thanks to the fact that an explicit dynamic model is used for prediction and the predicted control errors are minimised, the MPC algorithm is able to find the future control sequence which gives good control accuracy. As a result, unlike the PID controller, the MPC algorithms can be successfully applied to processes which are difficult to control (i.e. processes with a significant time-delay or with the so called inverse step-response). Of course, the model used for prediction should correspond with the properties of the process, the prediction horizon should be sufficiently long. The idea of MPC is very general, it does not impose any restrictions on the type and the structure of the model. That is why the MPC algorithm can be developed not only for processes with one input and one output (single-input single-output processes) but also for multivariable processes (multi-input multi-output processes). In such a case, thanks to using the model for prediction and optimisation of the future control errors, all existing cross-couplings are automatically taken into account.

This book discusses MPC algorithms for multi-input multi-output processes. The number of process inputs (the manipulated variables) is denoted by n_{u}, the number of process outputs (the controlled variables) is denoted by n_{y}. The vector notation is used for compactness of presentation: $u \in \mathbb{R}^{n_{\mathrm{u}}}$, $y \in \mathbb{R}^{n_{\mathrm{y}}}$. In some cases, however, it is necessary or more convenient to use individual signals: $u_1, \ldots, u_{n_{\mathrm{u}}}, y_1, \ldots, y_{n_{\mathrm{y}}}$. The vector of manipulated variables for the sampling instant $k+p$ calculated at the current instant k is then $u(k+p|k) = [u_1(k+p|k) \ldots u_{n_{\mathrm{u}}}(k+p|k)]^{\mathrm{T}}$, the corresponding vector of control increments is $\triangle u(k+p|k) = [\triangle u_1(k+p|k) \ldots \triangle u_{n_{\mathrm{u}}}(k+p|k)]^{\mathrm{T}}$.

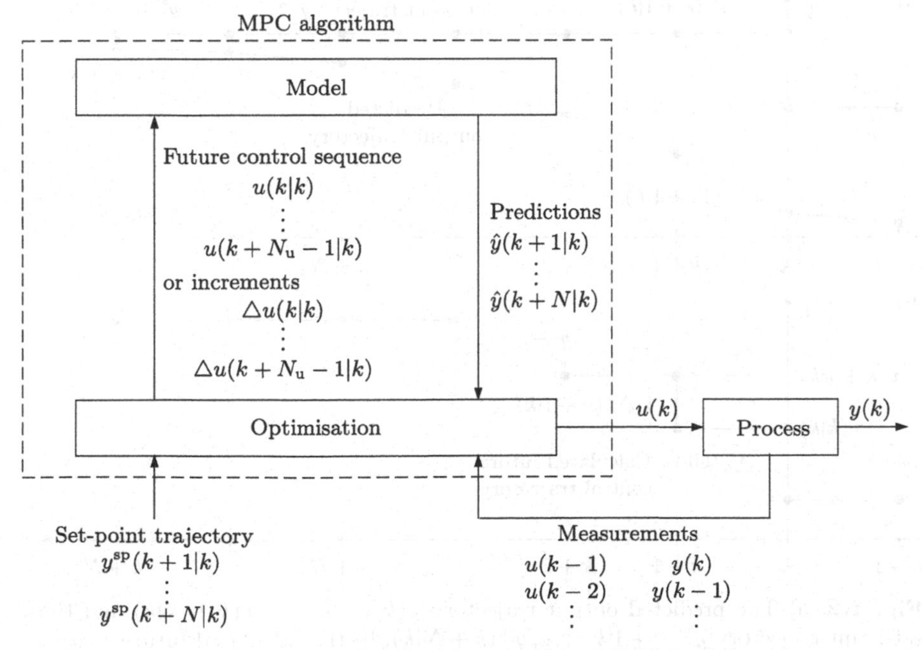

Fig. 1.1. The general structure of the MPC algorithm

The vector of decision variables of the algorithm (1.1) or (1.2) is of length $n_\mathrm{u}N_\mathrm{u}$. The minimised cost-function, similarly to that used in the case of the single-input single-output processes (1.3), is

$$\begin{aligned}J(k) &= \sum_{p=1}^{N}\sum_{m=1}^{n_\mathrm{y}} \mu_{p,m}\left(y_m^{\mathrm{sp}}(k+p|k) - \hat{y}_m(k+p|k)\right)^2 \\ &+ \sum_{p=0}^{N_\mathrm{u}-1}\sum_{n=1}^{n_\mathrm{u}} \lambda_{p,n}\left(\triangle u_n(k+p|k)\right)^2 \\ &= \sum_{p=1}^{N} \left\|y^{\mathrm{sp}}(k+p|k) - \hat{y}(k+p|k)\right\|_{\boldsymbol{M}_p}^2 + \sum_{p=0}^{N_\mathrm{u}-1} \left\|\triangle u(k+p|k)\right\|_{\boldsymbol{\Lambda}_p}^2 \end{aligned} \tag{1.4}$$

The weighting coefficients $\mu_{p,m} \geq 0$ and $\lambda_{p,n} > 0$ are tuning parameters of the MPC algorithm. The first ones make it possible to differentiate the influence of the predicted control errors of consecutive outputs of the process (over the prediction horizon). The second ones are used not only to differentiate the influence of the control increments of consecutive inputs of the process (over the control horizon), but, first of all, to establish the necessary scale between both parts of the cost-function. The weighting matrix

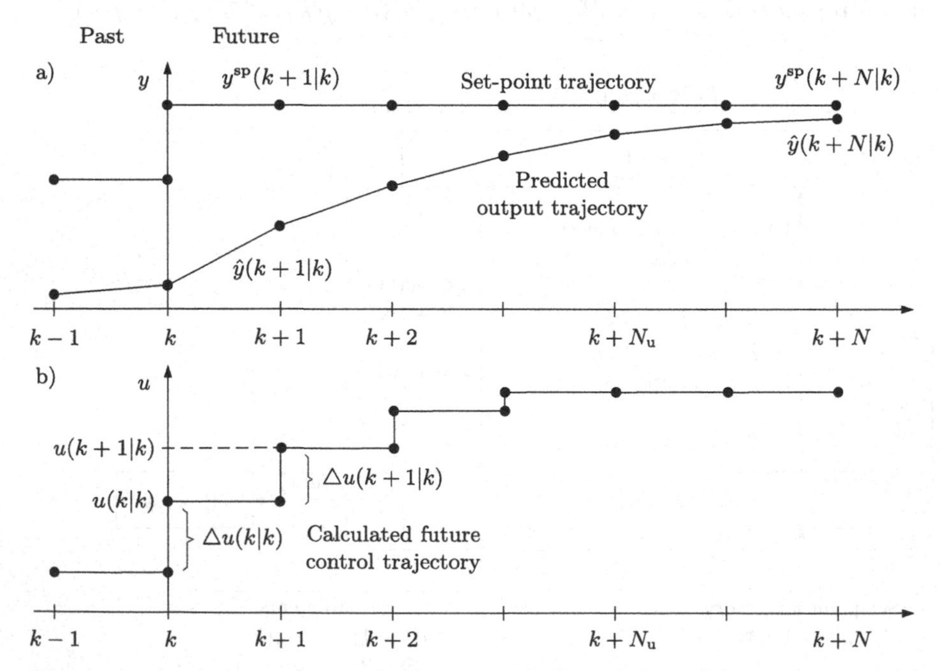

Fig. 1.2. a) The predicted output trajectory $\hat{y}(k+1|k), \ldots, \hat{y}(k+N|k)$ and the set-point trajectory $y^{\mathrm{sp}}(k+1|k), \ldots, y^{\mathrm{sp}}(k+N|k)$, b) the calculated future control trajectory $u(k|k), \ldots, u(k+N_\mathrm{u}-1|k)$

$\boldsymbol{M}_p = \text{diag}(\mu_{p,1}, \ldots, \mu_{p,n_y}) \geq 0$ is of dimensionality $n_y \times n_y$, the matrix $\boldsymbol{\Lambda}_p = \text{diag}(\lambda_{p,1}, \ldots, \lambda_{p,n_u}) > 0$ is of dimensionality $n_u \times n_u$. Typically, constant coefficients $\mu_{1,m} = \ldots = \mu_{N,m}$ are used for the whole prediction horizon and constant coefficients $\lambda_{0,n} = \ldots = \lambda_{N_u-1,n}$ for the whole control horizon (the matrices $\boldsymbol{M}_p$ and $\boldsymbol{\Lambda}_p$ are then constant, independent of p). The problem of tuning MPC algorithms (selection of horizons and weighting coefficients) is discussed elsewhere [53, 288, 316]. The minimised cost-function (1.4) takes into account predicted control errors from the sampling instant $k+1$ to $k+N$, but when the process is characterised by a significant time-delay, some initial components of the first part do not depend on the calculated future control sequence and can be omitted. For simplicity of presentation it is assumed that the prediction horizon N is constant for all outputs, the control horizon N_u is constant for all inputs. In the simplest approach, which is the most frequently used in practice, no future changes of the output set-point trajectory are known at the current sampling instant k. In such a case the set-point trajectory is assumed to be constant over the whole prediction horizon, i.e. $y^{sp}(k+p|k) = y^{sp}(k)$ for all $p = 1, \ldots, N$. In different words, only the current set-point vector $y^{sp}(k) \in \mathbb{R}^{n_y}$ is used. In place of the set-point trajectory, it is also possible to take into account the reference trajectory in the minimised cost-function [272, 273, 316]. It is usually described by the first-order dynamics

$$y^{ref}(k+p|k) = \gamma y^{ref}(k+p-1|k) + (1-\gamma) y^{sp}(k+p|k)$$

where $0 \leq \gamma < 1$, $y^{ref}(k|k) = y(k)$.

A great advantage of the MPC algorithm is the possibility of taking into account constraints imposed on process variables. Because the future control sequence is calculated on-line from an optimisation problem in which the cost-function (1.4) is minimised, all the necessary constraints can be taken into account in a natural way. In such a case the general MPC optimisation problem solved on-line at each sampling instant is

$$\begin{aligned} \min_{\boldsymbol{u}(k) \text{ or } \Delta \boldsymbol{u}(k)} & \left\{ J(k) = \sum_{p=1}^{N} \| y^{sp}(k+p|k) - \hat{y}(k+p|k) \|^2_{\boldsymbol{M}_p} \right. \\ & \left. + \sum_{p=0}^{N_u-1} \| \Delta u(k+p|k) \|^2_{\boldsymbol{\Lambda}_p} \right\} \\ & \text{subject to} \\ & u^{min} \leq u(k+p|k) \leq u^{max}, \ p = 0, \ldots, N_u - 1 \\ & -\Delta u^{max} \leq \Delta u(k+p|k) \leq \Delta u^{max}, \ p = 0, \ldots, N_u - 1 \\ & y^{min} \leq \hat{y}(k+p|k) \leq y^{max}, \ p = 1, \ldots, N \end{aligned} \tag{1.5}$$

where the vectors $u^{min} \in \mathbb{R}^{n_u}$ and $u^{max} \in \mathbb{R}^{n_u}$ define the constraints imposed on the magnitude of the manipulated variables, the vector $\Delta u^{max} \in \mathbb{R}^{n_u}$ defines the constraints imposed on the increments of the manipulated variables and the vectors $y^{\min} \in \mathbb{R}^{n_y}$ and $y^{\max} \in \mathbb{R}^{n_y}$ define the constraints imposed on the magnitude of the predicted output variables. The total number of decision variables of the MPC algorithm is $n_u N_u$, the number of constraints is $4n_u N_u + 2n_y N$.

It is possible to solve the MPC optimisation problem (1.5) provided that the feasible set (the set of possible solutions) is not empty. Any (but not conflicting) constraints imposed on the magnitude and on the increments of the manipulated variables may reduce the feasible set, but it is impossible that that set is empty. Unfortunately, depending on the current operating point of the process, unmeasured variables, the model-process mismatch and existing input constraints, the satisfaction of output constraints may be not possible. Hence, the feasible set of the MPC optimisation problem (1.5) is empty. In such a case a simple constraints' window method can be used [186, 316]. The output constraints are not imposed on the whole prediction horizon, the constraints for some first part of the horizon are neglected. A better idea is to use soft output constraints. In this approach it is assumed that the predicted values of the output variables may temporarily violate the original hard constraints $y^{\min} \leq \hat{y}(k+p|k) \leq y^{\max}$, which enforces the existence of the feasible set. In place of the MPC optimisation problem with hard output constraints (1.5), it is necessary to solve the following task

$$\begin{aligned}
\min_{\substack{\boldsymbol{u}(k) \text{ or } \Delta\boldsymbol{u}(k) \\ \varepsilon^{\min}(k+p),\ \varepsilon^{\max}(k+p)}} \Bigg\{ J(k) = & \sum_{p=1}^{N} \left\| y^{\mathrm{sp}}(k+p|k) - \hat{y}(k+p|k) \right\|_{\boldsymbol{M}_p}^2 \\
& + \sum_{p=0}^{N_u-1} \left\| \Delta u(k+p|k) \right\|_{\boldsymbol{\Lambda}_p}^2 \\
& + \rho^{\min} \sum_{p=1}^{N} \left\| \varepsilon^{\min}(k+p) \right\|^2 \\
& + \rho^{\max} \sum_{p=1}^{N} \left\| \varepsilon^{\max}(k+p) \right\|^2 \Bigg\}
\end{aligned}$$

$$\begin{aligned}
&\text{subject to} \qquad\qquad (1.6)\\
&u^{\min} \leq u(k+p|k) \leq u^{\max},\ p = 0, \ldots, N_u - 1 \\
&-\Delta u^{\max} \leq \Delta u(k+p|k) \leq \Delta u^{\max},\ p = 0, \ldots, N_u - 1 \\
&y^{\min} - \varepsilon^{\min}(k+p) \leq \hat{y}(k+p|k) \leq y^{\max} + \varepsilon^{\max}(k+p),\ p = 1, \ldots, N \\
&\varepsilon^{\min}(k+p) \geq 0,\ \varepsilon^{\max}(k+p) \geq 0,\ p = 1, \ldots, N
\end{aligned}$$

When it is necessary (i.e. when the feasible set is empty), the original hard output constraints are temporarily violated. The vectors $\varepsilon^{\min}(k+p) \in \mathbb{R}^{n_y}$ and $\varepsilon^{\max}(k+p) \in \mathbb{R}^{n_y}$, which determine the degree of constraint violation for consecutive sampling instants over the prediction horizon ($p = 1, \ldots, N$), are the additional decision variables of the MPC optimisation problem. They have positive values only when the corresponding hard constraints are violated.

The number of decision variables of the resulting MPC algorithm is $n_{\mathrm{u}}N_{\mathrm{u}}+2n_{\mathrm{y}}N$, the number of constraints is $4n_{\mathrm{u}}N_{\mathrm{u}}+4n_{\mathrm{y}}N$, $\rho^{\mathrm{min}}, \rho^{\mathrm{max}} > 0$ are penalty coefficients.

Defining the following set-point trajectory and the predicted output trajectory vectors of length $n_{\mathrm{y}}N$

$$\boldsymbol{y}^{\mathrm{sp}}(k) = \begin{bmatrix} y^{\mathrm{sp}}(k+1|k) \\ \vdots \\ y^{\mathrm{sp}}(k+N|k) \end{bmatrix}, \ \hat{\boldsymbol{y}}(k) = \begin{bmatrix} \hat{y}(k+1|k) \\ \vdots \\ \hat{y}(k+N|k) \end{bmatrix} \tag{1.7}$$

the additional variables vectors of length $n_{\mathrm{y}}N$

$$\boldsymbol{\varepsilon}^{\mathrm{min}}(k) = \begin{bmatrix} \varepsilon^{\mathrm{min}}(k+1) \\ \vdots \\ \varepsilon^{\mathrm{min}}(k+N) \end{bmatrix}, \ \boldsymbol{\varepsilon}^{\mathrm{max}}(k) = \begin{bmatrix} \varepsilon^{\mathrm{max}}(k+1) \\ \vdots \\ \varepsilon^{\mathrm{max}}(k+N) \end{bmatrix} \tag{1.8}$$

as well as the weighting matrices $\boldsymbol{M} = \mathrm{diag}(\boldsymbol{M}_1, \ldots, \boldsymbol{M}_N) \geq 0$ of dimensionality $n_{\mathrm{y}}N \times n_{\mathrm{y}}N$ and $\boldsymbol{\Lambda} = \mathrm{diag}(\boldsymbol{\Lambda}_0, \ldots, \boldsymbol{\Lambda}_{N_{\mathrm{u}}-1}) > 0$ of dimensionality $n_{\mathrm{u}}N_{\mathrm{u}} \times n_{\mathrm{u}}N_{\mathrm{u}}$, the minimised cost-function used in the optimisation task (1.6) can be expressed in the compact form

$$J(k) = \|\boldsymbol{y}^{\mathrm{sp}}(k) - \hat{\boldsymbol{y}}(k)\|^2_{\boldsymbol{M}} + \|\triangle \boldsymbol{u}(k)\|^2_{\boldsymbol{\Lambda}} + \rho^{\mathrm{min}} \left\|\boldsymbol{\varepsilon}^{\mathrm{min}}\right\|^2 + \rho^{\mathrm{max}} \left\|\boldsymbol{\varepsilon}^{\mathrm{max}}\right\|^2$$

Defining the input constraints vectors of length $n_{\mathrm{u}}N_{\mathrm{u}}$

$$\boldsymbol{u}^{\mathrm{min}} = \begin{bmatrix} u^{\mathrm{min}} \\ \vdots \\ u^{\mathrm{min}} \end{bmatrix}, \ \boldsymbol{u}^{\mathrm{max}} = \begin{bmatrix} u^{\mathrm{max}} \\ \vdots \\ u^{\mathrm{max}} \end{bmatrix}, \ \triangle \boldsymbol{u}^{\mathrm{max}} = \begin{bmatrix} \triangle u^{\mathrm{max}} \\ \vdots \\ \triangle u^{\mathrm{max}} \end{bmatrix}$$

and the output constraints vectors of length $n_{\mathrm{y}}N$

$$\boldsymbol{y}^{\mathrm{min}} = \begin{bmatrix} y^{\mathrm{min}} \\ \vdots \\ y^{\mathrm{min}} \end{bmatrix}, \ \boldsymbol{y}^{\mathrm{max}} = \begin{bmatrix} y^{\mathrm{max}} \\ \vdots \\ y^{\mathrm{max}} \end{bmatrix}$$

the general MPC optimisation problem with soft output constraints (1.6), which must be solved on-line at each sampling instant, can be expressed as

$$\begin{aligned} \min_{\substack{\boldsymbol{u}(k) \text{ or } \triangle \boldsymbol{u}(k) \\ \boldsymbol{\varepsilon}^{\mathrm{min}}(k),\ \boldsymbol{\varepsilon}^{\mathrm{max}}(k)}} & \Big\{ J(k) = \|\boldsymbol{y}^{\mathrm{sp}}(k) - \hat{\boldsymbol{y}}(k)\|^2_{\boldsymbol{M}} + \|\triangle \boldsymbol{u}(k)\|^2_{\boldsymbol{\Lambda}} \\ & \quad + \rho^{\mathrm{min}} \left\|\boldsymbol{\varepsilon}^{\mathrm{min}}(k)\right\|^2 + \rho^{\mathrm{max}} \left\|\boldsymbol{\varepsilon}^{\mathrm{max}}(k)\right\|^2 \Big\} \\ \text{subject to} & \\ & \boldsymbol{u}^{\mathrm{min}} \leq \boldsymbol{u}(k) \leq \boldsymbol{u}^{\mathrm{max}} \\ & -\triangle \boldsymbol{u}^{\mathrm{max}} \leq \triangle \boldsymbol{u}(k) \leq \triangle \boldsymbol{u}^{\mathrm{max}} \\ & \boldsymbol{y}^{\mathrm{min}} - \boldsymbol{\varepsilon}^{\mathrm{min}}(k) \leq \hat{\boldsymbol{y}}(k) \leq \boldsymbol{y}^{\mathrm{max}} + \boldsymbol{\varepsilon}^{\mathrm{max}}(k) \\ & \boldsymbol{\varepsilon}^{\mathrm{min}}(k) \geq 0,\ \boldsymbol{\varepsilon}^{\mathrm{max}}(k) \geq 0 \end{aligned} \tag{1.9}$$

If the future changes of the set-point are not known at the current sampling instant, the set-point vector is

$$\boldsymbol{y}^{\mathrm{sp}}(k) = \begin{bmatrix} y^{\mathrm{sp}}(k) \\ \vdots \\ y^{\mathrm{sp}}(k) \end{bmatrix}$$

In MPC optimisation problems (1.6) and (1.9) the number of the additional decision variables, the role of which is to prevent the feasible set of being empty, $(2n_{\mathrm{y}}N)$, may be significantly greater than the length of the calculated future control sequence $(n_{\mathrm{u}}N_{\mathrm{u}})$. It may happen because the prediction horizon is typically much longer than the control one. In order to reduce computational burden one may decrease the number of decision variables of the MPC optimisation problem by assuming that the same vectors $\varepsilon^{\min}(k) \in \mathbb{R}^{n_{\mathrm{y}}}$ and $\varepsilon^{\max}(k) \in \mathbb{R}^{n_{\mathrm{y}}}$ are used over the whole prediction horizon. Of course, it leads to some suboptimality, but the number of decision variables drops to $n_{\mathrm{u}}N_{\mathrm{u}} + 2n_{\mathrm{y}}$ and it is independent of the prediction horizon. The MPC optimisation problem is then

$$\begin{aligned}
&\min_{\substack{\boldsymbol{u}(k) \text{ or } \triangle\boldsymbol{u}(k) \\ \varepsilon^{\min}(k),\ \varepsilon^{\max}(k)}} \Bigg\{ J(k) = \sum_{p=1}^{N} \|y^{\mathrm{sp}}(k+p|k) - \hat{y}(k+p|k)\|^2_{\boldsymbol{M}_p} \\
&\qquad + \sum_{p=0}^{N_{\mathrm{u}}-1} \|\triangle u(k+p|k)\|^2_{\boldsymbol{\Lambda}_p} \\
&\qquad + \rho^{\min} \left\|\varepsilon^{\min}(k)\right\|^2 + \rho^{\max} \left\|\varepsilon^{\max}(k)\right\|^2 \Bigg\} \\
&\text{subject to} \\
&u^{\min} \leq u(k+p|k) \leq u^{\max},\ p = 0, \ldots, N_{\mathrm{u}} - 1 \\
&-\triangle u^{\max} \leq \triangle u(k+p|k) \leq \triangle u^{\max},\ p = 0, \ldots, N_{\mathrm{u}} - 1 \\
&y^{\min} - \varepsilon^{\min}(k) \leq \hat{y}(k+p|k) \leq y^{\max} + \varepsilon^{\max}(k),\ p = 1, \ldots, N \\
&\varepsilon^{\min}(k) \geq 0,\ \varepsilon^{\max}(k) \geq 0
\end{aligned} \tag{1.10}$$

The optimisation task (1.10) can be expressed in the compact way

$$\begin{aligned}
&\min_{\substack{\boldsymbol{u}(k) \text{ or } \triangle\boldsymbol{u}(k) \\ \varepsilon^{\min}(k),\ \varepsilon^{\max}(k)}} \Big\{ J(k) = \|\boldsymbol{y}^{\mathrm{sp}}(k) - \hat{\boldsymbol{y}}(k)\|^2_{\boldsymbol{M}} + \|\triangle\boldsymbol{u}(k)\|^2_{\boldsymbol{\Lambda}} \\
&\qquad + \rho^{\min} \left\|\varepsilon^{\min}(k)\right\|^2 + \rho^{\max} \left\|\varepsilon^{\max}(k)\right\|^2 \Big\} \\
&\text{subject to} \\
&\boldsymbol{u}^{\min} \leq \boldsymbol{u}(k) \leq \boldsymbol{u}^{\max} \\
&-\triangle\boldsymbol{u}^{\max} \leq \triangle\boldsymbol{u}(k) \leq \triangle\boldsymbol{u}^{\max} \\
&\boldsymbol{y}^{\min} - \boldsymbol{\varepsilon}^{\min}(k) \leq \hat{\boldsymbol{y}}(k) \leq \boldsymbol{y}^{\max} + \boldsymbol{\varepsilon}^{\max}(k) \\
&\varepsilon^{\min}(k) \geq 0,\ \varepsilon^{\max}(k) \geq 0
\end{aligned} \tag{1.11}$$

The additional decision variables of the MPC optimisation task (1.9) are the vectors $\boldsymbol{\varepsilon}^{\min}(k)$ and $\boldsymbol{\varepsilon}^{\max}(k)$ of length $n_{\mathrm{y}}N$, whereas in the optimisation problem (1.11) the additional variables are the vectors $\varepsilon^{\min}(k)$ and $\varepsilon^{\max}(k)$ of length n_{y}. The auxiliary vectors of length $n_{\mathrm{y}}N$ used in the optimisation problem (1.11) in the constraints imposed on the predicted output variables are

$$\boldsymbol{\varepsilon}^{\min}(k) = \begin{bmatrix} \varepsilon^{\min}(k) \\ \vdots \\ \varepsilon^{\min}(k) \end{bmatrix}, \ \boldsymbol{\varepsilon}^{\max}(k) = \begin{bmatrix} \varepsilon^{\max}(k) \\ \vdots \\ \varepsilon^{\max}(k) \end{bmatrix} \tag{1.12}$$

If the number of process input and output variables are equal ($n_{\mathrm{u}} = n_{\mathrm{y}}$) the optimisation problems formulated above have unique solutions. If $n_{\mathrm{u}} > n_{\mathrm{y}}$, in order to avoid non-uniqueness of the solution, the minimised cost-function should include an additional penalty term

$$\sum_{p=0}^{N_{\mathrm{u}}-1} \rho_p \left\| u(k+p|k) - u^{\mathrm{sp}}(k) \right\|^2$$

where the input set-point vector $u^{\mathrm{sp}}(k)$ corresponds to the output set-point vector $y^{\mathrm{sp}}(k)$. As an alternative, one can take into account the constraint $u(k + N_{\mathrm{u}} - 1|k) = u^{\mathrm{sp}}(k)$ or use the penalty term $\rho \left\| u(k+N_{\mathrm{u}}-1|k) - u^{\mathrm{sp}}(k) \right\|^2$), which is better from the numerical point of view. If $n_{\mathrm{u}} < n_{\mathrm{y}}$, which may be caused by an actuator malfunction, the MPC algorithm does not have the sufficient degree of freedom.

1.2 Prediction

The model is a very important part of any MPC algorithm. Its role is to calculate on-line the predicted values of the output variables which are next used to find the predicted control errors minimised in the MPC cost-function (1.3) or (1.4). During prediction calculation it is necessary to take into account the fact that the model is usually not perfect and the process may be affected by some unmeasured disturbances. The general vector prediction equation for the sampling instant $k+p$ is

$$\hat{y}(k+p|k) = y(k+p|k) + d(k) \tag{1.13}$$

where $p = 1, \ldots, N$. The scalar form of the prediction equation is

$$\hat{y}_m(k+p|k) = y_m(k+p|k) + d_m(k) \tag{1.14}$$

where $m = 1, \ldots, n_{\mathrm{y}}$, the mth output of the model for the sampling instant $k+p$ calculated at the current instant k is $y_m(k+p|k)$, the current estimation of the unmeasured disturbance acting on the mth process output is $d_m(k)$. In the most typical approach ("the DMC disturbance model") it is assumed that the disturbance is constant over the whole prediction horizon and its value is determined as the difference between the real (measured) value of the process output and the model output calculated using process input and output signals up the sampling instant $k-1$

$$d_m(k) = y_m(k) - y_m(k|k-1) \tag{1.15}$$

When the unmeasured disturbance estimation is used in the prediction equation, the MPC algorithm has the integral action (no steady-state error), even if the model is not perfect and the process is affected by some unmeasured disturbances.

The discrete-time dynamic model of the single-input single-output process is given by the following equation (in general the model is nonlinear)

$$y(k) = f(u(k-\tau), \ldots, u(k-n_{\mathrm{B}}), y(k-1), \ldots, y(k-n_{\mathrm{A}})) \tag{1.16}$$

where τ is the time-delay, the integer numbers n_{A} and n_{B} ($\tau \leq n_{\mathrm{B}}$) determine the order of dynamics and the function $f\colon \mathbb{R}^{n_{\mathrm{A}}+n_{\mathrm{B}}-\tau+1} \to \mathbb{R}$ describes the model. Using the general prediction equation (1.14) and recurrently the model (1.16), the consecutive output predictions over the prediction equation ($p = 1, \ldots, N$) can be expressed as

$$\begin{aligned}\hat{y}(k+p|k) = f(&\underbrace{u(k-\tau+p|k), \ldots, u(k|k)}_{I_{\mathrm{uf}}(p)}, \underbrace{u(k-1), \ldots, u(k-n_{\mathrm{B}}+p)}_{I_{\mathrm{u}}-I_{\mathrm{uf}}(p)},\\ &\underbrace{\hat{y}(k-1+p|k), \ldots, \hat{y}(k+1|k)}_{I_{\mathrm{yf}}(p)}, \underbrace{y(k), \ldots, y(k-n_{\mathrm{A}}+p)}_{n_{\mathrm{A}}-I_{\mathrm{yf}}(p)})\\ &+ d(k)\end{aligned} \tag{1.17}$$

The predictions $\hat{y}(k+p|k)$ depend on: $I_{\mathrm{uf}}(p) = \max(\min(p-\tau+1, I_{\mathrm{u}}), 0)$ (where $I_{\mathrm{u}} = n_{\mathrm{B}} - \tau + 1$) future values of the control signal (which are the decision variables of the MPC algorithm), $I_{\mathrm{u}} - I_{\mathrm{uf}}(p)$ values of the control signal used for control in previous iterations, $I_{\mathrm{yf}}(p) = \min(p-1, n_{\mathrm{A}})$ predicted values of the output signal and $n_{\mathrm{A}} - I_{\mathrm{yf}}(p)$ values of the process output measured in previous iterations. Using the equation (1.15), the unmeasured disturbance is estimated from

$$d(k) = y(k) - f(u(k-\tau), \ldots, u(k-n_{\mathrm{B}}), y(k-1), \ldots, y(k-n_{\mathrm{A}}))$$

1.3 Linear and Nonlinear MPC Algorithms

If a linear model is used for prediction, the predictions (1.17) are linear functions of the future control policy, which are the decision variables of the MPC algorithm. In such a case the minimised cost-function (1.4) is quadratic. In consequence, the MPC optimisation problem (1.9) or (1.11) is a quadratic optimisation task (the objective cost-function is quadratic, the constraints are linear). Its unique global minimum can be efficiently calculated by means of the active set method or the interior point method [226] in some limited number of iterations, it is possible to estimate the calculation time. The MPC approaches which use for prediction linear models are usually named, also in this book, the linear MPC algorithms. Although the name is used and accepted, it is not precise, because the real linear MPC algorithm is when no constraints are present which makes it possible to derive an explicit linear control law.

Typically, properties of the majority of technological processes are nonlinear. If nonlinearity is not significant, the linear model can be used and the linear MPC algorithm gives quite good control quality (the unavoidable prediction inaccuracy does not lead to the steady-state error because the algorithm has the integral action). If the process is significantly nonlinear, the linear model may be completely inadequate and the linear MPC algorithm is likely to give unsatisfactory control quality (the algorithm may be too slow or unstable, depending on the operating point). In such a case it is necessary to use for prediction a nonlinear model. Because the general idea of MPC is of course the same, one may expect that replacing the linear model by the nonlinear one does not make a big difference. Unfortunately, it is not true, the nonlinear MPC algorithm is completely different from the linear one. The output predictions (1.17) are nonlinear functions of the future control policy (the decisions variables of the algorithm). In consequence, the MPC optimisation problem becomes a nonlinear task: the minimised cost-function is nonlinear, the output constraints are also nonlinear. In general, the nonlinear MPC optimisation problem may be non-convex, it may have many local minima.

Despite huge progress in nonlinear optimisation, the universal algorithm which would guarantee finding the global solution to the nonlinear optimisation task has not been found yet. It is quite risky to use nonlinear optimisation on-line in MPC because it is impossible to check whether or not the solution found is the global one. Furthermore, it is impossible to guarantee that the solution is found at each iteration in the necessary time, which is particularly important when the process is fast and the sampling time is short. On the one hand, the gradient optimisation methods [226] may find local minima which give unsatisfactory control quality. On the other hand, global optimisation methods, such as evolutionary algorithms [283], are heuristic by nature. Although they can be successfully used off-line, for example to find the structure and parameters of nonlinear models [283, 332], but convergence and computational complexity issues seem to prevent from using them for on-line MPC optimisation, although such solutions are reported in the literature [72, 190, 233, 254]. There are also attempts to use simulated annealing for on-line nonlinear optimisation in MPC [2]. According to the authors of the cited publications heuristic nonlinear optimisation works, but, unfortunately, the necessity of using the heuristic approach to optimisation is not justified. For single-input single-output processes, provided that the control horizon $N_u = 1$ and no constraints are present, it is possible to use for optimisation the golden section search method [190]. Unfortunately, the mentioned limitations, in particular the lack of constraints, may turn out to be very important disadvantages.

The need to reduce computation complexity of the general nonlinear MPC approach with on-line nonlinear optimisation is an important issue. Any simplification of the optimisation task or even elimination of the necessity of solving it at each sampling instant is desirable. There are a few such methods in the literature.

In some specific cases it is possible to calculate the first element of the future control policy by means of nonlinear optimisation whereas the remaining ones are found from an explicit control law (similarly to the explicit linear MPC) [343]. Of course, the optimisation problem is still nonlinear, but the number of decision variables is reduced. An important reduction of computational burden may be possible when the model has a special structure. The nonlinear polynomial model can be transformed in such a way that the minimised cost-function is convex [305]. For the nonlinear second-order Volterra model the MPC optimisation problem is nonlinear, but the solution may be obtained by solving a number of quadratic optimisation tasks [64, 191]. An inverse of the steady-state part of the Hammerstein or Wiener model is usually used to compensate for nonlinearity of the process which makes it possible to formulate a quadratic optimisation MPC problem [6, 44, 73, 229, 230, 296].

If the following affine model

$$\begin{aligned} x(k+1) &= f(x(k)) + g(x(k))u(k) \\ y(k) &= h(x(k)) \end{aligned}$$

is used, it can be easily linearised using the feedback linearisation technique [112]. Thanks to linearisation, if no constraints are present, it is possible to derive an explicit control law [97]. Taking into account linear constraints imposed on magnitude of the manipulated variable, it leads to a nonlinear optimisation problem. It can be solved by successive linearisation, which, in turn, results in a quadratic optimisation problem [14].

An alternative method of reducing computational complexity of nonlinear MPC is to use the model in which the output values for consecutive sampling instants within the prediction horizon are linear functions of the future control policy but are nonlinear functions of the past (the quasi-linear model) [138]

$$\begin{aligned} y(k+1|k) &= f^1(\widetilde{\boldsymbol{x}}(k)) + g_0^1(\widetilde{\boldsymbol{x}}(k))u(k|k) \\ y(k+2|k) &= f^2(\widetilde{\boldsymbol{x}}(k)) + g_0^2(\widetilde{\boldsymbol{x}}(k))u(k|k) + g_1^2(\widetilde{\boldsymbol{x}}(k))u(k+1|k) \\ &\vdots \\ y(k+N|k) &= f^N(\widetilde{\boldsymbol{x}}(k)) + \sum_{i=0}^{N-1} g_i^N(\widetilde{\boldsymbol{x}}(k))u(k+i|k) \end{aligned} \tag{1.18}$$

where the vector $\widetilde{x}(k) = [u(k-1) \ldots u(k-n_{\mathrm{B}})\ y(k) \ldots y(k-n_{\mathrm{A}})]^{\mathrm{T}}$ consists of the signals recorded in the previous sampling instants, it defines the current operating point of the process. The functions $f^p(\widetilde{x}(k))$ and $g_i^p(\widetilde{x}(k))$ may be realised by neural networks. If the quasi-linear model is used for prediction, one obtains a quadratic optimisation MPC optimisation problem. Unfortunately, in general, prediction accuracy of the quasi-linear model may be lower than that of the general nonlinear model (1.16). The next model structure developed with the aim of reducing computational burden of MPC is [251]

$$\begin{aligned}\delta y(k) = & \, b_\tau(k)\delta u(k-\tau) + b_{\tau+1}(k)\delta u(k-\tau-1)\\ & \tilde{b}_\tau(k)(\delta u(k-\tau))^2 + \tilde{b}_{\tau+1}(k)(\delta u(k-\tau-1))^2\\ & - a_1\delta y(k-1) - a_2\delta y(k-2)\end{aligned}$$

where $\delta u = u(k) - u^{\mathrm{ss}}$ and $\delta y = y(k) - y^{\mathrm{ss}}$, the quantities u^{ss} and y^{ss} are found from a nonlinear (neural) steady-state model, i.e. $y^{\mathrm{ss}} = f^{\mathrm{ss}}(u^{\mathrm{ss}})$. The second-order dynamics is used, because it turns out to be sufficient in the majority of applications. The model is quadratic with respect to the input signal u and the time-varying coefficients $b_i(k)$ and $\tilde{b}_i(k)$ are calculated on-line using the nonlinear steady-state model and for a specific target set-point. In spite of the fact that the resulting MPC optimisation problem is not of quadratic optimisation type, it is far less demanding than the general nonlinear one (usually non-convex).

From the practical point of view, the MPC algorithms with successive on-line linearisation are the most important ones. The first solutions of this kind, analogously to the first linear MPC algorithms, were developed for chemical engineering [76, 78]. Experts in this field know nonlinear models of processes, hence, it seems straightforward to use them in MPC. In order to circumvent the necessity of repeating nonlinear optimisation at each sampling instant, a linear approximation of the nonlinear model is successively found on-line for the current operating point. Next, the linearised model is used for prediction. Thanks to linearisation, analogously to the linear MPC algorithms, the consecutive predictions are linear function of the future control policy and the minimised cost-function is quadratic which leads to a quadratic MPC optimisation problem. Because the linearised model is only an approximation of the original nonlinear one (and of the real process), the MPC algorithms with on-line model linearisation are suboptimal by nature. Nevertheless, they are appreciated by practitioners and used in many applications: *"Linearization is the only method which has found any wider use in industry beyond demonstration projects. For industry there has to be clear justification for solving non-linear programs on-line in a dynamic setting and there are no examples to bear that out in a convincing manner."* [215]. Although there are numerous theoretical investigations of the MPC approach with on-line nonlinear optimisation, but properties of the MPC algorithms with on-line linearisation are not discussed frequently in the literature (*"the theoretical purists tend to stay away from linearization approaches"* [215]). In fact, successive on-line linearisation makes stability and robustness investigations much more difficult than in the case of MPC with nonlinear optimisation.

To summarise, the main categories of MPC structures are:

a) the linear MPC algorithms with quadratic optimisation,
b) the nonlinear MPC algorithms with nonlinear optimisation,
c) the suboptimal MPC algorithms with successive on-line linearisation and quadratic optimisation.

It is necessary to point out that numerous applications of MPC algorithms motivated the development of many optimisation methods tailored especially for MPC, both quadratic [17, 211] and nonlinear ones [33, 195, 330]. An interesting review of the existing optimisation methods is given in [29], the possible directions of future research in this field are given in [86].

It is an interesting idea to use the fast MPC [329]. In this approach the MPC optimisation task is not solved precisely, but in an approximate way. Of course, it is likely to affect the quality of control, but computational burden can be significantly reduced. A yet another solution is to transform the MPC optimisation problem into a set of equations, which is next solved by means of the Newton method [246].

Explicit MPC algorithms with constraints have been extensively researched in recent years. It can be proven that for a linear model and for the typical quadratic cost-function, the optimal current value of the control signal is a function of the state [20, 322]. The whole state domain is divided into some sets, for each set the explicit control law is derived off-line. During on-line control it is necessary to find out to which set the current state of the process belongs and to use the corresponding precalculated control law. The general idea seems to be simple, but in practice the whole number of sets may be high. There are also attempts to develop similar MPC algorithms for nonlinear processes [105].

The next idea worth mentioning is to replace the numerical optimisation routine by a specially designed neural network. There are a few neural structures which solve the quadratic optimisation problem [139, 327]. The network described in [139] is used for optimisation in a linear MPC algorithm [238] and in a suboptimal MPC algorithm with on-line model linearisation [239].

1.4 History of MPC Algorithms

The well-known linear-quadratic regulator minimises the following cost-function

$$J(k) = \sum_{p=1}^{\infty} \left[\|x(k+p)\|^2_{\boldsymbol{Q}} + \|u(k+p)\|^2_{\boldsymbol{R}} \right]$$

Provided that a linear model is used, one obtains the explicit linear control law with the state feedback $u(k) = -\boldsymbol{K}x(k)$. Although linear-quadratic regulators were used in many applications, in particular in aviation and aerospace, they did not become popular in process control. For typical industrial processes, such as chemical reactors and distillation columns, the satisfaction of constraints is fundamentally important because the best operating point (which maximises the profit) is usually located in the vicinity of the constraints or even at the constraints.

The computer was used for control for the first time in 1959 (in a polymerisation reactor control system) [15]. That was a revolutionary breakthrough. The computer made it possible to use advanced control algorithms, not only PID controllers. The mathematical model of the process, previously used only for developing the controller, could be used directly on-line in the control algorithm. The new approach paved the way for control algorithms which could deal with multi-input multi-output nonlinear processes, in which the constraints may be taken into account. Moreover, the computer made it possible to design on-line optimisation systems.

The general idea of using a mathematical model of the process for prediction in the control algorithm is not completely new. Such an approach is used in the Internal Model Control (IMC) algorithm [77]. In spite of the fact that the idea of repetitive control with a moving horizon [71], which is practically used in all MPC algorithms, was formulated in the 1960s [127, 260], but it is difficult to precisely indicate when and where the first MPC structure was developed. It is because some preliminary versions of MPC algorithms were designed and implemented in the industry, reports of such works were published with a few years of delay [186]. It is an interesting fact that some of the very early computer control algorithms are actually archetypes of the suboptimal MPC algorithms used nowadays [263].

The first MPC algorithms were designed by engineers who worked in the industry, they were not invented by the university researches. That is why the MPC solutions were developed with applications in mind. The first works concerned with the Model Heuristic Predictive Control (MHPC) algorithm were published in 1976 and 1978 [272, 273], the algorithm was later known under the name Model Algorithmic Control (MAC) [282]. A unique feature of the algorithm is the fact that a linear finite impulse-response model is used for prediction. The minimised cost-function takes into account the sum of the predicted deviations from the reference trajectory. Such an approach is used to reduce excessive changes of the control signal because the typical penalty term (1.3) is not present in the MAC cost-function. According to the authors of the MAC algorithm, the simplicity of tuning and good application perspectives are its most important advantages. Because for optimisation a heuristic procedure is used, optimisation accuracy is not of crucial importance. Similarly, satisfaction of constraints is simply checked during calculations.

The Dynamic Matrix Control (DMC) algorithm was devised by the engineers who worked for Shell. It turned out to be very popular, in particular in chemical and petrochemical industry. The algorithm was applied for the first time in 1973 [8, 262], but its description was published a few years later [57].

A unique feature of the DMC algorithm is the fact that a linear finite step-response model is used for prediction. Unlike the MHPC (MAC) algorithm, the minimised cost-function takes into account not only the predicted control errors, but also the penalty term, which makes it possible to reduce excessive changes of the control signal and to improve the numerical properties of the optimisation problem. The vector of optimal increments of the manipulated variable can be calculated explicitly because optimisation of a quadratic function leads to the least-squares problem. It makes it possible to implement the DMC algorithm using relatively uncomplicated and inexpensive hardware.

The MHPC (MAC) and DMC algorithms belong to the first generation of MPC algorithms [8, 186, 262]. It is possible to state that their industrial applications made a revolutionary breakthrough in process control. In comparison with the classical PID controller (usually single loop ones), the MPC technique is able to efficiently control multi-input multi-output processes, such as chemical reactors and distillation columns. The principal idea of MPC and its tuning procedure are relatively simple, they can be understood by the process operating staff, which is of big importance when new ideas are introduced in the industry. The next advantage of MHPC (MAC) and DMC algorithms is the fact that impulse-response or step-response linear models are used. Although for describing property of a typical single-input single-output process it is necessary to use a few dozens of coefficients, but model interpretation is very intuitive. Moreover, model identification is quite simple, which is very important in the industry. It is necessary to add that the popularity of MHPC (MAC) and DMC algorithms is also caused by availability of software packages for model identification and algorithm development [95]. The most famous are the IDentification and COMmand (IDCOM) package for the MHPC (MAC) algorithm and the DMC software environment for the DMC scheme.

Approximate (heuristic) constraints handling, in particular imposed on the predicted output variables, is the fundamental disadvantage of the first generation of MPC algorithms. The problem was solved in the Quadratic Dynamic Matrix Control (QDMC) algorithm [75], which belongs to the second generation of MPC. It can be noticed that taking into account linear constraints imposed on the input signals and on the predicted output signals, provided that the minimised cost-function is quadratic which is true when the model is linear, leads to a quadratic optimisation problem. It must be solved on-line at each sampling instant. Although, on the one hand, the DMC and QDMC structures were developed many, many years ago, they are still very popular in the industry. They are usually available in the majority of professional MPC software packages [41, 186].

Rapid development of electronics, in particular microprocessor systems, greatly stimulated applications of MPC algorithms. Because computational burden of MPC is much higher than that of PID, MPC algorithms were initially used in supervisory control where the sampling time is relatively long, but the key issue is multivariable control in the presence of constraints.

After some years new, more powerful and cheaper generations of control hardware are available. As a result it is now possible to use MPC algorithms not only in supervisory control, but also in basic control.

Great industrial success of MPC algorithms was a factor which attracted the attention of university researches. In the 1980s the Generalized Predictive Control (GPC) algorithm [54] was developed. The algorithm uses a discrete-time input-output linear model. Unlike the step-response model used in DMC algorithm, such a model is able to describe a very broad class of processes, including unstable processes or processes with integration. Moreover, different disturbance models can be used. Interesting extensions of the GPC algorithm are the MPC structures with theoretically guaranteed stability. The most important and well-known such approaches are: the Stabilizing Input Output Receding Horizon Control (SIORHC) algorithm [23] and the Constrained Receding Horizon Predictive Control (CRHPC) algorithm [52]. In both control schemes some additional equality constraints imposed on the predicted outputs enforce stability. The algorithm GPC$^{\infty}$ [293] uses a different technique, since the prediction horizon is infinite, but the number of decision variables (determined by the control horizon) is finite.

The ideal of MPC can be also used in adaptive control, in which a time-varying model (typically a linear one) is successively identified on-line. Hence, the adaptive MPC approach can be applied for nonlinear processes [225]. Example members of this class are: the Predictor Based Self Tuning Controller [250], the Extended Horizon Adaptive Control (EHAC) algorithm [335], the Extended Prediction Self Adaptive Control (EPSAC) algorithm [111] and the Multistep Multivariable Adaptive Control (MUSMAR) algorithm [83].

Although in the second generation of MPC algorithms the constraints constitute an integral part of the optimisation problem, but some important issues are not addressed. If the output constraints are present, the problem of enlarging the feasible set may be solved by means of a simple constraints' window method (the output constraints for some first part of the prediction horizon are neglected) or by means of the soft constraints method (the constraints may be temporarily violated). A unique feature of the third generation of MPC algorithms is introduction of different kinds of constraints, which may have different priority. Moreover, the efficient methods for enlarging the feasible set are used. It is also possible to used different cost-functions and multicriterial optimisation. The Shell Multivariable Optimizing Controller (SMOC) [194] can be given as an example of such an algorithm. State-space linear models are used in the SMOC approach, which makes it possible to describe a very broad class of dynamic processes (the state-space approach is much more flexible than the input-output one). A state observer is used when necessary. Two categories of the output variables can be used: controlled ones and measured but uncontrolled ones.

Huge progress in control hardware platforms and development of new modelling and identification methods made it possible to consider practical applications of nonlinear structures which constitute the fourth generation of

MPC algorithms [21, 40, 203]. The most important issues in this field are stability and robustness problems. Of course, theoretical analysis of nonlinear control approaches is much more difficult than in the linear cases.

Nowadays the field of MPC blooms. The development is twofold: intensive research is carried out and the algorithms are applied to numerous processes. Both research and practical directions stimulate each other, which contribute to the progress of MPC. The most important scientific journals in the field of automatic control (e.g. Automatica, Control Engineering Practice, Journal of Process Control) publish each year many articles concerned with new algorithms, theoretical analysis and application reports. Example important books in this field include the works of: E. F. Camacho and C. Bordons [41], J. M. Maciejowski [186], J. B. Rawlings and D. Q. Mayne [268], J. A. Rossiter [278] and P. Tatjewski [316]. Example review articles (or book chapters) include the works of: F. Allgöwer, T. A. Badgwell, J. S. Qin, J. B. Rawlings and S. J. Wright [8], A. Bemporad and M. Morar [21], E. F. Camacho and C. Bordons [40], C. E. Garcia, D. M. Prett and M. Morari [74], M. A. Henson [95], L. Magni and R. Scattolini [187], D. Q. Mayne, J. B. Rawlings, C. V. Rao and P. O. M. Scokaert [203], M. Morari and J. H. Lee [215], S. J. Qin and T. A. Badgwell [262], J. B. Rawlings [269] and P. Tatjewski [315]. S. J. Qin and T. A. Badgwell mention in [262] a few thousands of applications of linear MPC algorithms, in particular in the refinery, petrochemical, chemical, paper and food processing industries. They noticed that the number of application doubled in the period of 5 years. Usually, the considered processes are multivariable, with a few dozens or even hundreds of input and output variables. It is also interesting to give less typical processes for which MPC technique can be used (real applications or simulation studies). They include robots [234], combustion engines [55, 286], airplane engines [218] and airplanes [110]. MPC algorithms can be used to control: a greenhouse [88], a solar power station [12, 25], a cement kiln [306], a warehouse in a supermarket [285], a waste water treatment plant [37], a fuel cell [89], traffic on a highway [19], a computer network [192], gas pipeline networks [345], a high energy accelerator [31], insulin delivery for diabetes [161, 323] or anaesthesia during operation [190].

There are a few professional software packages for modelling and developing MPC algorithms. Some typical examples are: np. DMC Plus from AspenTech, IDCOM-HEICON from Sherpa Engineering, SMOC from Shell Global Solutions, 3dMPC from ABB, Profit Controller and Profit NLC Controller from Honeywell, Pavilion8 from Pavilion Technologies, Connoisseur from Invensys, INCA MPC from IPCOS, DeltaV PredictPro from Emerson and MVC from GE Energy [186, 262]. Availability of such packages greatly contributed to the industrial success of MPC algorithms.

1.5 Process Models in MPC Algorithms

Because the model of the process is a key element in MPC algorithms, it is necessary to give desirable properties of a good model [248]:

a) approximation accuracy,
b) physical interpretation,
c) suitability for control,
d) easiness of development.

Model selection determines not only accuracy of control, but also computational complexity. The proper choice of the model structure affects reliability of the designed control system. One may notice that some of the above criteria are contradictory. Linear models may be found in a relatively simple way, the linear MPC algorithms are simple to design and are computationally uncomplicated. Unfortunately, the obtained control quality may be unsatisfactory for nonlinear systems, in particular when the operating point is changed significantly and fast. In such cases nonlinear models are straightforward, but their identification is more demanding. Furthermore, complexity of nonlinear MPC algorithms is higher than that of the classical linear ones.

The key issue is the type of the nonlinear model. There are three general classes of models [300]:

a) First-principle (fundamental) models (white-box models): they are developed using technological knowledge. Their parameters always have physical interpretation.
b) Empirical models (black-box models): for identification no technological knowledge is necessary, their structure is usually chosen arbitrarily and parameters, which have no physical interpretation, are found using some data set recorded during process operation.
c) Hybrid models (grey-box models) [261, 321, 333]: both technological knowledge and data sets are used during identifications. Usually, the general structure is motivated by technological considerations, but parameters are determined from data (the serial hybrid model). If the first-principle model exists, but it is not precise enough, one may use a parallel hybrid model which consists of the first-principle part and the black-box one.

Of course, one may attempt to develop first-principle models which are expected to precisely describe the processes [141, 193]. Physical interpretation is often the only advantage of such models and they have some important disadvantages. Firstly, the biggest difficulty, in particular for non-technologists, is to formulate all the necessary equations and validate the models. First-principle models usually have some parameters which must be determined experimentally. Secondly, such models may be too complicated to be used for control. For example, the first-principle model of the distillation column discussed in Example 3.2 consists of some 17 thousands of differential equations [331]. For prediction calculation in MPC they should be successively solved on-line. Obviously, it would be quite difficult (or even impossible) in practice. That is why empirical models are more interesting. They can be determined by non-technologists and are usually much less complicated than the first-principle ones. There are, however, processes, for which the first-principle models are quite uncomplicated and may be obtained quite easily, e.g. DC motors or robots.

1.5.1 Example Classes of Empirical Models of Nonlinear Processes

They are numerous types of nonlinear models described by the general equation (1.16). It is possible to use, among others, polynomials, bi-linear models, piece-wise structures, Volterra series models, cascade models, wavelets, fuzzy systems and neural networks [106, 223, 248, 300]. The mentioned models are different in terms of approximation accuracy and the number of parameters. Both these factors affect the possibility of using them in MPC algorithms. In the following part of the chapter some selected classes of models and their properties are shortly discussed.

Polynomial Models

Taking into account the list of arguments of the general nonlinear model (1.16), one may consider the following polynomial model [131]

$$\begin{aligned} y(k) = & \sum_{i=\tau}^{n_{\mathrm{B}}} \theta_{1,i} u(k-i) + \sum_{i=1}^{n_{\mathrm{A}}} \theta_{2,i} y(k-i) \\ & + \sum_{i=\tau}^{n_{\mathrm{B}}} \sum_{j=\tau}^{i} \theta_{3,i,j} u(k-i) u(k-j) + \sum_{i=1}^{n_{\mathrm{A}}} \sum_{j=1}^{i} \theta_{4,i,j} y(k-i) y(k-j) \\ & + \sum_{i=\tau}^{n_{\mathrm{B}}} \sum_{j=1}^{n_{\mathrm{A}}} \theta_{5,i,j} u(k-i) y(k-j) + \ldots \end{aligned} \tag{1.19}$$

The main advantage of the model is the fact that the output is a linear function of model parameters. As a result, model identification is a least-squares problem. A high (or very high) number of parameters is an important disadvantage of the model. It soars as the order of dynamics (defined by the integers n_{A} and n_{B}) and the polynomial degree are increased. High complexity of polynomial models hampers their applications. That is why low order models are used, e.g. second-order or third-order, provided that they are precise enough. If the output is only a function of some previous values of the input signal, the model is named a Volterra series model [64], which is also known as the Volterra neural network [235].

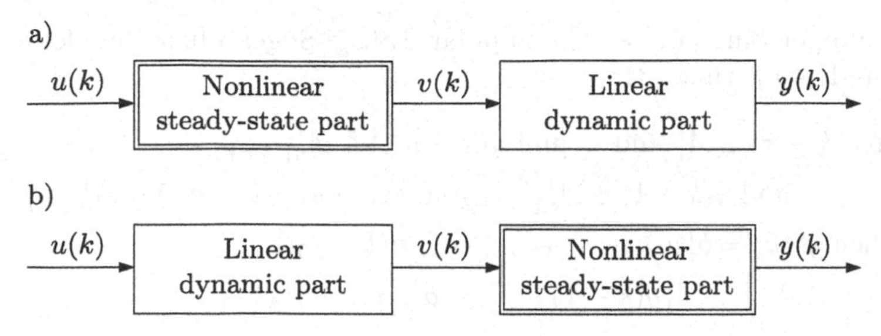

Fig. 1.3. Cascade (serial) models: a) Hammerstein model, b) Wiener model

Cascade (Serial) Models

For many processes one may notice that the steady-state properties are nonlinear (i.e. the steady-state characteristics of an actuator or of a sensor is nonlinear) whereas process dynamics is practically linear. In such cases cascade (serial) models can be used. There are two the most common types of such models: Hammerstein and Wiener structures depicted in Fig. 1.3. Both models consist of a nonlinear steady-state part and a linear dynamic part. In the Hammerstein structure the nonlinear part precedes the linear one, in the Wiener model the connection order is reversed. Although the cascade structures seem to be rather restrictive, they are quite frequently used in practice [103]. In the simplest case a polynomial is used in the steady-state nonlinear part, but it is also possible to use a potentially more flexible representation, e.g. a neural network.

In many cases a precise first-principle steady-state nonlinear model is known, but the full (dynamic) first-principle one is not known. It is because, typically, development of dynamic models is much more difficult than steady-state ones. In such cases the nonlinear part of the cascade model can be found using some data generated from the first-principle steady-state model whereas the linear part can be very efficiently determined using some recorded input-output process data [249].

If accuracy of the classic cascade models is not sufficient, one may use structures with 3 parts: the Hammerstein-Wiener model (2 nonlinear steady-state input and output parts, the linear dynamic part is in the middle) and the Wiener-Hammerstein models (2 linear dynamic input and output parts, the nonlinear steady-state part is in the middle). Unfortunately, such three-part cascade models may have many parameters, in particular in the case of multi-input multi-output processes.

Fuzzy Models

Fuzzy models are very intuitive [283]. One assumes that there are a few (or a few dozens or a few hundreds etc.) of simple models (usually linear) which quite precisely describe properties of the process in some neighbourhoods of typical operating points. The popular Takagi-Sugeno fuzzy model can be formulated as n_{R} rules [310]

$$\begin{aligned}
\mathrm{R}^1: \text{ if } & u(k-\tau) \in \mathcal{A}_1^1 \text{ and} \ldots \text{and } u(k-n_{\mathrm{B}}) \in \mathcal{A}_{n_{\mathrm{B}}-\tau+1}^1 \\
& \text{and } y(k-1) \in \mathcal{A}_{n_{\mathrm{B}}-\tau+2}^1 \text{ and} \ldots \text{and } y(k-n_{\mathrm{A}}) \in \mathcal{A}_{n_{\mathrm{A}}+n_{\mathrm{B}}-\tau+1}^1 \\
\text{then } & y^1(k) = b_\tau^1 u(k-\tau) + \ldots + b_{n_{\mathrm{B}}}^1 u(k-n_{\mathrm{B}}) \\
& \quad - a_1^1 y(k-1) - \ldots - a_{n_{\mathrm{A}}}^1 y(k-n_{\mathrm{A}}) \\
& \vdots \\
\mathrm{R}^{n_{\mathrm{R}}}: \text{ if } & u(k-\tau) \in \mathcal{A}_1^{n_{\mathrm{R}}} \text{ and} \ldots \text{and } u(k-n_{\mathrm{B}}) \in \mathcal{A}_{n_{\mathrm{B}}-\tau+1}^{n_{\mathrm{R}}} \\
& \text{and } y(k-1) \in \mathcal{A}_{n_{\mathrm{B}}-\tau+2}^{n_{\mathrm{R}}} \text{ and} \ldots \text{and } y(k-n_{\mathrm{A}}) \in \mathcal{A}_{n_{\mathrm{A}}+n_{\mathrm{B}}-\tau+1}^{n_{\mathrm{R}}} \\
\text{then } & y^{n_{\mathrm{R}}}(k) = b_\tau^{n_{\mathrm{R}}} u(k-\tau) + \ldots + b_{n_{\mathrm{B}}}^{n_{\mathrm{R}}} u(k-n_{\mathrm{B}}) \\
& \quad - a_1^{n_{\mathrm{R}}} y(k-1) - \ldots - a_{n_{\mathrm{A}}}^{n_{\mathrm{R}}} y(k-n_{\mathrm{A}})
\end{aligned}$$

where $\mathcal{A}_1^r, \ldots, \mathcal{A}_{n_{\mathrm{A}}+n_{\mathrm{B}}-\tau+1}^r$ are fuzzy sets ($r = 1, \ldots, n_{\mathrm{R}}$) which determine the regions where the consecutive local models can be activated. The output of the whole model is a weighted sum of all local models

$$y(k) = \frac{\sum_{r=1}^{n_{\mathrm{R}}} w^r(k) y^r(k)}{\sum_{r=1}^{n_{\mathrm{R}}} w^r(k)}$$

The activation level of the rth rule is usually calculated as an algebraic product of the membership function of the fuzzy sets, i.e. $w^r(k) = \mu_{\mathcal{A}_1^r}(k) \times \ldots \times \mu_{\mathcal{A}_{n_{\mathrm{A}}+n_{\mathrm{B}}-\tau+1}^r}(k)$. Different fuzzy sets can be use, e.g. triangle, trapezoid, Gussian, sigmoid.

In place of the fuzzy input-output model one may use a fuzzy step-response model [199] of a fuzzy state-space model [316]. Increasing the number of fuzzy rules and (or) of the fuzzy sets leads to significant increase of the total number of model parameters (the curse of dimensionality).

Multi-Layer Perceptron Feedforward Neural Networks

Development of artificial neural networks, shortly named neural networks [66, 94, 119, 235, 274, 283, 309], is the result of research of nervous systems of living organisms. The first significant model of a nervous cell was given in the work of W.S. McCulloch and W.H. Pitts [205]. The model of perceptron was discussed by F. Rosenblatt [277]. The cited works initiated a very intensive period of research. Publication of a book by M. Minsky and S. Papert [212], in which the disadvantages of the perceptron are pointed out (inability to act as a simple XOR function) resulted in significant reduction of funds spent on neural networks research and, in turn, decreased the scientific activity in that field. The 1980s witnessed a great popularity of neural networks. There were two reasons for that: invention of new structures (in particular multi-layer networks, which do not have disadvantages of the single-layer one) and rapid development in electronics, which made it possible to use neural structure in practice. The field of neural networks is now a very important, rapidly developing, interdisciplinary area of research. Theoretical developments motivate new applications and vice versa. Neural networks can be used, among others, in the following fields: approximation, modelling, prediction, classification, recognition (of patterns, sounds, voice), data compression and signal processing (e.g. filtration, separation). They can be used in numerous fields, not only in engineering, but also in physics, medicine, statistics and even in economics.

In spite of the fact that there is a great number of different types of neural networks, the Multi-Layer Perceptron (MLP) feedforward structure is the most popular. Usually, two layers are used. Fig. 1.4 depicts the structure of the Double-Layer Perceptron (DLP) network which corresponds to the general nonlinear dynamic model (1.16). The network has $n_{\mathrm{A}} + n_{\mathrm{B}} - \tau + 1$ inputs, K nonlinear hidden nodes (the nonlinear transfer function is denoted by $\varphi\colon \mathbb{R} \to \mathbb{R}$), one linear output element (sum) and one output $y(k)$. The unipolar sigmoid transfer function is usually used

$$\varphi(x) = \frac{1}{1 + \exp(-\beta x)}$$

or the bipolar one

$$\varphi(x) = \tanh(\beta x) = \frac{1 - \exp(-\beta x)}{1 + \exp(-\beta x)}$$

where β is a parameter, typically $\beta = 1$.

The DLP neural network has a feedforward structure which means that all the signals flow from the inputs to the output. The additional constant unitary inputs of the first and the second layers are biases. The weights of the first layer are denoted by $w_{i,j}^1$, where $i = 1, \ldots, K$, $j = 0, \ldots, n_{\mathrm{A}} + n_{\mathrm{B}} - \tau + 1$, the weights of the second layer are denoted by w_i^2, where $i = 0, \ldots, K$. The output signal can be expressed by the equation

$$y(k) = w_0^2 + \sum_{i=1}^{K} w_i^2 \varphi(z_i(k))$$

where the sum of input signals of the ith hidden node is

$$\begin{aligned} z_i(k) = {} & w_{i,0}^1 + w_{1,1}^1 u(k - \tau) + \ldots + w_{i,n_{\mathrm{B}}-\tau+1}^1 u(k - n_{\mathrm{B}}) \\ & + w_{i,n_{\mathrm{B}}-\tau+2}^1 y(k-1) + \ldots + w_{i,n_{\mathrm{A}}+n_{\mathrm{B}}-\tau+1}^1 y(k - n_{\mathrm{A}}) \end{aligned}$$

It can be proven theoretically that the MLP network with only one nonlinear hidden layer (i.e. the DLP network) is able to approximate any continuous function with an arbitrary degree of accuracy, provided that the number of hidden nodes is sufficient enough [98]. Hence the DLP network is a universal approximator of functions of many variables. Of course, it is also possible to use alternative approximation methods, e.g. polynomials. From the Stone-Weierstrass theorem it follows that any continuous function defined on some interval can be approximated arbitrarily closely by a polynomial. On the other hand, however, it may turn out that it is necessary to use polynomial models (1.19) of high orders to describe precisely enough properties of typical technological processes. The polynomial model is likely to have an excessive number of parameters, in particular when the process has many inputs and outputs. The polynomial models of high order may badly affect numerical properties of the identification procedure. Furthermore, the polynomial models may have oscillatory interpolation and extrapolation properties [103]. That is why the application of polynomial models is in practice limited. Typically, the neural models are likely to be much more precise and have significantly a lower number of parameters. Unlike the polynomials, the neural models have good interpolation and extrapolation properties. In many practical cases it also turns out that the neural models are more precise than the fuzzy ones and have a lower number of parameters.

The identification procedure of a neural model consists of three steps: selection of the model structure (i.e. the choice of model inputs and the number of hidden nodes), model training (i.e. optimisation of model parameters – weights) and model validation. A very simple, but in practice quite a good approach, is to train a series of neural models and choose the best one. Alternatively, one can use specialised algorithms for determination of the dynamic

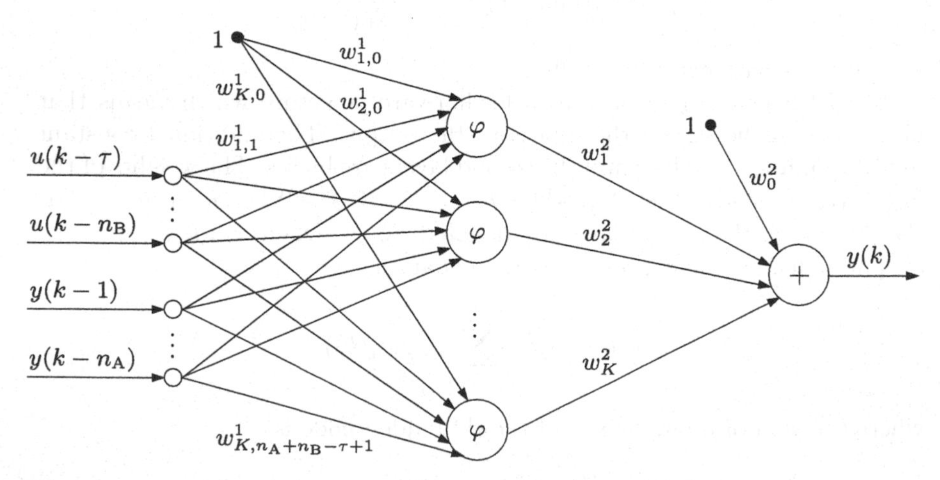

Fig. 1.4. The structure of the double-layer perceptron feedforward neural network

order of the model [34]. During network training the following Sum of Squared Errors (SSE) is minimised

$$\mathrm{SSE} = \sum_{k=1}^{n_\mathrm{p}} (y^{\mathrm{mod}}(k) - y(k))^2 \tag{1.20}$$

where $y^{\mathrm{mod}}(k)$ is the output signal calculated by the model for the sampling instant k, $y(k)$ is the real value of the recorded process output, n_p is the number of samples. During model training nonlinear gradient optimisation methods are usually used: the steepest descent method, the conjugated gradients methods, the quasi-Newton variable metrics methods (e.g. BFGS or DFP) or the Levenberg-Marquardt algorithm [235]. It also possible to use heuristic optimisation approaches (e.g. evolutionary algorithms, simulated annealing), in particular for initialisation [235]. During training two model configurations are possible: serial-parallel and parallel [222]. In the non-recurrent serial-parallel model the output signal (for consecutive sampling instants $k = 1, 2, \ldots, n_\mathrm{p}$) is a function of the process input and output signal values from previous instants

$$y^{\mathrm{mod}}(k) = f(u(k-\tau), \ldots, u(k-n_\mathrm{B}), y(k-1), \ldots, y(k-n_\mathrm{A}))$$

In the recurrent parallel model, which is also called the simulation model, the past process outputs are replaced by the model outputs calculated at previous sampling instants

$$y^{\mathrm{mod}}(k) = f(u(k-\tau), \ldots, u(k-n_\mathrm{B}), y^{\mathrm{mod}}(k-1), \ldots, y^{\mathrm{mod}}(k-n_\mathrm{A}))$$

It is also possible to minimise during model training the cost-function which takes into account the role of the model in MPC [154, 298]

$$\mathrm{SSE} = \sum_{k=1}^{n_\mathrm{p}-N} \sum_{p=1}^{N} (y^{\mathrm{mod}}(k+p) - y(k+p))^2 \tag{1.21}$$

During training all model errors over the whole prediction horizon for all available data samples are taken into account. Of course, such training of a recurrent network is much more computationally demanding in comparison with training a classical serial-parallel model.

A great number of very efficient algorithms for finding the structure of neural models have been developed [66, 235]. As the examples it is necessary to mention Optimal Brain Damage (OBD) [126] and Optimal Brain Surgeon (OBS) [93] reduction (pruning) algorithms, network growing methods [66] as well as algorithms useful for searching the best structure [61].

Radial Basis Neural Networks

Sigmoid transfer functions are used in the hidden layer (or layers) of the most popular perceptron neural networks. All neurons participate in generating the output signal (they are active for the whole data set). That is why the perceptron networks are sometimes named "global approximators" [235]. Local approximation is as alternative. In this approach the hidden nodes are characterised by local transfer functions, i.e. Radial Basis Functions (RBF). Consecutive hidden elements are active only for some chosen part of the data. The transfer functions of hidden nodes change radially around their centres and they have non-zero values only in some neighbourhood of the centres.

Fig. 1.5 depicts the structure of the radial basis neural network which corresponds to the general nonlinear dynamic model (1.16). It is similar to the perceptron neural network shown in Fig. 1.4 because both networks are feedforward, they have the same inputs, K hidden nodes characterised by the nonlinear transfer function $\varphi\colon \mathbb{R} \to \mathbb{R}$ and one output $y(k)$. Unlike the double-layer perceptron network, the radial basis structure has only one layer of weights, they are denoted by w_i, where $i = 0, \ldots, K$, the hidden nodes do not have the bias inputs. For the most popular Gaussian transfer function, the output of the network is

$$y(k) = w_0 + \sum_{i=1}^{K} w_i \exp\left(-\frac{1}{2\sigma_i^2} z_i(k)\right)$$

where

$$\begin{aligned} z_i(k) = &(u(k-\tau) - c_{i,1})^2 + \ldots + (u(k-n_\mathrm{B}) - c_{i,n_\mathrm{B}-\tau+1})^2 \\ &+ (y(k-1) - c_{i,n_\mathrm{B}-\tau+2})^2 + \ldots + (y(k-n_\mathrm{A}) - c_{i,n_\mathrm{A}+n_\mathrm{B}-\tau+1})^2 \end{aligned}$$

The vectors $\left[c_{i,1} \ldots c_{i,n_\mathrm{A}+n_\mathrm{B}-\tau+1}\right]^\mathrm{T}$ define the centres of hidden nodes and the spread parameters σ_i determine the neighbourhood of the centre in which the transfer functions give non-zero values, $i = 1, \ldots, K$.

The radial basis network with a sufficient number of nodes, similarly to the double-layer perceptron one, is a universal approximator [243].

Unlike the multi-layer perceptron neural networks, in which the transfer functions give non-zero values only in some neighbourhood of their centres,

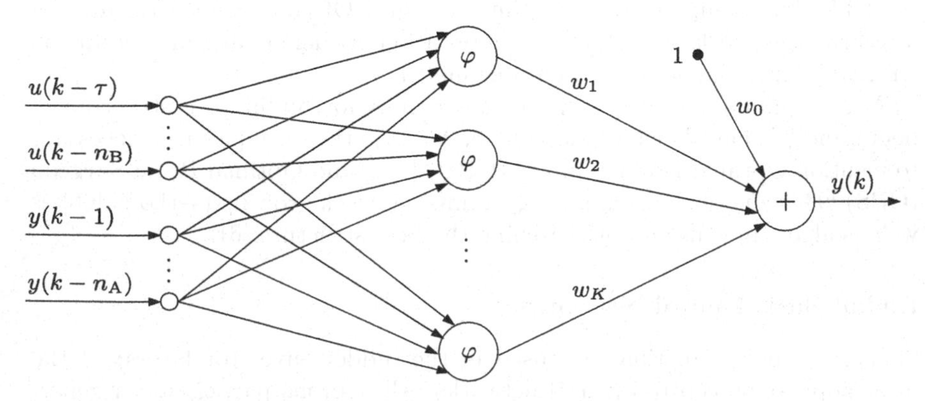

Fig. 1.5. The structure of the radial basis neural network

the radial basis structures are local approximators. The perceptron networks, because of a global character of their transfer functions, do not have any mechanism which would make it possible to define the regions in which activity of consecutive hidden elements is the highest. In such a case it is very difficult to associate activity of the hidden neurons with the region of the training data set. It means that it is difficult to find the initial values of weights. Training of the multi-layer perceptron neural network is a nonlinear optimisation problem, in which the main difficulty is to avoid shallow local minima of the minimised cost-function. Conversely, because the transfer functions used in radial basis neural networks have a local character, it is possible to link their parameters with the training data set. In consequence, it is possible to obtain the initial values of model parameters in a relatively simple way. For radial basis neural network training a hybrid algorithm is usually used, which has two separate stages: determination of the parameters of the transfer functions and optimisation of weights [235]. Such an approach makes the training process quite fast and efficient. Additionally, it is possible to use the Gram-Schmidt ortogonalisation algorithm [235], which is a very powerful method of controlling the number of hidden nodes. In practice the structure of the double-layer perceptron neural networks is chosen experimentally by the trial and error approach whereas finding the structure of radial basis network is an integral part of training. Unfortunately, because of a local character of hidden nodes, radial basis networks typically have more (sometimes much more) parameters than the multi-layer perceptron structures of similar accuracy.

Both double-layer perceptron and radial basis neural networks are frequently used in practice as models of different processes [101]. They can be also used in control structures, e.g. in the IMC algorithm or in the control system with the inverse model [101, 227]. Unfortunately, unlike MPC algorithms, both mentioned control structures can be mainly used for single-input single-output processes and they do not take into account any constraints.

Recurrent Neural Networks

In MPC algorithms the double-layer perceptron neural model (or the radial basis structure) is in fact used recurrently because consecutive predictions of the output variable over the prediction horizon are functions of some previous predictions (Eq. (1.17)). Hence, the parallel training configuration seems to be more appropriate than the simpler serial-parallel one. In such a case feedback signals from some previous model outputs are used as the inputs. The feedforward network becomes a recurrent structure.

Both described classes of neural networks can be successfully used as models of numerous processes. The problems appear if the state is not a function of a fixed set of a finite number of past values of the input and output [244, 340]. In such cases it is necessary to use recurrent networks which have a special structure with feedbacks. The feedback signal may be taken from the hidden

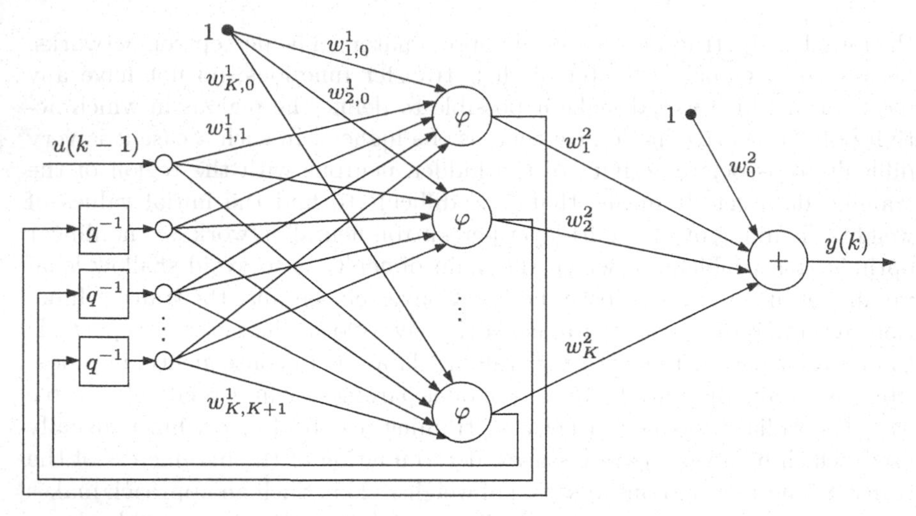

Fig. 1.6. The structure of the recurrent Elman neural network

layer or from the output (or outputs) and be delivered to the input (or inputs) of the network. Alternatively, cross-feedbacks between hidden elements of the same layer can be used. Fig. 1.6 depicts the example recurrent neural network of the Elman type [235]. It has a partially recurrent architecture, because the feedback from the hidden layer and inputs is used, the output signal of the network is not used recurrently. If in the Elman neural network the second layer (the output layer) is not present, one obtains the Real Time Recurrent Network network [235]. In comparison with the classical feedforward neural networks, training of the recurrent structures is much more demanding.

The state-space neural model [340]

$$x(k+1) = f(x(k), u(k))$$
$$y(k) = g(x(k))$$

also has a recurrent structure. The model consists of two neural networks: the first one is responsible for the state equation, the second one for the output relation. The model has some feedbacks, because the outputs of the first network (i.e. the state variables) are delivered through delay block to its inputs. In comparison with recurrent networks, i.e. the Elman structure, the number of state variables is independent of the number of hidden nodes.

Feedforward Neural Networks with Dynamic Neurons

The Locally Recurrent Globally Feedforward (LRGF) neural network [244] is an interesting alternative to the recurrent structures. The structure of the LRGF networks is fully feedforward and it is similar to that of the MLP networks because they consist of a few layers and all signals flow into only one direction, i.e. from the inputs to the outputs. The are no feedbacks.

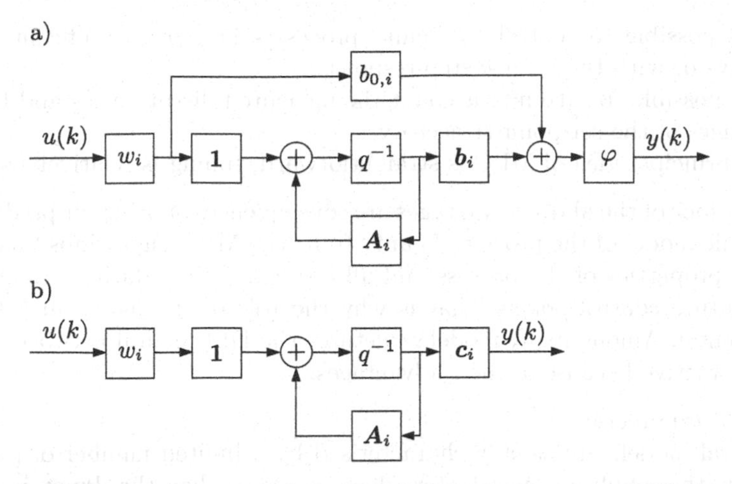

Fig. 1.7. The dynamic neurons: a) the Infinite Impulse Response (IIR) filter, b) the Finite Impulse Response (FIR) filter; $u(k)$ – the neuron input, $y(k)$ – the neuron input, $\boldsymbol{A}_i$, $b_{0,i}$, $\boldsymbol{b}_i$, $\boldsymbol{c}_i$ – neuron parameters, φ – the nonlinear transfer function

Dynamic properties are obtained because the dynamic neurons with some internal feedbacks are used. Two the most popular neuron types are: the Infinite Impulse Response (IIR) filter and the Finite Impulse Response (FIR) filter. The structure of both dynamic neurons is depicted in Fig. 1.7.

In the first layer of the cascade LRGF neural network the IIR filters are used, in the second layer – the FIR filters, there are also some additional connections between the inputs and the hidden layer. It can be proven than the cascade LRGF network is able to approximate the trajectory generated by any function of the C^1 class, provided that the number of nodes is sufficient enough [244]. The network can be used for modelling of "difficult" processes, for which the classical multi-layer perceptron or radial basis function neural networks is unable to give required accuracy. Training of the LRGF network is much less demanding than training of the typical recurrent structures, i.e. the Elman one.

1.6 Summary

The MPC algorithm is the only one among the advanced control techniques (more advanced than the classical PID controller) which is successfully used in numerous applications [262, 316]. It is because the MPC technique has a few important advantages:

a) it is possible to take into account constraints imposed on the input variables, on the predicted output variables and (or) on the predicted state variables,
b) it is possible to control multi-input multi-output processes,

c) it is possible to control "difficult" processes i.e. with significant time-delays or with the inverse step-response,
d) it is possible to take into account the measured disturbances and future changes of the set-point trajectory,
e) the principal idea of MPC is straightforward, tuning is relatively easy.

The first four of the above advantages are consequences of using for prediction a dynamic model of the process. Thanks to it, the MPC algorithms take into account properties of the process and all the constraints during calculation of the future control policy. That is why the role of the model in MPC is fundamental. Among many model structures the DLP neural networks seem to be attractive because of their advantages:

a) They are precise.
b) Neural models are usually characterised by a limited number of parameters, the number of parameters does not soar when the dynamic order of the model or the number of input, output and (or) state variables are increased.
c) A great number of efficient neural network training and structure optimisation algorithms are available. For training some data sets are only used, no technological knowledge is necessary.
d) Unlike the first-principle (fundamental) models, the neural models do not consist of any differential (or algebraic) equations which should be repeatedly solved on-line in MPC algorithms. Such an approach would be difficult, not only because of significant computational complexity, but also because numerical problems (e.g. ill-conditioning) are possible.
e) The neural models may be used in nonlinear MPC algorithms in a relative simple way, their implementation is not difficult.

A family of MPC algorithms based on different types of neural models (based on the DLP networks) are detailed in the following chapters.

As discussed in this book and in many publications cited in literature reviews, the neural models are very efficient. On the other hand, the choice of the model class is, of course, subjective. On may notice that the existing research groups have their own preferred model types and they concentrate on developing identification and control algorithms for chosen models. As examples one may mention cascade Hammerstein and Wiener models [103], fuzzy models [199] and Support Vector Machines (SVM) [289], for example Least Squares Support Vector Machines (LS-SVM) [308].

2

MPC Algorithms Based on Double-Layer Perceptron Neural Models: the Prototypes

This chapter discusses a family of nonlinear MPC algorithms (the prototypes) and their implementation details for the DLP feedforward neural models. The "ideal" MPC algorithm with nonlinear optimisation and a few suboptimal MPC algorithms with different on-line linearisation methods are discussed. In order to illustrate properties of the considered MPC algorithms they are compared in two control systems: a yeast fermentation reactor and a high-pressure high-purity ethylene-ethane distillation column are considered. In particular, good control accuracy and low computational complexity of the suboptimal MPC approaches are emphasised.

2.1 Double-Layer Perceptron Neural Models

The dynamic process has n_{u} inputs $u_1, \ldots, u_{n_{\mathrm{u}}}$, n_{h} measured disturbances (uncontrolled inputs) $h_1, \ldots, h_{n_{\mathrm{h}}}$ and n_{y} outputs $y_1, \ldots, y_{n_{\mathrm{y}}}$. Its model is

$$\begin{aligned}
y_1(k) = f_1(\boldsymbol{x}_1(k)) = f_1(&u_1(k-\tau^{1,1}), \ldots, u_1(k-n_{\mathrm{B}}^{1,1}), \ldots, \\
&u_{n_{\mathrm{u}}}(k-\tau^{1,n_{\mathrm{u}}}), \ldots, u_{n_{\mathrm{u}}}(k-n_{\mathrm{B}}^{1,n_{\mathrm{u}}}), \\
&h_1(k-\tau_{\mathrm{h}}^{1,1}), \ldots, h_1(k-n_{\mathrm{C}}^{1,1}), \ldots, \\
&h_{n_{\mathrm{h}}}(k-\tau_{\mathrm{h}}^{1,n_{\mathrm{h}}}), \ldots, h_{n_{\mathrm{h}}}(k-n_{\mathrm{C}}^{1,n_{\mathrm{h}}}), \\
&y_1(k-1), \ldots, y_1(k-n_{\mathrm{A}}^{1}))
\end{aligned}$$

$$\vdots$$

$$\begin{aligned}
y_{n_{\mathrm{y}}}(k) = f_{n_{\mathrm{y}}}(\boldsymbol{x}_{n_{\mathrm{y}}}(k)) = f_{n_{\mathrm{y}}}(&u_1(k-\tau^{n_{\mathrm{y}},1}), \ldots, u_1(k-n_{\mathrm{B}}^{n_{\mathrm{y}},1}), \ldots, \\
&u_{n_{\mathrm{u}}}(k-\tau^{n_{\mathrm{y}},n_{\mathrm{u}}}), \ldots, u_{n_{\mathrm{u}}}(k-n_{\mathrm{B}}^{n_{\mathrm{y}},n_{\mathrm{u}}}), \\
&h_1(k-\tau_{\mathrm{h}}^{n_{\mathrm{y}},1}), \ldots, h_1(k-n_{\mathrm{C}}^{n_{\mathrm{y}},1}), \ldots, \\
&h_{n_{\mathrm{h}}}(k-\tau_{\mathrm{h}}^{n_{\mathrm{y}},n_{\mathrm{h}}}), \ldots, h_{n_{\mathrm{h}}}(k-n_{\mathrm{C}}^{n_{\mathrm{y}},n_{\mathrm{h}}}), \\
&y_{n_{\mathrm{y}}}(k-1), \ldots, y_{n_{\mathrm{y}}}(k-n_{\mathrm{A}}^{n_{\mathrm{y}}}))
\end{aligned} \tag{2.1}$$

M. Ławryńczuk, *Computationally Efficient Model Predictive Control Algorithms*,
Studies in Systems, Decision and Control 3,
DOI: 10.1007/978-3-319-04229-9_2, © Springer International Publishing Switzerland 2014

where integer numbers n_{A}^m, $n_{\mathrm{B}}^{m,n}$, $\tau^{m,n}$ (where $m = 1, \ldots, n_{\mathrm{y}}$, $n = 1, \ldots, n_{\mathrm{u}}$) and $n_{\mathrm{C}}^{m,n}$, $\tau_{\mathrm{h}}^{m,n}$ (where $m = 1, \ldots, n_{\mathrm{y}}$, $n = 1, \ldots, n_{\mathrm{h}}$) determine the order of dynamics, $\tau^{m,n} \leq n_{\mathrm{B}}^{m,n}$ and $\tau_{\mathrm{h}}^{m,n} \leq n_{\mathrm{C}}^{m,n}$. The model consists of n_{y} independent neural networks, which are multi-input single-output models. They are trained independently. It is also possible to use only one neural network which has n_{y} outputs, but all outputs depend on the same input arguments whereas in practice the order of dynamics of consecutive parts of the process is different. Furthermore, the multi-input multi-output model usually has many parameters (weights), its training is difficult. Such a model may be useful only in some specific cases.

The structure of the DLP feedforward neural network used as the model of the mth output is depicted in Fig. 2.1. The network has I^m inputs, one hidden layer containing K^m nodes with the nonlinear transfer function $\varphi\colon \mathbb{R} \to \mathbb{R}$ and one output $y_m(k)$. The weights of the first layer are denoted by $w_{i,j}^{1,m}$,

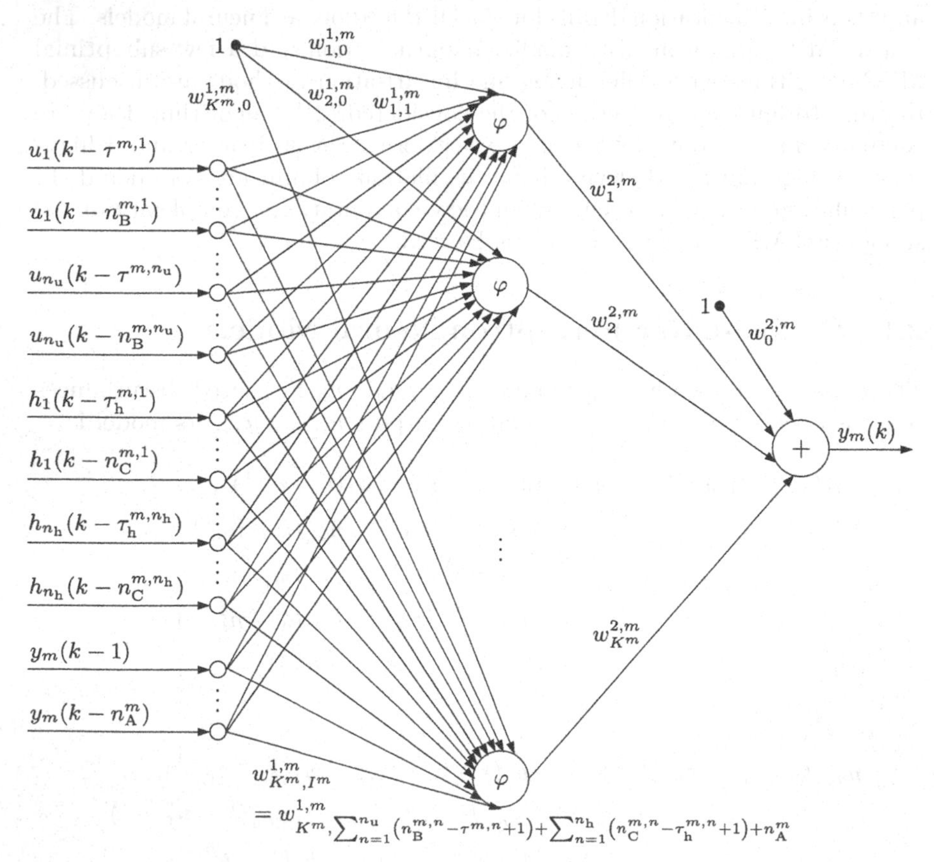

Fig. 2.1. The structure of the multi-input single-output neural model of the mth output, $m = 1, \ldots, n_{\mathrm{y}}$

where $i = 1, \ldots, K^m$, $j = 0, \ldots, I^m$, $m = 1, \ldots, n_{\mathrm{y}}$, the weights of the second layer are denoted by $w_i^{2,m}$, where $i = 0, \ldots, K$, $m = 1, \ldots, n_{\mathrm{y}}$. Consecutive multi-input single-output models can be described by the functions $f_m \colon \mathbb{R}^{I^m} \to \mathbb{R}$ of the general form (2.1). For the mth output the number of model inputs which depend on the process input signal u_n ($n = 1, \ldots, n_{\mathrm{u}}$) is $I_{\mathrm{u}}^{m,n} = n_{\mathrm{B}}^{m,n} - \tau^{m,n} + 1$, the number of model inputs which depend on the measured disturbance h_n ($n = 1, \ldots, n_{\mathrm{h}}$) is $I_{\mathrm{h}}^{m,n} = n_{\mathrm{C}}^{m,n} - \tau_{\mathrm{h}}^{m,n} + 1$ and the number of model inputs which depend on the process output signal y_m is n_{A}^m. Hence, the number of model inputs which depend on all process input signals is $I_{\mathrm{u}}^m = \sum_{n=1}^{n_{\mathrm{u}}} I_{\mathrm{u}}^{m,n}$, the number of model inputs which depend on all measured disturbances is $I_{\mathrm{h}}^m = \sum_{n=1}^{n_{\mathrm{h}}} I_{\mathrm{h}}^{m,n}$. The total number of model inputs is then

$$\begin{aligned} I^m &= I_{\mathrm{u}}^m + I_{\mathrm{h}}^m + n_{\mathrm{A}}^m = \sum_{n=1}^{n_{\mathrm{u}}} I_{\mathrm{u}}^{m,n} + \sum_{n=1}^{n_{\mathrm{h}}} I_{\mathrm{h}}^{m,n} + n_{\mathrm{A}}^m \\ &= \sum_{n=1}^{n_{\mathrm{u}}} (n_{\mathrm{B}}^{m,n} - \tau^{m,n} + 1) + \sum_{n=1}^{n_{\mathrm{h}}} (n_{\mathrm{C}}^{m,n} - \tau_{\mathrm{h}}^{m,n} + 1) + n_{\mathrm{A}}^m \end{aligned}$$

The neural network shown in Fig. 2.1 is described by the following equation

$$y_m(k) = w_0^{2,m} + \sum_{i=1}^{K^m} w_i^{2,m} \varphi(z_i^m(k)) \tag{2.2}$$

where

$$\begin{aligned} z_i^m(k) = w_{i,0}^{1,m} &+ \sum_{n=1}^{n_{\mathrm{u}}} \sum_{j=1}^{I_{\mathrm{u}}^{m,n}} w_{i,J_{\mathrm{u}}^{m,n}+j}^{1,m} u_n(k - \tau^{m,n} + 1 - j) \\ &+ \sum_{n=1}^{n_{\mathrm{h}}} \sum_{j=1}^{I_{\mathrm{h}}^{m,n}} w_{i,I_{\mathrm{u}}^m + J_{\mathrm{h}}^{m,n}+j}^{1,m} h_n(k - \tau_{\mathrm{h}}^{m,n} + 1 - j) \\ &+ \sum_{j=1}^{n_{\mathrm{A}}^m} w_{i,I_{\mathrm{u}}^m + I_{\mathrm{h}}^m + j}^{1,m} y_m(k - j) \end{aligned} \tag{2.3}$$

is the sum of the input signals connected to the ith hidden node and

$$J_{\mathrm{u}}^{m,n} = \begin{cases} 0 & \text{if } n = 1 \\ \sum_{i=1}^{n-1} I_{\mathrm{u}}^{m,i} & \text{if } n = 2, \ldots, n_{\mathrm{u}} \end{cases}, \quad J_{\mathrm{h}}^{m,n} = \begin{cases} 0 & \text{if } n = 1 \\ \sum_{i=1}^{n-1} I_{\mathrm{h}}^{m,i} & \text{if } n = 2, \ldots, n_{\mathrm{h}} \end{cases}$$

Because the model takes into account the measured disturbances, the MPC algorithms discussed next have the ability to compensate for such disturbances. If disturbance measurements are not available, they are not arguments of the model.

2.2 MPC Algorithm with Nonlinear Optimisation (MPC-NO)

The most straightforward solution is to use for prediction the neural model directly, without any simplifications. Using the rudimentary prediction equation (1.14) and recurrently the model (2.1), one obtains consecutive predictions over the prediction horizon ($p = 1, \ldots, N$) for all outputs ($m = 1, \ldots, n_{\mathrm{y}}$)

$$\begin{aligned}\hat{y}_m(k+p|k) = f_m(&\underbrace{u_n(k-\tau^{m,n}+p|k), \ldots, u_n(k|k)}_{I_{\mathrm{uf}}^{m,n}(p),\ n=1,\ldots,n_{\mathrm{u}}},\\ &\underbrace{u_n(k-1), \ldots, u_n(k-n_{\mathrm{B}}^{m,n}+p)}_{I_{\mathrm{u}}^{m,n}-I_{\mathrm{uf}}^{m,n}(p),\ n=1,\ldots,n_{\mathrm{u}}}\\ &\underbrace{h_n(k-\tau_{\mathrm{h}}^{m,n}+p|k), \ldots, h_n(k+1|k)}_{I_{\mathrm{hf}}^{m,n}(p),\ n=1,\ldots,n_{\mathrm{h}}},\\ &\underbrace{h_n(k), \ldots, h_n(k-n_{\mathrm{C}}^{m,n}+p)}_{I_{\mathrm{h}}^{m,n}-I_{\mathrm{hf}}^{m,n}(p),\ n=1,\ldots,n_{\mathrm{h}}},\\ &\underbrace{\hat{y}_m(k-1+p|k), \ldots, \hat{y}_m(k+1|k)}_{I_{\mathrm{yf}}^{m}(p)},\\ &\underbrace{y_m(k), \ldots, y_m(k-n_{\mathrm{A}}+p)}_{n_{\mathrm{A}}^{m}-I_{\mathrm{yf}}^{m}(p)}) + d_m(k)\end{aligned} \tag{2.4}$$

where $I_{\mathrm{uf}}^{m,n}(p) = \max(\min(p-\tau^{m,n}+1, I_{\mathrm{u}}^{m,n}), 0)$, $I_{\mathrm{hf}}^{m,n}(p) = \max(\min(p-\tau_{\mathrm{h}}^{m,n}, I_{\mathrm{h}}^{m,n}), 0)$, $I_{\mathrm{yf}}^{m}(p) = \min(p-1, n_{\mathrm{A}}^{m})$. The prediction equation (2.4) is similar to that used in the case of single-input single-output processes (Eq. (1.17)). From Eq. (1.15), the unmeasured disturbance acting on the mth output of the process is estimated from

$$\begin{aligned}d_m(k) = y_m(k) - f_m(&u_1(k-\tau^{m,1}), \ldots, u_1(k-n_{\mathrm{B}}^{m,1}), \ldots,\\ &u_{n_{\mathrm{u}}}(k-\tau^{m,n_{\mathrm{u}}}), \ldots, u_{n_{\mathrm{u}}}(k-n_{\mathrm{B}}^{m,n_{\mathrm{u}}}),\\ &h_1(k-\tau_{\mathrm{h}}^{m,1}), \ldots, h_1(k-n_{\mathrm{C}}^{m,1}), \ldots,\\ &h_{n_{\mathrm{h}}}(k-\tau_{\mathrm{h}}^{m,n_{\mathrm{h}}}), \ldots, h_{n_{\mathrm{h}}}(k-n_{\mathrm{C}}^{m,n_{\mathrm{h}}})\\ &y_{n_{\mathrm{y}}}(k-1), \ldots, y_{n_{\mathrm{y}}}(k-n_{\mathrm{A}}^{m}))\end{aligned}$$

Because the predicted output trajectory (2.4) is a nonlinear function of the future control signals, it is necessary to use a nonlinear optimisation routine for minimisation of the cost-function (1.4). That is why one obtains the MPC algorithm with Nonlinear Optimisation (MPC-NO). Its structure is depicted in Fig. 2.2. During optimisation, for the currently calculated future control sequence $\boldsymbol{u}(k)$, the predicted output trajectory $\hat{\boldsymbol{y}}(k)$ is calculated.

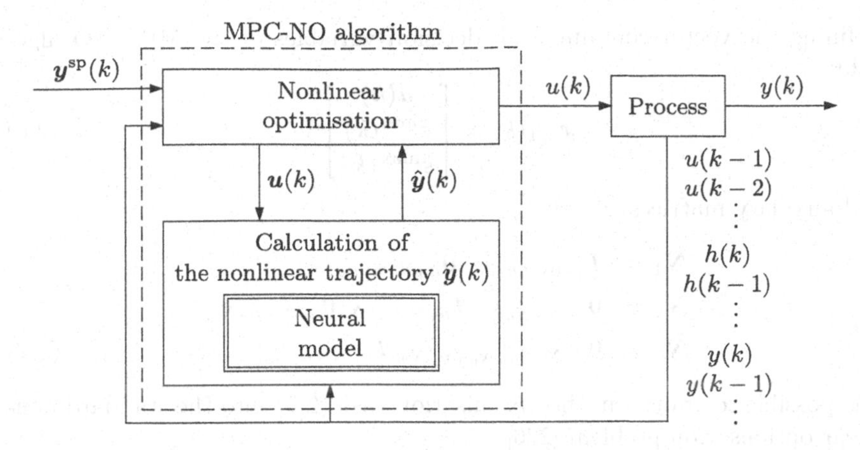

Fig. 2.2. The structure of the MPC algorithm with Nonlinear Optimisation (MPC-NO)

In the MPC-NO algorithm the values of the future control signals $\boldsymbol{u}(k)$ are calculated, rather than the increments $\triangle \boldsymbol{u}(k)$. Using the relation

$$\triangle \boldsymbol{u}(k) = \boldsymbol{J}^{\mathrm{NO}} \boldsymbol{u}(k) + \boldsymbol{u}^{\mathrm{NO}}(k)$$

where the matrix

$$\boldsymbol{J}^{\mathrm{NO}} = \begin{bmatrix} \boldsymbol{I}_{n_u \times n_u} & \boldsymbol{0}_{n_u \times n_u} & \boldsymbol{0}_{n_u \times n_u} & \cdots & \boldsymbol{0}_{n_u \times n_u} \\ -\boldsymbol{I}_{n_u \times n_u} & \boldsymbol{I}_{n_u \times n_u} & \boldsymbol{0}_{n_u \times n_u} & \cdots & \boldsymbol{0}_{n_u \times n_u} \\ \boldsymbol{0}_{n_u \times n_u} & -\boldsymbol{I}_{n_u \times n_u} & \boldsymbol{I}_{n_u \times n_u} & \cdots & \boldsymbol{0}_{n_u \times n_u} \\ \vdots & \vdots & \vdots & \ddots & \vdots \\ \boldsymbol{0}_{n_u \times n_u} & \boldsymbol{0}_{n_u \times n_u} & \boldsymbol{0}_{n_u \times n_u} & \cdots & \boldsymbol{I}_{n_u \times n_u} \end{bmatrix}$$

is of dimensionality $n_u N_u \times n_u N_u$ and the vector

$$\boldsymbol{u}^{\mathrm{NO}}(k) = \begin{bmatrix} -u(k-1) \\ 0_{n_u(N_u-1)} \end{bmatrix}$$

is of length $n_u N_u$, the optimisation problem (1.9) can be expressed as

$$\min_{\substack{\boldsymbol{u}(k) \\ \boldsymbol{\varepsilon}^{\min}(k),\ \boldsymbol{\varepsilon}^{\max}(k)}} \Big\{ J(k) = \|\boldsymbol{y}^{\mathrm{sp}}(k) - \hat{\boldsymbol{y}}(k)\|^2_{\boldsymbol{M}} + \left\| \boldsymbol{J}^{\mathrm{NO}} \boldsymbol{u}(k) + \boldsymbol{u}^{\mathrm{NO}}(k) \right\|^2_{\boldsymbol{\Lambda}} + \rho^{\min} \left\| \boldsymbol{\varepsilon}^{\min}(k) \right\|^2 + \rho^{\max} \left\| \boldsymbol{\varepsilon}^{\max}(k) \right\|^2 \Big\}$$

subject to

$$\boldsymbol{u}^{\min} \leq \boldsymbol{u}(k) \leq \boldsymbol{u}^{\max}$$
$$-\triangle \boldsymbol{u}^{\max} \leq \boldsymbol{J}^{\mathrm{NO}} \boldsymbol{u}(k) + \boldsymbol{u}^{\mathrm{NO}}(k) \leq \triangle \boldsymbol{u}^{\max}$$
$$\boldsymbol{y}^{\min} - \boldsymbol{\varepsilon}^{\min}(k) \leq \hat{\boldsymbol{y}}(k) \leq \boldsymbol{y}^{\max} + \boldsymbol{\varepsilon}^{\max}(k)$$
$$\boldsymbol{\varepsilon}^{\min}(k) \geq 0,\ \boldsymbol{\varepsilon}^{\max}(k) \geq 0 \tag{2.5}$$

Defining the vector containing all decision variables of the MPC-NO algorithm

$$\boldsymbol{x}_{\mathrm{opt}}(k) = \begin{bmatrix} \boldsymbol{u}(k) \\ \boldsymbol{\varepsilon}^{\min}(k) \\ \boldsymbol{\varepsilon}^{\max}(k) \end{bmatrix} \tag{2.6}$$

and auxiliary matrices

$$\begin{aligned} \boldsymbol{N}_1 &= \begin{bmatrix} \boldsymbol{I}_{n_{\mathrm{u}}N_{\mathrm{u}} \times n_{\mathrm{u}}N_{\mathrm{u}}} & \boldsymbol{0}_{n_{\mathrm{u}}N_{\mathrm{u}} \times 2n_{\mathrm{y}}N} \end{bmatrix} \\ \boldsymbol{N}_2 &= \begin{bmatrix} \boldsymbol{0}_{n_{\mathrm{y}}N \times n_{\mathrm{u}}N_{\mathrm{u}}} & \boldsymbol{I}_{n_{\mathrm{y}}N \times n_{\mathrm{y}}N} & \boldsymbol{0}_{n_{\mathrm{y}}N \times n_{\mathrm{y}}N} \end{bmatrix} \\ \boldsymbol{N}_3 &= \begin{bmatrix} \boldsymbol{0}_{n_{\mathrm{y}}N \times (n_{\mathrm{u}}N_{\mathrm{u}}+n_{\mathrm{y}}N)} & \boldsymbol{I}_{n_{\mathrm{y}}N \times n_{\mathrm{y}}N} \end{bmatrix} \end{aligned} \tag{2.7}$$

it is possible to transform the optimisation task (2.5) into the standard nonlinear optimisation problem [226]

$$\begin{aligned} &\min_{\boldsymbol{x}_{\mathrm{opt}}(k)} \{f_{\mathrm{opt}}(\boldsymbol{x}_{\mathrm{opt}}(k))\} \\ &\text{subject to} \\ &\boldsymbol{g}_{\mathrm{opt}}(\boldsymbol{x}_{\mathrm{opt}}(k)) \leq 0 \end{aligned} \tag{2.8}$$

where

$$\begin{aligned} f_{\mathrm{opt}}(\boldsymbol{x}_{\mathrm{opt}}(k)) = {} & \left\| \boldsymbol{y}^{\mathrm{sp}}(k) - \hat{\boldsymbol{y}}(k) \right\|_{\boldsymbol{M}}^2 + \left\| \boldsymbol{J}^{\mathrm{NO}} \boldsymbol{N}_1 \boldsymbol{x}_{\mathrm{opt}}(k) + \boldsymbol{u}^{\mathrm{NO}}(k) \right\|_{\boldsymbol{\Lambda}}^2 \\ & + \rho^{\min} \left\| \boldsymbol{N}_2 \boldsymbol{x}_{\mathrm{opt}}(k) \right\|^2 + \rho^{\max} \left\| \boldsymbol{N}_3 \boldsymbol{x}_{\mathrm{opt}}(k) \right\|^2 \end{aligned} \tag{2.9}$$

and

$$\boldsymbol{g}_{\mathrm{opt}}(\boldsymbol{x}_{\mathrm{opt}}(k)) = \begin{bmatrix} -\boldsymbol{N}_1 \boldsymbol{x}_{\mathrm{opt}}(k) + \boldsymbol{u}^{\min} \\ \boldsymbol{N}_1 \boldsymbol{x}_{\mathrm{opt}}(k) - \boldsymbol{u}^{\max} \\ -\boldsymbol{J}^{\mathrm{NO}} \boldsymbol{N}_1 \boldsymbol{x}_{\mathrm{opt}}(k) - \boldsymbol{u}^{\mathrm{NO}}(k) - \triangle \boldsymbol{u}^{\max} \\ \boldsymbol{J}^{\mathrm{NO}} \boldsymbol{N}_1 \boldsymbol{x}_{\mathrm{opt}}(k) + \boldsymbol{u}^{\mathrm{NO}}(k) - \triangle \boldsymbol{u}^{\max} \\ -\hat{\boldsymbol{y}}(k) - \boldsymbol{N}_2 \boldsymbol{x}_{\mathrm{opt}}(k) + \boldsymbol{y}^{\min} \\ \hat{\boldsymbol{y}}(k) - \boldsymbol{N}_3 \boldsymbol{x}_{\mathrm{opt}}(k) - \boldsymbol{y}^{\max} \\ -\boldsymbol{N}_2 \boldsymbol{x}_{\mathrm{opt}}(k) \\ -\boldsymbol{N}_3 \boldsymbol{x}_{\mathrm{opt}}(k) \end{bmatrix} \tag{2.10}$$

If the constraints imposed on values of the control signals are only present, one may use the L-BFGS-B algorithm [39]. If all constraints are linear, i.e. when no output constraints are taken into account, it is possible to use the Rosen's gradient projection method [276]. If the nonlinear constraints imposed on the output predictions are present, one may use the optimisation methods with the penalty function (in particular with a shifted penalty function) [226], the SQP algorithm [226, 257] or the Ipopt algorithm [330].

Analytically calculated derivatives of the minimised cost-function and of the constraints are used (it is conceptually better than numerical differentiation). Differentiating the cost-function (2.9) with respect to the vector of decision variables of the algorithm gives the vector of length $n_{\mathrm{u}}N_{\mathrm{u}} + 2n_{\mathrm{y}}N$

$$\begin{aligned}\frac{\mathrm{d}f_{\mathrm{opt}}(\boldsymbol{x}_{\mathrm{opt}}(k))}{\mathrm{d}\boldsymbol{x}_{\mathrm{opt}}(k)} &= 2\left(\frac{\mathrm{d}\hat{\boldsymbol{y}}(k)}{\mathrm{d}\boldsymbol{x}_{\mathrm{opt}}(k)}\right)^{\mathrm{T}} \boldsymbol{M}(\hat{\boldsymbol{y}}(k) - \boldsymbol{y}^{\mathrm{sp}}(k)) \\ &+ 2\boldsymbol{N}_1^{\mathrm{T}}(\boldsymbol{J}^{\mathrm{NO}})^{\mathrm{T}}\boldsymbol{\Lambda}(\boldsymbol{J}^{\mathrm{NO}}\boldsymbol{N}_1\boldsymbol{x}_{\mathrm{opt}}(k) + \boldsymbol{u}^{\mathrm{NO}}(k)) \\ &+ 2\rho^{\min}\boldsymbol{N}_2^{\mathrm{T}}\boldsymbol{N}_2\boldsymbol{x}_{\mathrm{opt}}(k) + 2\rho^{\max}\boldsymbol{N}_3^{\mathrm{T}}\boldsymbol{N}_3\boldsymbol{x}_{\mathrm{opt}}(k) \end{aligned} \quad (2.11)$$

The matrix of the derivatives of the predicted output trajectory is of dimensionality $n_{\mathrm{y}}N \times (n_{\mathrm{u}}N_{\mathrm{u}} + 2n_{\mathrm{y}}N)$ and has the structure

$$\frac{\mathrm{d}\hat{\boldsymbol{y}}(k)}{\mathrm{d}\boldsymbol{x}_{\mathrm{opt}}(k)} = \begin{bmatrix} \frac{\mathrm{d}\hat{\boldsymbol{y}}(k)}{\mathrm{d}\boldsymbol{u}(k)} & \frac{\mathrm{d}\hat{\boldsymbol{y}}(k)}{\mathrm{d}\boldsymbol{\varepsilon}^{\min}(k)} & \frac{\mathrm{d}\hat{\boldsymbol{y}}(k)}{\mathrm{d}\boldsymbol{\varepsilon}^{\max}(k)} \end{bmatrix} \quad (2.12)$$

where the derivatives of the predicted trajectory with respect to $\boldsymbol{\varepsilon}^{\min}(k)$ and $\boldsymbol{\varepsilon}^{\max}(k)$ are all zeros matrices of dimensionality $n_{\mathrm{y}}N \times n_{\mathrm{y}}N$. The matrix of the derivatives of the constraints (2.10) is of dimensionality $(4n_{\mathrm{u}}N_{\mathrm{u}} + 4n_{\mathrm{y}}N) \times (n_{\mathrm{u}}N_{\mathrm{u}} + 2n_{\mathrm{y}}N)$ and has the structure

$$\frac{\mathrm{d}\boldsymbol{g}_{\mathrm{opt}}(\boldsymbol{x}_{\mathrm{opt}}(k))}{\mathrm{d}\boldsymbol{x}_{\mathrm{opt}}(k)} = \begin{bmatrix} -\boldsymbol{N}_1 \\ \boldsymbol{N}_1 \\ -\boldsymbol{J}^{\mathrm{NO}}\boldsymbol{N}_1 \\ \boldsymbol{J}^{\mathrm{NO}}\boldsymbol{N}_1 \\ -\frac{\mathrm{d}\hat{\boldsymbol{y}}(k)}{\mathrm{d}\boldsymbol{x}_{\mathrm{opt}}(k)} - \boldsymbol{N}_2 \\ \frac{\mathrm{d}\hat{\boldsymbol{y}}(k)}{\mathrm{d}\boldsymbol{x}_{\mathrm{opt}}(k)} - \boldsymbol{N}_3 \\ -\boldsymbol{N}_2 \\ -\boldsymbol{N}_3 \end{bmatrix} \quad (2.13)$$

It is possible to significantly reduce the number of decision variables of the MPC-NO algorithm. The optimisation problem (1.6) has as many as $n_{\mathrm{u}}N_{\mathrm{u}}+2n_{\mathrm{y}}N$ decision variables. One may assume that the variables $\varepsilon^{\min}(k+p)$ and $\varepsilon^{\min}(k+p)$, respectively, are constant on the whole prediction horizon, which leads to the general optimisation problem (1.10). In such a case the vector of the decision variables

$$\boldsymbol{x}_{\mathrm{opt}}(k) = \begin{bmatrix} \boldsymbol{u}(k) \\ \varepsilon^{\min}(k) \\ \varepsilon^{\max}(k) \end{bmatrix} \quad (2.14)$$

has length $n_{\mathrm{u}}N_{\mathrm{u}}+2n_{\mathrm{y}}$, independent of the prediction horizon, and the number of constraints drops to $4n_{\mathrm{u}}N_{\mathrm{u}} + 2n_{\mathrm{y}}N + 2n_{\mathrm{y}}$. The auxiliary matrices are then

$$\begin{aligned} \boldsymbol{N}_1 &= \begin{bmatrix} \boldsymbol{I}_{n_{\mathrm{u}}N_{\mathrm{u}} \times n_{\mathrm{u}}N_{\mathrm{u}}} & \boldsymbol{0}_{n_{\mathrm{u}}N_{\mathrm{u}} \times 2n_{\mathrm{y}}} \end{bmatrix} \\ \boldsymbol{N}_2 &= \begin{bmatrix} \boldsymbol{0}_{n_{\mathrm{y}} \times n_{\mathrm{u}}N_{\mathrm{u}}} & \boldsymbol{I}_{n_{\mathrm{y}} \times n_{\mathrm{y}}} & \boldsymbol{0}_{n_{\mathrm{y}} \times n_{\mathrm{y}}} \end{bmatrix} \\ \boldsymbol{N}_3 &= \begin{bmatrix} \boldsymbol{0}_{n_{\mathrm{y}} \times (n_{\mathrm{u}}N_{\mathrm{u}}+n_{\mathrm{y}})} & \boldsymbol{I}_{n_{\mathrm{y}} \times n_{\mathrm{y}}} \end{bmatrix} \end{aligned} \quad (2.15)$$

In consequence, the length of the derivatives of the cost-function (2.11) is also reduced in comparison with the original formulation. Analogously, the number of columns of the predicted output trajectory matrix (2.12) and dimensionality of the constraints derivatives matrices (2.13) are decreased.

The nonlinear optimisation problem (2.8) is solved using the analytical derivatives of the cost-function (Eq. (2.11)). The constraints' derivatives of the structure (2.13) are also used, where the predicted output derivatives have the structure (2.12). The matrix of the predicted output derivatives with respect of the future control signals, which is used in the derivative vector of the cost-function (2.12) and in the constraints derivative matrix (2.13), is of dimensionality $n_\mathrm{y}N \times n_\mathrm{u}N_\mathrm{u}$ and has the general structure

$$\frac{\mathrm{d}\hat{\boldsymbol{y}}(k)}{\mathrm{d}\boldsymbol{u}(k)} = \begin{bmatrix} \frac{\partial \hat{y}(k+1|k)}{\partial u(k|k)} & \cdots & \frac{\partial \hat{y}(k+1|k)}{\partial u(k+N_\mathrm{u}-1|k)} \\ \vdots & \ddots & \vdots \\ \frac{\partial \hat{y}(k+N|k)}{\partial u(k|k)} & \cdots & \frac{\partial \hat{y}(k+N|k)}{\partial u(k+N_\mathrm{u}-1|k)} \end{bmatrix} \tag{2.16}$$

where the matrices

$$\frac{\partial \hat{y}(k+p|k)}{\partial u(k+r|k)} = \begin{bmatrix} \frac{\partial \hat{y}_1(k+p|k)}{\partial u_1(k+r|k)} & \cdots & \frac{\partial \hat{y}_1(k+p|k)}{\partial u_{n_\mathrm{u}}(k+r|k)} \\ \vdots & \ddots & \vdots \\ \frac{\partial \hat{y}_{n_\mathrm{y}}(k+p|k)}{\partial u_1(k+r|k)} & \cdots & \frac{\partial \hat{y}_{n_\mathrm{y}}(k+p|k)}{\partial u_{n_\mathrm{u}}(k+r|k)} \end{bmatrix}$$

of dimensionality $n_\mathrm{y} \times n_\mathrm{u}$ are calculated for all $p = 1, \ldots, N$ and $r = 0, \ldots, N_\mathrm{u} - 1$.

Computational burden of the MPC-NO algorithm is seriously affected by the initial solution of the optimisation problem, the control sequence $\boldsymbol{u}^0(k)$. Of course, $\boldsymbol{\varepsilon}^{\min}(k)$ and $\boldsymbol{\varepsilon}^{\max}(k)$ are at first zeros vectors. The simplest initialisation method, but also the worst one, it is to use a constant vector $\boldsymbol{u}^0(k)$. A better approach is to take into account the current operating point of the process

$$\boldsymbol{u}^0(k) = \boldsymbol{u}(k-1) = \begin{bmatrix} u(k-1) \\ \vdots \\ u(k-1) \end{bmatrix}$$

or the last $n_\mathrm{u}(N_\mathrm{u} - 1)$ elements of the optimal control sequence calculated at the previous sampling instant

$$\boldsymbol{u}^0(k) = \boldsymbol{u}(k|k-1) = \begin{bmatrix} u^0(k|k) \\ \vdots \\ u^0(k+N_\mathrm{u}-3|k) \\ u^0(k+N_\mathrm{u}-2|k) \\ u^0(k+N_\mathrm{u}-1|k) \end{bmatrix} = \begin{bmatrix} u(k|k-1) \\ \vdots \\ u(k+N_\mathrm{u}-3|k-1) \\ u(k+N_\mathrm{u}-2|k-1) \\ u(k+N_\mathrm{u}-2|k-1) \end{bmatrix} \tag{2.17}$$

The vector $u(k+N_{\mathrm{u}}-2|k-1)$ is deliberately used twice because the vector $u(k+N_{\mathrm{u}}-1|k-1)$ is not calculated at the sampling instant $k-1$.

Implementation of MPC-NO Algorithm for Neural Model

All the description given so far is universal, practically any differentiable nonlinear model can be used in the MPC-NO algorithm. The key implementation elements are: calculation of the predicted output trajectory $\hat{\boldsymbol{y}}(k)$ and determination of the matrix (2.16), which consists of derivatives of this trajectory with respect to the future control sequence $\boldsymbol{u}(k)$. For the DLP neural model depicted in Fig. 2.1 and described by Eqs. (2.2) and (2.3), using the general prediction equation (1.14), one obtains predictions

$$\hat{y}_m(k+p|k) = w_0^{2,m} + \sum_{i=1}^{K^m} w_i^{2,m}\varphi(z_i^m(k+p|k)) + d_m(k) \tag{2.18}$$

where, using Eq. (1.15), the unmeasured disturbances are estimated from

$$d_m(k) = y_m(k) - w_0^{2,m} - \sum_{i=1}^{K^m} w_i^{2,m}\varphi(z_i^m(k)) \tag{2.19}$$

The sum of the input signals of the ith hidden node ($i=1,\ldots,K^m$), which results from the structure of the model, is calculated from Eq. (2.3). According to Eq. (2.4), the output predictions are functions of past and future signals, hence

$$\begin{aligned}
z_i^m(k+p|k) = w_{i,0}^{1,m} &+ \sum_{n=1}^{n_{\mathrm{u}}} \sum_{j=1}^{I_{\mathrm{uf}}^{m,n}(p)} w_{i,J_{\mathrm{u}}^{m,n}+j}^{1,m} u_n(k-\tau^{m,n}+1-j+p|k) \\
&+ \sum_{n=1}^{n_{\mathrm{u}}} \sum_{j=I_{\mathrm{uf}}^{m,n}(p)+1}^{I_{\mathrm{u}}^{m,n}} w_{i,J_{\mathrm{u}}^{m,n}+j}^{1,m} u_n(k-\tau^{m,n}+1-j+p) \\
&+ \sum_{n=1}^{n_{\mathrm{h}}} \sum_{j=1}^{I_{\mathrm{hf}}^{m,n}(p)} w_{i,I_{\mathrm{u}}^m+J_{\mathrm{h}}^{m,n}+j}^{1,m} h_n(k-\tau_{\mathrm{h}}^{m,n}+1-j+p|k) \\
&+ \sum_{n=1}^{n_{\mathrm{h}}} \sum_{j=I_{\mathrm{hf}}^{m,n}(p)+1}^{I_{\mathrm{h}}^{m,n}} w_{i,I_{\mathrm{u}}^m+J_{\mathrm{h}}^{m,n}+j}^{1,m} h_n(k-\tau_{\mathrm{h}}^{m,n}+1-j+p) \\
&+ \sum_{j=1}^{I_{\mathrm{yf}}^m(p)} w_{i,I_{\mathrm{u}}^m+I_{\mathrm{h}}^m+j}^{1,m} \hat{y}_m(k-j+p|k) \\
&+ \sum_{j=I_{\mathrm{yf}}^m(p)+1}^{n_{\mathrm{A}}^m} w_{i,I_{\mathrm{u}}^m+I_{\mathrm{h}}^m+j}^{1,m} y_m(k-j+p)
\end{aligned} \tag{2.20}$$

If future values of the measured disturbances are known, they can be used for prediction calculation. Usually it is not possible. In such a case it is necessary to assume that $h_n(k+p|k) = h_n(k)$ for $p \geq 1$.

The entries of the derivatives matrix of the predicted output trajectory with respect to the future control signals are calculated by differentiating Eq. (2.18)

$$\frac{\partial \hat{y}_m(k+p|k)}{\partial u_n(k+r|k)} = \sum_{i=1}^{K^m} w_i^{2,m} \frac{\mathrm{d}\varphi(z_i^m(k+p|k))}{\mathrm{d}z_i^m(k+p|k)} \frac{\partial z_i^m(k+p|k)}{\partial u_n(k+r|k)} \tag{2.21}$$

It may be noticed that

$$\frac{\partial z_i^m(k+p|k)}{\partial u_n(k+r|k)} = \frac{\partial \hat{y}_m(k+p|k)}{\partial u_n(k+r|k)} = 0 \text{ if } r \geq p - \tau^{m,n} + 1 \tag{2.22}$$

The first partial derivatives in the right side of Eq. (2.21) depend on the nonlinear transfer function φ used in the hidden layer of the neural model. If the hyperbolic tangent is used ($\varphi = \tanh(\cdot)$), one has

$$\frac{\mathrm{d}\varphi(z_i^m(k+p|k))}{\mathrm{d}z_i^m(k+p|k)} = 1 - \tanh^2\left(z_i^m(k+p|k)\right)$$

Differentiating Eq. (2.20), it is possible to obtain the second partial derivatives in the right side of Eq. (2.21)

$$\begin{aligned}\frac{\partial z_i^m(k+p|k)}{\partial u_n(k+r|k)} = & \sum_{j=1}^{I_{\mathrm{uf}}^{m,n}(p)} w_{i,J_{\mathrm{u}}^{m,n}+j}^{1,m} \frac{\partial u_n(k-\tau^{m,n}+1-j+p|k)}{\partial u_n(k+r|k)} \\ & + \sum_{j=1}^{I_{\mathrm{yf}}^{m}(p)} w_{i,I_{\mathrm{u}}^m+I_{\mathrm{h}}^m+j}^{1,m} \frac{\partial \hat{y}_m(k-j+p|k)}{\partial u_n(k+r|k)}\end{aligned} \tag{2.23}$$

Because $u_n(k+p|k) = u_n(k+N_{\mathrm{u}}-1|k)$ for $p \geq N_{\mathrm{u}}$, the first partial derivatives in the right part of the above equation may have only two values

$$\frac{\partial u_n(k+p|k)}{\partial u_n(k+r|k)} = \begin{cases} 1 & \text{if } p = r,\ p > r \text{ i } r = N_{\mathrm{u}} - 1 \\ 0 & \text{otherwise} \end{cases} \tag{2.24}$$

whereas the second partial derivatives in the right part of Eq. (2.23) must be calculated recurrently.

2.3 Suboptimal MPC Algorithms with On-Line Linearisation

The MPC-NO algorithm is the "ideal" MPC approach, because the full nonlinear model is used for prediction. As a result, the predicted output trajectory is a nonlinear function of the future control policy, which is successively calculated on-line. Unfortunately, the optimisation problem (2.5) is nonlinear, it may be non-convex. In general, computational complexity of nonlinear optimisation is usually high. Quite frequently the available hardware platform is insufficient or on-line calculations are too slow (in comparison with the sampling time). Moreover, despite huge progress in nonlinear optimisation, the universal algorithm which would guarantee finding the global solution to the nonlinear optimisation task at each sampling instant and within an assumed time has not been found yet.

This book is concerned with computationally efficient suboptimal MPC algorithms. A whole family of such algorithms is discussed, the basic feature of which is successive model or predicted trajectory on-line linearisation. Thanks to linearisation, the predicted trajectory is a linear function of the calculated future control sequence. Hence, linearisation makes it possible to calculate the decision variables from a quadratic optimisation problem, the necessity of on-line nonlinear optimisation is eliminated. It is necessary to emphasise the fact that quadratic optimisation is much more efficient than nonlinear optimisation in two respects: quantitative and qualitative. Firstly, quadratic optimisation tasks can be solved by means of the active set method or the interior point method [226] in some limited number of iterations, it is possible to estimate the calculation time. The quadratic optimisation problem always has a unique global minimum (provided that the weighting matrices $\boldsymbol{M} \geq 0$ and $\boldsymbol{\Lambda} > 0$). Furthermore, quadratic optimisation is much less computationally demanding than nonlinear optimisation. The suboptimal MPC algorithms may be applied for controlling fast processes.

The main idea of the simplest suboptimal MPC algorithms with on-line model linearisation is to some extent similar to the general idea of adaptive MPC techniques [225]. In both approaches the model used for prediction is a linear function of the calculated future input sequence. The difference is the fact that development of the suboptimal MPC algorithms is preceded by identification and validation of a nonlinear model whereas in adaptive MPC parameters of the linear model are successively calculated on-line. The time-varying linear model used in adaptive MPC is usually not verified in any way. Successive on-line identification is always risky, in particular when variability of process signals is not significant (the parameter blow-up phenomenon). In such a case properties of the model may by very different from those of the real process and the resulting adaptive MPC algorithm is likely to give unsatisfactory results. That is why the adaptive MPC techniques are not popular in the industry, simulation results are mainly published in the literature.

The Taylor series expansion formula is used for linearisation throughout the book. The linear approximation of a nonlinear function $y(x)$ performed for the linearisation point $\bar{x}$ is

$$y(x) = y(\bar{x}) + \left.\frac{\mathrm{d}y(x)}{\mathrm{d}x}\right|_{x=\bar{x}} (x - \bar{x}) \tag{2.25}$$

The above formula is universal, it can be used for linearisation of scalar and vector functions.

2.3.1 MPC Algorithm with Nonlinear Prediction and Linearisation for the Current Operating Point (MPC-NPL)

Using the Taylor series expansion formula (2.25), the linear approximation of the nonlinear model (2.1) is

$$y_m(k) = f_m(\bar{\boldsymbol{x}}_m(k)) + \sum_{n=1}^{n_{\mathrm{u}}} \sum_{l=1}^{n_{\mathrm{B}}^{m,n}} b_l^{m,n}(\bar{\boldsymbol{x}}_m(k))(u_n(k-l) - \bar{u}_n(k-l))$$
$$- \sum_{l=1}^{n_{\mathrm{A}}^m} a_l^m(\bar{\boldsymbol{x}}_m(k))(y_m(k-l) - \bar{y}_m(k-l)) \quad (2.26)$$

where $m = 1, \ldots, n_{\mathrm{y}}$. The coefficients of the linearised model are calculated analytically as

$$a_l^m(k) = -\left.\frac{\partial f_m(\boldsymbol{x}_m(k))}{\partial y_m(k-l)}\right|_{\boldsymbol{x}_m(k)=\bar{\boldsymbol{x}}_m(k)} \quad (2.27)$$

for all $l = 1, \ldots, n_{\mathrm{A}}$, $m = 1, \ldots, n_{\mathrm{y}}$ and

$$b_l^{m,n}(k) = \left.\frac{\partial f_m(\boldsymbol{x}_m(k))}{\partial u_n(k-l)}\right|_{\boldsymbol{x}_m(k)=\bar{\boldsymbol{x}}_m(k)} \quad (2.28)$$

for all $l = 1, \ldots, n_{\mathrm{B}}$, $m = 1, \ldots, n_{\mathrm{y}}$, $n = 1, \ldots, n_{\mathrm{u}}$. Linearisation is carried out locally, in some neighbourhood of the current operating point defined by the vectors $\bar{\boldsymbol{x}}_1(k), \ldots, \bar{\boldsymbol{x}}_{n_{\mathrm{y}}}(k)$ which are arguments of the nonlinear model (2.1). They consist of the control signals used in previous sampling instants as well as the measured disturbances and the output values recorded previously, i.e. the signals $\bar{u}_n(k-l)$, $\bar{h}_n(k-l)$ and $\bar{y}_m(k-l)$). The input signals $u_n(k-l)$ and the output signals $y_m(k-l)$ are the only arguments of the linearised model whereas the measured disturbances influence the time-varying values of model coefficients. In order to simplify the notation, the time delay is not present in the linearised model, although it is used in the nonlinear model (2.1). The linearised model (2.26) may be expressed in its incremental form

$$\delta y_m(k) = \sum_{n=1}^{n_{\mathrm{u}}} \sum_{l=1}^{n_{\mathrm{B}}^{m,n}} b_l^{m,n}(k)\delta u_n(k-l) - \sum_{l=1}^{n_{\mathrm{A}}^m} a_l^m(k)\delta y_m(k-l)$$

where $\delta y_m(k) = y_m(k) - f_m(\bar{\boldsymbol{x}}_m(k))$, $\delta u_n(k-l) = u_n(k-l) - \bar{u}_n(k-l)$ and $\delta y_m(k-l) = y_m(k-l) - \bar{y}_m(k-l)$, the notation $a_l^m(k) = a_l^m(\bar{\boldsymbol{x}}_m(k))$ and $b_l^{m,n}(k) = b_l^{m,n}(\bar{\boldsymbol{x}}_m(k))$ is used for short. In order to simplify further derivation, the incremental notation is neglected. The linearised model is

$$y_m(k) = \sum_{n=1}^{n_{\mathrm{u}}} \sum_{l=1}^{n_{\mathrm{B}}^{m,n}} b_l^{m,n}(k)u_n(k-l) - \sum_{l=1}^{n_{\mathrm{A}}^m} a_l^m(k)y_m(k-l) \quad (2.29)$$

Using the linearised model (2.29) recurrently, from the general prediction equation (1.14) one can calculate the predictions for consecutive outputs ($m = 1, \ldots, n_{\mathrm{y}}$) over the whole prediction horizon ($p = 1, \ldots, N$)

$$\begin{aligned}
\hat{y}_m(k+1|k) = \sum_{n=1}^{n_{\mathrm{u}}} & (b_1^{m,n}(k)u_n(k|k) + b_2^{m,n}(k)u_n(k-1) \\
& + b_3^{m,n}(k)u_n(k-2) + \ldots + b_{n\mathrm{B}}^{m,n}(k)u_n(k - n_{\mathrm{B}}^{m,n} + 1)) \\
& - a_1^m(k)y_m(k) - a_2^m(k)y_m(k-1) \\
& - a_3^m(k)y_m(k-2) - \ldots - a_{n\mathrm{A}}^m(k)y_m(k - n_{\mathrm{A}}^m + 1) + d_m(k) \\
\hat{y}_m(k+2|k) = \sum_{n=1}^{n_{\mathrm{u}}} & (b_1^{m,n}(k)u_n(k+1|k) + b_2^{m,n}(k)u_n(k|k) \\
& + b_3^{m,n}(k)u_n(k-1) + \ldots + b_{n\mathrm{B}}^{m,n}(k)u_n(k - n_{\mathrm{B}}^{m,n} + 2)) \\
& - a_1^m(k)\hat{y}_m(k+1|k) - a_2^m(k)y_m(k) \\
& - a_3^m(k)y_m(k-1) - \ldots - a_{n\mathrm{A}}^m(k)y_m(k - n_{\mathrm{A}}^m + 2) + d_m(k) \\
\hat{y}_m(k+3|k) = \sum_{n=1}^{n_{\mathrm{u}}} & (b_1^{m,n}(k)u_n(k+2|k) + b_2^{m,n}(k)u_n(k+1|k) \\
& + b_3^{m,n}(k)u_n(k|k) + \ldots + b_{n\mathrm{B}}^{m,n}(k)u_n(k - n_{\mathrm{B}}^{m,n} + 3)) \\
& - a_1^m(k)\hat{y}_m(k+2|k) - a_2^m(k)\hat{y}_m(k+1|k) \\
& - a_3^m(k)y_m(k) - \ldots - a_{n\mathrm{A}}^m(k)y_m(k - n_{\mathrm{A}}^m + 3) + d_m(k) \\
& \vdots
\end{aligned}$$

Using the very convenient vector-matrix notation for multi-input multi-output processes, the predictions can be expressed as functions of the future control increments (the influence of the past is not taken into account)

$$\begin{aligned}
\hat{y}(k+1|k) &= \boldsymbol{S}_1(k)\triangle u(k|k) + \ldots \\
\hat{y}(k+2|k) &= \boldsymbol{S}_2(k)\triangle u(k|k) + \boldsymbol{S}_1(k)\triangle u(k+1|k) + \ldots \\
\hat{y}(k+3|k) &= \boldsymbol{S}_3(k)\triangle u(k|k) + \boldsymbol{S}_2(k)\triangle u(k+1|k) + \boldsymbol{S}_1(k)\triangle u(k+2|k) + \ldots \\
&\vdots
\end{aligned} \tag{2.30}$$

where the matrices containing step-response coefficients of the linearised model are of dimensionality of $n_{\mathrm{y}} \times n_{\mathrm{u}}$ and have the structure

$$\boldsymbol{S}_p(k) = \begin{bmatrix} s_p^{1,1}(k) & \ldots & s_p^{1,n_{\mathrm{u}}}(k) \\ \vdots & \ddots & \vdots \\ s_p^{n_{\mathrm{y}},1}(k) & \ldots & s_p^{n_{\mathrm{y}},n_{\mathrm{u}}}(k) \end{bmatrix} \tag{2.31}$$

Scalar step-response coefficients are calculated recurrently using the current parameters of the linearised model (i.e. for the current sampling instant k)

over the whole prediction horizon ($p = 1, \ldots, N$) and for all inputs and outputs ($m = 1, \ldots, n_y$, $n = 1, \ldots, n_u$) from the formula

$$s_p^{m,n}(k) = \sum_{i=1}^{\min(p,n_B)} b_i^{m,n}(k) - \sum_{i=1}^{\min(p-1,n_A)} a_i^m(k) s_{p-i}^{m,n}(k) \tag{2.32}$$

Taking into account Eq. (2.30), the vector of output signals predicted over the prediction horizon (1.7), of length $n_y N$, can be expressed as

$$\hat{\boldsymbol{y}}(k) = \underbrace{\boldsymbol{G}(k)\triangle\boldsymbol{u}(k)}_{\text{future}} + \underbrace{\boldsymbol{y}^0(k)}_{\text{past}} \tag{2.33}$$

The first part of the above sum depends only on the future, because it is a function of the currently calculated control increments. The dynamic matrix

$$\boldsymbol{G}(k) = \begin{bmatrix} \boldsymbol{S}_1(k) & \boldsymbol{0}_{n_y \times n_u} & \ldots & \boldsymbol{0}_{n_y \times n_u} \\ \boldsymbol{S}_2(k) & \boldsymbol{S}_1(k) & \ldots & \boldsymbol{0}_{n_y \times n_u} \\ \vdots & \vdots & \ddots & \vdots \\ \boldsymbol{S}_N(k) & \boldsymbol{S}_{N-1}(k) & \ldots & \boldsymbol{S}_{N-N_u+1}(k) \end{bmatrix} \tag{2.34}$$

is of dimensionality $n_y N \times n_u N_u$ and consists of the step-response coefficients of the linear approximation of the nonlinear model obtained for the current operating point of the process. The expression $\boldsymbol{G}(k)\triangle\boldsymbol{u}(k)$ is sometimes called the forced part of the predicted output trajectory. The free trajectory vector

$$\boldsymbol{y}^0(k) = \begin{bmatrix} y^0(k+1|k) \\ \vdots \\ y^0(k+N|k) \end{bmatrix}$$

has length $n_y N$ and depends only on the past. The free trajectory is calculated using the full nonlinear model (not the linearised one) and taking into account only the influence of the past (free trajectory calculation for the neural model is discussed on p. 48).

Eq. (2.33) is the suboptimal prediction equation. Naturally, the locally linearised model (2.29) only approximates the properties of the original nonlinear model. That is why one may expect some differences between the predicted nonlinear output trajectory (which is calculated by means of the full nonlinear model in the MPC-NO algorithm from Eqs (2.18) and (2.20)) and the suboptimal trajectory, calculated from the locally linearised model. Intuitively, the greater the future control increments $\triangle\boldsymbol{u}(k)$, the bigger discrepancy between those trajectories. On the other hand, using the suboptimal prediction equation (2.33) and the relation

$$\boldsymbol{u}(k) = \boldsymbol{J}\triangle\boldsymbol{u}(k) + \boldsymbol{u}(k-1) \tag{2.35}$$

where the vector

$$\boldsymbol{u}(k-1) = \begin{bmatrix} u(k-1) \\ \vdots \\ u(k-1) \end{bmatrix}$$

is of length $n_{\mathrm{u}}N_{\mathrm{u}}$ and the matrix

$$\boldsymbol{J} = \begin{bmatrix} \boldsymbol{I}_{n_{\mathrm{u}}\times n_{\mathrm{u}}} & \boldsymbol{0}_{n_{\mathrm{u}}\times n_{\mathrm{u}}} & \boldsymbol{0}_{n_{\mathrm{u}}\times n_{\mathrm{u}}} & \ldots & \boldsymbol{0}_{n_{\mathrm{u}}\times n_{\mathrm{u}}} \\ \boldsymbol{I}_{n_{\mathrm{u}}\times n_{\mathrm{u}}} & \boldsymbol{I}_{n_{\mathrm{u}}\times n_{\mathrm{u}}} & \boldsymbol{0}_{n_{\mathrm{u}}\times n_{\mathrm{u}}} & \ldots & \boldsymbol{0}_{n_{\mathrm{u}}\times n_{\mathrm{u}}} \\ \vdots & \vdots & \vdots & \ddots & \vdots \\ \boldsymbol{I}_{n_{\mathrm{u}}\times n_{\mathrm{u}}} & \boldsymbol{I}_{n_{\mathrm{u}}\times n_{\mathrm{u}}} & \boldsymbol{I}_{n_{\mathrm{u}}\times n_{\mathrm{u}}} & \ldots & \boldsymbol{I}_{n_{\mathrm{u}}\times n_{\mathrm{u}}} \end{bmatrix}$$

is of dimensionality $n_{\mathrm{u}}N_{\mathrm{u}} \times n_{\mathrm{u}}N_{\mathrm{u}}$, the general MPC optimisation problem (1.9) can be transformed to the following quadratic optimisation task

$$\begin{aligned} &\min_{\substack{\triangle\boldsymbol{u}(k) \\ \boldsymbol{\varepsilon}^{\min}(k),\ \boldsymbol{\varepsilon}^{\max}(k)}} \Big\{ J(k) = \left\|\boldsymbol{y}^{\mathrm{sp}}(k) - \boldsymbol{G}(k)\triangle\boldsymbol{u}(k) - \boldsymbol{y}^{0}(k)\right\|_{\boldsymbol{M}}^{2} + \left\|\triangle\boldsymbol{u}(k)\right\|_{\boldsymbol{\Lambda}}^{2} \\ &\qquad\qquad + \rho^{\min}\left\|\boldsymbol{\varepsilon}^{\min}(k)\right\|^{2} + \rho^{\max}\left\|\boldsymbol{\varepsilon}^{\max}(k)\right\|^{2} \Big\} \\ &\text{subject to} \\ &\boldsymbol{u}^{\min} \leq \boldsymbol{J}\triangle\boldsymbol{u}(k) + \boldsymbol{u}(k-1) \leq \boldsymbol{u}^{\max} \\ &-\triangle\boldsymbol{u}^{\max} \leq \triangle\boldsymbol{u}(k) \leq \triangle\boldsymbol{u}^{\max} \\ &\boldsymbol{y}^{\min} - \boldsymbol{\varepsilon}^{\min}(k) \leq \boldsymbol{G}(k)\triangle\boldsymbol{u}(k) + \boldsymbol{y}^{0}(k) \leq \boldsymbol{y}^{\max} + \boldsymbol{\varepsilon}^{\max}(k) \\ &\boldsymbol{\varepsilon}^{\min}(k) \geq 0,\ \boldsymbol{\varepsilon}^{\max}(k) \geq 0 \end{aligned} \tag{2.36}$$

The optimisation problem (2.36) can be expressed in the standard form typical of quadratic optimisation [226]

$$\begin{aligned} &\min_{\boldsymbol{x}_{\mathrm{opt}}(k)} \left\{ \frac{1}{2}\boldsymbol{x}_{\mathrm{opt}}^{\mathrm{T}}(k)\boldsymbol{H}_{\mathrm{opt}}(k)\boldsymbol{x}_{\mathrm{opt}}(k) + \boldsymbol{f}_{\mathrm{opt}}^{\mathrm{T}}(k)\boldsymbol{x}_{\mathrm{opt}}(k) \right\} \\ &\text{subject to} \\ &\boldsymbol{A}_{\mathrm{opt}}(k)\boldsymbol{x}_{\mathrm{opt}}(k) \leq \boldsymbol{b}_{\mathrm{opt}}(k) \end{aligned} \tag{2.37}$$

where the vector of decision variables has length of $n_{\mathrm{u}}N_{\mathrm{u}}+2Nn_{\mathrm{y}}$ and is similar to the vector (2.6) used in the MPC-NO algorithm, but in place of the future values of the control signals the corresponding increments are calculated in the MPC-NPL algorithm

$$\boldsymbol{x}_{\mathrm{opt}}(k) = \begin{bmatrix} \triangle\boldsymbol{u}(k) \\ \boldsymbol{\varepsilon}^{\min}(k) \\ \boldsymbol{\varepsilon}^{\max}(k) \end{bmatrix}$$

The minimised cost function is defined by

$$\boldsymbol{H}_{\mathrm{opt}}(k) = 2(\boldsymbol{N}_1^{\mathrm{T}}\boldsymbol{G}^{\mathrm{T}}(k)\boldsymbol{M}\boldsymbol{G}(k)\boldsymbol{N}_1 + \boldsymbol{N}_1^{\mathrm{T}}\boldsymbol{\Lambda}\boldsymbol{N}_1 + \rho^{\min}\boldsymbol{N}_2^{\mathrm{T}}\boldsymbol{N}_2 + \rho^{\max}\boldsymbol{N}_3^{\mathrm{T}}\boldsymbol{N}_3)$$
$$\boldsymbol{f}_{\mathrm{opt}}(k) = -2\boldsymbol{N}_1^{\mathrm{T}}\boldsymbol{G}^{\mathrm{T}}(k)\boldsymbol{M}(\boldsymbol{y}^{\mathrm{sp}}(k) - \boldsymbol{y}^{0}(k))$$

and the constraints are defined by

$$\boldsymbol{A}_{\text{opt}}(k)=\begin{bmatrix}-\boldsymbol{J}\boldsymbol{N}_1\\ \boldsymbol{J}\boldsymbol{N}_1\\ -\boldsymbol{N}_1\\ \boldsymbol{N}_1\\ -\boldsymbol{G}(k)\boldsymbol{N}_1-\boldsymbol{N}_2\\ \boldsymbol{G}(k)\boldsymbol{N}_1-\boldsymbol{N}_3\\ -\boldsymbol{N}_2\\ -\boldsymbol{N}_3\end{bmatrix},\ \boldsymbol{b}_{\text{opt}}(k)=\begin{bmatrix}-\boldsymbol{u}^{\min}+\boldsymbol{u}(k-1)\\ \boldsymbol{u}^{\max}-\boldsymbol{u}(k-1)\\ \triangle\boldsymbol{u}^{\max}\\ \triangle\boldsymbol{u}^{\max}\\ -\boldsymbol{y}^{\min}+\boldsymbol{y}^0(k)\\ \boldsymbol{y}^{\max}-\boldsymbol{y}^0(k)\\ 0_{n_{\text{y}}N}\\ 0_{n_{\text{y}}N}\end{bmatrix} \tag{2.38}$$

The auxiliary matrices $\boldsymbol{N}_1$, $\boldsymbol{N}_2$ and $\boldsymbol{N}_3$ are defined by Eqs. (2.7). In order to reduce the number of decision variables of the algorithm, which should lead to reducing computational burden, one may assume that the same vectors $\varepsilon^{\min}(k)$ and $\varepsilon^{\max}(k)$ of length n_{y} are used for all sampling instants of the prediction horizon (in the same way it is done in the optimisation problems (1.10) and (1.11)). Length of the decision variables vector

$$\boldsymbol{x}_{\text{opt}}(k)=\begin{bmatrix}\triangle\boldsymbol{u}(k)\\ \varepsilon^{\min}(k)\\ \varepsilon^{\max}(k)\end{bmatrix} \tag{2.39}$$

is then independent of the prediction horizon and is $n_{\text{u}}N_{\text{u}}+2n_{\text{y}}$. In such a case the matrices $\boldsymbol{N}_1$, $\boldsymbol{N}_2$ and $\boldsymbol{N}_3$ are given by Eqs. (2.15), the all zeros vectors used in the vector $\boldsymbol{b}_{\text{opt}}(k)$ of the structure defined by Eq. (2.38) are of length n_{y}.

The discussed suboptimal algorithm is named the MPC algorithm with Nonlinear Prediction and Linearisation (MPC-NPL). Its structure is shown in Fig. 2.3. At each sampling instant (algorithm iteration k) the following steps are repeated on-line:

1. The local linear approximation (2.29) of the nonlinear (e.g. neural) model (2.1) is obtained for the current operating point of the process, i.e. the coefficients $a_l^m(k)$ for $l=1,\ldots,n_{\text{A}}$, $m=1,\ldots,n_{\text{y}}$ and the coefficients $b_l^{m,n}(k)$ for $l=1,\ldots,n_{\text{B}}$, $m=1,\ldots,n_{\text{y}}$, $n=1,\ldots,n_{\text{u}}$ are calculated. The linearised model is next used to find the step-response coefficients $s_p^{m,n}(k)$ from Eq. (2.32), which comprise the dynamic matrix $\boldsymbol{G}(k)$ given by Eq. (2.34).
2. The nonlinear model of the process is used to estimate the unmeasured disturbances and to find the nonlinear free trajectory $\boldsymbol{y}^0(k)$.
3. The quadratic optimisation problem (2.36) is solved to calculate the future control increments vector $\triangle\boldsymbol{u}(k)$.
4. The first n_{u} elements of the determined sequence are applied to the process, i.e. $u(k)=\triangle u(k|k)+u(k-1)$.
5. The iteration of the algorithm is increased, i.e. $k:=k+1$, the algorithm goes to step 1.

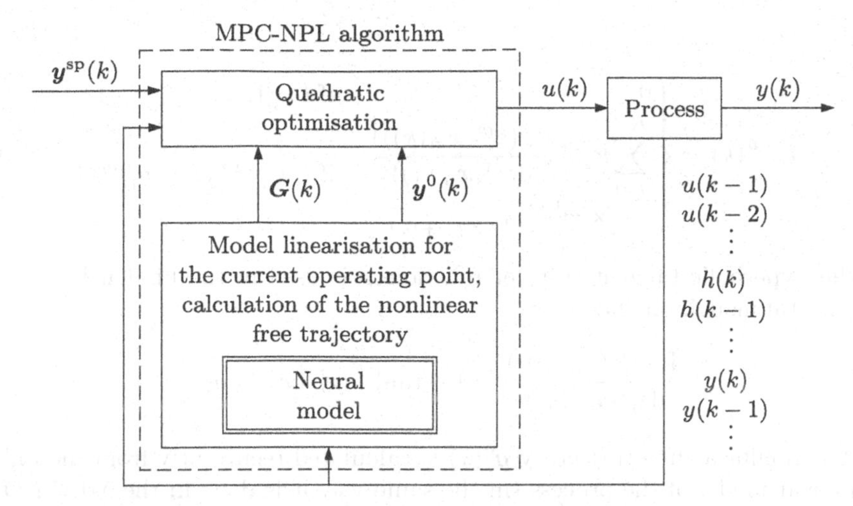

Fig. 2.3. The structure of the MPC algorithm with Nonlinear Prediction and Linearisation (MPC-NPL)

Because the output signals $(y_1(k), \ldots, y_{n_y}(k))$ may be measured at the current sampling instant, one may use for linearisation the operating point defined by the vectors $\bar{\boldsymbol{x}}_m(k+1)$. If the control time-delay is 1, one may assume for linearisation that the unknown (currently calculated) control signals which define the current operating point are equal to the signals from the previous iteration, i.e. $u_1(k) = u_1(k-1), \ldots, u_{n_u}(k) = u_{n_u}(k-1)$. Alternatively, one may use the control signals calculated at the previous iteration for the current one, i.e. $u_1(k) = u_1(k|k-1), \ldots, u_{n_u}(k) = u_{n_u}(k|k-1)$. When the disturbance time-delay is 1, the signals $h_1(k), \ldots, h_{n_h}(k)$ are available for measurement.

If the process is mildly nonlinear or if the set-point is not changed fast and significantly, linearisation may be repeated not at each iteration, but less frequently.

Implementation of MPC-NPL Algorithm for Neural Model

The MPC-NPL algorithm is universal, there are no restrictions imposed on the model type. For the chosen model it is necessary to derive the parameters of its linear approximation given by the general formula (2.29) and the nonlinear free trajectory. For the DLP neural model shown in Fig. 2.1, using Eqs. (2.2), (2.3), (2.27) and (2.28), the coefficients of the linearised model are

$$a_l^m(k) = -\sum_{i=1}^{K^m} w_l^{2,m} \frac{\mathrm{d}\varphi(z_i^m(\bar{\boldsymbol{x}}_m(k)))}{\mathrm{d}z_i^m(\bar{\boldsymbol{x}}_m(k))} w_{i,I_\mathrm{u}^m+I_\mathrm{h}^m+l}^{1,m}$$

and

$$b_l^{m,n}(k) = \begin{cases} 0 & \text{if } l = 1, \ldots, \tau^{m,n} - 1 \\ \displaystyle\sum_{i=1}^{K^m} w_i^{2,m} \frac{\mathrm{d}\varphi(z_i^m(\bar{\boldsymbol{x}}_m(k)))}{\mathrm{d}z_i^m(\bar{\boldsymbol{x}}_m(k))} & \text{if } l = \tau^{m,n}, \ldots, n_\mathrm{B}^{m,n} \\ \quad \times w_{i,J_\mathrm{u}^{m,n} - \tau^{m,n} + 1 + l}^{1,m} & \end{cases}$$

If the hyperbolic tangent transfer function φ is used in the hidden layer of the neural model, one has

$$\frac{\mathrm{d}\varphi(z_i^m(\bar{\boldsymbol{x}}_m(k)))}{\mathrm{d}z_i^m(\bar{\boldsymbol{x}}_m(k))} = 1 - \tanh^2\left(z_i^m(\bar{\boldsymbol{x}}_m(k))\right)$$

The nonlinear free trajectory $\boldsymbol{y}^0(k)$ is calculated recurrently from the full nonlinear model of the process (in the same way it is done in the MPC-NO algorithm), using the nonlinear prediction equation (2.18), but assuming no influence of the future

$$y_m^0(k+p|k) = w_0^{2,m} + \sum_{i=1}^{K^m} w_i^{2,m} \varphi(z_i^{m,0}(k+p|k)) + d_m(k)$$

for all $m = 1, \ldots, n_\mathrm{y}$ and $p = 1, \ldots, N$. The unmeasured disturbances are calculated from Eq. (2.19), in the same way it is done in the MPC-NO approach. Using Eq. (2.20) and considering only the past one has

$$\begin{aligned} z_i^{m,0}(k+p|k) = w_{i,0}^{1,m} &+ \sum_{n=1}^{n_\mathrm{u}} \sum_{j=1}^{I_\mathrm{uf}^{m,n}(p)} w_{i,J_\mathrm{u}^{m,n}+j}^{1,m} u_n(k-1) \\ &+ \sum_{n=1}^{n_\mathrm{u}} \sum_{j=I_\mathrm{uf}^{m,n}(p)+1}^{I_\mathrm{u}^{m,n}} w_{i,J_\mathrm{u}^{m,n}+j}^{1,m} u_n(k-\tau^{m,n}+1-j+p) \\ &+ \sum_{n=1}^{n_\mathrm{h}} \sum_{j=1}^{I_\mathrm{hf}^{m,n}(p)} w_{i,I_\mathrm{u}^m+J_\mathrm{h}^{m,n}+j}^{1,m} h_n(k-\tau_\mathrm{h}^{m,n}+1-j+p|k) \\ &+ \sum_{n=1}^{n_\mathrm{h}} \sum_{j=I_\mathrm{hf}^{m,n}(p)+1}^{I_\mathrm{h}^{m,n}} w_{i,I_\mathrm{u}^m+J_\mathrm{h}^{m,n}+j}^{1,m} h_n(k-\tau_\mathrm{h}^{m,n}+1-j+p) \\ &+ \sum_{j=1}^{I_\mathrm{yf}^m(p)} w_{i,I_\mathrm{u}^m+I_\mathrm{h}^m+j}^{1,m} y_m^0(k-j+p|k) \\ &+ \sum_{j=I_\mathrm{yf}^m(p)+1}^{n_\mathrm{A}^m} w_{i,I_\mathrm{u}^m+I_\mathrm{h}^m+j}^{1,m} y_m(k-j+p) \end{aligned} \tag{2.40}$$

In comparison with Eq. (2.20), for the free trajectory calculation all control increments from the current sampling instant k are assumed to be 0, i.e. $u_n(k+p|k) = u_n(k-1)$ for $p \geq 0$ (the second part of the right side of Eq. (2.40)). In place of the nonlinear predictions $\hat{y}_m(k+p|k)$ used in the second last part of Eq. (2.20), which depend on the future control policy, the free trajectory is used in Eq. (2.40).

2.3.2 MPC Algorithm with Successive Linearisation for the Current Operating Point (MPC-SL)

One may notice that the suboptimal prediction equation (2.33) used in the MPC-NPL algorithm is very similar to that of the classical MPC algorithms (e.g. DMC or GPC), which are based on linear models with constant parameters. There are, however, two fundamental differences. Firstly, the dynamic matrix $\boldsymbol{G}$ used in the linear MPC algorithms is constant whereas in the MPC-NPL algorithm the nonlinear model is successively linearised on-line and the operating-point dependent matrix $\boldsymbol{G}(k)$ is updated accordingly. Secondly, the linear model is also used in the linear MPC algorithm to find the dynamic matrix and the free trajectory. In the suboptimal MPC algorithm it is possible to use the linear approximation of the nonlinear model not only to calculate the step-response coefficients which comprise the dynamic matrix, but also to determine the free trajectory (and, naturally, to estimate the unmeasured disturbances which act on the process). Such an approach is known under the name the MPC algorithm with Successive Linearisation (MPC-SL). Its structure is shown in Fig. 2.4. In both MPC-SL and MPC-NPL algorithms the same quadratic optimisation problem (2.36) is solved

Using the current linear approximation (2.26) of the general nonlinear model (2.1), the general prediction equation (1.14) and Eq. (2.4), the predicted output trajectory over the prediction horizon ($p = 1, \ldots, N$) for consecutive outputs ($m = 1, \ldots, n_{\mathrm{y}}$) can be expressed as

$$\begin{aligned}\hat{y}_m(k+p|k) = f_m(\bar{\boldsymbol{x}}_m(k)) &+ \sum_{n=1}^{n_{\mathrm{u}}} \sum_{l=1}^{I_{\mathrm{uf}}^{m,n}(p)} b_l^{m,n}(k)(u_n(k-l+p|k) - \bar{u}_n(k-l)) \\ &+ \sum_{n=1}^{n_{\mathrm{u}}} \sum_{l=I_{\mathrm{uf}}^{m,n}(p)+1}^{n_{\mathrm{B}}^{m,n}} b_l^{m,n}(k)(u_n(k-l+p) - \bar{u}_n(k-l)) \\ &+ \sum_{l=1}^{I_{\mathrm{yf}}^{m}(p)} a_l^m(k)(\hat{y}_m(k-l+p|k) - \bar{y}_m(k-l)) \\ &+ \sum_{l=I_{\mathrm{yf}}^{m}(p)+1}^{n_{\mathrm{A}}^{m}} a_l^m(k)(y_m(k-l+p) - \bar{y}_m(k-l)) + d_m(k)\end{aligned}$$

Analogously to the MPC-NPL algorithm (Eq (2.40)), for the free trajectory calculation the influence of the past is taken into account, hence

$$\begin{aligned} y_m^0(k+p|k) &= f_m(\bar{\boldsymbol{x}}_m(k)) + \sum_{n=1}^{n_{\mathrm{u}}} \sum_{l=1}^{I_{\mathrm{uf}}^{m,n}(p)} b_l^{m,n}(k)(u_n(k-1) - \bar{u}_n(k-l)) \\ &+ \sum_{n=1}^{n_{\mathrm{u}}} \sum_{l=I_{\mathrm{uf}}^{m,n}(p)+1}^{n_{\mathrm{B}}^{m,n}} b_l^{m,n}(k)(u_n(k-l+p) - \bar{u}_n(k-l)) \\ &+ \sum_{l=1}^{I_{\mathrm{yf}}^{m}(p)} a_l^m(k)(y_m^0(k-l+p|k) - \bar{y}_m(k-l)) \\ &+ \sum_{l=I_{\mathrm{yf}}^{m}(p)+1}^{n_{\mathrm{A}}^{m}} a_l^m(k)(y_m(k-l+p) - \bar{y}_m(k-l)) + d_m(k) \end{aligned} \tag{2.41}$$

Using Eq. (1.15) and the linearised model (2.26), the unmeasured disturbances are estimated from

$$\begin{aligned} d_m(k) = y_m(k) - f_m(\bar{\boldsymbol{x}}_m(k)) - \sum_{n=1}^{n_{\mathrm{u}}} \sum_{l=1}^{n_{\mathrm{B}}^{m,n}} b_l^{m,n}(\bar{\boldsymbol{x}}_m(k))(u_n(k-l) - \bar{u}_n(k-l)) \\ + \sum_{l=1}^{n_{\mathrm{A}}^{m}} a_l^m(\bar{\boldsymbol{x}}_m(k))(y_m(k-l) - \bar{y}_m(k-l)) \end{aligned}$$

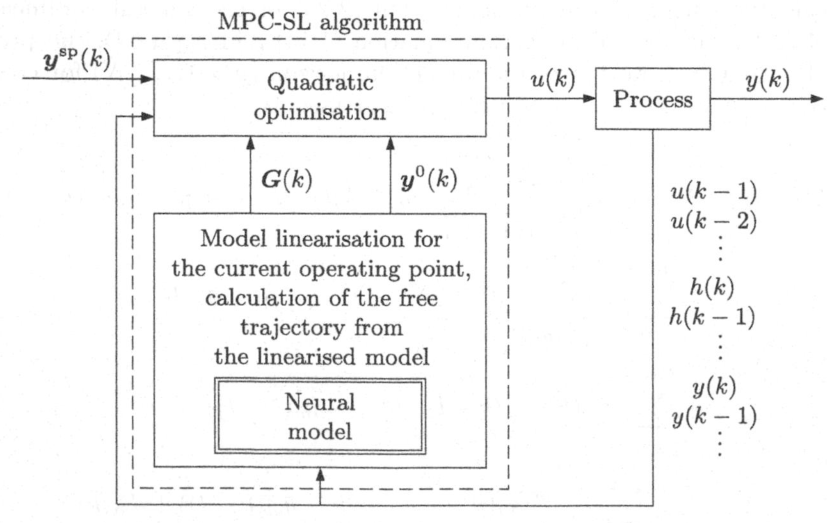

Fig. 2.4. The structure of the MPC algorithm with Successive Linearisation (MPC-SL)

The free trajectory (2.41) does not depend explicitly on the measured disturbances $h_1, \ldots, h_{\mathrm{h}}$, but they determine the current operating point of the process used for linearisation. Hence, the coefficients of the linearised model ($a_j^m(k)$ and $b_j^{m,n}(k)$) depend on the measured disturbances.

Presentation of the MPC-SL algorithm is motivated by historic reasons. Nowadays there are no arguments against using the nonlinear model for finding the free trajectory [316]. Computational burden of the whole MPC algorithm is mainly influenced by quadratic optimisation. The MPC-SL algorithm may be used in place of the MPC-NPL one only for some simple processes, when the set-point changes are moderate.

2.3.3 MPC Algorithm with Nonlinear Prediction and Linearisation along the Trajectory (MPC-NPLT)

A unique feature of MPC-NPL and MPC-SL algorithms is the fact a linear approximation of the nonlinear model is calculated on-line for the current operating point and used for prediction over the whole prediction horizon. If the nonlinear model is precise enough, the MPC-NPL algorithm should give better control than the MPC-SL one because the nonlinear free trajectory is used. It may happen, however, that for the same nonlinear model the MPC-NPL strategy gives much worse results than the "ideal" MPC-NO approach. The problem is then the inaccuracy of the suboptimal prediction, i.e. it is significantly different from the nonlinear prediction used in the MPC-NO approach. In the suboptimal MPC algorithms described so far model linearisation is carried out for the current operating point, which is defined by some previous signals. When the set-point changes significantly and fast or when strong disturbances affect the process, the obtained linear approximation is likely to describe precisely enough properties of the nonlinear model (and properties of the real process) only for the first part of the prediction horizon. For long horizons the discrepancy between the suboptimal predicted trajectory and the nonlinear one (which should be very similar to the trajectory of the process) may be significant.

The discussed approach is named the MPC algorithm with Nonlinear Prediction and Linearisation along the Trajectory (MPC-NPLT). A straightforward solution is carry out linearisation not for the current operating point, but along some future input trajectory

$$\boldsymbol{u}^{\mathrm{traj}}(k) = \begin{bmatrix} u^{\mathrm{traj}}(k|k) \\ \vdots \\ u^{\mathrm{traj}}(k+N_{\mathrm{u}}-1|k) \end{bmatrix}$$

of length $n_{\mathrm{u}}N_{\mathrm{u}}$, where, of course, $u^{\mathrm{traj}}(k+p|k) = u^{\mathrm{traj}}(k+N_{\mathrm{u}}-1|k)$ for $p = N_{\mathrm{u}}, \ldots, N$. Using the model, for the assumed input trajectory $\boldsymbol{u}^{\mathrm{traj}}(k)$ the predicted output trajectory

$$\hat{\boldsymbol{y}}^{\mathrm{traj}}(k) = \begin{bmatrix} \hat{y}^{\mathrm{traj}}(k+1|k) \\ \vdots \\ \hat{y}^{\mathrm{traj}}(k+N|k) \end{bmatrix}$$

of length $n_{\mathrm{y}}N$ may be calculated. Ideally, it would be best to use for linearisation the optimal control trajectory calculated at the current sampling instant. Of course, it is unknown before solving the MPC optimisation problem. A straightforward approach is to use for linearisation the control signals calculated and applied to the process at the previous sampling instant

$$\boldsymbol{u}^{\mathrm{traj}}(k) = \boldsymbol{u}(k-1) = \begin{bmatrix} u(k-1) \\ \vdots \\ u(k-1) \end{bmatrix} \tag{2.42}$$

In such a case the predicted output trajectory $\hat{\boldsymbol{y}}^{\mathrm{traj}}(k)$ is in fact the nonlinear free trajectory. The MPC-NPLT algorithm with linearisation along the trajectory $\boldsymbol{u}(k-1)$ is named MPC-NPLT$_{\boldsymbol{u}(k-1)}$. It is also possible to use for linearisation the last $n_{\mathrm{u}}(N_{\mathrm{u}}-1)$ elements of the optimal input trajectory calculated at the previous iteration

$$\boldsymbol{u}^{\mathrm{traj}}(k) = \boldsymbol{u}(k|k-1) = \begin{bmatrix} u(k|k-1) \\ \vdots \\ u(k+N_{\mathrm{u}}-3|k-1) \\ u(k+N_{\mathrm{u}}-2|k-1) \\ u(k+N_{\mathrm{u}}-2|k-1) \end{bmatrix} \tag{2.43}$$

The second version of the MPC-NPLT algorithm is named MPC-NPLT$_{\boldsymbol{u}(k|k-1)}$.

If linearisation is carried out along the input trajectory $\boldsymbol{u}^{\mathrm{traj}}(k)$, using the Taylor series expansion formula (2.25), the linear approximation of the nonlinear predicted output trajectory $\hat{\boldsymbol{y}}(k)$, i.e. approximation of the function $\hat{\boldsymbol{y}}(\boldsymbol{u}(k))\colon \mathbb{R}^{n_{\mathrm{u}}N_{\mathrm{u}}} \to \mathbb{R}^{n_{\mathrm{y}}N}$, can be expressed as

$$\hat{\boldsymbol{y}}(k) = \hat{\boldsymbol{y}}^{\mathrm{traj}}(k) + \boldsymbol{H}(k)(\boldsymbol{u}(k) - \boldsymbol{u}^{\mathrm{traj}}(k)) \tag{2.44}$$

where the matrix

$$\boldsymbol{H}(k) = \left.\frac{\mathrm{d}\hat{\boldsymbol{y}}(k)}{\mathrm{d}\boldsymbol{u}(k)}\right|_{\substack{\hat{\boldsymbol{y}}(k)=\hat{\boldsymbol{y}}^{\mathrm{traj}}(k) \\ \boldsymbol{u}(k)=\boldsymbol{u}^{\mathrm{traj}}(k)}} = \frac{\mathrm{d}\hat{\boldsymbol{y}}^{\mathrm{traj}}(k)}{\mathrm{d}\boldsymbol{u}^{\mathrm{traj}}(k)}$$

$$= \begin{bmatrix} \dfrac{\partial \hat{y}^{\mathrm{traj}}(k+1|k)}{\partial u^{\mathrm{traj}}(k|k)} & \cdots & \dfrac{\partial \hat{y}^{\mathrm{traj}}(k+1|k)}{\partial u^{\mathrm{traj}}(k+N_{\mathrm{u}}-1|k)} \\ \vdots & \ddots & \vdots \\ \dfrac{\partial \hat{y}^{\mathrm{traj}}(k+N|k)}{\partial u^{\mathrm{traj}}(k|k)} & \cdots & \dfrac{\partial \hat{y}^{\mathrm{traj}}(k+N|k)}{\partial u^{\mathrm{traj}}(k+N_{\mathrm{u}}-1|k)} \end{bmatrix} \tag{2.45}$$

is of dimensionality $n_{\text{y}}N \times n_{\text{u}}N_{\text{u}}$. In the discussed approach, similarly to MPC-SL and MPC-NPL algorithms, control increments are calculated rather than the values of the control signals. Using Eqs. (2.35) and (2.44), the linear approximation of the nonlinear predicted output trajectory becomes

$$\hat{\boldsymbol{y}}(k) = \boldsymbol{H}(k)\boldsymbol{J}\triangle\boldsymbol{u}(k) + \hat{\boldsymbol{y}}^{\text{traj}}(k) + \boldsymbol{H}(k)(\boldsymbol{u}(k-1) - \boldsymbol{u}^{\text{traj}}(k)) \tag{2.46}$$

Thanks to linearisation, the predicted output trajectory is a linear function of the future control increments. The obtained formula is similar the prediction equation (2.33) used in the MPC-NPL algorithm with model linearisation at the current operating point. In the MPC-NPL scheme, however, the forced part $\boldsymbol{G}(k)\triangle\boldsymbol{u}(k)$ of the predicted output trajectory depends only on the future values of the control signals (i.e. the decision variables of the algorithm) and the free trajectory $\boldsymbol{y}^0(k)$ depends only on the past whereas in the MPC-NPLT algorithm the expression $\boldsymbol{H}(k)\boldsymbol{J}\triangle\boldsymbol{u}(k)$ explicitly depends on the future control moves, but the part $\hat{\boldsymbol{y}}^{\text{traj}}(k)+\boldsymbol{H}(k)(\boldsymbol{u}(k-1)-\boldsymbol{u}^{\text{traj}}(k))$ depends on some past measurements and also on the future, i.e. on the assumed input trajectory $\boldsymbol{u}^{\text{traj}}(k)$.

Using the suboptimal prediction equation (2.46), it is possible to transform the general MPC optimisation problem (1.9) into the MPC-NPLT quadratic optimisation task

$$\min_{\substack{\triangle\boldsymbol{u}(k)\\ \boldsymbol{\varepsilon}^{\min}(k),\ \boldsymbol{\varepsilon}^{\max}(k)}} \Big\{J(k) = \Big\|\boldsymbol{y}^{\text{sp}}(k) - \boldsymbol{H}(k)\boldsymbol{J}\triangle\boldsymbol{u}(k) - \hat{\boldsymbol{y}}^{\text{traj}}(k) - \boldsymbol{H}(k)(\boldsymbol{u}(k-1) - \boldsymbol{u}^{\text{traj}}(k))\Big\|_{\boldsymbol{M}}^2 + \|\triangle\boldsymbol{u}(k)\|_{\boldsymbol{\Lambda}}^2 + \rho^{\min}\left\|\boldsymbol{\varepsilon}^{\min}(k)\right\|^2 + \rho^{\max}\left\|\boldsymbol{\varepsilon}^{\max}(k)\right\|^2\Big\}$$

subject to (2.47)

$$\boldsymbol{u}^{\min} \leq \boldsymbol{J}\triangle\boldsymbol{u}(k) + \boldsymbol{u}(k-1) \leq \boldsymbol{u}^{\max}$$
$$-\triangle\boldsymbol{u}^{\max} \leq \triangle\boldsymbol{u}(k) \leq \triangle\boldsymbol{u}^{\max}$$
$$\boldsymbol{y}^{\min} - \boldsymbol{\varepsilon}^{\min}(k) \leq \boldsymbol{H}(k)\boldsymbol{J}\triangle\boldsymbol{u}(k) + \hat{\boldsymbol{y}}^{\text{traj}}(k) + \boldsymbol{H}(k)(\boldsymbol{u}(k-1) - \boldsymbol{u}^{\text{traj}}(k)) \leq \boldsymbol{y}^{\max} + \boldsymbol{\varepsilon}^{\max}(k)$$
$$\boldsymbol{\varepsilon}^{\min}(k) \geq 0,\ \boldsymbol{\varepsilon}^{\max}(k) \geq 0$$

The structure of the MPC-NPLT algorithm is depicted in Fig. 2.5. At each sampling instant (algorithm iteration k) the following steps are repeated on-line:

1. The predicted output trajectory $\hat{\boldsymbol{y}}^{\text{traj}}(k)$ is calculated for the assumed future input trajectory $\boldsymbol{u}^{\text{traj}}(k)$ using the nonlinear (e.g. neural) model of the process.
2. The nonlinear model of the process is also used to find the linear approximation of the predicted trajectory $\hat{\boldsymbol{y}}(k)$ along the trajectory $\boldsymbol{u}^{\text{traj}}(k)$ given by Eq. (2.44), i.e. the matrix $\boldsymbol{H}(k)$ given by Eq. (2.45) is found.
3. The quadratic optimisation problem (2.47) is solved to calculate the future control increments vector $\triangle\boldsymbol{u}(k)$.

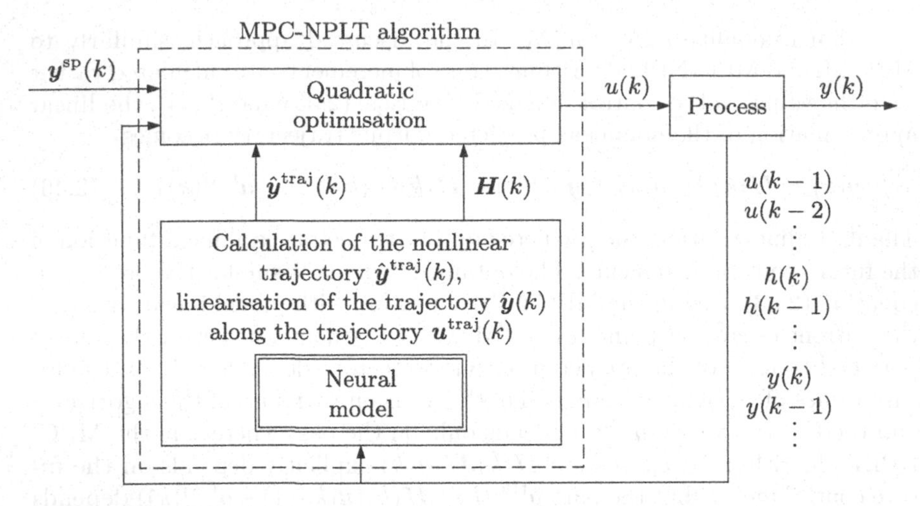

Fig. 2.5. The structure of the MPC algorithm with Nonlinear Prediction and Linearisation along the Trajectory (MPC-NPLT)

4. The first n_{u} elements of the determined sequence are applied to the process, i.e. $u(k) = \triangle u(k|k) + u(k-1)$.
5. The iteration of the algorithm is increased, i.e. $k\colon = k+1$, the algorithm goes to step 1.

It is possible to design hybrid MPC-NPL-NPLT and MPC-SL-NPLT algorithms. In their first phase the future control increments $\triangle \boldsymbol{u}(k)$ are found by means of MPC-NPL or MPC-SL approaches. In the second phase the MPC-NPLT approach is used, the future input trajectory $\boldsymbol{u}^{\text{traj}}(k)$ used for linearisation corresponds to the future control increments found in the first phase. From Eq. (2.35) one has $\boldsymbol{u}^{\text{traj}}(k) = \boldsymbol{J}\triangle \boldsymbol{u}(k) + \boldsymbol{u}(k-1)$.

Implementation of MPC-NPLT Algorithm for Neural Model

The equations used in implementation of the MPC-NPLT algorithm are quite similar to those used in the MPC-NO approach. For the DLP neural model described by Eqs. (2.2) and (2.3), using the general prediction equation (1.14), similarly to Eq. (2.18), it is possible to obtain the predicted output trajectory in the following way

$$\hat{y}_m^{\text{traj}}(k+p|k) = w_0^{2,m} + \sum_{i=1}^{K^m} w_i^{2,m} \varphi(z_i^{m,\text{traj}}(k+p|k)) + d_m(k) \tag{2.48}$$

where the unmeasure disturbances are calculated from Eq. (2.19), in the same way it is done in MPC-NO and MPC-NPL algorithms. Similarly to Eq. (2.20), one has

$$\begin{aligned}
z_i^{m,\mathrm{traj}}(k+p|k) = w_{i,0}^{1,m} &+ \sum_{n=1}^{n_\mathrm{u}} \sum_{j=1}^{I_\mathrm{uf}^{m,n}(p)} w_{i,J_\mathrm{u}^{m,n}+j}^{1,m} u_n^{\mathrm{traj}}(k-\tau^{m,n}+1-j+p|k) \\
&+ \sum_{n=1}^{n_\mathrm{u}} \sum_{j=I_\mathrm{uf}^{m,n}(p)+1}^{I_\mathrm{u}^{m,n}} w_{i,J_\mathrm{u}^{m,n}+j}^{1,m} u_n(k-\tau^{m,n}+1-j+p) \\
&+ \sum_{n=1}^{n_\mathrm{h}} \sum_{j=1}^{I_\mathrm{hf}^{m,n}(p)} w_{i,I_\mathrm{u}^m+J_\mathrm{h}^{m,n}+j}^{1,m} h_n(k-\tau_\mathrm{h}^{m,n}+1-j+p|k) \\
&+ \sum_{n=1}^{n_\mathrm{h}} \sum_{j=I_\mathrm{hf}^{m,n}(p)+1}^{I_\mathrm{h}^{m,n}} w_{i,I_\mathrm{u}^m+J_\mathrm{h}^{m,n}+j}^{1,m} h_n(k-\tau_\mathrm{h}^{m,n}+1-j+p) \\
&+ \sum_{j=1}^{I_\mathrm{yf}^m(p)} w_{i,I_\mathrm{u}^m+I_\mathrm{h}^m+j}^{1,m} \hat{y}_m^{\mathrm{traj}}(k-j+p|k) \\
&+ \sum_{j=I_\mathrm{yf}^m(p)+1}^{n_\mathrm{A}^m} w_{i,I_\mathrm{u}^m+I_\mathrm{h}^m+j}^{1,m} y_m(k-j+p)
\end{aligned} \tag{2.49}$$

In comparison with Eq. (2.20) used in the MPC-NO algorithm, in the MPC-NPLT approach the assumed input trajectory (the second part of the right side of Eq. (2.49)) and the corresponding output trajectory (the second last part of the right side of Eq. (2.49)) must be used.

The linear approximation of the predicted output trajectory along some assumed trajectory $\hat{\boldsymbol{y}}^{\mathrm{traj}}(k)$ is given by Eq. (2.44) and characterised by the matrix $\boldsymbol{H}(k)$ given by Eq. (2.45). One may notice that the matrix $\boldsymbol{H}(k)$ is the same as the derivatives matrix of the predicted trajectory with respect to the future control sequence (2.16). The entries of that matrix, analogously to Eq. (2.21), are

$$\frac{\partial \hat{y}_m^{\mathrm{traj}}(k+p|k)}{\partial u_n^{\mathrm{traj}}(k+r|k)} = \sum_{i=1}^{K^m} w_i^{2,m} \frac{\mathrm{d}\varphi(z_i^{m,\mathrm{traj}}(k+p|k))}{\mathrm{d}z_i^{m,\mathrm{traj}}(k+p|k)} \frac{\partial z_i^{m,\mathrm{traj}}(k+p|k)}{\partial u_n^{\mathrm{traj}}(k+r|k)} \tag{2.50}$$

where, similarly to Eq. (2.23), one has

$$\begin{aligned}
\frac{\partial z_i^{m,\mathrm{traj}}(k+p|k)}{\partial u_n^{\mathrm{traj}}(k+r|k)} = &\sum_{j=1}^{I_\mathrm{uf}^{m,n}(p)} w_{i,J_\mathrm{u}^{m,n}+j}^{1,m} \frac{\partial u_n^{\mathrm{traj}}(k-\tau^{m,n}+1-j+p|k)}{\partial u_n^{\mathrm{traj}}(k+r|k)} \\
&+ \sum_{j=1}^{I_\mathrm{yf}^m(p)} w_{i,I_\mathrm{u}^m+I_\mathrm{h}^m+j}^{1,m} \frac{\partial \hat{y}_m^{\mathrm{traj}}(k-j+p|k)}{\partial u_n^{\mathrm{traj}}(k+r|k)}
\end{aligned} \tag{2.51}$$

2.3.4 MPC Algorithm with Nonlinear Prediction and Linearisation along the Predicted Trajectory (MPC-NPLPT)

The MPC-NPLT algorithm is recommended when control quality obtained in the MPC-NPL algorithm with model linearisation for the current operating point is not sufficient. In the MPC-NPLT approach a linear approximation of the predicted output trajectory $\hat{\boldsymbol{y}}(k)$ is calculated along some assumed future input trajectory $\boldsymbol{u}^{\mathrm{traj}}(k)$. It is also possible to repeat iteratively nonlinear prediction, trajectory linearisation and calculation of the future control policy a few times at each sampling instant. Intuitively, such an approach is likely to increase prediction accuracy and, in consequence, quality of control. The discussed method is used in the MPC algorithm with Nonlinear Prediction and Linearisation along the Predicted Trajectory (MPC-NPLPT). Its structure is depicted in Fig. 2.6. Let t be the index of internal iterations ($t = 1, 2, \ldots$). In the tth internal iteration the predicted output trajectory is linearised along the input trajectory $\boldsymbol{u}^{t-1}(k)$ found at the previous internal iteration $(t-1)$, which corresponds to the predicted output trajectory $\hat{\boldsymbol{y}}^{t-1}(k)$. The initial trajectory $\boldsymbol{u}^{0}(k)$ (initialisation of the internal loop) may be defined using the control signals applied at the previous sampling instant $\boldsymbol{u}(k-1)$ or the signals $\boldsymbol{u}(k|k-1)$ calculated for the current sampling instant at the previous one (Eqs. (2.42) and (2.43)). The first internal iteration of the MPC-NPLPT algorithm is in fact the same as in the MPC-NPLT one, in the following internal iterations linearisation is carried out for the successively updated input trajectory.

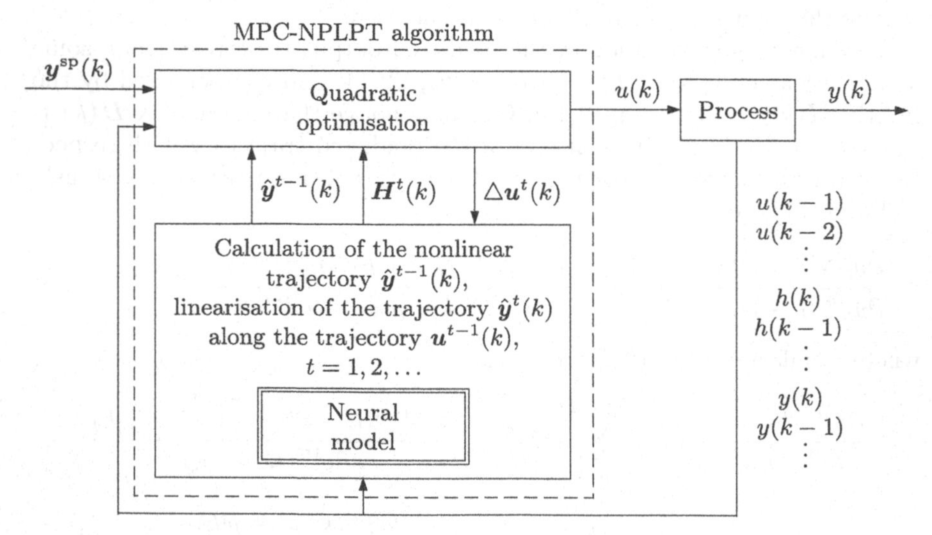

Fig. 2.6. The structure of the MPC algorithm with Nonlinear Prediction and Linearisation along the Predicted Trajectory (MPC-NPLPT)

If linearisation is carried out along the input trajectory $\boldsymbol{u}^{t-1}(k)$, using the Taylor series expansion formula (2.25), the linear approximation of the nonlinear predicted output trajectory $\hat{\boldsymbol{y}}^t(k)$, i.e. approximation of the function $\hat{\boldsymbol{y}}^t(\boldsymbol{u}^t(k))\colon \mathbb{R}^{n_\mathrm{u}N_\mathrm{u}} \to \mathbb{R}^{n_\mathrm{y}N}$, can be expressed as

$$\hat{\boldsymbol{y}}^t(k) = \hat{\boldsymbol{y}}^{t-1}(k) + \boldsymbol{H}^t(k)(\boldsymbol{u}^t(k) - \boldsymbol{u}^{t-1}(k)) \tag{2.52}$$

where the matrix

$$\boldsymbol{H}^t(k) = \left.\frac{\mathrm{d}\hat{\boldsymbol{y}}(k)}{\mathrm{d}\boldsymbol{u}(k)}\right|_{\substack{\hat{\boldsymbol{y}}(k)=\hat{\boldsymbol{y}}^{t-1}(k)\\ \boldsymbol{u}(k)=\boldsymbol{u}^{t-1}(k)}} = \frac{\mathrm{d}\hat{\boldsymbol{y}}^{t-1}(k)}{\mathrm{d}\boldsymbol{u}^{t-1}(k)}$$
$$= \begin{bmatrix} \dfrac{\partial \hat{y}^{t-1}(k+1|k)}{\partial u^{t-1}(k|k)} & \cdots & \dfrac{\partial \hat{y}^{t-1}(k+1|k)}{\partial u^{t-1}(k+N_\mathrm{u}-1|k)} \\ \vdots & \ddots & \vdots \\ \dfrac{\partial \hat{y}^{t-1}(k+N|k)}{\partial u^{t-1}(k|k)} & \cdots & \dfrac{\partial \hat{y}^{t-1}(k+N|k)}{\partial u^{t-1}(k+N_\mathrm{u}-1|k)} \end{bmatrix} \tag{2.53}$$

is of dimensionality of $n_\mathrm{y}N \times n_\mathrm{u}N_\mathrm{u}$. Using the relation

$$\boldsymbol{u}^t(k) = \boldsymbol{J}\triangle\boldsymbol{u}^t(k) + \boldsymbol{u}(k-1) \tag{2.54}$$

which is a counterpart of Eqs. (2.35), and Eq. (2.52), the linear approximation of the nonlinear predicted output trajectory becomes

$$\hat{\boldsymbol{y}}^t(k) = \boldsymbol{H}^t(k)\boldsymbol{J}\triangle\boldsymbol{u}^t(k) + \hat{\boldsymbol{y}}^{t-1}(k) + \boldsymbol{H}^t(k)(\boldsymbol{u}(k-1) - \boldsymbol{u}^{t-1}(k)) \tag{2.55}$$

Thanks to linearisation, the predicted output trajectory $\hat{\boldsymbol{y}}^t(k)$ is a linear function of the future control increments $\triangle\boldsymbol{u}^t(k)$ calculated at the internal iteration t.

Using the suboptimal prediction equation (2.55) and the relation (2.54), it is possible to transform the general MPC optimisation problem (1.9) into the following MPC-NPLPT quadratic optimisation task

$$\min_{\substack{\triangle\boldsymbol{u}^t(k)\\ \boldsymbol{\varepsilon}^{\min}(k),\ \boldsymbol{\varepsilon}^{\max}(k)}} \Big\{J(k) = \big\|\boldsymbol{y}^{\mathrm{sp}}(k) - \boldsymbol{H}^t(k)\boldsymbol{J}\triangle\boldsymbol{u}^t(k) - \hat{\boldsymbol{y}}^{t-1}(k) - \boldsymbol{H}^t(k)(\boldsymbol{u}(k-1) - \boldsymbol{u}^{t-1}(k))\big\|_{\boldsymbol{M}}^2 + \|\triangle\boldsymbol{u}^t(k)\|_{\boldsymbol{\Lambda}}^2 + \rho^{\min}\left\|\boldsymbol{\varepsilon}^{\min}(k)\right\|^2 + \rho^{\max}\left\|\boldsymbol{\varepsilon}^{\max}(k)\right\|^2\Big\}$$

subject to (2.56)

$$\boldsymbol{u}^{\min} \leq \boldsymbol{J}\triangle\boldsymbol{u}^t(k) + \boldsymbol{u}(k-1) \leq \boldsymbol{u}^{\max}$$
$$-\triangle\boldsymbol{u}^{\max} \leq \triangle\boldsymbol{u}^t(k) \leq \triangle\boldsymbol{u}^{\max}$$
$$\boldsymbol{y}^{\min} - \boldsymbol{\varepsilon}^{\min}(k) \leq \boldsymbol{H}^t(k)\boldsymbol{J}\triangle\boldsymbol{u}^t(k) + \hat{\boldsymbol{y}}^{t-1}(k) + \boldsymbol{H}^t(k)(\boldsymbol{u}(k-1) - \boldsymbol{u}^{t-1}(k)) \leq \boldsymbol{y}^{\max} + \boldsymbol{\varepsilon}^{\max}(k)$$
$$\boldsymbol{\varepsilon}^{\min}(k) \geq 0,\ \boldsymbol{\varepsilon}^{\max}(k) \geq 0$$

If the set-point changes in the consecutive iterations are not significant, one may expect that one internal iteration is sufficient (which leads to the MPC-NPLT algorithm). The internal iterations are continued if

$$\sum_{p=0}^{N_0}(y^{\mathrm{sp}}(k-p)-y(k-p))^2 \geq \delta_{\mathrm{y}} \tag{2.57}$$

where N_0 is a time horizon and the quantity $\delta_{\mathrm{y}} > 0$ is found experimentally. If the difference between the future control increments calculated in two consecutive internal iterations is not significant, it means when

$$\left\|\triangle \boldsymbol{u}^t(k)-\triangle \boldsymbol{u}^{t-1}(k)\right\|^2 < \delta_{\mathrm{u}} \tag{2.58}$$

where the quantity $\delta_{\mathrm{u}} > 0$ is adjusted experimentally, the internal iterations are terminated.

At each sampling instant (algorithm iteration k) of the MPC-NPLPT algorithm shown in Fig. 2.6 the following steps are repeated on-line:

1. The first internal iteration ($t = 1$): The predicted output trajectory $\hat{\boldsymbol{y}}^0(k)$ is calculated for the assumed input trajectory $\boldsymbol{u}^0(k)$ using the nonlinear (e.g. neural) model of the process.
2. The nonlinear model of the process is also used to find the linear approximation of the predicted trajectory $\hat{\boldsymbol{y}}^1(k)$ along the trajectory $\boldsymbol{u}^0(k)$ given by Eq. (2.52), i.e. the matrix $\boldsymbol{H}^1(k)$ given by Eq. (2.53) is found.
3. The quadratic optimisation problem (2.56) is solved to calculate the future control increments vector $\triangle\boldsymbol{u}^1(k)$.
4. If the condition (2.57) is satisfied, the internal iterations are continued for $t = 2, \ldots, t_{\max}$.
   4.1. The predicted output trajectory $\hat{\boldsymbol{y}}^{t-1}(k)$ is calculated for the input trajectory $\boldsymbol{u}^{t-1}(k) = \boldsymbol{J}\triangle\boldsymbol{u}^{t-1}(k)+\boldsymbol{u}(k-1)$ using the nonlinear model of the process.
   4.2. The nonlinear model of the process is also used to find the linear approximation of the predicted trajectory $\hat{\boldsymbol{y}}^t(k)$ along the trajectory $\boldsymbol{u}^{t-1}(k)$, i.e. the matrix $\boldsymbol{H}^t(k)$ is found.
   4.3. The quadratic optimisation problem (2.56) is solved to calculate the future control increments vector $\triangle\boldsymbol{u}^t(k)$ for the current internal iteration.
   4.4. If the condition (2.58) is satisfied or $t > t_{\max}$, the internal iterations are terminated. Otherwise, the internal iteration index is increased, i.e. $t\colon = t+1$, the algorithms goes to step 4.1.
5. The first n_{u} elements of the determined sequence are applied to the process, i.e. $u(k) = \triangle u^t(k|k) + u(k-1)$.
6. The iteration of the algorithm is increased, i.e. $k\colon = k+1$, the algorithm goes to step 1.

It is possible to design hybrid MPC-NPL-NPLPT and MPC-SL-NPLPT algorithms. In the first phase the future control sequence is found by means of the MPC-NPL or MPC-SL approach. Next, the MPC-NPLPT algorithm is used, the input trajectory $\boldsymbol{u}^0(k)$ used for linearisation corresponds to the future control increments found in the first phase. When the set-point changes are not significant, the second phase may be not activated or only one internal iteration may be sufficient (Eq. (2.57)).

Implementation of MPC-NPLPT Algorithm for Neural Model

Implementation details of the MPC-NPLPT are very similar to those of the MPC-NPLT one. Similarly to Eq. (2.48) the predicted output trajectory in the tth internal iteration is

$$\hat{y}_m^t(k+p|k) = w_0^{2,m} + \sum_{i=1}^{K^m} w_i^{2,m}\varphi(z_i^{m,t}(k+p|k)) + d_m(k) \tag{2.59}$$

where the unmeasured disturbances are calculated from (2.19), in the same way it is done in MPC-NO, MPC-NPL and MPC-NPLT algorithm. Similarly to Eq. (2.49), one has

$$\begin{aligned} z_i^{m,t}(k+p|k) = w_{i,0}^{1,m} &+ \sum_{n=1}^{n_{\text{u}}} \sum_{j=1}^{I_{\text{uf}}^{m,n}(p)} w_{i,J_{\text{u}}^{m,n}+j}^{1,m} u_n^t(k-\tau^{m,n}+1-j+p|k) \\ &+ \sum_{n=1}^{n_{\text{u}}} \sum_{j=I_{\text{uf}}^{m,n}(p)+1}^{I_{\text{u}}^{m,n}} w_{i,J_{\text{u}}^{m,n}+j}^{1,m} u_n(k-\tau^{m,n}+1-j+p) \\ &+ \sum_{n=1}^{n_{\text{h}}} \sum_{j=1}^{I_{\text{hf}}^{m,n}(p)} w_{i,I_{\text{u}}^m+J_{\text{h}}^{m,n}+j}^{1,m} h_n(k-\tau_{\text{h}}^{m,n}+1-j+p|k) \\ &+ \sum_{n=1}^{n_{\text{h}}} \sum_{j=I_{\text{hf}}^{m,n}(p)+1}^{I_{\text{h}}^{m,n}} w_{i,I_{\text{u}}^m+J_{\text{h}}^{m,n}+j}^{1,m} h_n(k-\tau_{\text{h}}^{m,n}+1-j+p) \\ &+ \sum_{j=1}^{I_{\text{yf}}^m(p)} w_{i,I_{\text{u}}^m+I_{\text{h}}^m+j}^{1,m} \hat{y}_m^t(k-j+p|k) \\ &+ \sum_{j=I_{\text{yf}}^m(p)+1}^{n_{\text{A}}^m} w_{i,I_{\text{u}}^m+I_{\text{h}}^m+j}^{1,m} y_m(k-j+p) \end{aligned} \tag{2.60}$$

Linearisation is carried out in the MPC-NPLPT algorithm along the trajectory $\boldsymbol{u}^{t-1}(k)$. The entries of the matrix $\boldsymbol{H}^t(k)$ defined by Eq. (2.53) are calculated using some formulae derived for the MPC-NO algorithm, in a similar way it is done in the MPC-NPLT algorithm. Similarly to Eqs. (2.21) and (2.50), they are found from

$$\frac{\partial \hat{y}_m^{t-1}(k+p|k)}{\partial u_n^{t-1}(k+r|k)} = \sum_{i=1}^{K^m} w_i^{2,m} \frac{\mathrm{d}\varphi(z_i^{m,t-1}(k+p|k))}{\mathrm{d}z_i^{m,t-1}(k+p|k)} \frac{\partial z_i^{m,t-1}(k+p|k)}{\partial u_n^{t-1}(k+r|k)}$$

where, using Eqs. (2.59) and (2.60), similarly to Eqs. (2.23) and (2.51), one has

$$\frac{\partial z_i^{m,t-1}(k+p|k)}{\partial u_n^{t-1}(k+r|k)} = \sum_{j=1}^{I_{\mathrm{uf}}^{m,n}(p)} w_{i,J_{\mathrm{u}}^{m,n}+j}^{1,m} \frac{\partial u_n^{t-1}(k-\tau^{m,n}+1-j+p|k)}{\partial u_n^{t-1}(k+r|k)}$$
$$+ \sum_{j=1}^{I_{\mathrm{yf}}^{m}(p)} w_{i,I_{\mathrm{u}}^{m}+I_{\mathrm{h}}^{m}+j}^{1,m} \frac{\partial \hat{y}_m^{t-1}(k-j+p|k)}{\partial u_n^{t-1}(k+r|k)}$$

2.3.5 MPC Newton-Like Algorithm with Nonlinear Prediction and Approximation along the Predicted Trajectory (MPC-NNPAPT)

Although simulations of the MPC-NPLPT algorithm clearly indicate its usefulness, convergence of internal iterations may be an issue. In the simplest case one may compare values of the minimised cost-function in consecutive internal iterations and terminate them if those values increase. A fundamentally better idea is to use the MPC Newton-like algorithm with Nonlinear Prediction and Approximation along the Predicted Trajectory (MPC-NNPAPT). A quadratic approximation of the minimised cost-function is used in the considered algorithm, which is a characteristic feature of Newton and Newton-like optimisation algorithms [226]. In general, quadratic approximation is much better than the linear one. As a result, the quasi-Newton variable metrics unconstrained optimisation methods are very efficient, e.g. DFP or BFGS algorithms. In constrained optimisation the quadratic approximation approach is used in the SQP algorithm. The solution of a nonlinear optimisation problem is obtained by means of solving a series of quadratic optimisation subproblems. Convergence of the SQP algorithm is superlinear; when the Hessian matrix is calculated precisely (not approximated), convergence is quadratic. Thanks to its efficiency (relative low computational complexity) and robustness, the SQP optimisation method is very popular. The rudimentary version of the SQP algorithm is described in [257], its convergence is proven in [256].

Taking into account the general optimisation problem used in the MPC-NO algorithm (2.8), the Lagrange function is defined

$$\mathcal{L}(\boldsymbol{x}_{\mathrm{opt}}^t(k), \boldsymbol{\varrho}^t(k)) = f_{\mathrm{opt}}(\boldsymbol{x}_{\mathrm{opt}}^t(k)) + (\boldsymbol{\varrho}^t(k))^{\mathrm{T}} \boldsymbol{g}_{\mathrm{opt}}(\boldsymbol{x}_{\mathrm{opt}}^t(k)) \tag{2.61}$$

where t is the index of internal iterations, the decision variables vector $\boldsymbol{x}_{\mathrm{opt}}^t(k)$ of length $n_{\mathrm{u}}N_{\mathrm{u}} + 2n_{\mathrm{y}}N$ is defined in the same way it is done in the MPC-NO algorithm (Eq. (2.6)), the number of constraints is $4n_{\mathrm{u}}N_{\mathrm{u}} + 4n_{\mathrm{y}}N$. The vector of Lagrange multipliers is denoted by $\boldsymbol{\varrho}^t(k)$. During calculations the derivatives vector of the minimised cost-function is used

$$\boldsymbol{f}^t(k) = \frac{\mathrm{d}f_{\mathrm{opt}}(\boldsymbol{x}_{\mathrm{opt}}^t(k))}{\mathrm{d}\boldsymbol{x}_{\mathrm{opt}}^t(k)}$$

which has length $n_{\mathrm{u}}N_{\mathrm{u}} + 2n_{\mathrm{y}}N$, the matrix of the second-order derivatives of the Lagrange function

$$\boldsymbol{B}^t(k) = \frac{\mathrm{d}^2\mathcal{L}(\boldsymbol{x}_{\mathrm{opt}}^t(k), \boldsymbol{\varrho}^t(k))}{\mathrm{d}(\boldsymbol{x}_{\mathrm{opt}}^t(k))^2}$$

of dimensionality $(n_{\mathrm{u}}N_{\mathrm{u}} + 2n_{\mathrm{y}}N) \times (n_{\mathrm{u}}N_{\mathrm{u}} + 2n_{\mathrm{y}}N)$ and the matrix of derivatives of the constraints

$$\boldsymbol{G}^t(k) = \frac{\mathrm{d}\boldsymbol{g}_{\mathrm{opt}}(\boldsymbol{x}_{\mathrm{opt}}^t(k))}{\mathrm{d}\boldsymbol{x}_{\mathrm{opt}}^t(k)}$$

of dimensionality $(4n_{\mathrm{u}}N_{\mathrm{u}} + 4n_{\mathrm{y}}N) \times (n_{\mathrm{u}}N_{\mathrm{u}} + 2n_{\mathrm{y}}N)$. The index t indicates the internal iteration for the current main iteration (sampling instant) k. At each iteration k the following steps are repeated:

1. Initialisation: the initial point $\boldsymbol{x}_{\mathrm{opt}}^0(k)$ and the Lagrange multipliers $\boldsymbol{\varrho}^0(k)$ are chosen, the internal iteration index indicates the first iteration ($t = 1$).
2. The derivatives vector $\boldsymbol{f}^t(k)$ of the minimised function, the Hessian matrix $\boldsymbol{B}^t(k)$ of the Lagrange function and the matrix of the derivatives of the constraints $\boldsymbol{G}^t(k)$ are calculated for the latest available solution $\boldsymbol{x}_{\mathrm{opt}}^t(k)$ (the solution from the previous internal iteration).
3. The direction vector $\boldsymbol{d}^t(k)$ and the vector of Lagrange multipliers $\boldsymbol{\varrho}^t(k)$ are calculated from the following quadratic optimisation task

$$\min_{\boldsymbol{d}^t(k)} \left\{ \frac{1}{2}(\boldsymbol{d}^t(k))^{\mathrm{T}} \boldsymbol{B}^t(k)\boldsymbol{d}^t(k) + \boldsymbol{f}^{\mathrm{T}}(k)\boldsymbol{d}^t(k) \right\}$$

subject to

$$\boldsymbol{G}^t(k)\boldsymbol{d}^t(k) + \boldsymbol{g}_{\mathrm{opt}}(\boldsymbol{x}_{\mathrm{opt}}^t(k)) \leq 0$$

4. The solution vector for the next internal iteration is calculated

$$\boldsymbol{x}_{\mathrm{opt}}^{t+1}(k) = \boldsymbol{x}_{\mathrm{opt}}^t(k) + \alpha \boldsymbol{d}^t(k)$$

where $0 < \alpha \leq 1$ is the step length into the optimisation direction $\boldsymbol{d}^t(k)$.
5. If the stopping criterion (e.g. a norm of the gradient of the minimised cost-function or the difference of its value in two consecutive internal iterations) is satisfied, the optimisation algorithm is terminated. The first n_{u} elements of the determined sequence $\boldsymbol{x}_{\mathrm{opt}}^{t+1}(k)$ are applied to the process, i.e. $u(k) = u^{t+1}(k|k)$.
6. In the next internal iteration $(t + 1)$ the algorithm goes to step 2.

In comparison with the general formulation of the SQP algorithm, there are no equality constraints in the considered algorithm. It is because in the MPC algorithms discussed so far they are not necessary.

Theoretically proven convergence, "inherited" from the SQP algorithm, is the fundamental advantage of the MPC-NNPAPT algorithm. A quadratic approximation of the minimised cost-function and a linear approximation of the constraints are determined at each iteration of the algorithm. All the calculations are carried out for the vector $\boldsymbol{x}^t_{\text{opt}}(k)$ which means that linearisation along the future control trajectory is used. The structure of the algorithm is depicted in Fig. 2.7.

The nonlinear (e.g. neural) model of the process and some equations used in the MPC-NO scheme are used in the second step of the MPC-NNPAPT algorithm. The minimised cost-function is similar to that given by Eq. (2.9) and the constraints are similar to those given by Eq. (2.10), the only difference is the fact that the vector $\boldsymbol{x}^t_{\text{opt}}(k)$ must be used in place of $\boldsymbol{x}_{\text{opt}}(k)$. The derivatives vector $\boldsymbol{f}^t(k)$ are calculated from Eqs. (2.11) and (2.12), the matrix of derivatives of the constraints $\boldsymbol{G}^t(k)$ is found from Eq. (2.13). Eq. (2.18) is used to find the predicted trajectory, Eqs. (2.21) and (2.23) together with the conditions (2.22) and (2.24) are used in order to calculate the matrix of the derivatives of the predicted output trajectory with respect to the future control sequence (2.16). In all the mentioned equations it is necessary to take into account the decision variables of the algorithm for the internal iteration t. In the MPC-NNPAPT algorithm the matrix of the second-order derivatives $\boldsymbol{B}^t(k)$ is calculated analytically using the nonlinear model of the process whereas in the rudimentary SQP algorithm it is typically approximated by

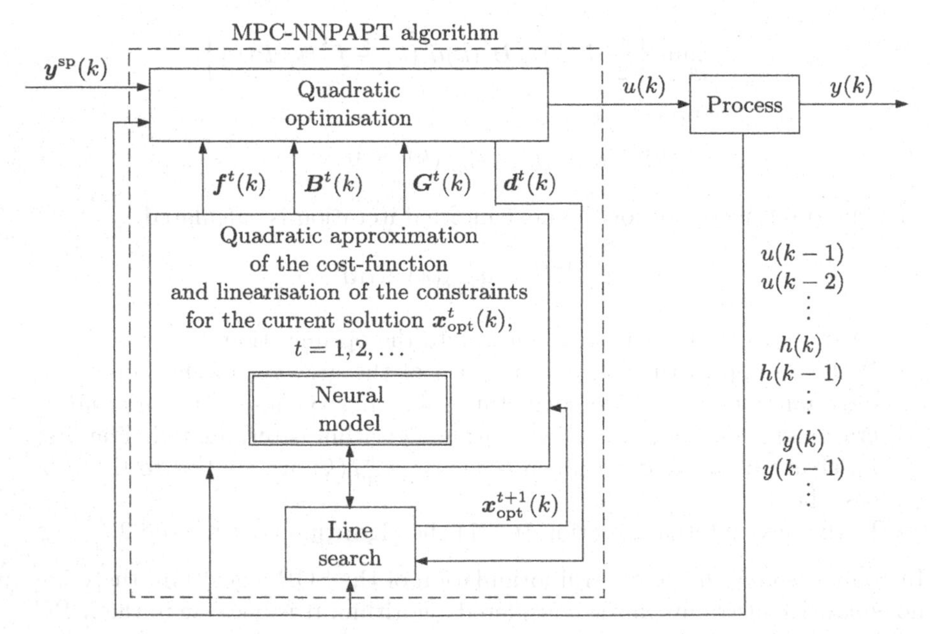

Fig. 2.7. The structure of the MPC Newton-like algorithm with Nonlinear Prediction and Approximation along the Predicted Trajectory (MPC-NNPAPT)

means of the BFGS method [257]. Differentiating twice the Lagrange function (2.61), the matrix $\boldsymbol{B}^t(k)$ is calculated from

$$\boldsymbol{B}^t(k) = \frac{\mathrm{d}^2 f_{\mathrm{opt}}(\boldsymbol{x}_{\mathrm{opt}}^t(k))}{\mathrm{d}(\boldsymbol{x}_{\mathrm{opt}}^t(k))^2} + \sum_{i=1}^{4n_{\mathrm{u}}N_{\mathrm{u}}+4n_{\mathrm{y}}N} \varrho_i^t(k) \frac{\mathrm{d}^2 g_{\mathrm{opt},i}(\boldsymbol{x}_{\mathrm{opt}}^t(k))}{\mathrm{d}(\boldsymbol{x}_{\mathrm{opt}}^t(k))^2} \tag{2.62}$$

Differentiating Eq. (2.11), one has

$$\frac{\mathrm{d}^2 f_{\mathrm{opt}}(\boldsymbol{x}_{\mathrm{opt}}^t(k))}{\mathrm{d}(\boldsymbol{x}_{\mathrm{opt}}^t(k))^2} = 2\begin{bmatrix} \boldsymbol{Q}(\boldsymbol{x}_{\mathrm{opt}}^t(k)) & \boldsymbol{0}_{n_{\mathrm{u}}N_{\mathrm{u}}\times 2n_{\mathrm{y}}N} \\ \boldsymbol{0}_{2n_{\mathrm{y}}N\times n_{\mathrm{u}}N_{\mathrm{u}}} & \boldsymbol{0}_{2n_{\mathrm{y}}N\times 2n_{\mathrm{y}}N} \end{bmatrix}$$
$$+ 2\boldsymbol{N}_1^{\mathrm{T}}(\boldsymbol{J}^{\mathrm{NO}})^{\mathrm{T}}\boldsymbol{\Lambda}\boldsymbol{J}^{\mathrm{NO}}\boldsymbol{N}_1 + 2\rho^{\min}\boldsymbol{N}_2^{\mathrm{T}}\boldsymbol{N}_2 + 2\rho^{\max}\boldsymbol{N}_3^{\mathrm{T}}\boldsymbol{N}_3$$

where the matrix

$$\boldsymbol{Q}(\boldsymbol{x}_{\mathrm{opt}}^t(k)) = \frac{\mathrm{d}}{\mathrm{d}\boldsymbol{u}^t(k)}\left[\left(\frac{\mathrm{d}\hat{\boldsymbol{y}}(k)}{\mathrm{d}\boldsymbol{u}^t(k)}\right)^{\mathrm{T}} \boldsymbol{M}(\hat{\boldsymbol{y}}(k) - \boldsymbol{y}^{\mathrm{sp}}(k))\right]$$

is of dimensionality $n_{\mathrm{u}}N_{\mathrm{u}} \times n_{\mathrm{u}}N_{\mathrm{u}}$. Using the notation $\boldsymbol{e}(k) = \hat{\boldsymbol{y}}(k) - \boldsymbol{y}^{\mathrm{sp}}(k)$, one has

$$\boldsymbol{Q}(\boldsymbol{x}_{\mathrm{opt}}^t(k)) = \begin{bmatrix} \boldsymbol{d}_{1,0,1,0}^{\mathrm{T}}(k)\boldsymbol{M}\boldsymbol{e}(k) & \cdots & \boldsymbol{d}_{n_{\mathrm{u}},N_{\mathrm{u}}-1,1,0}^{\mathrm{T}}(k)\boldsymbol{M}\boldsymbol{e}(k) \\ \vdots & \ddots & \vdots \\ \boldsymbol{d}_{1,0,n_{\mathrm{u}},N_{\mathrm{u}}-1}^{\mathrm{T}}(k)\boldsymbol{M}\boldsymbol{e}(k) & \cdots & \boldsymbol{d}_{n_{\mathrm{u}},N_{\mathrm{u}}-1,n_{\mathrm{u}},N_{\mathrm{u}}-1}^{\mathrm{T}}(k)\boldsymbol{M}\boldsymbol{e}(k) \end{bmatrix}$$
$$+ \left(\frac{\mathrm{d}\hat{\boldsymbol{y}}(k)}{\mathrm{d}\boldsymbol{u}^t(k)}\right)^{\mathrm{T}} \boldsymbol{M} \frac{\mathrm{d}\hat{\boldsymbol{y}}(k)}{\mathrm{d}\boldsymbol{u}^t(k)}$$

where the vectors

$$\boldsymbol{d}_{n_1,r_1,n_2,r_2}(k) = \frac{\partial^2 \hat{\boldsymbol{y}}(k)}{\partial u_{n_1}^t(k+r_1|k)\partial u_{n_2}^t(k+r_2|k)}$$

are of length $n_{\mathrm{y}}N$. Taking into account the structure of the vector $\hat{\boldsymbol{y}}(k)$, defined by Eq. (1.7), they may be expressed in the following form

$$\boldsymbol{d}_{n_1,r_1,n_2,r_2}(k) = \begin{bmatrix} \dfrac{\partial^2 \hat{y}(k+1|k)}{\partial u_{n_1}^t(k+r_1|k)\partial u_{n_2}^t(k+r_2|k)} \\ \vdots \\ \dfrac{\partial^2 \hat{y}(k+N|k)}{\partial u_{n_1}^t(k+r_1|k)\partial u_{n_2}^t(k+r_2|k)} \end{bmatrix}$$

where the vectors

$$\frac{\partial^2 \hat{y}(k+p|k)}{\partial u_{n_1}^t(k+r_1|k)\partial u_{n_2}^t(k+r_2|k)} = \begin{bmatrix} \dfrac{\partial^2 \hat{y}_1(k+p|k)}{\partial u_{n_1}^t(k+r_1|k)\partial u_{n_2}^t(k+r_2|k)} \\ \vdots \\ \dfrac{\partial^2 \hat{y}_{n_{\mathrm{y}}}(k+p|k)}{\partial u_{n_1}^t(k+r_1|k)\partial u_{n_2}^t(k+r_2|k)} \end{bmatrix} \tag{2.63}$$

are of length n_{y}. The matrix of the derivatives of the predicted output trajectory with respect to the future control policy has the structure given by Eq. (2.16), but during calculations it is necessary to take into account the decision variables for the internal iteration t.

Taking into consideration the constraints (2.10), the second-order derivatives occurring in the second part of the right side of Eq. (2.62) are the matrices of dimensionality $(n_{\mathrm{u}}N_{\mathrm{u}}+2n_{\mathrm{y}}N)\times(n_{\mathrm{u}}N_{\mathrm{u}}+2n_{\mathrm{y}}N)$ and of the structure

$$\frac{\mathrm{d}^2 g_{\mathrm{opt},i}(\boldsymbol{x}_{\mathrm{opt}}^t(k))}{\mathrm{d}(\boldsymbol{x}_{\mathrm{opt}}^t(k))^2}=\begin{bmatrix}\boldsymbol{R}_i(\boldsymbol{x}_{\mathrm{opt}}^t(k)) & \mathbf{0}_{n_{\mathrm{u}}N_{\mathrm{u}}\times 2n_{\mathrm{y}}N}\\ \mathbf{0}_{2n_{\mathrm{y}}N\times n_{\mathrm{u}}N_{\mathrm{u}}} & \mathbf{0}_{2n_{\mathrm{y}}N\times 2n_{\mathrm{y}}N}\end{bmatrix}$$

where the index $i=1,\ldots,4n_{\mathrm{u}}N_{\mathrm{u}}+4n_{\mathrm{y}}N$ indicates the constraint, the matrix

$$\boldsymbol{R}_i(\boldsymbol{x}_{\mathrm{opt}}^t(k))=\begin{bmatrix} r_{1,0,1,0}(\boldsymbol{x}_{\mathrm{opt}}^t(k)) & \cdots & r_{n_{\mathrm{u}},N_{\mathrm{u}}-1,1,0}(\boldsymbol{x}_{\mathrm{opt}}^t(k))\\ \vdots & \ddots & \vdots\\ r_{1,0,n_{\mathrm{u}},N_{\mathrm{u}}-1}(\boldsymbol{x}_{\mathrm{opt}}^t(k)) & \cdots & r_{n_{\mathrm{u}},N_{\mathrm{u}}-1,n_{\mathrm{u}},N_{\mathrm{u}}-1}(\boldsymbol{x}_{\mathrm{opt}}^t(k))\end{bmatrix} \tag{2.64}$$

consists of the second-order partial derivatives of the predicted output trajectory

$$\begin{aligned} r_{n_1,r_1,n_2,r_2}(\boldsymbol{x}_{\mathrm{opt}}^t(k)) &= \frac{\partial^2 g_{\mathrm{opt},i}(\boldsymbol{x}_{\mathrm{opt}}^t(k))}{\partial u_{n_1}^t(k+r_1|k)\partial u_{n_2}^t(k+r_2|k)}\\ &= z(i)\frac{\partial^2 \hat{y}_m(k+p|k)}{\partial u_{n_1}^t(k+r_1|k)\partial u_{n_2}^t(k+r_2|k)} \end{aligned} \tag{2.65}$$

where

$$z(i)=\begin{cases}-1 & \text{if } 4n_{\mathrm{u}}N_{\mathrm{u}}+1\le i\le 4n_{\mathrm{u}}N_{\mathrm{u}}+n_{\mathrm{y}}N\\ 1 & \text{if } 4n_{\mathrm{u}}N_{\mathrm{u}}+n_{\mathrm{y}}N+1\le i\le 4n_{\mathrm{u}}N_{\mathrm{u}}+2n_{\mathrm{y}}N\end{cases}$$

The index m occurring in the right part of Eq. (2.65) is the integer part of the division of $i-4n_{\mathrm{u}}N_{\mathrm{u}}$ by n_{y}, p is the rest of the division. If the calculated matrix $\boldsymbol{B}^t(k)$ is not positive-definite, it must be modified to guarantee such a feature [26].

To sum up, in the discussed MPC-NNPAPT algorithm the derivatives vector $\boldsymbol{f}^t(k)$, the Hessian matrix $\boldsymbol{B}^t(k)$ and the derivatives vector $\boldsymbol{G}^t(k)$ are calculated on-line analytically, during calculations the structure of the model and the predicted trajectory are taken into account.

The step length α is found in the fourth step of the algorithm by a line search procedure in such a way that the merit function

$$\Psi(\boldsymbol{x}_{\mathrm{opt}}^{t+1}(k),\boldsymbol{\varrho}(k))=f(\boldsymbol{x}_{\mathrm{opt}}^t(k))+\sum_{i=1}^{4n_{\mathrm{u}}N_{\mathrm{u}}+4n_{\mathrm{y}}N} r_i\min(0,g_{\mathrm{opt},i}(\boldsymbol{x}_{\mathrm{opt}}^t(k)))$$

is minimised. Typically, the Armijo's method [257] is used for line search. The initial values of the penalty parameters r_i (for $t = 0$) are the same as those of the Lagrange multipliers ($r_i^0 = \varrho_i^0(k)$), in the following iterations they are

$$r_i^t = \max(|\varrho_i^t(k)|, 0.5(r_i^{t-1} + |\varrho_i^t(k)|))\Psi(\boldsymbol{x}_{\mathrm{opt}}^{t+1}(k), \boldsymbol{\varrho}(k))$$
$$= \boldsymbol{x}_{\mathrm{opt}}^t(k) + \sum_{i=1}^{4n_{\mathrm{u}}N_{\mathrm{u}}+4n_{\mathrm{y}}N} r_i$$

It is also possible to use the alternative merit function proposed by K. Schittkowski in [295]. In order to reduce the number of the decision variables it is also possible to assume that the same vectors $\varepsilon^{\min}(k)$ and $\varepsilon^{\max}(k)$ of length n_{y} are used for the whole prediction horizon. In such a case the decision variables vector (2.39) is of length $n_{\mathrm{u}}N_{\mathrm{u}}+2n_{\mathrm{y}}$, the number of constraints is $4n_{\mathrm{u}}N_{\mathrm{u}} + 2n_{\mathrm{y}}N + 2n_{\mathrm{y}}$.

Implementation of MPC-NNPAPT Algorithm for Neural Model

The description of the MPC-NNPAPT algorithm presented so far is universal, no restrictions on the model class are imposed (but it must be differentiable). For the DLP neural model, differentiating Eq. (2.18), one obtains the elements of the vector (2.63) and the entries of the matrix (2.64)

$$\frac{\partial^2 \hat{y}_m(k+p|k)}{u_{n_1}^t(k+r_1|k)\partial u_{n_2}^t(k+r_2|k)} = \tag{2.66}$$
$$\sum_{i=1}^{K^m} w_i^{2,m}\left[\frac{\mathrm{d}^2\varphi(z_i^m(k+p|k))}{\mathrm{d}(z_i^m(k+p|k))^2}\frac{\partial z_i^m(k+p|k)}{\partial u_{n_1}^t(k+r_1|k)}\frac{\partial z_i^m(k+p|k)}{\partial u_{n_2}^t(k+r_2|k)}\right.$$
$$\left.+\frac{\partial\varphi(z_i^m(k+p|k))}{\partial z_i^m(k+p|k)}\frac{\partial^2 z_i^m(k+p|k)}{\partial u_{n_1}^t(k+r_1|k)\partial u_{n_2}^t(k+r_2|k)}\right]$$

For the transfer function $\varphi = \tanh(\cdot)$, one has

$$\frac{\mathrm{d}^2\varphi(z_i^m(k+p|k))}{\mathrm{d}(z_i^m(k+p|k))^2} = -2\tanh\left(z_i^m(k+p|k)\right)\left(1-\tanh^2\left(z_i^m(k+p|k)\right)\right)$$

The second and the third derivatives occurring in the right side of Eq. (2.66) are calculated from Eq. (2.23). Differentiating Eq. (2.23), one obtains the second-order derivatives

$$\frac{\partial^2 z_i^m(k+p|k)}{\partial u_{n_1}^t(k+r_1|k)\partial u_{n_2}^t(k+r_2|k)} = \sum_{j=1}^{I_{\mathrm{yf}}^{m,n}(p)} w_{i,I_{\mathrm{u}}^m+I_{\mathrm{h}}^m+j}^{1,m}$$
$$\times\frac{\partial^2 \hat{y}_m(k-j+p|k)}{\partial u_{n_1}^t(k+r_1|k)\partial u_{n_2}^t(k+r_2|k)}$$

2.3.6 Suboptimal Explicit MPC Algorithm

In order to reduce computational burden of calculations of the MPC algorithms presented so far, one may use their suboptimal versions in which no on-line optimisation is used. Removing the constraints, the MPC-NPL optimisation task (2.36) reduces to minimisation of the quadratic cost-function

$$\min_{\Delta \boldsymbol{u}(k)} \left\{ J(k) = \left\| \boldsymbol{y}^{\mathrm{sp}}(k) - \boldsymbol{G}(k)\Delta \boldsymbol{u}(k) - \boldsymbol{y}^{0}(k) \right\|_{\boldsymbol{M}}^{2} + \left\| \Delta \boldsymbol{u}(k) \right\|_{\boldsymbol{\Lambda}}^{2} \right\} \tag{2.67}$$

By equating its first-order derivatives with respect to the vector $\Delta \boldsymbol{u}(k)$ to the zeros vector of length $n_{\mathrm{u}}N_{\mathrm{u}}$, one obtains the optimal vector of the decision variables of the algorithm (i.e. the vector of future control increments)

$$\Delta \boldsymbol{u}(k) = \boldsymbol{K}(k)(\boldsymbol{y}^{\mathrm{sp}}(k) - \boldsymbol{y}^{0}(k)) \tag{2.68}$$

where the matrix

$$\boldsymbol{K}(k) = (\boldsymbol{G}^{\mathrm{T}}(k)\boldsymbol{M}\boldsymbol{G}(k) + \boldsymbol{\Lambda})^{-1}\boldsymbol{G}^{\mathrm{T}}(k)\boldsymbol{M} \tag{2.69}$$

is of dimensionality $n_{\mathrm{u}}N_{\mathrm{u}} \times n_{\mathrm{y}}N$. The obtained solution is the unique global minimum of the optimisation problem (2.67) because the second-order derivatives matrix $\frac{\mathrm{d}^2 J(k)}{\mathrm{d}(\Delta \boldsymbol{u}(k))^2} = 2(\boldsymbol{G}^{\mathrm{T}}(k)\boldsymbol{M}\boldsymbol{G}(k) + \boldsymbol{\Lambda})$ is positive-definite if $\boldsymbol{M} \geq 0$ and $\boldsymbol{\Lambda} > 0$.

In the explicit algorithm it is not necessary to calculate the whole control increments vector $\Delta \boldsymbol{u}(k)$, but only its first n_{u} elements, i.e. the control increments for the current sampling instant. In place of Eq. (2.68) the control law is

$$\Delta u(k|k) = \boldsymbol{K}^{n_{\mathrm{u}}}(k)(\boldsymbol{y}^{\mathrm{sp}}(k) - \boldsymbol{y}^{0}(k)) \tag{2.70}$$

where the matrix $\boldsymbol{K}_{n_{\mathrm{u}}}(k)$ consists of the first n_{u} rows of the matrix $\boldsymbol{K}(k)$. The matrix $\boldsymbol{K}^{n_{\mathrm{u}}}(k)$ depends on the dynamic matrix $\boldsymbol{G}(k)$, which, in turn, is repeatedly calculated on-line and is composed of the step-response coefficients of the locally linearised model. It means that model linearisation (at each sampling instant or less frequently) entails the necessity of finding the matrix $\boldsymbol{K}^{n_{\mathrm{u}}}(k)$. It can be done by means of any numerically efficient method, e.g. by the Low-Upper (LU) matrix factorisation and solution of a set of linear equations [81, 259]. Details of such an approach are thoroughly discussed in [162].

The calculated control moves may not satisfy the constraints imposed on the values and on the rates of change of the current control signals

$$u^{\min} \leq u(k|k) \leq u^{\max}, \ -\Delta u^{\max} \leq \Delta u(k|k) \leq \Delta u^{\max}$$

That is why the calculated vector $\triangle u(k|k)$ is next projected onto the set of feasible solutions (determined by the constraints). The projection procedure is

$$\begin{aligned}
&\text{if } \triangle u_n(k|k) < -\triangle u_n^{\max}, \text{ set } \triangle u_n(k|k) = -\triangle u_n^{\max}\\
&\text{if } \triangle u_n(k|k) > \triangle u_n^{\max}, \text{ set } \triangle u_n(k|k) = \triangle u_n^{\max}\\
&\text{calculate } u_n(k|k) = \triangle u_n(k|k) + u_n(k-1)\\
&\text{if } u_n(k|k) < u_n^{\min}, \text{ set } u_n(k|k) = u_n^{\min}\\
&\text{if } u_n(k|k) > u_n^{\max}, \text{ set } u_n(k|k) = u_n^{\max}
\end{aligned}$$

for $n = 1, \ldots, n_{\mathrm{u}}$. The constraints are imposed only on the control signals for the current iteration, because, unlike the MPC algorithms with on-line linearisation, the future control policy over the whole control horizon is not calculated. When necessary, the constraints imposed on the values of the predicted outputs may be transformed into some constraints imposed on the future control signals [202].

The structure of the explicit MPC-NPL algorithm is depicted in Fig. 2.8, it is very similar to the structure of the classical MPC-NPL algorithm with on-line optimisation shown in Fig. 2.3. Model linearisation and nonlinear free trajectory calculation are done in the same way in both approaches whereas in the explicit algorithm quadratic optimisation is not necessary and the matrix $\boldsymbol{K}_{n_{\mathrm{u}}}(k)$, which defines the control law (2.70), is calculated from a matrix decomposition task and linear equations.

Of course, the explicit MPC-SL algorithm can be used in place of the MPC-NPL approach, the only difference is the fact that the linearised model, not

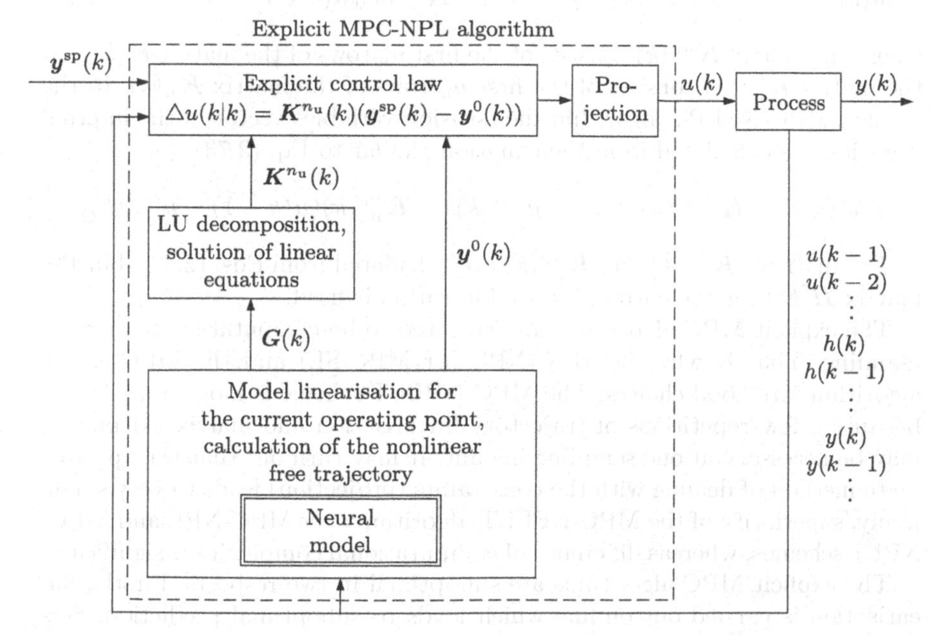

Fig. 2.8. The structure of the explicit MPC-NPL algorithm

the nonlinear one, is used for free trajectory calculation. It is also possible to use explicit versions of the MPC algorithms with more advanced trajectory linearisation (i.e. MPC-NPLT and MPC-NPLPT ones). Removing the constraints from the MPC-NPLT problem (2.47), one obtains

$$\min_{\triangle \boldsymbol{u}(k)} \Big\{ J(k) = \Big\| \boldsymbol{y}^{\mathrm{sp}}(k) - \boldsymbol{H}(k)\boldsymbol{J}\triangle\boldsymbol{u}(k) - \hat{\boldsymbol{y}}^{\mathrm{traj}}(k) - \boldsymbol{H}(k)(\boldsymbol{u}(k-1) - \boldsymbol{u}^{\mathrm{traj}}(k)) \Big\|_{\boldsymbol{M}}^2 + \|\triangle\boldsymbol{u}(k)\|_{\boldsymbol{\Lambda}}^2 \Big\} \tag{2.71}$$

The optimal vector of future control increments is

$$\triangle\boldsymbol{u}(k) = \boldsymbol{K}(k)(\boldsymbol{y}^{\mathrm{sp}}(k) - \hat{\boldsymbol{y}}^{\mathrm{traj}}(k)) + \boldsymbol{K}_{\mathrm{u}}(k)(\boldsymbol{u}(k-1) - \boldsymbol{u}^{\mathrm{traj}}(k))$$

where the matrices

$$\begin{aligned} \boldsymbol{K}(k) &= (\boldsymbol{J}^{\mathrm{T}}\boldsymbol{H}^{\mathrm{T}}(k)\boldsymbol{M}\boldsymbol{H}(k)\boldsymbol{J} + \boldsymbol{\Lambda})^{-1}\boldsymbol{J}^{\mathrm{T}}\boldsymbol{H}^{\mathrm{T}}(k)\boldsymbol{M} \\ \boldsymbol{K}_{\mathrm{u}}(k) &= -\boldsymbol{K}(k)\boldsymbol{H}(k) \end{aligned} \tag{2.72}$$

are of dimensionality $n_{\mathrm{u}}N_{\mathrm{u}} \times n_{\mathrm{y}}N$ and $n_{\mathrm{u}}N_{\mathrm{u}} \times n_{\mathrm{u}}N_{\mathrm{u}}$, respectively. The obtained solution is the global minimum of the optimisation problem (2.71) because the matrix $\frac{\mathrm{d}^2 J(k)}{\mathrm{d}(\triangle\boldsymbol{u}(k))^2} = 2(\boldsymbol{J}^{\mathrm{T}}\boldsymbol{H}^{\mathrm{T}}(k)\boldsymbol{M}\boldsymbol{H}(k)\boldsymbol{J} + \boldsymbol{\Lambda})$ is positive-definite if $\boldsymbol{M} \geq 0$ and $\boldsymbol{\Lambda} > 0$. Of course, it is not necessary to calculate the whole vector $\triangle\boldsymbol{u}(k)$ but only its first n_{u} elements, i.e. the control increments for the current sampling instant

$$\triangle u(k|k) = \boldsymbol{K}^{n_{\mathrm{u}}}(k)(\boldsymbol{y}^{\mathrm{sp}}(k) - \hat{\boldsymbol{y}}^{\mathrm{traj}}(k)) + \boldsymbol{K}_{\mathrm{u}}^{n_{\mathrm{u}}}(k)(\boldsymbol{u}(k-1) - \boldsymbol{u}^{\mathrm{traj}}(k)) \tag{2.73}$$

where the matrix $\boldsymbol{K}^{n_{\mathrm{u}}}(k)$ consists of the first n_{u} rows of the matrix $\boldsymbol{K}(k)$ and the matrix $\boldsymbol{K}_{\mathrm{u}}^{n_{\mathrm{u}}}(k)$ consists of the first n_{u} rows of the matrix $\boldsymbol{K}_{\mathrm{u}}(k)$. In the explicit MPC-NPLPT algorithm the decision variables vector in the internal iteration t is calculated from the equation similar to Eq. (2.73)

$$\triangle u^t(k|k) = \boldsymbol{K}^{n_{\mathrm{u}}}(k)(\boldsymbol{y}^{\mathrm{sp}}(k) - \hat{\boldsymbol{y}}^{t-1}(k)) + \boldsymbol{K}_{\mathrm{u}}^{n_{\mathrm{u}}}(k)(\boldsymbol{u}(k-1) - \boldsymbol{u}^{t-1}(k))$$

The matrices $\boldsymbol{K}^{n_{\mathrm{u}}}(k)$ and $\boldsymbol{K}_{\mathrm{u}}^{n_{\mathrm{u}}}(k)$ are calculated from Eqs. (2.72), but the matrix $\boldsymbol{H}^t(k)$ for the current internal iteration is used.

The explicit MPC algorithms are expected to be computationally not demanding. That is why the MPC-NPL (or MPC-SL) and the MPC-NPLT algorithms are good choices. The MPC-NPLPT scheme is more complicated because a few repetitions of trajectory linearisation and matrix calculation may be necessary at one sampling instant. It may turn out that the approximate method of dealing with the constraints (projection) leads to very small, if any, superiority of the MPC-NPLPT algorithm over MPC-NPL and MPC-NPLT schemes, whereas difference of computational complexity is significant.

The explicit MPC algorithms are suboptimal in two respects. Firstly, linearisation is carried out on-line which leads to suboptimal prediction. Secondly, for simple calculation of the decision variables all the constraints are neglected, but the obtained solution in next projected onto the set of feasible solutions. As the practical experience indicates, for many processes they give control accuracy not much worse than that of the classical (numerical) MPC algorithms with on-line quadratic optimisation. It may be not true for all processes, in particular when the process has many inputs and outputs.

2.4 Hybrid MPC Algorithms with Linearisation and Nonlinear Optimisation

It is possible to combine the suboptimal MPC approaches and the MPC-NO algorithm. In the first phase the suboptimal scheme is used. If the process is close to the desired set-point or if changes of the set-point are not significant, the obtained control policy is applied to the process. Otherwise, the second phase is activated in which nonlinear optimisation is used, the control policy found in the first phase is the initial point. In the first phase it is possible to use any suboptimal MPC algorithm, in the numerical version (with quadratic optimisation) or in the explicit version.

2.5 Example 2.1

The first process under consideration is the yeast fermentation reactor (*Saccharomyces cerevisiae*) whose control system structure is shown in Fig. 2.9. Yeast is commonly used in many branches of the food industry, in particular in: bakeries, breweries, wineries and distilleries.

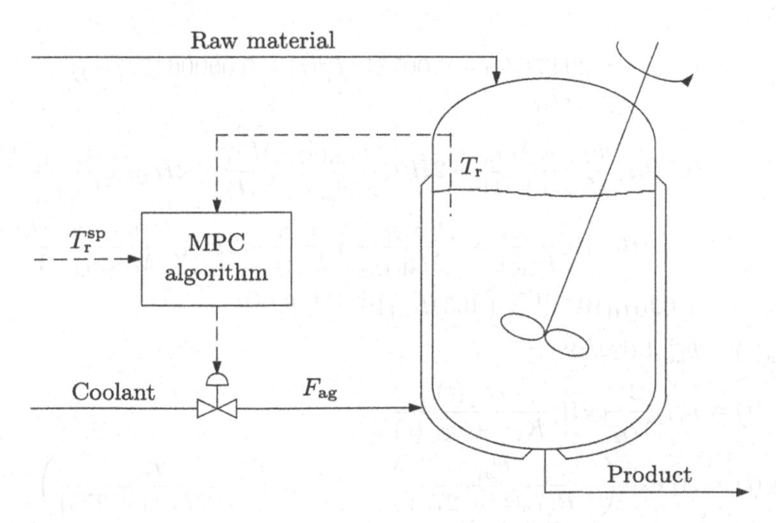

Fig. 2.9. The yeast fermentation reactor control system structure

The first-principle model of the yeast fermentation process consists of the following nonlinear differential equations [221]

$$
\begin{aligned}
\frac{\mathrm{d}V(t)}{\mathrm{d}t} &= F_{\mathrm{i}}(t) - F_{\mathrm{e}}(t)\\
\frac{\mathrm{d}c_{\mathrm{X}}(t)}{\mathrm{d}t} &= \mu_{\mathrm{X}}(t)c_{\mathrm{X}}(t)\frac{c_{\mathrm{S}}(t)}{K_{\mathrm{S}}+c_{\mathrm{S}}(t)}\exp(-K_{\mathrm{P}}c_{\mathrm{P}}(t)) - \frac{F_{\mathrm{e}}(t)}{V(t)}c_{\mathrm{X}}(t)\\
\frac{\mathrm{d}c_{\mathrm{P}}(t)}{\mathrm{d}t} &= \mu_{\mathrm{P}}c_{\mathrm{X}}(t)\frac{c_{\mathrm{S}}(t)}{K_{\mathrm{S}_1}+c_{\mathrm{S}}(t)}\exp(-K_{\mathrm{P}_1}c_{\mathrm{P}}(t)) - \frac{F_{\mathrm{e}}(t)}{V(t)}c_{\mathrm{P}}(t)\\
\frac{\mathrm{d}c_{\mathrm{S}}(t)}{\mathrm{d}t} &= -\frac{1}{R_{\mathrm{SX}}}\mu_{\mathrm{X}}(t)c_{\mathrm{X}}(t)\frac{c_{\mathrm{S}}(t)}{K_{\mathrm{S}}+c_{\mathrm{S}}(t)}\exp(-K_{\mathrm{P}}c_{\mathrm{P}}(t))\\
&\quad - \frac{1}{R_{\mathrm{SP}}}\mu_{\mathrm{P}}c_{\mathrm{X}}(t)\frac{c_{\mathrm{S}}(t)}{K_{\mathrm{S}_1}+c_{\mathrm{S}}(t)}\exp(-K_{\mathrm{P}_1}c_{\mathrm{P}}(t))\\
&\quad + \frac{F_{\mathrm{i}}(t)}{V(t)}c_{\mathrm{S,in}}(t) - \frac{F_{\mathrm{e}}(t)}{V(t)}c_{\mathrm{S}}(t)\\
\frac{\mathrm{d}c_{\mathrm{O}_2}(t)}{\mathrm{d}t} &= k_{\mathrm{la}}(t)(c^*_{\mathrm{O}_2}(t) - c_{\mathrm{O}_2}(t)) - r_{\mathrm{O}_2}(t) - \frac{F_{\mathrm{e}}(t)}{V(t)}c_{\mathrm{O}_2}(t)\\
\frac{\mathrm{d}T_{\mathrm{r}}(t)}{\mathrm{d}t} &= \frac{F_{\mathrm{i}}(t)}{V(t)}(T_{\mathrm{in}}(t)+273) - \frac{F_{\mathrm{e}}(t)}{V(t)}(T_{\mathrm{r}}(t)+273) + \frac{r_{\mathrm{O}_2}(t)\triangle H_{\mathrm{r}}}{32\rho_{\mathrm{r}}C_{\mathrm{heat,r}}}\\
&\quad - \frac{K_{\mathrm{T}}A_{\mathrm{T}}(T_{\mathrm{r}}(t) - T_{\mathrm{ag}}(t))}{V\rho_{\mathrm{r}}C_{\mathrm{heat,r}}}\\
\frac{\mathrm{d}T_{\mathrm{ag}}(t)}{\mathrm{d}t} &= \frac{F_{\mathrm{ag}}(t)}{V_{\mathrm{j}}}(T_{\mathrm{in,ag}}(t) - T_{\mathrm{ag}}(t)) + \frac{K_{\mathrm{T}}A_{\mathrm{T}}(T_{\mathrm{r}}(t) - T_{\mathrm{ag}}(t))}{V_j\rho_{\mathrm{ag}}C_{\mathrm{heat,ag}}}
\end{aligned}
$$

where

$$
\begin{aligned}
c^*_{\mathrm{O}_2}(t) &= (14.6 - 0.3943T_{\mathrm{r}}(t) + 0.007714T_{\mathrm{r}}^2(t) - 0.0000646T_{\mathrm{r}}^3(t))\\
&\quad \times 10^{-\sum(H_iI_i)(t)}\\
(H_iI_i)(t) &= 0.5H_{\mathrm{Na}}\frac{m_{\mathrm{NaCl}}}{M_{\mathrm{NaCl}}}\frac{M_{\mathrm{Na}}}{V(t)} + 2H_{\mathrm{Ca}}\frac{m_{\mathrm{CaCO_3}}}{M_{\mathrm{CaCO_3}}}\frac{M_{\mathrm{Ca}}}{V(t)} + 2H_{\mathrm{Mg}}\frac{m_{\mathrm{MgCl_2}}}{M_{\mathrm{MgCl_2}}}\frac{M_{\mathrm{Mg}}}{V(t)}\\
&\quad + 0.5H_{\mathrm{Cl}}\left(\frac{m_{\mathrm{NaCl}}}{M_{\mathrm{NaCl}}} + 2\frac{m_{\mathrm{MgCl_2}}}{M_{\mathrm{MgCl_2}}}\right)\frac{M_{\mathrm{Cl}}}{V(t)} + 2H_{\mathrm{CO_3}}\frac{m_{\mathrm{CaCO_3}}}{M_{\mathrm{CaCO_3}}}\frac{M_{\mathrm{CO_3}}}{V(t)}\\
&\quad + 0.5H_{\mathrm{H}}10^{-\mathrm{pH}(t)} + 0.5H_{\mathrm{OH}}10^{-(14-\mathrm{pH}(t))}\\
k_{\mathrm{la}}(t) &= k_{\mathrm{la0}}1.024^{T_{\mathrm{r}}(t)-20}\\
r_{\mathrm{O}_2}(t) &= \mu_{\mathrm{O}_2}\frac{1}{Y_{\mathrm{O}_2}}c_{\mathrm{X}}(t)\frac{c_{\mathrm{O}_2}(t)}{K_{\mathrm{O}_2}+c_{\mathrm{O}_2}(t)}\\
\mu_{\mathrm{X}}(t) &= A_1\exp\left(-\frac{E_{\mathrm{a}_1}}{R(T_{\mathrm{r}}(t)+273)}\right) - A_2\exp\left(-\frac{E_{\mathrm{a}_2}}{R(T_{\mathrm{r}}(t)+273)}\right)
\end{aligned}
$$

Model parameters are given in Table 2.1, the nominal operating point is given in Table 2.2.

Table 2.1. The parameters of the first-principle model of the yeast fermentation reactor

$A_1 = 9.5 \times 10^8$	$k_{\mathrm{la0}} = 38$ 1/h	$M_{\mathrm{Mg}} = 24$ g/mol
$A_2 = 2.55 \times 10^{33}$	$K_{\mathrm{O_2}} = 8.86$ mg/l	$M_{\mathrm{MgCl_2}} = 95$ g/mol
$A_{\mathrm{T}} = 1$ m^2	$K_{\mathrm{P}} = 0.139$ g/l	$M_{\mathrm{Na}} = 23$ g/mol
$C_{\mathrm{heat,ag}} = 4.18$ J/(g K)	$K_{\mathrm{P_1}} = 0.07$ g/l	$M_{\mathrm{NaCl}} = 58.5$ g/mol
$C_{\mathrm{heat,r}} = 4.18$ J/(g K)	$K_{\mathrm{S}} = 1.03$ g/l	$R = 8.31$ J/(mol K)
$E_{\mathrm{a_1}} = 55000$ J/mol	$K_{\mathrm{S_1}} = 1.68$ g/l	$R_{\mathrm{SP}} = 0.435$
$E_{\mathrm{a_2}} = 220000$ J/mol	$K_{\mathrm{T}} = 3.6 \times 10^5$ J/(h m^2 K)	$R_{\mathrm{SX}} = 0.607$
$H_{\mathrm{Ca}} = -0.303$	$m_{\mathrm{CaCO_3}} = 100$ g	$V_{\mathrm{j}} = 50$ l
$H_{\mathrm{Cl}} = 0.844$	$m_{\mathrm{MgCl_2}} = 100$ g	$Y_{\mathrm{O_2}} = 0.97$ mg/mg
$H_{\mathrm{CO_3}} = 0.485$	$m_{\mathrm{NaCl}} = 500$ g	$\Delta H_{\mathrm{r}} = 518$ kJ/mol O$_2$
$H_{\mathrm{H}} = -0.774$	$M_{\mathrm{Ca}} = 40$ g/mol	$\mu_{\mathrm{O_2}} = 0.5$ 1/h
$H_{\mathrm{Mg}} = -0.314$	$M_{\mathrm{CaCO_3}} = 90$ g/mol	$\mu_{\mathrm{P}} = 1.79$ 1/h
$H_{\mathrm{Na}} = -0.550$	$M_{\mathrm{Cl}} = 35.5$ g/mol	$\rho_{\mathrm{ag}} = 1000$ g/l
$H_{\mathrm{OH}} = 0.941$	$M_{\mathrm{CO_3}} = 60$ g/mol	$\rho_{\mathrm{r}} = 1080$ g/l

Table 2.2. The nominal operating point of the yeast fermentation reactor

$c_{\mathrm{O_2}} = 3.106953$ mg/l	$F_{\mathrm{i}} = 51$ l/h
$c_{\mathrm{P}} = 12.515241$ g/l	pH $= 6$
$c_{\mathrm{S}} = 29.738924$ g/l	$T_{\mathrm{ag}} = 27.053939$ °C
$c_{\mathrm{S,in}} = 60$ g/l	$T_{\mathrm{in}} = 25$ °C
$c_{\mathrm{X}} = 0.904677$ g/l	$T_{\mathrm{in,ag}} = 15$ °C
$F_{\mathrm{ag}} = 18$ l/h	$T_{\mathrm{r}} = 29.573212$ °C
$F_{\mathrm{e}} = 51$ l/h	$V = 1000$ l

From the perspective of a control algorithm the reactor is a single-input single-output process: the coolant flow rate (F_{ag}) is the input (the manipulated variable), the reactor temperature (T_{r}) is the output (the controlled variable). Both steady-state and dynamic properties of the process are nonlinear. Fig. 2.10 shows the steady-state process characteristics $T_{\mathrm{r}}(F_{\mathrm{ag}})$ and the dependence of the steady-state gain on the operating point. For the assumed range of the input signal the gain changes more than 75 times: from -5.76×10^{-1} to -7.62×10^{-3}. Due to a nonlinear nature of the process it cannot be controlled efficiently by the PID algorithm [221].

Yeast Fermentation Reactor Modelling

During the research carried out the first-principle model is treated as the "real" process. It has been found out experimentally that the differential equations of the model are stiff. The classical methods used for solving differential equations are very inefficient, because very small step size is necessary (which leads to a great number of iterations). That is why for simulations the method designed for solving stiff differential equations is used [297].

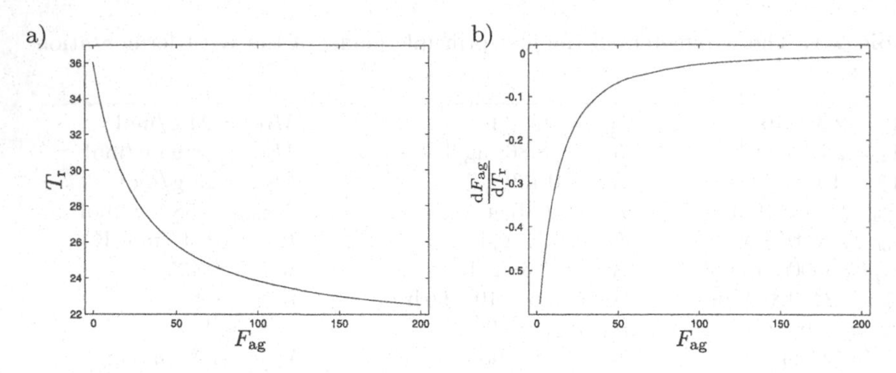

Fig. 2.10. a) The steady-state characteristics $T_r(F_{ag})$ of the yeast fermentation reactor, b) the dependence of the steady-state gain on the operating point

Every empirical model, including a neural one, must be able to generalise. The neural model is trained using some data set, possibly large and containing signals from the whole operating domain. When the model is used in the control algorithm, it should mimic the process, i.e. react correctly to different data sequences (but from the operating domain). In order to make the model have the generalisation ability, assessment of model quality is of key importance. Usually, three data sets are used for model identification [235, 274]:

a) the training data set,
b) the validation data set,
c) the test data set.

For training (optimisation of the model parameters) the training data set is used, i.e. the model error is minimised for that set (the training error). Additionally, at each iteration of the training algorithm, the model error for the validation data set (the validation error) is calculated. If the validation error increases, training is terminated to avoid overtraining (overfitting). Initially during training both training and validation errors decreases, but the validation error grows from some iteration. It means that the model fits to the training data set only and the generalisation ability deteriorates. If such an overtrained model is used in MPC, model predictions may be significantly different from the real behaviour of the process. Usually, numerous models are found, i.e. of a different structure and training is repeated a few times with different initial parameters (the multi-start approach), model selection is performed taking into account the validation error only. Finally, to independently assess generalisation of the chosen model, the third set is used to calculate the test error. It is important that the test data set is used neither for training nor for model selection. Therefore, the test error gives an unbiased estimate of the generalisation error. If the test error is comparable with training and validation errors one may conclude that the chosen model has

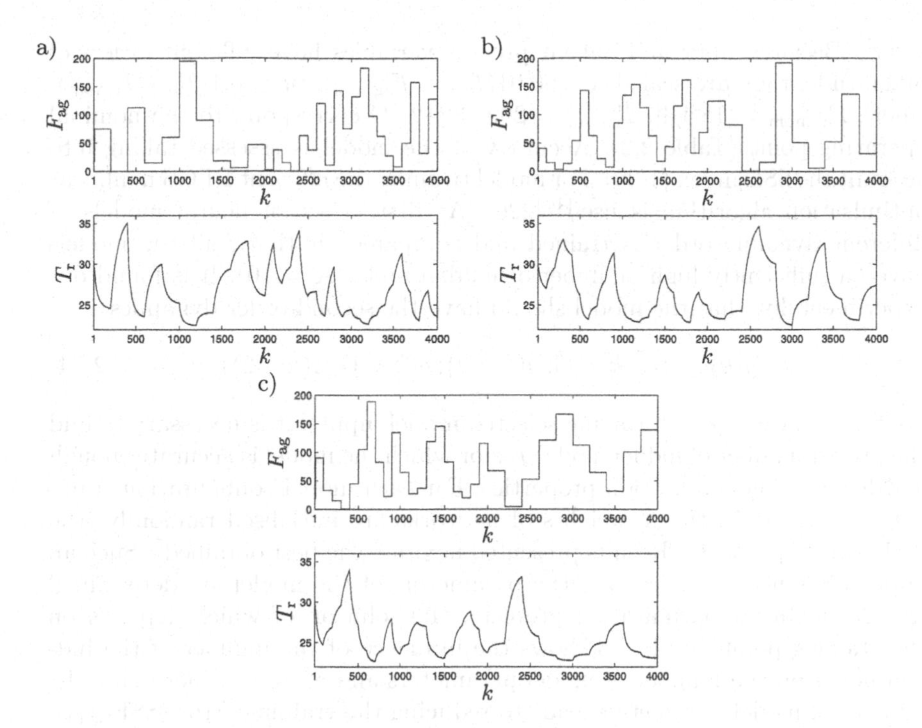

Fig. 2.11. Three data sets used for identification of empirical models of the yeast fermentation reactor: a) the training data set, b) the validation data set, c) the test data set

good generalisation ability[1]. In the simplest case the model is trained until the training error has an assumed value, the validation error is not evaluated during training. The validation error is found after training for different model structures and the model is selected taking into account the validation error. Unfortunately, such an approach does not prevent overtraining from occurring and the number of training iterations must be adjusted experimentally.

Fig. 2.11 depicts three data sets used for identification of empirical models of the yeast fermentation reactor. Each set has 4000 samples. They are obtained as a result of open-loop simulation (without any controller) of the first-principle model[2]. Because process dynamics is slow, the sampling period of all models (and hence of MPC algorithms) is 30 minutes [221]. The output signal contains small measurement noise.

The DLP neural network with one hidden layer is used for modelling, the hyperbolic tangent transfer function $\varphi = \tanh(\cdot)$ is used in the hidden

[1] The same identification method is used for obtaining all empirical models described in the book.

[2] Open-loop simulations are used to generate data in all examples described in the book.

layer[3]. Because input and output process variables have different orders of magnitude, they are scaled: $u = 0.01(F_{\rm ag} - F_{\rm ag,nom})$, $y = 0.1(T_r - T_{\rm r,nom})$, where $F_{\rm ag,nom} = 18$ l/h, $T_{\rm r,nom} = 29.573212$ °C correspond to the nominal operating point (Table 2.2). Accuracy of the model is assessed taking into account the SSE index (1.20). For model training the efficient BFGS nonlinear optimisation algorithm is used[4] [226]. At first, a set of neural models of different dynamic order is trained and compared (initially, all the models have "a sufficiently high" number of hidden nodes $K = 10$). It is found out experimentally, that the model should have the second-order dynamics

$$y(k) = f(u(k-1), u(k-2), y(k-1), y(k-2)) \tag{2.74}$$

i.e. $\tau = 1$, $n_{\rm A} = n_{\rm B} = 2$. Having selected model inputs, it is necessary to find the proper number of hidden nodes K, for which the model is accurate enough and has good generalisation properties. For each model configuration training is repeated 10 times, weights of networks are initialised randomly (the multi-start approach). Results presented next are the best obtained[5]. Such an approach is necessary because the parameters of the model are determined from a nonlinear optimisation problem, the solution of which depends on the starting point. Table 2.3 shows the influence of the number of the hidden nodes on the number of model parameters and accuracy. Increasing the number of model parameters leads to reducing the training error ($\rm SSE_{train}$). At the same time, the models with too many weights (for $K > 3$) do not have good generalisation ability, because increasing the number of weights entails increasing the validation error ($\rm SSE_{wer}$). A straightforward option is to choose the model with only 3 hidden nodes, which has 19 weights. For the chosen model the test error ($\rm SSE_{test}$) is also calculated. Its value is low, comparable with training and validation errors.

In order to emphasise high accuracy of the neural model the linear model

$$y(k) = b_1 u(k-1) + b_2 u(k-2) - a_1 y(k-1) - a_2 y(k-2) \tag{2.75}$$

is also found. It has the same inputs as the neural one. Unfortunately, due to strong nonlinear nature of the process, the linear model is very inaccurate. The linear model errors for the training and validation data sets are given in Table 2.3, the test error is of course not calculated. For the linear model the validation error is more than 420 times greater when compared with the error for the neural model. Fig. 2.12 compares the validation data set and the outputs of the neural and linear models, the differences between the data and model outputs are also shown.

[3] The transfer function $\varphi = \tanh(\cdot)$ is used in the hidden layer of all DLP neural models described in the book.

[4] The BFGS algorithm is used for identification of all nonlinear models discussed in the book. If not specified, training is carried out in the parallel (recurrent) configuration.

[5] The multi-start approach is used for identification of all nonlinear models discussed in the book.

Table 2.3. Comparison of empirical models of the yeast fermentation reactor in terms of the number of parameters (NP) and accuracy (SSE)

Model	NP	SSE_{train}	SSE_{val}	SSE_{test}
Linear	4	9.1815×10^{1}	7.7787×10^{1}	–
Neural, $K = 1$	7	1.1649×10^{1}	1.3895×10^{1}	–
Neural, $K = 2$	13	3.2821×10^{-1}	3.2568×10^{-1}	–
Neural, $K = 3$	19	2.0137×10^{-1}	1.8273×10^{-1}	1.4682×10^{-1}
Neural, $K = 4$	25	1.9868×10^{-1}	1.9063×10^{-1}	–
Neural, $K = 5$	31	1.3642×10^{-1}	1.9712×10^{-1}	–
Neural, $K = 6$	37	1.3404×10^{-1}	2.0440×10^{-1}	–
Neural, $K = 7$	43	1.2801×10^{-1}	2.9391×10^{-1}	–

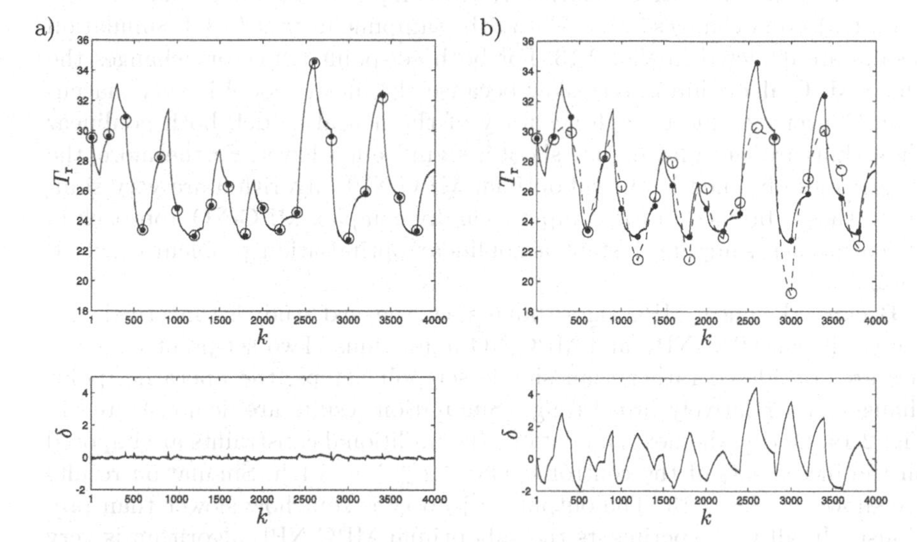

Fig. 2.12. The validation data set (*solid line with dots*) vs. the output of the model of the yeast fermentation reactor (*dashed line with circles*): a) the neural model, b) the linear model; δ – differences between the data and model outputs

Yeast Fermentation Reactor Predictive Control

The following MPC algorithms are used in the control system of the yeast fermentation reactor:

a) the linear MPC algorithm based on the linear model (2.75),
b) the "ideal" nonlinear MPC-NO algorithm based on the neural model (2.74) with 3 hidden nodes,
c) the nonlinear MPC-NPL algorithm based on the same neural model.

All simulations are carried out in Matlab[6]. The GPC algorithm with the DMC disturbance model is used as the linear MPC technique[7] [316]. The GPC algorithm, similarly to the MPC-NPL one, needs solving on-line, at each sampling instant, a quadratic optimisation problem (the active-set method [226] is used for quadratic optimisation[8]). The SQP algorithm [226] is used for nonlinear optimisation necessary in the MPC-NO approach, the initial point is defined by the last $n_{\mathrm{u}}(N_{\mathrm{u}}-1)$ elements of the optimal control sequence calculated at the previous sampling instant (Eq. (2.17))[9]. All MPC algorithms have the same parameters: $N=10$, $N_{\mathrm{u}}=3$, $\lambda_p=1.5$ for $p=1,\ldots,N$ (values of the tuning parameters are chosen experimentally[10]). The following constraints are imposed on the value of the control signal: $F_{\mathrm{ag}}^{\min}=0$ l/h, $F_{\mathrm{ag}}^{\max}=200$ l/h.

At first, all three MPC algorithms are compared assuming that the set-point trajectory changes by ± 0.5 °C at the sampling instant $k=3$. Simulation results are depicted in Fig. 2.13. For both set-point trajectory changes the linear MPC algorithm is very slow because the linear model is very inaccurate. Conversely, due to high accuracy of the neural model, both nonlinear algorithms are much faster, overshoot is significantly lower. Furthermore, the trajectories obtained in the suboptimal MPC-NPL algorithm are very similar to those obtained in the computationally complex MPC-NO approach in which at each sampling instant a nonlinear optimisation problem is solved on-line.

Because the linear MPC algorithm is slow, the next study is concerned with the nonlinear MPC-NPL and MPC-NO algorithms. Two set-point trajectories are considered rather than simple set-point steps, the operating point changes in a relatively broad range. Simulation results are demonstrated in Fig. 2.14. During the next simulations the additional constraints are imposed on the increments of the control signal: $\triangle F_{\mathrm{ag}}^{\max}=3$ l/h. Simulation results are shown in Fig. 2.15. The output trajectory is somehow slower than previously. In all the experiments the suboptimal MPC-NPL algorithm is very precise, its trajectories are very similar to the trajectories obtained in the MPC-NO approach.

Table 2.4 compares control accuracy of all considered MPC algorithms in terms of the sum of squared errors over the simulation horizon ($\mathrm{SSE_{sim}}$), which is calculated after simulations. The constraints imposed on the increments of the control signal do not have any effect in the case of the linear

---

[6] All simulations considered in this book are carried out in Matlab.

[7] The GPC algorithm with the DMC disturbance model is used as the linear MPC technique in all examples discussed in the book.

[8] The active-set method is used for quadratic optimisation in all examples discussed in the book.

[9] The SQP algorithm is used for nonlinear optimisation in the MPC-NO approach in all examples discussed in the book. As the initial solution of the optimisation task the last $n_{\mathrm{u}}(N_{\mathrm{u}}-1)$ elements of the optimal control sequence calculated at the previous sampling instant are used.

[10] The values of the tuning parameters of all MPC algorithms discussed in the book are chosen experimentally.

MPC algorithm. It is because the linear model is very imprecise and the resulting MPC algorithm generates relatively small increments of the control signal, the additional constraints are never active, the output trajectory is slow. Conversely, the neural model is very precise, when it is used in the nonlinear MPC algorithms, the variability of the control signal is much bigger and the output trajectories are fast.

Because trajectories obtained in both nonlinear MPC algorithms which use the same neural model are very similar, it is interesting to compare their computational complexity. Table 2.5 shows computational burden in terms of Million of FLOating Point operationS, MFLOPS) for MPC-NPL and MPC-NO algorithms for two prediction horizons ($N = 5, 10$) and for six control horizons ($N_{\mathrm{u}} = 1, 2, 3, 4, 5, 10$). Additionally, results for the MPC-NO algorithm with numerically approximated gradients are also given. For all combinations of horizons summarised results for both set-point trajectories are given. Of course, the control horizon has a predominant impact on computational burden because the MPC-optimisation problem has N_{u} decision variables. Differences of computational complexity are very significant. For example, for $N = 10$ and $N_{\mathrm{u}} = 3$, the suboptimal MPC-NPL algorithm

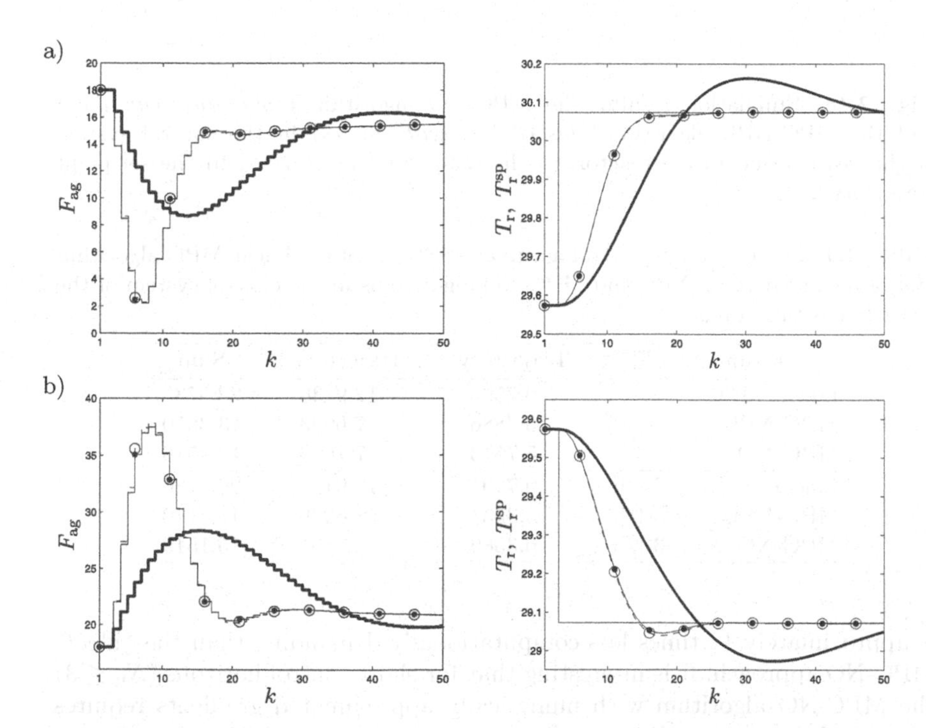

Fig. 2.13. Simulation results: the linear MPC algorithm (*thick solid line*), the MPC-NO algorithm (*solid line with dots*) and the MPC-NPL algorithm (*dashed line with circles*) in the control system of the yeast fermentation reactor: a) the set point changes by 0.5 °C, b) the set point changes by −0.5 °C

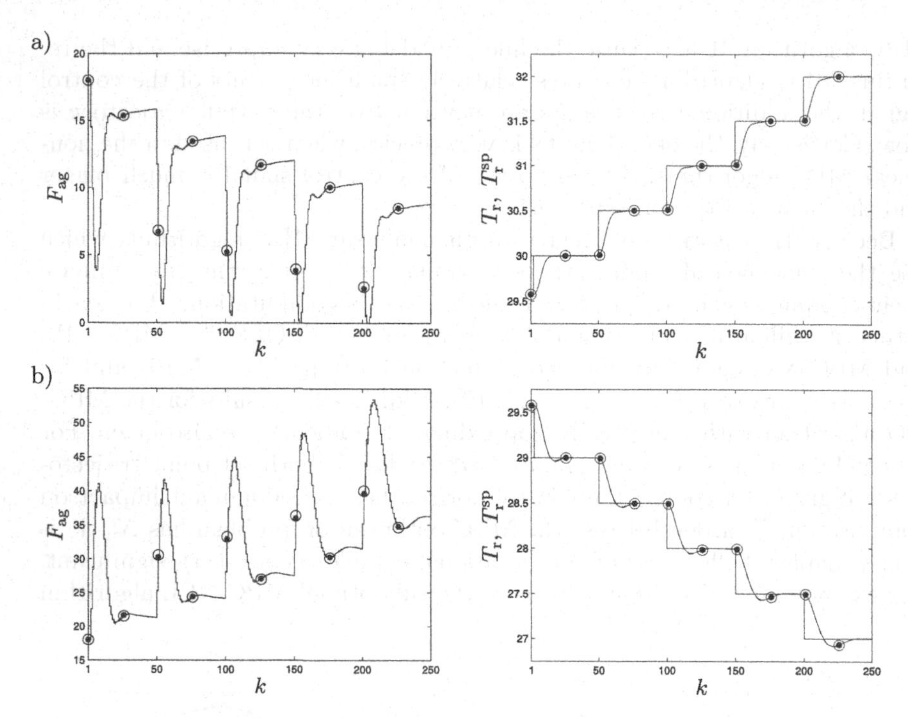

Fig. 2.14. Simulation results: the MPC-NO algorithm (*solid line with dots*) and the MPC-NPL algorithm (*dashed line with circles*) in the control system of the yeast fermentation reactor: a) the set-point trajectory 1, b) the set-point trajectory 2

Table 2.4. Comparison of control accuracy (SSE_{sim}) of the linear MPC algorithm and of nonlinear MPC-NPL and MPC-NO algorithms in the control system of the yeast fermentation reactor

Algorithm	$\triangle F_{ag}^{max}$	Trajectory 1	Trajectory 2	Sum
Linear MPC	–	9.7595	12.9136	22.6730
MPC-NPL	–	5.7886	7.6063	13.3950
MPC-NO	–	5.7513	7.7002	13.4518
Linear	3 l/h	9.7595	12.9136	22.6730
MPC-NPL	3 l/h	6.3531	8.8239	15.1770
MPC-NO	3 l/h	6.3389	8.8130	15.1519

is approximately 12 times less computationally demanding than the "ideal" MPC-NO approach. It is interesting that for short control horizons ($N_u \leq 3$) the MPC-NO algorithm with numerically approximated gradients requires less calculations than the algorithm with analytically calculated gradients. Advantages of the analytical approach are evident for longer horizons.

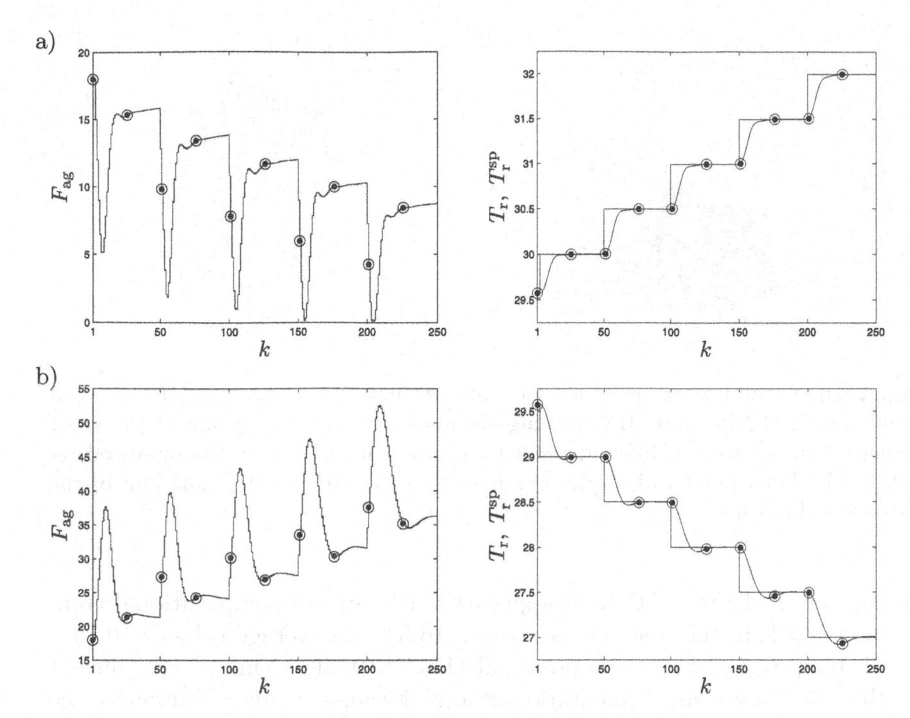

Fig. 2.15. Simulation results: the MPC-NO algorithm (*solid line with dots*) and the MPC-NPL algorithm (*dashed line with circles*) with the constraints imposed on the increments of control signal $\triangle F_{ag}^{max} = 3$ l/h in the control system of the yeast fermentation reactor: a) the set-point trajectory 1, b) the set-point trajectory 2

Table 2.5. Comparison of computational complexity (in MFLOPS) of nonlinear MPC-NPL, MPC-NO and MPC-NO$_{apr}$ (with numerically approximated gradients) algorithms in the control system of the yeast fermentation reactor for different prediction and control horizons (summarised results for both set-point trajectories are given), $N = 10$

Algorithm	N	$N_u = 1$	$N_u = 2$	$N_u = 3$	$N_u = 4$	$N_u = 5$	$N_u = 10$
NPL	5	0.40	0.53	0.85	1.29	1.92	–
NO	5	2.61	5.04	8.00	12.65	18.37	–
NO$_{apr}$	5	2.47	4.32	7.98	15.25	26.53	–
NPL	10	0.63	0.79	1.14	1.62	2.31	9.13
NO	10	5.20	9.04	13.56	19.17	26.26	76.50
NO$_{apr}$	10	4.38	7.58	12.63	20.09	31.77	154.15

Fig. 2.16 compares graphically computational burden of both nonlinear MPC algorithms for different combination of horizons ($N = 5, \ldots, 15$, $N_u = 1, \ldots, 10$). In general, the MPC-NPL algorithm is many times less demanding than the MPC-NO one. In the worst case (from the perspective of

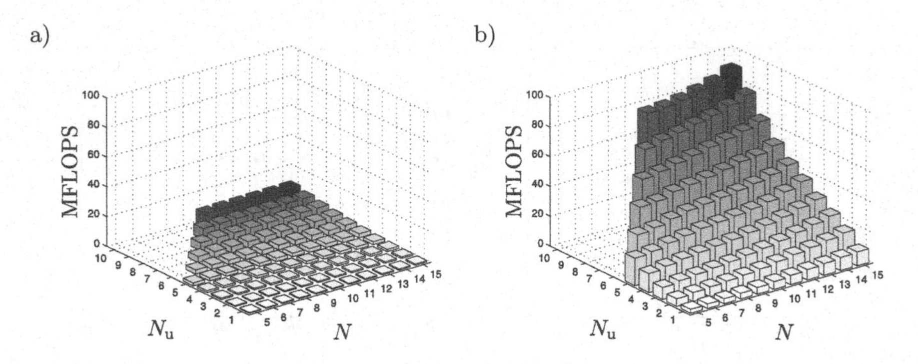

Fig. 2.16. Graphical comparison of computational complexity (in MFLOPS) of nonlinear MPC-NPL and MPC-NO algorithms in the control system of the yeast fermentation reactor for different prediction and control horizons (summarised results for both set-point trajectories are given): a) the MPC-NPL algorithm, b) the MPC-NO algorithm

the efficiency of the MPC-NPL approach) the ratio of computational complexity is 6.61, in the best case it rises to 16.64, the average value is 10.59.

All things considered, the suboptimal MPC-NPL algorithm is very efficient for the discussed yeast fermentation reactor because it gives practically the same trajectories as the MPC-NO approach and, at the same time, it is much less computationally demanding. It is not necessary to use more advanced on-line linearisation methods (i.e. along the trajectory), the rudimentary model linearisation for the current operating point is sufficient.

2.6 Example 2.2

The second process under consideration is a high-purity, high-pressure (1.93 MPa) ethylene-ethane distillation column shown in Fig. 2.17. It is used in one of the leading Polish refineries (Orlen, refinery in Płock). The feed stream consists of ethylene (approximately 80%), ethane (approximately 20%) and traces of hydrogen, methane and propylene. The vapour phase is transmitted from the bottom part of the process to the top whereas the liquid phase moves from the top to the bottom. When the vapour moves through the liquid, composition of both phases changes (the higher the tray of the column, the higher the composition, which means purification). The column has 121 trays, the feed stream is delivered to the tray number 37. The liquid from the bottom product tank is used to produce the vapour stream V in the evaporator and the excessive amount of the liquid is removed from the process as the bottom product B. The vapour is condensed into liquid in the top part of the process. It is next removed from the tray number 110 as the top product stream P and some part of the obtained liquid is fed back to the process (to the tray number 121) as the reflux stream R.

In order to stabilise levels in reflux and bottom product tanks, two classical PID controllers are used (denoted as LC). They manipulate the flow rate of the bottom product stream B and the flow rate of the reflux stream R, respectively. The third PID controller (denoted as TC) is used to stabilise the temperature on the tray number 13 (the so called controlled tray), which entails stabilisation of the flow rate of the vapour stream V. The TC controller adjusts the amount of heat delivered to the evaporator. All three PID controllers, which comprise the basic control layer, are necessary, they guarantee safe process operation.

The supervisory MPC control loop has one manipulated variable r which is the reflux ratio $r = R/P$, and one controlled variable z which represents the impurity of the product. The product of distillation is ethylene, but, according to technological requirements, it may contain up to 1000 ppm (parts per million) of ethane. The objective is to develop a supervisory MPC algorithm which could be able to increase relatively fast the impurity level when it is possible, i.e. when the composition changes in the feed stream are insignificant (the decision is taken by a process operator). Reducing the purity of the

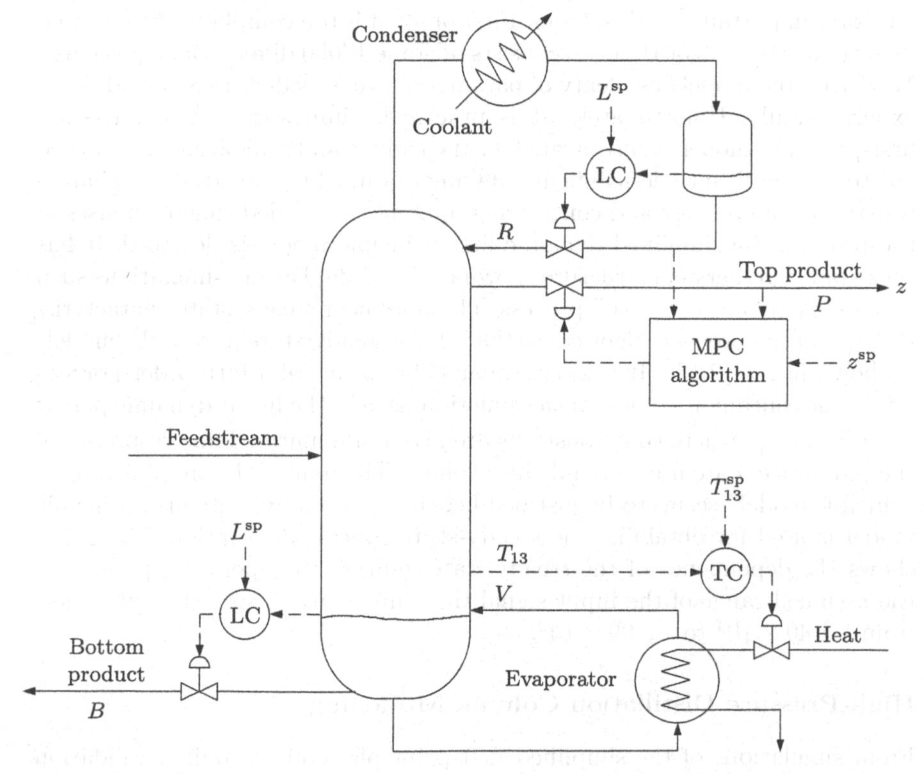

Fig. 2.17. The high-pressure high-purity ethylene-ethane distillation column control system structure

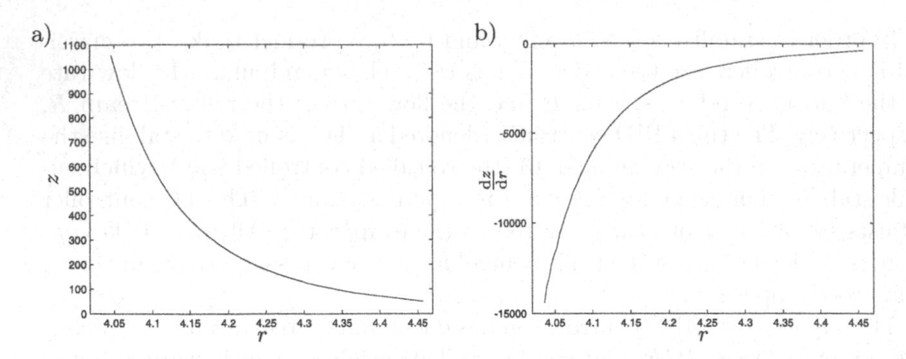

Fig. 2.18. a) The steady-state characteristics $z(r)$ of the high-pressure distillation column, b) the dependence of the steady-state gain on the operating point

product, of course taking into account the technological limit, results in decreasing energy consumption. The production scale is very big, the nominal value of the product stream flow rate is 43 tons/h.

The dynamic first-principle model of the process is available [331], but it has some important disadvantages. First of all, it is too complicated to be used directly in MPC algorithms (it consists of some 17000 differential equations). Moreover, the model has plenty of parameters whose values must be adjusted experimentally. Unfortunately, it is practically impossible. A steady-state first-principle model is also available. Its tuning and validation has turned out to be significantly less difficult. Having obtained the steady-state characteristics of the process and considering numerous recorded time-responses of the process, the simplified dynamic first-principle model is designed. It has the serial Hammerstein structure shown in Fig. 1.3a. During simulations such a model is used as the "real" process. The nonlinear steady-state characteristics $z(r)$ of the process (identical to that of the steady-state part of the model) is shown in Fig. 2.18a. It is approximated by means of a fifth order polynomial. The continuous-time transfer function used in the linear dynamic part is $G(s) = \frac{e^{-80s}}{150s+1}$, where time constants are given in minutes. Time-responses of the real process are used to find their values. The name "the simplified first-principle model" seems to be justified because the steady-state first-principle model is used for obtaining the steady-state process description. Fig. 2.18b shows the dependence of the steady-state gain on the operating point. For the assumed range of the input signal the gain changes more than 35 times: from -1.40×10^4 to -3.99×10^2.

High-Pressure Distillation Column Modelling

From simulations of the simplified first-principle model training, validation and test data sets are obtained. Each set has 1500 samples. Sample time is 40 minutes (enforced by a slow composition analyser). The output signal contains small measurement noise. Because input and output process variables have different orders of magnitude, they are scaled: $u = 10(r - r_{\text{nom}})/3$, $y = 0.001(z - z_{\text{nom}})$, where in the nominal operating point $r_{\text{nom}} = 4.3399$, $z_{\text{nom}} = 100.2275$. Accuracy of the model is assessed taking into account the SSE index (1.20).

For the assumed simplified first-principle model, the neural model of the same order of dynamics is

$$y(k) = f(u(k-3), y(k-1)) \tag{2.76}$$

and the corresponding linear model is

$$y(k) = b_3 u(k-3) - a_1 y(k-1) \tag{2.77}$$

Table 2.6 compares empirical models of the distillation column in terms of the number of parameters and accuracy. Due to significant nonlinearity of the process, the linear model is very inaccurate. Considering the neural models, for small networks increasing the number of hidden nodes results in reducing SSE_{val}, but for networks with $K = 7$ and $K = 8$ hidden nodes the validation error increases. Hence, there are two good candidates: the networks with $K = 5$ or $K = 6$ nodes, because the error for the network with $K = 4$ nodes is significantly higher. Finally, as a reasonable compromise between accuracy and complexity, the model with $K = 5$ hidden nodes is chosen (the model with $K = 6$ nodes gives the SSE_{val} index only 3% lower, but it has 4 weights more, which is 16%). Finally, for the chosen model, the test error is calculated. Its value is comparable with those for training and validation sets, which means that the model generalises well. The chosen neural model is very precise: its validation error is more than 4650 times lower when compared with the error for the linear one. Fig. 2.19 compares the validation data set and the outputs of the neural and linear models, the differences between the data and model outputs are also shown. For some operating points the linear model gives negative impurity level, which is not physically possible.

Table 2.6. Comparison of empirical models of the distillation column in terms of the number of parameters (NP) and accuracy (SSE)

Model	NP	$\text{SSE}_{\text{train}}$	SSE_{val}	SSE_{test}
Linear	2	2.3230×10^{1}	1.6099×10^{1}	–
Neural, $K = 1$	5	1.4855×10^{0}	1.6528×10^{0}	–
Neural, $K = 2$	9	9.4224×10^{-3}	1.1372×10^{-2}	–
Neural, $K = 3$	13	4.6645×10^{-3}	5.3305×10^{-3}	–
Neural, $K = 4$	17	3.7614×10^{-3}	4.3407×10^{-3}	–
Neural, $K = 5$	21	3.1962×10^{-3}	3.4611×10^{-3}	3.4669×10^{-3}
Neural, $K = 6$	25	3.1483×10^{-3}	3.3515×10^{-3}	–
Neural, $K = 7$	29	3.1316×10^{-3}	3.4612×10^{-3}	–
Neural, $K = 8$	33	3.0909×10^{-3}	3.4819×10^{-3}	–

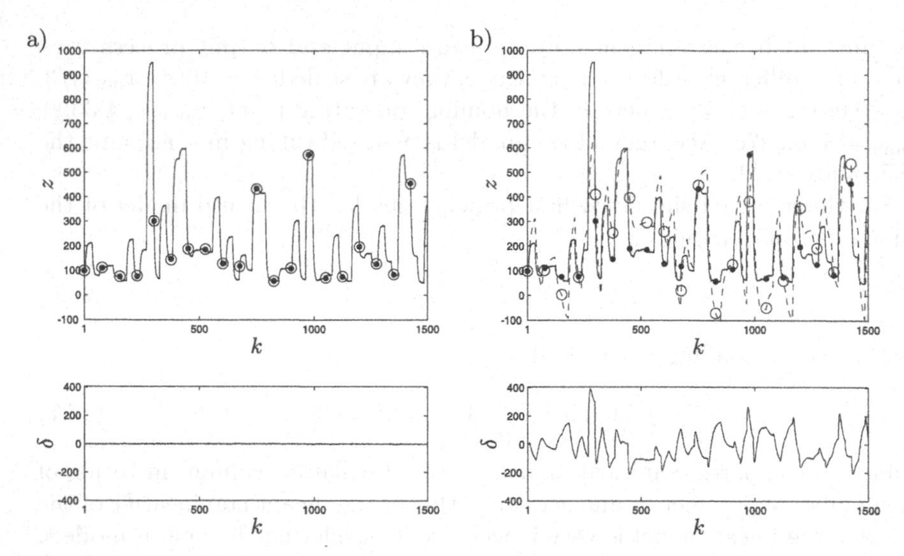

Fig. 2.19. The validation data set (*solid line with dots*) vs. the output of the model of the high-pressure distillation column (*dashed line with circles*): a) the neural model, b) the linear model; δ – differences between the data and model outputs

High-Pressure Distillation Column Predictive Control

At first, the linear MPC algorithm is developed in which the linear model (2.77) is used. Tuning parameters of the algorithm are: $N = 10$, $N_{\mathrm{u}} = 3$, $\lambda_p = 2$. The constraints are imposed on the value of the control signal: $r^{\min} = 4.051$, $r^{\max} = 4.4571$. Three set-point trajectories z^{ref} are considered: the set-point changes at the sampling instant $k = 1$ from its nominal value (100 ppm) to 350 ppm, 600 ppm and 850 ppm, respectively.

Because the distillation process is significantly nonlinear, the linear model is very imprecise. In consequence, the linear MPC algorithm does not work properly which is depicted in Fig. 2.20. The algorithm works only for the first set-point trajectory (the low purity level), strong oscillations occur for two other trajectories. Inefficiency of the linear MPC approaches applied to the considered distillation process is also discussed in [316, 317], where two linear MPC algorithms are used: the first one uses a model for the nominal operating point (100 ppm), the second one uses a model for the neighbourhood of the point 850 ppm. Both MPC algorithms work only locally, for operating points for which the models are good. For other operating regions they do not work (there are oscillations or the output responses are too slow).

The following nonlinear MPC algorithms are used in the control system of the high-pressure distillation column:

a) the classical MPC-NPL algorithm with model linearisation for the current operating point,

b) the MPC-NPLT$_{\boldsymbol{u}(k-1)}$ algorithm with the predicted output trajectory linearisation along the future input trajectory defined by the control signal used in the previous sampling instant,
c) the MPC-NPLT$_{\boldsymbol{u}(k|k-1)}$ algorithm with the predicted output trajectory linearisation along the $N_{\mathrm{u}}-1$ elements of the future input trajectory calculated in the previous sampling instant,
d) the hybrid MPC-NPL-NPLPT algorithm with iteratively repeated nonlinear prediction and trajectory linearisation at each sampling instant of the second phase,
e) the MPC-NO algorithm.

All nonlinear MPC algorithms use the same neural model (2.76) with 5 hidden nodes. Tuning parameters of all algorithms are the same as those of the linear one. In the first phase of the hybrid MPC-NPL-NPLPT algorithm, the classical MPC-NPL approach is used. In the second phase the MPC-NPLPT algorithm is used, the solution obtained in the first phase if used for initialisation of the future control trajectory along which output trajectory linearisation is carried out. Of course, one may imagine using the MPC-NPLPT scheme without the first phase, but in such a case trajectory linearisation is always performed at each sampling instant, even when the output signal is close to the constant set-point. Intuitively, in such a case simple model linearisation used in the MPC-NPL approach seems to be sufficient.

In order to demonstrate the role of tuning parameters of the MPC-NPLPT algorithm, its two versions are considered:

a) version 1: the parameter of the internal loop continuation is $\delta_{\mathrm{y}}=5\times10^{-2}$, the parameter of the internal loop termination is $\delta_{\mathrm{u}}=10^{-3}$,
b) version 2: $\delta_{\mathrm{y}}=10^{-4}$, $\delta_{\mathrm{u}}=10^{-5}$.

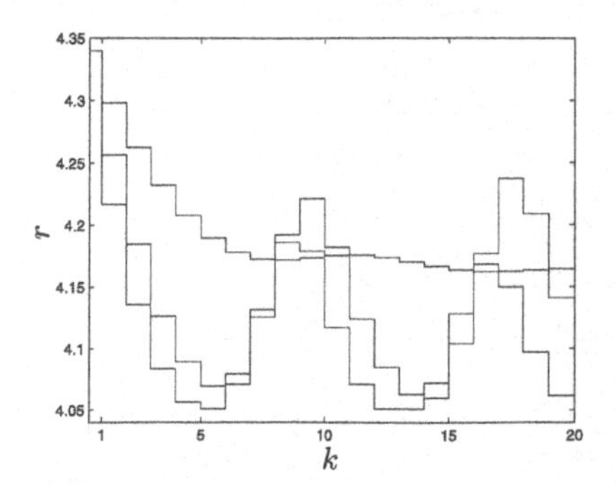

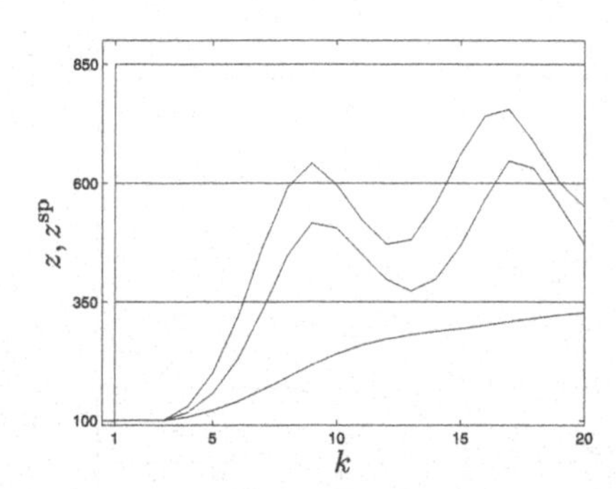

Fig. 2.20. Simulation results of the linear MPC algorithm in the control system of the high-pressure distillation column

In both cases $N_0 = 2$, $n_{\max} = 10$. The parameters δ_{u} and δ_{y} of the first version are chosen to obtain a good compromise between closed-loop accuracy and computational complexity, the second version gives excellent control performance.

At first, the classical MPC-NPL algorithm and the "ideal" MPC-NO approach are compared in Fig. 2.21. It is worth remembering that the MPC-NPL algorithm works very well (similarly to the MPC-NO one) in the control system of the yeast fermentation process (Figs. 2.13, 2.14 and 2.15). It is not true in the control system of the high-pressure distillation column. Although the MPC-NPL algorithm works much better than the linear MPC scheme, it is much slower than the MPC-NO one. Both algorithms use the same neural model, low performance of the suboptimal algorithm is caused by using for prediction over the whole prediction horizon the same local approximation of the model obtained for the current operating point. Intuitively, when the set-point changes fast and significantly, the linearised model approximates well the nonlinear model (and the real process) only for the beginning of the prediction horizon.

Fig. 2.22 depicts simulation results of two MPC algorithms with trajectory linearisation carried out once at each sampling instant (MPC-NPLT$_{\boldsymbol{u}(k-1)}$ and MPC-NPLT$_{\boldsymbol{u}(k|k-1)}$), the trajectories of the MPC-NO algorithms are also shown. The MPC-NPLT$_{\boldsymbol{u}(k-1)}$ algorithm with the predicted output trajectory linearisation along the control signal used in the previous sampling instant is faster than the MPC-NPL algorithm, but the MPC-NPLT$_{\boldsymbol{u}(k|k-1)}$ algorithm with trajectory linearisation along the input trajectory calculated in the previous sampling instant is significantly faster. In both cases changes of the control signal are bigger than in the MPC-NPL approach. Although the MPC algorithms with trajectory linearisation are faster than the MPC-NPL one, they are still slower than the MPC-NO approach.

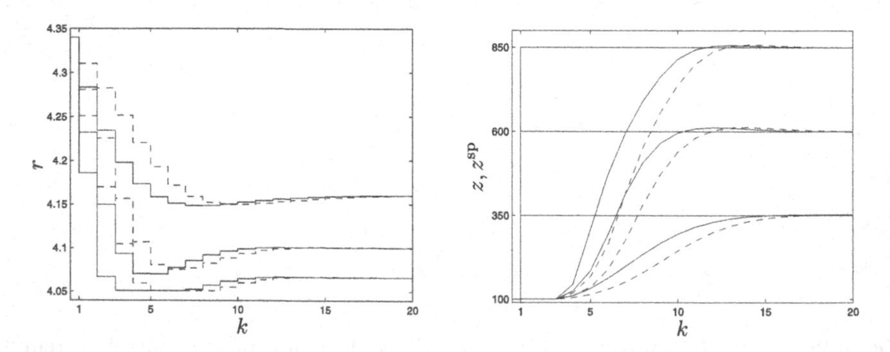

Fig. 2.21. Simulation results: the MPC-NO algorithm (*solid line*) and the MPC-NPL algorithm (*dashed line*) in the control system of the high-pressure distillation column

Fig. 2.23 depicts simulation results obtained in the first version of the MPC-NPL-NPLPT algorithm, Fig. 2.24 shows results of the second version. The MPC algorithm with iteratively repeated nonlinear prediction and trajectory linearisation at each sampling instant is very precise: in its first version one may decipher some differences from the trajectories obtained in the MPC-NO approach while in its second version the suboptimal algorithm gives practically the same results as the MPC-NO one.

Table 2.7 compares control accuracy of all considered nonlinear MPC algorithms in terms of the sum of squared errors over the simulation horizon (SSE_{sim}). The algorithms in which trajectory linearisation is performed once at each sampling instant (MPC-NPLT$_{\boldsymbol{u}(k-1)}$ and MPC-NPLT$_{\boldsymbol{u}(k|k-1)}$) are more precise than the MPC-NPL approach. An appropriate choice of the parameters δ_{y} and δ_{u} makes it possible to obtain in the suboptimal MPC-NPL-NPLPT algorithm practically the same accuracy as in the MPC-NO

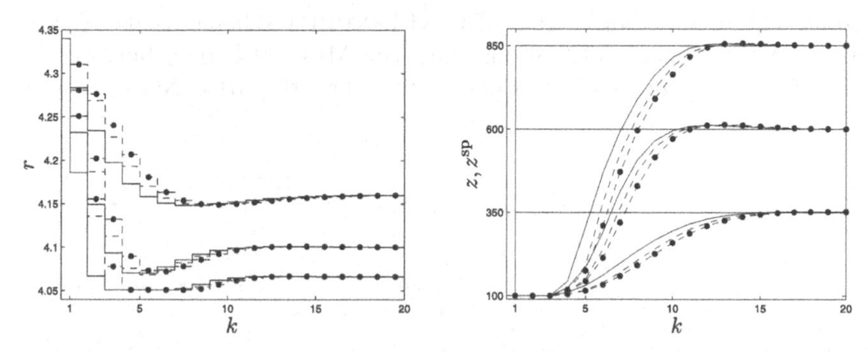

Fig. 2.22. Simulation results: the MPC-NO algorithm (*solid line*), the MPC-NPLT$_{\boldsymbol{u}(k-1)}$ algorithm (*solid line with dots*) and the MPC-NPLT$_{\boldsymbol{u}(k|k-1)}$ (*dashed line*) in the control system of the high-pressure distillation column

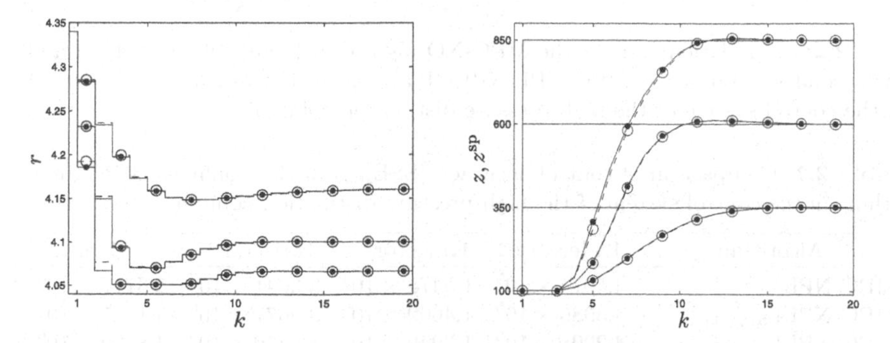

Fig. 2.23. Simulation results: the MPC-NO algorithm (*solid line with dots*) and the first version of the MPC-NPL-NPLPT algorithm (*dashed line with circles*) in the control system of the high-pressure distillation column

one. The simple model linearisation used in the MPC-NPL approach turns out to be insufficient in the control system of the high-pressure distillation column, trajectory linearisation is necessary.

Table 2.8 shows computational burden of compared nonlinear MPC algorithms. MPC-NPLT$_{\boldsymbol{u}(k-1)}$ and MPC-NPLT$_{\boldsymbol{u}(k|k-1)}$ algorithms are approximately twice more complicated in comparison with the simplest MPC-NPL one, the MPC-NPLT$_{\boldsymbol{u}(k-1)}$ algorithm is a little less complicated than the MPC-NPLT$_{\boldsymbol{u}(k|k-1)}$ one because the relation $\boldsymbol{u}^{\text{traj}}(k) = \boldsymbol{u}(k-1)$ makes calculations a little bit simpler. Since the parameters δ_{y} and δ_{u} determine continuation and termination of the internal loop of the second phase of the MPC-NPL-NPLPT algorithm, they have a predominant impact on computational burden. In the first version of the hybrid algorithm computational burden is only some two times higher than in the MPC-NPL algorithm and as much as some five times lower than in the MPC-NO one.

Table 2.9 shows summarised (for three set-point trajectories) computational burden of compared nonlinear MPC algorithms for different control horizons. The first version of the MPC-NPL-NPLPT scheme needs approximately 2.5 times more calculations than the MPC-NPL one, but, on the other hand, it is significantly less demanding than the MPC-NO approach.

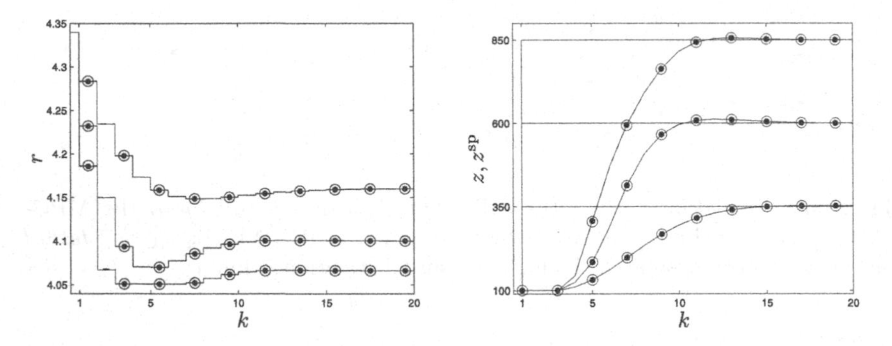

Fig. 2.24. Simulation results: the MPC-NO algorithm (*solid line with dots*) and the second version of the MPC-NPL-NPLPT algorithm (*dashed line with circles*) in the control system of the high-pressure distillation column

Table 2.7. Comparison of control accuracy (SSE_{sim}) of the nonlinear MPC algorithms in the control system of the high-pressure distillation column

Algorithm	Trajectory 1	Trajectory 2	Trajectory 3	Sum
MPC-NPL	4.5974×10^5	1.5478×10^6	3.3641×10^6	5.3717×10^6
MPC-NPLT$_{\boldsymbol{u}(k-1)}$	4.3986×10^5	1.4609×10^6	3.2077×10^6	5.1085×10^6
MPC-NPLT$_{\boldsymbol{u}(k\|k-1)}$	4.2008×10^5	1.3949×10^6	3.0450×10^6	4.8599×10^6
MPC-NPL-NPLPT v. 1	3.7830×10^5	1.2945×10^6	2.7776×10^6	4.4504×10^6
MPC-NPL-NPLPT v. 2	3.7582×10^5	1.2886×10^6	2.7298×10^6	4.3943×10^6
NO	3.7540×10^5	1.2876×10^6	2.7239×10^6	4.3869×10^6

The quantitative reduction of computational burden is in the range 3.2–5.8. The MPC-NO algorithm, in which the SQP algorithm is used for nonlinear optimisation, is to some extent conceptually similar to the successive linearisation used in the second phase of the hybrid MPC-NPL-NPLPT algorithms (a sequence of quadratic optimisation problems are solved to obtain the solution to a nonlinear one in both approaches). The MPC-NPL-NPLPT algorithm is yet significantly less complicated, it it characterised by lower computational burden.

Fig. 2.25 shows the number of internal iterations in two considered versions of the MPC-NPL-NPLPT algorithm. Obviously, increasing accuracy (version 2) leads to increasing the number of internal iterations and computational complexity.

In order to demonstrate iterative linearisation and optimisation of the future control policy, behaviour of the MPC-NPL-NPLPT algorithm is demonstrated for the first iteration (the sampling instant $k = 1$). At first, a linear approximation of the neural model is obtained for the current operating point. Its coefficients are $b_3(k) = -0.0406$, $a_1(k) = -0.7649$. Step-response coefficients of the linearised model are calculated. They can be used

Table 2.8. Comparison of computational complexity (in MFLOPS) of nonlinear MPC algorithms in the control system of the high-pressure distillation column

Algorithm	Trajectory 1	Trajectory 2	Trajectory 3	Sum
MPC-NPL	0.05	0.05	0.05	0.15
MPC-NPLT$_{\boldsymbol{u}(k-1)}$	0.11	0.11	0.11	0.33
MPC-NPLT$_{\boldsymbol{u}(k\|k-1)}$	0.11	0.11	0.12	0.34
MPC-NPL-NPLPT v. 1	0.13	0.12	0.14	0.39
MPC-NPL-NPLPT v. 2	0.22	0.21	0.22	0.65
NO	0.77	0.71	0.65	2.13

Table 2.9. Comparison of computational complexity (in MFLOPS) of nonlinear MPC algorithms in the control system of the high-pressure distillation column for different control horizons (summarised results for three set-point trajectories are given), $N = 10$

Algorithm	$N_u = 1$	$N_u = 2$	$N_u = 3$	$N_u = 4$	$N_u = 5$	$N_u = 10$
MPC-NPL	0.09	0.11	0.15	0.22	0.31	1.28
MPC-NPLT$_{\boldsymbol{u}(k-1)}$	0.14	0.22	0.33	0.46	0.61	1.92
MPC-NPLT$_{\boldsymbol{u}(k\|k-1)}$	0.14	0.23	0.34	0.47	0.62	1.87
MPC-NPL-NPLPT v. 1	0.18	0.25	0.39	0.55	0.77	3.01
MPC-NPL-NPLPT v. 2	0.25	0.38	0.65	0.95	1.33	4.69
NO	0.87	1.47	2.13	3.04	3.54	9.69

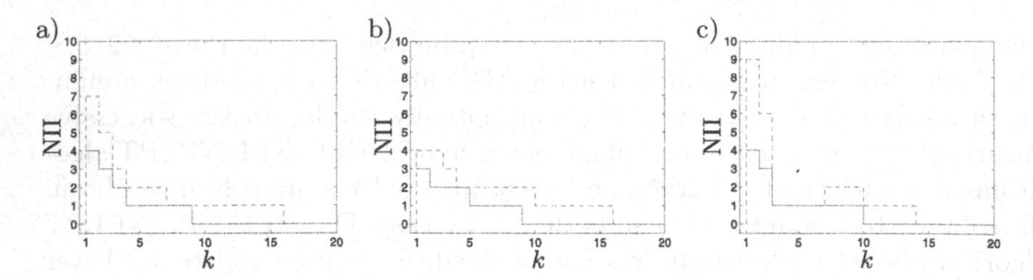

Fig. 2.25. The number of internal iterations (NII) in version 1 (*solid line*) and in version 2 (*dashed line*) of the MPC-NPL-NPLPT algorithm in the control system of the high-pressure distillation column: a) set-point trajectory 1, b) set-point trajectory 2, c) set-point trajectory 3

to find the impulse-response coefficients matrix of the linear approximation of the nonlinear model for the current operating point [41]

$$\boldsymbol{H}^0(k) = \begin{bmatrix} 0 & 0 & 0 \\ 0 & 0 & 0 \\ -0.0406 & 0 & 0 \\ -0.0310 & -0.0406 & 0 \\ -0.0238 & -0.0310 & -0.0406 \\ -0.0182 & -0.0238 & -0.0310 \\ -0.0139 & -0.0182 & -0.0238 \\ -0.0106 & -0.0139 & -0.0182 \\ -0.0081 & -0.0106 & -0.0139 \\ -0.0063 & -0.0081 & -0.0106 \end{bmatrix}$$

Because the same linear approximation of the neural model is used for the entire prediction horizon, the entries in the consecutive columns of the matrix $\boldsymbol{H}^0(k)$ are the same (but shifted vertically). The calculated (scaled) future control policy over the control horizon is $\boldsymbol{r}^0(k) = [4.3104\ 4.2860\ 4.2667]^{\mathrm{T}}$. In the first internal iteration a linear approximation of the predicted output trajectory is calculated along the obtained future control trajectory $\boldsymbol{r}^0(k)$. The linearised trajectory is characterised by the matrix

$$\boldsymbol{H}^1(k) = \begin{bmatrix} 0 & 0 & 0 \\ 0 & 0 & 0 \\ -0.0531 & 0 & 0 \\ -0.0407 & -0.0656 & 0 \\ -0.0312 & -0.0503 & -0.0766 \\ -0.0239 & -0.0385 & -0.1354 \\ -0.0183 & -0.0295 & -0.1806 \\ -0.0140 & -0.0226 & -0.2152 \\ -0.0108 & -0.0173 & -0.2418 \\ -0.0082 & -0.0133 & -0.2622 \end{bmatrix}$$

The matrix $\boldsymbol{H}^1(k)$ is significantly different from the initial matrix $\boldsymbol{H}^0(k)$. One may notice that the corresponding entries are different, the entries in the consecutive columns are not the same. The calculated future control policy is $\boldsymbol{r}^1(k) = [4.2966\ 4.2596\ 4.2297]^{\mathrm{T}}$. In the fourth internal iteration the linearised output trajectory is characterised by the matrix

$$\boldsymbol{H}^4(k) = \begin{bmatrix} 0 & 0 & 0 \\ 0 & 0 & 0 \\ -0.0651 & 0 & 0 \\ -0.0499 & -0.0934 & 0 \\ -0.0383 & -0.0717 & -0.1271 \\ -0.0294 & -0.0551 & -0.2248 \\ -0.0226 & -0.0423 & -0.2998 \\ -0.0173 & -0.0324 & -0.3575 \\ -0.0133 & -0.0249 & -0.4017 \\ -0.0102 & -0.0191 & -0.4356 \end{bmatrix}$$

and $\boldsymbol{r}^4(k) = [4.2853\ 4.2377\ 4.1991]^{\mathrm{T}}$. In the seventh internal iteration

$$\boldsymbol{H}^7(k) = \begin{bmatrix} 0 & 0 & 0 \\ 0 & 0 & 0 \\ -0.0666 & 0 & 0 \\ -0.0511 & -0.0976 & 0 \\ -0.0393 & -0.0750 & -0.1370 \\ -0.0301 & -0.0575 & -0.2423 \\ -0.0231 & -0.0442 & -0.3231 \\ -0.0178 & -0.0339 & -0.3851 \\ -0.0136 & -0.0260 & -0.4327 \\ -0.0105 & -0.0199 & -0.4692 \end{bmatrix}$$

and $\boldsymbol{r}^7(k) = [4.2838\ 4.2347\ 4.1948]^{\mathrm{T}}$. Because differences between the linear approximations of the predicted output trajectory calculated in the fourth and the seventh internal iterations are small, differences between the calculated vectors $\boldsymbol{r}^4(k)$ and $\boldsymbol{r}^7(k)$ are also small, the trajectories obtained in two versions of the MPC-NPL-NPLPT algorithm are very similar (Fig. 2.23 and Fig. 2.24). Evolution of the future control policy calculated in consecutive internal iterations is shown in Fig. 2.26.

2.7 Literature Review

2.7.1 MPC-NO Algorithm

The discussed implementation method of the MPC-NO algorithm based on the DLP neural model is an extended version of the algorithm presented in [177]. The processes for which the algorithm has been simulated include:

a) a bioreactor [163],

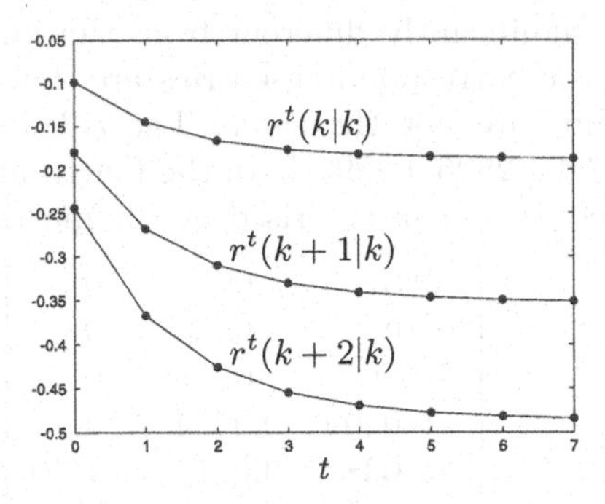

Fig. 2.26. Evolution of the future control policy calculated in consecutive internal iterations of the MPC-NPL-NPLPT algorithm (the sampling instant $k = 1$, the first set-point trajectory) in the control system of the high-pressure distillation column

b) a low-pressure cyclohexane-heptane distillation column [178],
c) a low-pressure methanol-water distillation column [149, 177],
d) an insulin administration process for the diabetics [161],
e) the yeast fermentation reactor discussed in this chapter [169],
f) a polymerisation reactor [149, 165, 177, 316],
g) the high-pressure ethylene-ethane distillation column discussed in this chapter [185, 316, 317].

A comparison of the influence of the initial point on computational complexity of the MPC-NO algorithm is given in [149, 178].

Derivation of the MPC-NO algorithm, quite similar to that discussed in this book, but for single-input single-output processes and without taking into account the measured disturbances, is given in [227]. Because the BFGS and Levenberg-Marquardt algorithms are used for nonlinear optimisation, no constraints are taken into account. Unfortunately, in such a case the ability of constraints handling, which is one of the most important advantages of MPC algorithms, is lost. A wrong prediction equation is used in [227] because, unlike Eq. (1.14), no estimation of the unmeasured disturbance is added to the model output which means that the resulting algorithm does not have the integral action. Although the authors have noticed the steady-state error and interpreted it as a result of the model-process mismatch, they have not taken into account this fact in the prediction equation. An application of the MPC-NO algorithm to a laboratory servo is described in [227].

The literature concerned with the MPC-NO algorithm is very rich, but application-orientated works dominate (simulation results and reports of applications to real processes). An early example of using neural networks in the MPC-NO algorithm is described in [311]. Only a single-input single-output process is considered, both prediction and control horizons are unitary. The steepest descent algorithm is used for nonlinear optimisation (the derivatives are calculated analytically), no constraints are considered.

The article [320] presents an another MPC-NO approach in which a numerically "strong" nonlinear optimisation routine with numerical approximation of the derivatives. The neural model is treated as a "black-box" one, its specific structure is not used during prediction and derivatives calculation. The algorithm is applied to a real chemical process with 3 inputs and 3 outputs. The BFGS algorithm with a penalty function is used for nonlinear optimisation. It has been found out empirically that it would be better to use analytical calculation of derivatives because it is much more numerically efficient than numerical approximation. On the average, the time needed for solving the nonlinear optimisation problem is 15 seconds whereas the sampling time is 10 minutes.

The literature concerned with applications of the MPC-NO algorithm with numerical approximation of derivatives in which the neural model is used for prediction is very rich. The applications include:

a) a distillation column [51] (a real process),
b) a catalytic cracking process [326] (a simulated process),
c) a crystallisation process [281] (a simulated process),
d) a neutralisation process (the pH reactor) [337, 336] (a real process),
e) a steel pickling process [116] (a simulated process),
f) a flotation column [214] (a real process),
g) the yeast fermentation process discussed in this chapter [115, 221] (the simulated process),
h) a fuel-ethanol fermentation process [56] (a simulated process),
i) a polystyrene batch polymerisation reactor [99] (a real process),
j) an aircraft gas turbine engine [218] (a simulated process).

The SQP algorithm is usually used for nonlinear optimisation. The Feasible Sequential Quadratic Programming (FSQP) algorithm is used in [281]. In consecutive iterations the FSQP algorithm finds only the feasible solutions, which satisfy all the constraints. One may guess that the FSQP algorithm is used rather than the rudimentary SQP one because the model is unable to correctly calculate the output predicted trajectory for some future input sequence which does not satisfy the constraints. The golden search method is used in [214], which is possible when the control horizon is unitary. An evolutionary algorithm is used for nonlinear optimisation in [254]. Although in the MPC-NO algorithm the numerical approximation of the derivatives can be justified in practice, but when a non-deterministic heuristic optimisation method is used for on-line optimisation, one is likely to question reliability of such an approach.

The rudimentary formulation of the MPC-NO is very universal, different model structures can be used for prediction, including alternative neural models, not only the DLP network. In such cases the predicted output trajectory and the matrix containing derivatives of that trajectory with respect to the future control policy given by Eq.(2.16) are different. The MPC-NO algorithm based on the RBF neural model is used in [323] to control an insulin administration process for the diabetics (a simulated process). No constraints are taken into account, but it makes it possible to use the DFP variable metrics algorithm for nonlinear optimisation. The MPC-NO algorithm for the SVM model is discussed in [102]. The unconstrained Newton method is used for optimisation, the derivatives and the Hessian matrix are calculated analytically.

The MPC-NO algorithm can be also designed in its adaptive version. Such a control strategy based on the RBF neural model which is successively updated on-line by means of a specially developed training algorithm is discussed in [7]. The control algorithm is applied to a chemical plant (a simulated process). A similar MPC-NO strategy is discussed in [328]. The model is trained on-line by a recurrent least-squares procedure. The algorithm is used to control the air-fuel ratio in an engine (a simulated process). In both cited works the SQP algorithm is used for nonlinear optimisation.

The literature concerned with the MPC-NO algorithm is very rich, but it is necessary to point out that many works have two important disadvantages. Firstly, the majority of authors treat the necessity of on-line nonlinear optimisation with no criticism. The SQP algorithm with numerically approximated derivatives is a predominant solution. The computational complexity issues are usually not discussed, no attempt is made to reduce on-line computational burden. Secondly, frequently the MPC-NO is not compared to any alternative control technique, for the majority of authors it is sufficient to show that the neural model can be used in MPC. Quite seldom the MPC-NO algorithm is compared with the PID controller or with the linear MPC approach, which is more appropriate.

2.7.2 Suboptimal MPC Algorithms with On-Line Linearisation

The general idea of the simplest suboptimal MPC algorithms with on-line model linearisation (the MPC-SL and MPC-NPL approaches) has been known for years [24, 76, 78, 316]. They are used in various industrial sectors [215]. The discussed implementation methods for the DLP neural model are extended versions of the algorithms presented in [177]. The applications of the MPC-SL strategy include:

a) a chemical extractor column [213] (a simulated process),
b) an aircraft gas turbine engine [218] (a simulated process),
c) a spark ignition engine [55] (a real process).

The applications of the MPC-NPL strategy include:

a) a bioreactor [163] (a simulated process),
b) a solar power plant [12, 25] (a real process),
c) a low-pressure cyclohexane-heptane distillation column [178] (a simulated process),

d) a low-pressure methanol-water distillation column [149, 177] (a simulated process),
e) an insulin administration process for the diabetics [161] (a simulated process),
f) the yeast fermentation reactor discussed in this chapter [169] (the simulated process),
g) a polymerisation reactor [149, 165, 177, 316] (a simulated process),
h) a spark ignition engine [286] (a simulated process),
i) the high-pressure ethylene-ethane distillation column discussed in this chapter [185, 316, 317] (the simulated process).

In the majority of the works cited above it is demonstrated that accuracy of the suboptimal MPC algorithms is very close to that obtained in the computationally demanding MPC-NO approach. Comparisons of computational complexity of both algorithms classes are given in [149, 165, 163, 169, 178]. Some works are mainly concerned with neural modelling and obtained system trajectories, but implementation details of the suboptimal MPC algorithm are not given [12, 25].

The explicit MPC-NPL algorithm based on the DLP neural model is discussed in [162]. The control signals for the current sampling instant are only calculated, not the whole future control policy. For this purpose the computationally efficient LU matrix decomposition procedure is used. For a simulated low-pressure methanol-water distillation column the explicit MPC-NPL algorithm gives the trajectories very similar to those obtained in the classical MPC-NPL algorithm with quadratic optimisation, which, in turn, gives results similar to those generated by the "ideal" MPC-NO approach. An alternative derivation of the MPC-SL algorithm for the DLP neural model, similar to the one presented in this book but for a single-input single-output process without measured disturbances, is given in [227]. The algorithm is applied to a laboratory servo.

There are relatively few publications concerned with more complicated suboptimal MPC algorithms (more advanced than the MPC-SL and MPC-NPL approaches). In the NPL+ algorithm discussed in [316] nonlinear prediction and quadratic optimisation is repeated a few times at one iteration when the set-point changes significantly. In order to simplify calculations only the control increments for the current sampling instant (i.e. the signals $\triangle u(k|k)$) are obtained in that manner, the remaining decision variables are calculated in the same way it is done in the classical MPC-NPL algorithm. The MPC-NPLT algorithm based on the DLP neural model is discussed in [145] whereas the MPC-NPL-NPLPT approach is described in [146]; in both cases the algorithms are applied to the high-pressure distillation column considered in this chapter.

The most advanced suboptimal MPC algorithm, the MPC-NNPAPT strategy, the main advantage of which over the MPC-NPLPT one is the fact that convergence of the internal iterations is guaranteed, is a natural use of the SQP approach [257]. A similar Newton-like IMC algorithm is presented in [67] (in the IMC structure the horizon is unitary). A new version of that algorithm, in which the constraints are taken into account, is described in [133]. The extended version of that algorithm is discussed in [132, 231], where the unitary horizon is not longer necessary. The core idea of the last cited approach is similar to that of the MPC-NNPAPT algorithm discussed in this book.

It is necessary to mention relatively infrequently used adaptive suboptimal MPC techniques. For example, the adaptive MPC-SL algorithm is discussed in [140]. The neural model is trained on-line and estimation of parameters of its linear approximation for the current operating point of the process is obtained by means of the recurrent least-squares method. Unfortunately, no constraints are taken into account. The constraints imposed on the control signal are introduced in the MPC-SL algorithm discussed in [4] (at each sampling instant the neural model is also retrained by the recurrent least-squares method).

Because the basic formulation of the discussed suboptimal MPC algorithms is very universal, numerous alternative models can be used, in particular different neural models. The MPC-NPL algorithm based on the RBF network and its application to a simulated polymerisation reactor are presented in [184]. The results are practically the same as those obtained in the MPC-NPL algorithm based on the DLP model [149, 165, 177, 316], but the RBF models usually have more hidden nodes and hence more parameters. The MPC-SL and MPC-NPL algorithms based on Takagi-Sugeno fuzzy models are discussed in [200, 317, 316].

The literature concerned with neural modelling is very rich. The neural modelling methodology is detailed in the majority of works cited in this literature review. A long list of publications devoted to neural modelling is given in an excellent review paper [101].

2.8 Chapter Summary

Two general approaches to nonlinear predictive control are discussed in this chapter: the MPC-NO algorithm, in which the full nonlinear model is used for prediction without any simplifications, and the suboptimal MPC techniques with on-line model or trajectory linearisation. In the first case it is necessary to use on-line nonlinear optimisation which is computationally expensive and not reliable (it is impossible to guarantee finding the global solution at each sampling instant, it is difficult to assess the time needed for calculations). That is why 5 suboptimal MPC techniques are discussed: MPC-SL, MPC-NPL, MPC-NPLT, MPC-NPLPT and MPC-NNPAPT. The suboptimal algorithms differ in terms of the linearisation method: in the first two approaches the model is locally linearised for the current operating point, in the MPC-NPLT algorithm the predicted output trajectory is linearised along an assumed future input trajectory, in the MPC-NPLPT approach nonlinear prediction and trajectory linearisation may be repeated a few times at each sampling instant. In the MPC-NNPAPT algorithm a quadratic approximation of the minimised cost-function and a linear approximation of the constraints are used, convergence of internal iterations is guaranteed, but the algorithm is quite complicated, much more than the other suboptimal MPC ones. To sum up, in all the discussed suboptimal approaches quadratic optimisation rather than nonlinear optimisation is used on-line, which means that the global solution is always found, it is possible to assess the calculation time. At each sampling instant of the first 3 suboptimal MPC algorithms only one quadratic optimisation problem is solved, in the last 2 algorithms a sequence of such problems is necessary.

It is possible to combine some of the presented algorithms. As a result one obtains the hybrid (two-phase) MPC-SL-NPLT, MPC-SL-NPLPT, MPC-NPL-NPLT and MPC-NPL-NPLPT algorithms. It is also possible to combine the suboptimal algorithms with the MPC-NO approach, the solution obtained in the first phase is likely to be quite close to that of the nonlinear optimisation problem (the MPC-SL-NO, MPC-NPL-NO and MPC-NPLT-NO algorithms).

Almost all suboptimal MPC algorithms can be formulated in their explicit versions (excluding the most advanced MPC-NNPAPT approach) in which for calculation of the current values of the control signals the constraints are not taken into account. It is done intentionally, because a simple matrix decomposition task is carried out and linear equations are solved in place of quadratic optimisation. Finally, the obtained solution is projected onto the feasible set determined by the constraints.

The discussed formulation of the MPC algorithms are very universal, for prediction various model structures can be used (provided that the model is differentiable). Implementation details for the DLP neural models with one hidden layer are given. The resulting MPC-SL and MPC-NPL algorithms are available in the DiaSter software platform which can be used for modelling, process diagnostics and control [118].

In order to demonstrate the discussed MPC algorithms two technological processes are considered: the yeast fermentation reactor and the high-pressure high-purity ethylene-ethane distillation column. In the first case the rudimentary MPC-NPL algorithm turns out to give practically the same accuracy as the MPC-NO one and, at the same time, it is several times or even more than 10 times less computationally demanding (computational complexity depends on horizons, in particular on the control horizon). The distillation process is much more difficult, because the MPC-NPL algorithm is significantly slower than the MPC-NO one, the simple model linearisation for the current operating point turns out to be insufficient. The MPC-NPLT$_{\boldsymbol{u}(k-1)}$ and MPC-NPLT$_{\boldsymbol{u}(k|k-1)}$ algorithms with trajectory linearisation are better.

Finally, the MPC-NPL-NPLPT algorithm with iterative trajectory linearisation makes it possible to obtain the same control accuracy as the MPC-NO one. The suboptimal algorithm is a few times less computationally demanding.

3

MPC Algorithms Based on Neural Hammerstein and Wiener Models

This chapter is concerned with nonlinear MPC algorithms based on cascade Hammerstein and Wiener neural models. A few different structures of such multi-input multi-output models are discussed and implementation details of three algorithms introduced in the previous chapter are given (MPC-NO, MPC-NPL and MPC-NPLPT schemes are considered). Additionally, the MPC algorithms with simplified linearisation, which is possible due to special structures of the models, are discussed (MPC-NPSL and MPC-SSL structures). Apart from the MPC-NPL algorithm for the Wiener model, all other discussed algorithms do not need the inverse of the steady-state part. Modelling abilities of cascade neural models are demonstrated for a polymerisation process, properties of the presented MPC algorithms are compared in the control systems of two processes.

3.1 Hammerstein and Wiener Models

The neural models used in the MPC algorithms discussed in the previous chapter are trained using some data sets. The technological knowledge may be used in some limited way to determine the order of model dynamics. The classical neural models are entirely the black-box structures, their parameters (the weights) do not have any physical interpretation, their structure is always the same but the number of hidden nodes and the model inputs are chosen in such a way that the model error for some available data set (sets) are minimised. Quite frequently, on the basis of theoretical considerations and analysis of process time-responses, one may conclude that the steady-state properties of the process are nonlinear (the sensors or actuators are usually nonlinear) whereas dynamic behaviour is practically linear. In such cases one may use the cascade Hammerstein or Wiener models shown in Fig. 1.3 which consists of a nonlinear steady-state part and a linear dynamic part.

It is also necessary to mention additional reasons why Hammerstein and Wiener models are sound alternatives to the black-box neural model. Firstly,

M. Ławryńczuk, *Computationally Efficient Model Predictive Control Algorithms*, Studies in Systems, Decision and Control 3,
DOI: 10.1007/978-3-319-04229-9_3, © Springer International Publishing Switzerland 2014

although it is usually difficult to develop the full first-principle dynamic model of the process, it is usually much simpler to obtain the corresponding steady-state description. Such a steady-state model may be used to generate the data to find the steady-state part of the cascade model. The linear dynamic part may be found on the basis of the recorded process time-responses. The resulting cascade model seems to be conceptually better than the empirical black-box model. Secondly, in practice it may turn out that the cascade model has a lower number of parameters than the black-box one.

In the nonlinear steady-state part one may use polynomials. Such an approach can be used primarily for modelling the simplest single-input single-output processes. Unfortunately, for modelling a broad class of processes it would be necessary to use polynomials of high orders. Such an approach has two fundamental disadvantages: the number of model parameters is likely to be significant (in particular for multi-input multi-output processes) and, which is very important, numerical properties of the model may be bad (in particular the identification process may be ill-conditioned) [103]. That is why different alternatives may be used in the steady-state nonlinear part: splines [59], wavelets [302], SVM approximators [80], piece-wise linear approximators [296], neural networks [103] and fuzzy systems [1]. In particular, thanks to their advantages, the neural networks can be successfully used in the cascade Hammerstein and Wiener structures. Such an approach has the following advantages:

1. Neural networks are universal approximators [98], when compared with alternative structures they usually give better accuracy and have a lower number of parameters.
2. Unlike polynomials, neural networks have good numerical properties: approximation is smooth, they have good interpolation and extrapolation properties. Neural networks are recommended for cascade models of processes with many inputs and many outputs.
3. A lot of very efficient identification algorithms for cascade neural models are available [103].

3.2 MPC Algorithms Based on Neural Hammerstein Models

3.2.1 Neural Hammerstein Models

The Hammerstein model consists of two parts connected in series: a nonlinear steady-state one and a linear one. The structure of the model of a single-input single-output process is depicted in Fig. 1.3. For processes with n_{u} inputs, n_{h} measured disturbances and n_{y} outputs it is possible to use a few variants of the model. Fig. 3.1 shows two the most popular configurations, namely the model with one nonlinear steady-state part and the model with separate steady-state parts. In both cases the number of auxiliary signals (between nonlinear and linear parts) is n_{v}. In the first case the nonlinear steady-state part has n_{v} outputs, in the second case as many as n_{v} independent multi-input single-output nonlinear steady-state parts are used. Of course, the model structure must be chosen for the particular process. It is natural to assume that $n_{\mathrm{v}} = n_{\mathrm{u}} + n_{\mathrm{h}}$. Alternatively, the value of n_{v} may be also chosen on the basis of some technological knowledge, but it is usually treated as an additional parameter which has a strong impact on model accuracy and complexity, in the same way the number of hidden nodes in neural networks is chosen. Some simplifications and combinations of the Hammerstein structures shown in Fig. 3.1 are possible. For example, Fig. 3.2 depicts the model with simplified nonlinear steady-state parts. They have only one input and one output, the number of auxiliary signals is always $n_{\mathrm{v}} = n_{\mathrm{u}} + n_{\mathrm{h}}$. An alternative may be to use a few simplified steady-state parts and a few multi-input single-output ones. It is always beneficial to evaluate and compare in practice numerous structures, some technological knowledge or observations concerning process behaviour may be very useful.

The description given in the following part of the chapter is concerned with the Hammerstein model with n_{v} separate steady-state parts shown in Fig. 3.1b. The nonlinear steady-state parts are described by

$$v(k) = g\left(\begin{bmatrix} u(k) \\ h(k) \end{bmatrix}\right) \tag{3.1}$$

where the steady-state parts are characterised by the function $g\colon \mathbb{R}^{n_{\mathrm{u}}+n_{\mathrm{h}}} \to \mathbb{R}^{n_{\mathrm{v}}}$. As the steady-state parts n_{v} DLP neural networks are used. The structure of such networks is depicted in Fig. 3.3. Each network has $n_{\mathrm{u}}+n_{\mathrm{h}}$ inputs $(u_1, \ldots, u_{n_{\mathrm{u}}}, h_1, \ldots, h_{n_{\mathrm{h}}})$, K^s hidden nodes with the nonlinear transfer function $\varphi\colon \mathbb{R} \to \mathbb{R}$ and one output $v_s(k)$. The output signals of the consecutive networks ($s = 1, \ldots, n_{\mathrm{v}}$) are

$$v_s(k) = w_0^{2,s} + \sum_{i=1}^{K^s} w_i^{2,s} \varphi(z_i^s(k)) \tag{3.2}$$

where

$$z_i^s(k) = w_{i,0}^{1,s} + \sum_{j=1}^{n_{\mathrm{u}}} w_{i,j}^{1,s} u_j(k) + \sum_{j=1}^{n_{\mathrm{h}}} w_{i,n_{\mathrm{u}}+j}^{1,s} h_j(k) \tag{3.3}$$

is the sum of the input signals connected to the ith hidden nodes ($i = 1, \ldots, K^s$). The weights of the first layer are denoted by $w_{i,j}^{1,s}$, where $i = 1, \ldots, K^s$, $j = 0, \ldots, n_{\mathrm{u}} + n_{\mathrm{h}}$, $s = 1, \ldots, n_{\mathrm{v}}$, the weights of the second layer are denoted by $w_i^{2,s}$, where $i = 0, \ldots, K^s$, $s = 1, \ldots, n_{\mathrm{v}}$. From Eqs. (3.2) and (3.3) one obtains

$$v_s(k) = w_0^{2,s} + \sum_{i=1}^{K^s} w_i^{2,s} \varphi\left(w_{i,0}^{1,s} + \sum_{j=1}^{n_{\mathrm{u}}} w_{i,j}^{1,s} u_j(k) + \sum_{j=1}^{n_{\mathrm{h}}} w_{i,n_{\mathrm{u}}+j}^{1,s} h_j(k)\right) \tag{3.4}$$

The linear dynamic part is defined by the difference equation

$$\boldsymbol{A}(q^{-1})y(k) = \boldsymbol{B}(q^{-1})v(k) \tag{3.5}$$

where the matrices

$$\boldsymbol{A}(q^{-1}) = \begin{bmatrix} A_{1,1}(q^{-1}) & \ldots & 0 \\ \vdots & \ddots & \vdots \\ 0 & \ldots & A_{n_y,n_y}(q^{-1}) \end{bmatrix}$$

$$\boldsymbol{B}(q^{-1}) = \begin{bmatrix} B_{1,1}(q^{-1}) & \ldots & B_{1,n_v}(q^{-1}) \\ \vdots & \ddots & \vdots \\ B_{n_y,1}(q^{-1}) & \ldots & B_{n_y,n_v}(q^{-1}) \end{bmatrix} \tag{3.6}$$

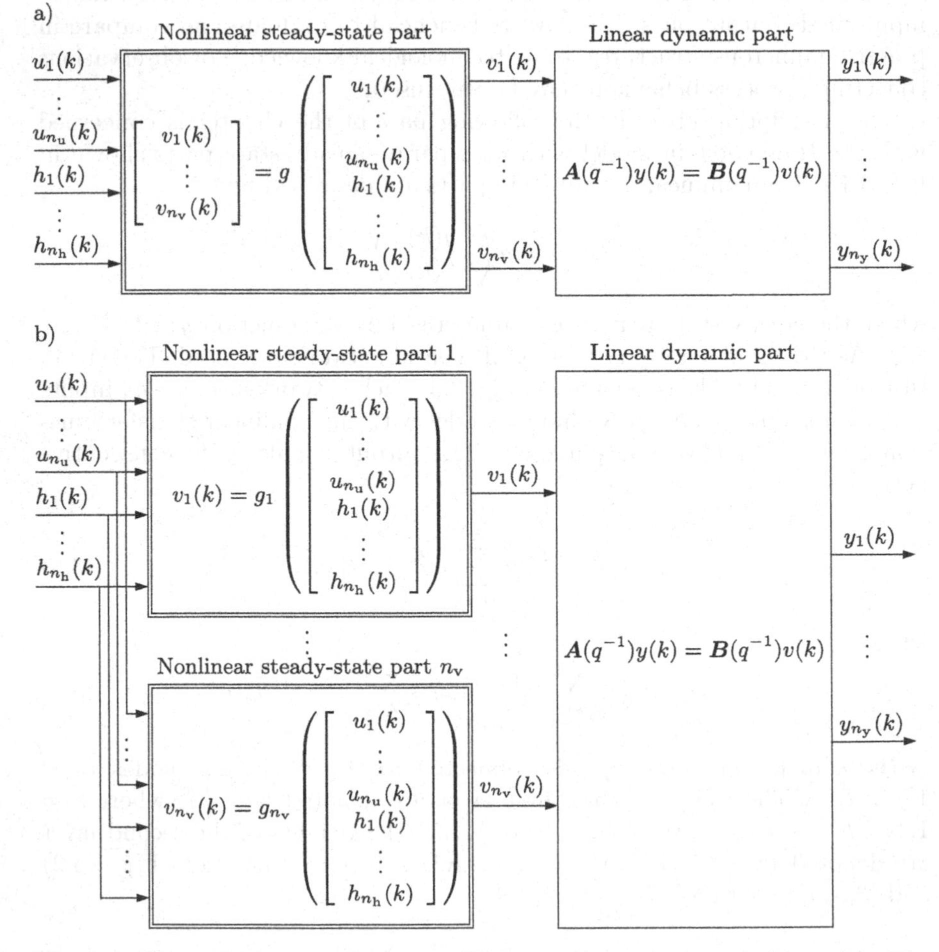

Fig. 3.1. The multi-input multi-output Hammerstein structures: a) the model with one steady-state part, b) the model with n_v steady-state parts

consist of polynomials of the time-delay operator q^{-1}

$$A_{m,m}(q^{-1}) = 1 + a_1^m q^{-1} + \ldots + a_{n_A}^m q^{-n_A}$$
$$B_{m,s}(q^{-1}) = b_1^{m,s} q^{-1} + \ldots + b_{n_B}^{m,s} q^{-n_B} \tag{3.7}$$

From Eqs. (3.5), (3.6) and (3.7) one can calculate consecutive outputs of the Hammerstein model ($m = 1, \ldots, n_y$)

$$y_m(k) = \sum_{s=1}^{n_v} \sum_{l=1}^{n_B} b_l^{m,s} v_s(k-l) - \sum_{l=1}^{n_A} a_l^m y_m(k-l) \tag{3.8}$$

In the above equation the outputs depend on the auxiliary signals $v_1, \ldots, v_{n_v}$. They can be easily eliminated. From Eqs. (3.1) and (3.5) one has

$$\boldsymbol{A}(q^{-1})y(k) = \boldsymbol{B}(q^{-1})g\left(\begin{bmatrix} u(k) \\ h(k) \end{bmatrix}\right)$$

Using Eqs. (3.4) and (3.8), one obtains

$$y_m(k) = \sum_{s=1}^{n_v} \sum_{l=1}^{n_B} b_l^{m,s} \left[w_0^{2,s} + \sum_{i=1}^{K^s} w_i^{2,s} \varphi\left(w_{i,0}^{1,s} + \sum_{j=1}^{n_u} w_{i,j}^{1,s} u_j(k-l) + \sum_{j=1}^{n_h} w_{i,n_u+j}^{1,s} h_j(k-l) \right) \right] - \sum_{l=1}^{n_A} a_l^m y_m(k-l) \tag{3.9}$$

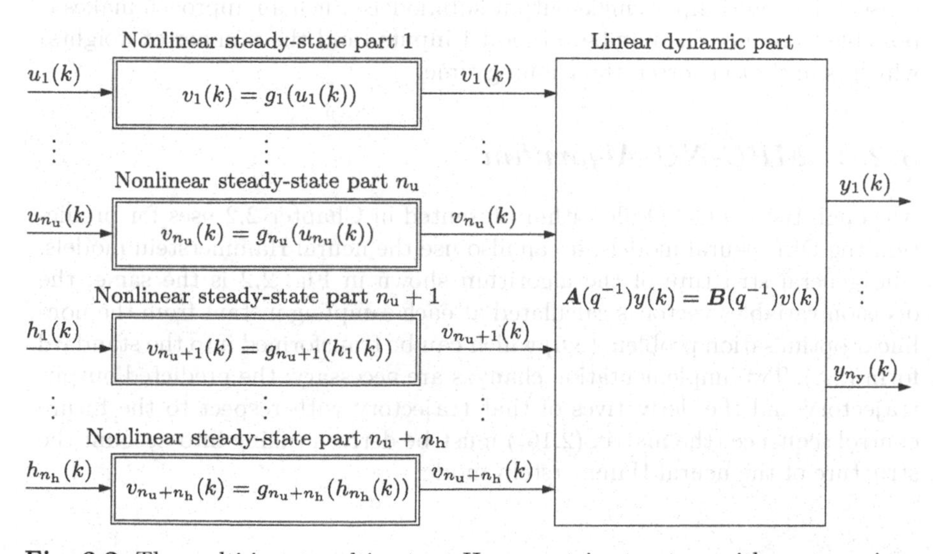

Fig. 3.2. The multi-input multi-output Hammerstein structure with $n_u + n_h$ simplified steady-state parts

In the current iteration the model outputs depend in Eq. (3.9) on the input, measured disturbance and output signals from previous sampling instants, there is no dependence on the auxiliary signals. The output signals can be expressed as general functions

$$\begin{aligned} y_m(k) = f_m(&u_1(k-1), \ldots, u_1(k-n_{\mathrm{B}}), \ldots, u_{n_{\mathrm{u}}}(k-1), \ldots, u_{n_{\mathrm{u}}}(k-n_{\mathrm{B}}), \\ &h_1(k-1), \ldots, h_1(k-n_{\mathrm{B}}), \ldots, h_{n_{\mathrm{h}}}(k-1), \ldots, h_{n_{\mathrm{h}}}(k-n_{\mathrm{B}}), \\ &y_m(k-1), \ldots, y_m(k-n_{\mathrm{A}})) \end{aligned} \tag{3.10}$$

where the consecutive multi-input single-output models are characterised by the functions $f_m \colon \mathbb{R}^{n_{\mathrm{A}}+(n_{\mathrm{u}}+n_{\mathrm{h}})n_{\mathrm{B}}} \to \mathbb{R}$, $m = 1, \ldots, n_{\mathrm{y}}$. For the neural Hammerstein model it is possible to obtain the general equation (3.10) similar to Eq. (2.1) used in the case of the DLP neural model shown in Fig. 2.1 and described by Eqs. (2.2) and (2.3). The fundamental difference is the fact that in the Hammerstein model (3.9) the model outputs for the current sampling instant depend in a linear way on output signals measured at the previous sampling instants, nonlinear properties of the model are obtained by assuming a nonlinear influence of only previous inputs and measured disturbances on the current outputs. No explicit time-delay is present in the linear dynamic part, the order of dynamics of all input-output channels (defined by the integer numbers n_{A} and n_{B}) is the same. Such an approach does not result in any significant model complication because complexity of the neural network is independent of the order of dynamics, but simplifies the further derivations. If the time-delay is present, some of the coefficients $b_i^{m,s}$ are 0. In the classical black-box neural model shown in Fig. 2.1 explicit time-delays are taken into account and the order of dynamics is defined separately for consecutive multi-input single-output submodels. Such an approach makes it possible to reduce the number of model inputs (and the number of weights) which is likely to shorten the training time.

3.2.2 MPC-NO Algorithm

Although the MPC-NO algorithm presented in Chapter 2.2 uses for prediction the DLP neural models, it can also use the neural Hammerstein models. The general structure of the algorithm shown in Fig. 2.2 is the same, the decision variables vector is calculated at each sampling instant from the nonlinear optimisation problem (2.5) which can be transformed into the standard form (2.8). Two implementation changes are necessary: the predicted output trajectory and the derivatives of that trajectory with respect to the future control sequence (the matrix (2.16)) must be derived taking into account the structure of the neural Hammerstein model.

Using the general prediction equation (1.14) and Eq. (3.9), the predicted output trajectory for consecutive outputs ($m = 1, \ldots, n_{\mathrm{y}}$) is

$$\begin{aligned}
\hat{y}_m(k+p|k) = &\sum_{s=1}^{n_{\mathrm{v}}} \sum_{l=1}^{I_{\mathrm{uf}}(p)} b_l^{m,s} \left[w_0^{2,s} + \sum_{i=1}^{K^s} w_i^{2,s} \varphi\left(w_{i,0}^{1,s} + \sum_{j=1}^{n_{\mathrm{u}}} w_{i,j}^{1,s} u_j(k-l+p|k) \right.\right. \\
&\left.\left. + \sum_{j=1}^{n_{\mathrm{h}}} w_{i,n_{\mathrm{u}}+j}^{1,s} h_j(k-l+p|k) \right) \right] \\
&+ \sum_{s=1}^{n_{\mathrm{v}}} \sum_{l=I_{\mathrm{uf}}(p)+1}^{n_{\mathrm{B}}} b_l^{m,s} \left[w_0^{2,s} + \sum_{i=1}^{K^s} w_i^{2,s} \varphi\left(w_{i,0}^{1,s} + \sum_{j=1}^{n_{\mathrm{u}}} w_{i,j}^{1,s} u_j(k-l+p) \right.\right. \\
&\left.\left. + \sum_{j=1}^{n_{\mathrm{h}}} w_{i,n_{\mathrm{u}}+j}^{1,s} h_j(k-l+p) \right) \right] \\
&- \sum_{l=1}^{I_{\mathrm{yf}}(p)} a_l^m \hat{y}_m(k-l+p|k) - \sum_{l=I_{\mathrm{yf}}(p)+1}^{n_{\mathrm{A}}} a_l^m y_m(k-l+p) \\
&+ d_m(k)
\end{aligned} \tag{3.11}$$

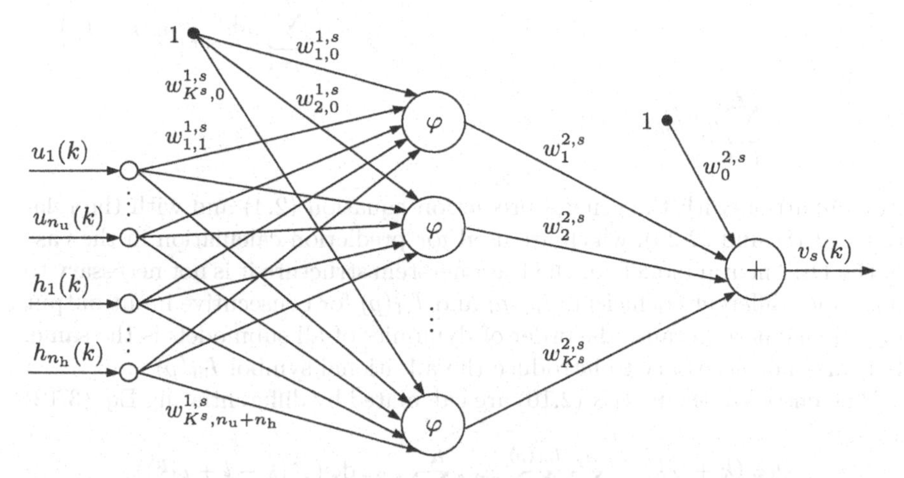

Fig. 3.3. The structure of the neural network used in the nonlinear part of the multi-input multi-output Hammerstein model with n_{v} steady-state parts, $s = 1, \ldots, n_{\mathrm{v}}$

where $I_{\mathrm{uf}}(p) = \min(p, n_{\mathrm{B}})$, $I_{\mathrm{yf}}(p) = \min(p-1, n_{\mathrm{A}})$. Eq. (3.11) can be rewritten in a compact form

$$\begin{aligned}\hat{y}_m(k+p|k) = &\sum_{s=1}^{n_{\mathrm{v}}} \sum_{l=1}^{I_{\mathrm{uf}}(p)} b_l^{m,s} \left[w_0^{2,s} + \sum_{i=1}^{K^s} w_i^{2,s} \varphi(z_i^s(k-l+p|k)) \right] \\ &+ \sum_{s=1}^{n_{\mathrm{v}}} \sum_{l=I_{\mathrm{uf}}(p)+1}^{n_{\mathrm{B}}} b_l^{m,s} \left[w_0^{2,s} + \sum_{i=1}^{K^s} w_i^{2,s} \varphi(z_i^s(k-l+p)) \right] \\ &- \sum_{l=1}^{I_{\mathrm{yf}}(p)} a_l^m \hat{y}_m(k-l+p|k) - \sum_{l=I_{\mathrm{yf}}(p)+1}^{n_{\mathrm{A}}} a_l^m y_m(k-l+p) \\ &+ d_m(k) \end{aligned} \tag{3.12}$$

where the signals $z_i^s(k-l+p|k)$ and $z_i^s(k-l+p)$ are calculated from Eq. (3.3). Taking into account Eqs. (1.15) and (3.9), the unmeasured disturbances are estimated from

$$\begin{aligned}d_m(k) = y_m(k) - &\sum_{s=1}^{n_{\mathrm{v}}} \sum_{l=1}^{n_{\mathrm{B}}} b_l^{m,s} \left[w_0^{2,s} + \sum_{i=1}^{K^s} w_i^{2,s} \varphi\left(w_{i,0}^{1,s} + \sum_{j=1}^{n_{\mathrm{u}}} w_{i,j}^{1,s} u_j(k-l) \right.\right. \\ &\left.\left. + \sum_{j=1}^{n_{\mathrm{h}}} w_{i,n_{\mathrm{u}}+j}^{1,r} h_j(k-l) \right) \right] \\ &+ \sum_{l=1}^{n_{\mathrm{A}}} a_l^m y_m(k-l)\end{aligned}$$

In comparison with the general prediction equation (2.4) and with the relations (2.18) and (2.20), which are used for prediction calculation in the case of the DLP neural model, for the Hammerstein structure it is not necessary to introduce different coefficients $I_{\mathrm{uf}}(p)$ and $I_{\mathrm{yf}}(p)$ for consecutive input-output model channels, because the order of dynamics of all submodels is the same. It is also not necessary to introduce the additional symbol $I_{\mathrm{hf}}(p)$.

The entries of the matrix (2.16) are calculated by differentiating Eq. (3.12)

$$\begin{aligned}\frac{\partial \hat{y}_m(k+p|k)}{\partial u_n(k+r|k)} = &\sum_{s=1}^{n_{\mathrm{v}}} \sum_{l=1}^{I_{\mathrm{uf}}(p)} b_l^{m,s} \sum_{i=1}^{K^s} w_i^{2,s} \frac{\mathrm{d}\varphi(z_i^s(k-l+p|k))}{\mathrm{d}z_i^s(k-l+p|k)} \\ &\times w_{i,n}^{1,s} \frac{\partial u_n(k-l+p|k)}{\partial u_n(k+r|k)} \\ &- \sum_{l=1}^{I_{\mathrm{yf}}(p)} a_l^m \frac{\partial \hat{y}_m(k-l+p|k)}{\partial u_n(k+r|k)}\end{aligned} \tag{3.13}$$

3.2.3 Suboptimal MPC Algorithms with On-Line Linearisation

It is also possible to derive the suboptimal MPC algorithms introduced in Chapter 2.3 for the neural Hammerstein model. Their general structures are the same, the same quadratic optimisation problems are solved. For the particular structure of the model it is necessary to derive the coefficients of the linearised model (or the linearised output trajectory) and the predicted trajectory, including the free trajectory. Implementation details of the most popular MPC-NPL algorithm and of the MPC-NPLPT one are given next. Additionally, an alternative simplified approach to on-line linearisation possible due to the specific structure of the model and the resulting MPC-NPSL algorithm are described.

MPC-NPL Algorithm

The general structure of the MPC-NPL algorithm, shown in Fig. 2.3, is universal, independent of the model used. The decision variables vector is calculated at each sampling instant from the quadratic optimisation task (2.36), which can be transformed into the standard form (2.37). The linear approximation (2.29) of the nonlinear Hammerstein model is calculated from the universal equations (2.27) and (2.28). Nota bene, the coefficients of the linearised model $a_l^m(k)$ and $b_l^{m,n}(k)$ depend on the current operating point of the process whereas the constants a_l^m are $b_l^{m,n}$ denote the parameters of the linear dynamic part of the Hammerstein model. Differentiating Eq. (3.9), one obtains the coefficients of the linearised model

$$a_l^m(k) = a_l^m$$

for $m = 1, \ldots, n_{\mathrm{y}}$ and

$$b_l^{m,n}(k) = \sum_{s=1}^{n_{\mathrm{v}}} b_l^{m,s} \sum_{i=1}^{K^s} w_i^{2,s} \frac{\mathrm{d}\varphi(z_i^s(k-l))}{\mathrm{d}z_i^s(k-l)} w_{i,n}^{1,s}$$

for all $l = 1, \ldots, n_{\mathrm{B}}$, $m = 1, \ldots, n_{\mathrm{y}}$, $n = 1, \ldots, n_{\mathrm{u}}$.

The nonlinear free trajectory is calculated recurrently for all $m = 1, \ldots, n_{\mathrm{y}}$ and $p = 1, \ldots, N$ from the predicted output trajectory used in the MPC-NO algorithm, defined by Eq. (3.11), but taking into account only the influence of the past

$$
\begin{aligned}
y_m^0(k+p|k) = &\sum_{s=1}^{n_\mathrm{v}} \sum_{l=1}^{I_\mathrm{uf}(p)} b_l^{m,s} \Bigg[w_0^{2,s} + \sum_{i=1}^{K^s} w_i^{2,s} \varphi \Bigg(w_{i,0}^{1,s} + \sum_{j=1}^{n_\mathrm{u}} w_{i,j}^{1,s} u_j(k-1) \\
&+ \sum_{j=1}^{n_\mathrm{h}} w_{i,n_\mathrm{u}+j}^{1,s} h_j(k-l+p|k) \Bigg) \Bigg] \\
&+ \sum_{s=1}^{n_\mathrm{v}} \sum_{l=I_\mathrm{uf}(p)+1}^{n_\mathrm{B}} b_l^{m,s} \Bigg[w_0^{2,s} + \sum_{i=1}^{K^s} w_i^{2,s} \varphi \Bigg(w_{i,0}^{1,s} + \sum_{j=1}^{n_\mathrm{u}} w_{i,j}^{1,s} u_j(k-l+p) \\
&+ \sum_{j=1}^{n_\mathrm{h}} w_{i,n_\mathrm{u}+j}^{1,s} h_j(k-l+p) \Bigg) \Bigg] \\
&- \sum_{l=1}^{I_\mathrm{yf}(p)} a_l^m y_m^0(k-l+p|k) - \sum_{l=I_\mathrm{yf}(p)+1}^{n_\mathrm{A}} a_l^m y_m(k-l+p) + d_m(k)
\end{aligned}
$$

MPC-NPLPT Algorithm

The general structure of the MPC-NPLPT algorithm, shown in Fig. 2.6, is universal, the decision variables vector is calculated from the quadratic optimisation task (2.56). The linear approximation of the nonlinear predicted output trajectory is defined by Eq. (2.52). For the neural Hammerstein model it is necessary to derive the predicted output trajectory and the derivatives of that trajectory with respect to the future control policy (the matrix (2.53)). From Eq. (3.11), the predicted output trajectory for the tth internal iteration is

$$
\begin{aligned}
\hat{y}_m^t(k+p|k) = &\sum_{s=1}^{n_\mathrm{v}} \sum_{l=1}^{I_\mathrm{uf}(p)} b_l^{m,s} \Bigg[w_0^{2,s} + \sum_{i=1}^{K^s} w_i^{2,s} \varphi \Bigg(w_{i,0}^{1,s} + \sum_{j=1}^{n_\mathrm{u}} w_{i,j}^{1,s} u_j^t(k-l+p|k) \\
&+ \sum_{j=1}^{n_\mathrm{h}} w_{i,n_\mathrm{u}+j}^{1,s} h_j(k-l+p|k) \Bigg) \Bigg] \\
&+ \sum_{s=1}^{n_\mathrm{v}} \sum_{l=I_\mathrm{uf}(p)+1}^{n_\mathrm{B}} b_l^{m,s} \Bigg[w_0^{2,s} + \sum_{i=1}^{K^s} w_i^{2,s} \varphi \Bigg(w_{i,0}^{1,s} + \sum_{j=1}^{n_\mathrm{u}} w_{i,j}^{1,s} u_j(k-l+p) \\
&+ \sum_{j=1}^{n_\mathrm{h}} w_{i,n_\mathrm{u}+j}^{1,s} h_j(k-l+p) \Bigg) \Bigg] \\
&- \sum_{l=1}^{I_\mathrm{yf}(p)} a_l^m \hat{y}_m^t(k-l+p|k) - \sum_{l=I_\mathrm{yf}(p)+1}^{n_\mathrm{A}} a_l^m y_m(k-l+p) \\
&+ d_m(k)
\end{aligned}
\tag{3.14}
$$

The only difference between the obtained trajectory and Eq. (3.11) used in the MPC-NO algorithm is the fact that in the first case the internal iteration index t is taken into account for the input and output trajectories.

The matrix (2.53) contains the derivatives of the predicted output trajectory calculated in the previous internal iteration $(t-1)$ with respect to the future control sequence found in the same internal iteration. Using the trajectory (3.14) and remembering that the index t must be replaced by $t-1$, one has

$$\frac{\partial \hat{y}_m^{t-1}(k+p|k)}{\partial u_n^{t-1}(k+r|k)} = \sum_{s=1}^{n_{\text{v}}} \sum_{l=1}^{I_{\text{uf}}(p)} b_l^{m,s} \sum_{i=1}^{K^s} w_i^{2,s} \frac{\mathrm{d}\varphi(z_i^{s,t-1}(k-l+p|k))}{\mathrm{d}z_i^{s,t-1}(k-l+p|k)} \times w_{i,n}^{1,s} \frac{\partial u_n^{t-1}(k-l+p|k)}{\partial u_n^{t-1}(k+r|k)} - \sum_{l=1}^{I_{\text{yf}}(p)} a_l^m \frac{\partial \hat{y}_m^{t-1}(k-l+p|k)}{\partial u_n^{t-1}(k+r|k)}$$

Basing on the equations given above it is also possible to implement the MPC-NPLT algorithm, in which trajectory linearisation is performed once at each sampling instant.

MPC Algorithms with Simplified Linearisation (MPC-NPSL, MPC-SSL)

Thanks to the specific cascade structure of the Hammerstein model it is possible to derive some suboptimal MPC algorithms with simplified model linearisation. In the MPC-NPL algorithm a precise linear approximation of the Hammerstein model (3.9) for the current operating point of the process is calculated analytically using the Taylor series expansion formula. Similarly, in the MPC-NPLPT algorithm a linear approximation of the nonlinear predicted output trajectory (3.11) is analytically calculated. The simplified linearisation approach is much simpler: the gain of the nonlinear steady-state part of the model is calculated for the current operating point and the obtained value is used to update the gain of the linear dynamic part. Using the structure of the multi-input multi-output Hammerstein model with separate n_{v} steady-state parts shown in Fig. 3.1b, for the current operating conditions the steady-state part may be described by

$$v(k) = \boldsymbol{K}^{\text{H}}(k)u(k) \tag{3.15}$$

where $\boldsymbol{K}^{\mathrm{H}}(k)$ denotes the gain matrix which depends on the current operating point of the process. It has the following structure

$$\boldsymbol{K}^{\mathrm{H}}(k) = \begin{bmatrix} k^{\mathrm{H}}_{1,1}(k) & \ldots & k^{\mathrm{H}}_{1,n_{\mathrm{u}}}(k) \\ \vdots & \ddots & \vdots \\ k^{\mathrm{H}}_{n_{\mathrm{v}},1}(k) & \ldots & k^{\mathrm{H}}_{n_{\mathrm{v}},n_{\mathrm{u}}}(k) \end{bmatrix} = \begin{bmatrix} \dfrac{\partial v_1(k)}{\partial u_1(k)} & \cdots & \dfrac{\partial v_1(k)}{\partial u_{n_{\mathrm{u}}}(k)} \\ \vdots & \ddots & \vdots \\ \dfrac{\partial v_{n_{\mathrm{v}}}(k)}{\partial u_1(k)} & \cdots & \dfrac{\partial v_{n_{\mathrm{v}}}(k)}{\partial u_{n_{\mathrm{u}}}(k)} \end{bmatrix}$$

Inserting Eq. (3.15) to Eq. (3.5), which describes the linear part of the model, one has

$$\boldsymbol{A}(q^{-1})y(k) = \boldsymbol{B}(q^{-1})\boldsymbol{K}^{\mathrm{H}}(k)u(k)$$

Taking into account the structure of the polynomial matrices $\boldsymbol{A}(q^{-1})$ and $\boldsymbol{B}(q^{-1})$, defined by Eq. (3.6), one has

$$\begin{bmatrix} A_{1,1}(q^{-1})y_1(k) \\ \vdots \\ A_{n_{\mathrm{y}},n_{\mathrm{y}}}(q^{-1})y_{n_{\mathrm{y}}}(k) \end{bmatrix} =$$

$$\begin{bmatrix} \sum_{s=1}^{n_{\mathrm{v}}} B_{1,s}(q^{-1})k^{\mathrm{H}}_{s,1}(k) & \ldots & \sum_{s=1}^{n_{\mathrm{v}}} B_{1,s}(q^{-1})k^{\mathrm{H}}_{s,n_{\mathrm{u}}}(k) \\ \vdots & \ddots & \vdots \\ \sum_{s=1}^{n_{\mathrm{v}}} B_{n_{\mathrm{y}},s}(q^{-1})k^{\mathrm{H}}_{s,1}(k) & \ldots & \sum_{s=1}^{n_{\mathrm{v}}} B_{n_{\mathrm{y}},s}(q^{-1})k^{\mathrm{H}}_{s,n_{\mathrm{u}}}(k) \end{bmatrix} \begin{bmatrix} u_1(k) \\ \vdots \\ u_{n_{\mathrm{u}}}(k) \end{bmatrix}$$

which leads to the coefficients of the linearised model

$$a_l^m(k) = a_l^m, \ b_l^{m,n}(k) = \sum_{s=1}^{n_{\mathrm{v}}} b_l^{m,s} k^{\mathrm{H}}_{s,n}(k)$$

Taking into account the neural structure of the nonlinear steady-state part of the model and Eq. (3.4), the entries of the gain matrix $\boldsymbol{K}^{\mathrm{H}}(k)$ are

$$k^{\mathrm{H}}_{s,n}(k) = \sum_{i=1}^{K^s} w_i^{2,s} \frac{\mathrm{d}\varphi(z_i^s(k))}{\mathrm{d}z_i^s(k)} w_{i,n}^{1,s}$$

for all $n = 1, \ldots, n_{\mathrm{u}}$, $s = 1, \ldots, n_{\mathrm{v}}$. According to Eq. (3.3), the measured disturbances have an impact on the signals $z_i^s(k)$ and on the coefficients $k_{s,n}(k)$.

The matrices (2.31) consisting of the step-response coefficients of the multi-input multi-output linearised model are

$$\boldsymbol{S}_p(k) = \boldsymbol{S}_p\boldsymbol{K}^{\mathrm{H}}(k) = \begin{bmatrix} \sum_{r=1}^{n_{\mathrm{v}}} s_p^{1,r} k^{\mathrm{H}}_{r,1}(k) & \ldots & \sum_{r=1}^{n_{\mathrm{v}}} s_p^{r,n_{\mathrm{u}}} k^{\mathrm{H}}_{r,n_{\mathrm{u}}}(k) \\ \vdots & \ddots & \vdots \\ \sum_{r=1}^{n_{\mathrm{v}}} s_p^{n_{\mathrm{y}},r} k^{\mathrm{H}}_{r,1}(k) & \ldots & \sum_{r=1}^{n_{\mathrm{v}}} s_p^{n_{\mathrm{y}},r} k^{\mathrm{H}}_{r,n_{\mathrm{u}}}(k) \end{bmatrix} \quad (3.16)$$

where $\boldsymbol{S}_p$ for $p = 1, \ldots, N$ denote constant matrices (of dimensionality $n_{\mathrm{y}} \times n_{\mathrm{v}}$) containing step-response coefficients of the linear part of the Hammerstein model, the scalar coefficients of the linear part are denoted by $s_p^{m,s}$. The dynamic matrix of the linearised model (of the structure (2.34)) depends on constant step-response coefficients of the linear part of the model and on the entries of the gain matrix $\boldsymbol{K}^{\mathrm{H}}(k)$. The linear approximation of the model obtained in this way can be of course used in the MPC-SL approach, which leads to the MPC algorithm with Simplified Successive Linearisation (MPC-SSL). When the free trajectory is calculated from the full nonlinear Hammerstein model one obtains the MPC algorithm with Nonlinear Prediction and Simplified Linearisation (MPC-NPSL). In both cases the same quadratic optimisation problem (2.36) is solved.

Naturally, in the described MPC-NPSL algorithm it is also possible to use an alternative Hammerstein model, for example the structure with $n_{\mathrm{u}} + n_{\mathrm{h}}$ simplified steady-state parts (with one input and one output) shown in Fig. 3.2. In such a case the gain matrix $\boldsymbol{K}^{\mathrm{H}}(k)$ is diagonal.

It is necessary to once again emphasise the fact that the simplified linearisation is different than the precise one used in the MPC-SL and MPC-NPL algorithms.

3.3 MPC Algorithms Based on Neural Wiener Models

3.3.1 Neural Wiener Models

In comparison with the Hammerstein model, in the Wiener one the order of model parts is reversed: the linear dynamic part precedes the nonlinear steady-state one. The structure of the single-input single-output Wiener model is shown in Fig. 1.3. For multi-input multi-output processes a few variants are possible. Fig. 3.4 shows two the most popular configurations. In the first case only one steady-state part is used, it has n_{v} inputs and n_{y} outputs. The number of auxiliary signals can be treated as an additional model parameter, but it is straightforward to assume that $n_{\mathrm{v}} = n_{\mathrm{y}}$. In the second model configuration as many as n_{y} separate steady-state parts are used, each of them has n_{v} inputs and one output. Of course, some simplifications and combinations of the structures shown in Fig. 3.4 are possible. For example, Fig. 3.5 shows the model with n_{y} simplified steady-state parts with one input and one output. For such a structure the condition $n_{\mathrm{v}} = n_{\mathrm{y}}$ is straightforward.

The description given in the following part of the chapter is concerned with the Wiener model with n_{y} separate steady-state parts shown in Fig. 3.4b (MPC-NO and MPC-NPLPT algorithms) and with the simplified model

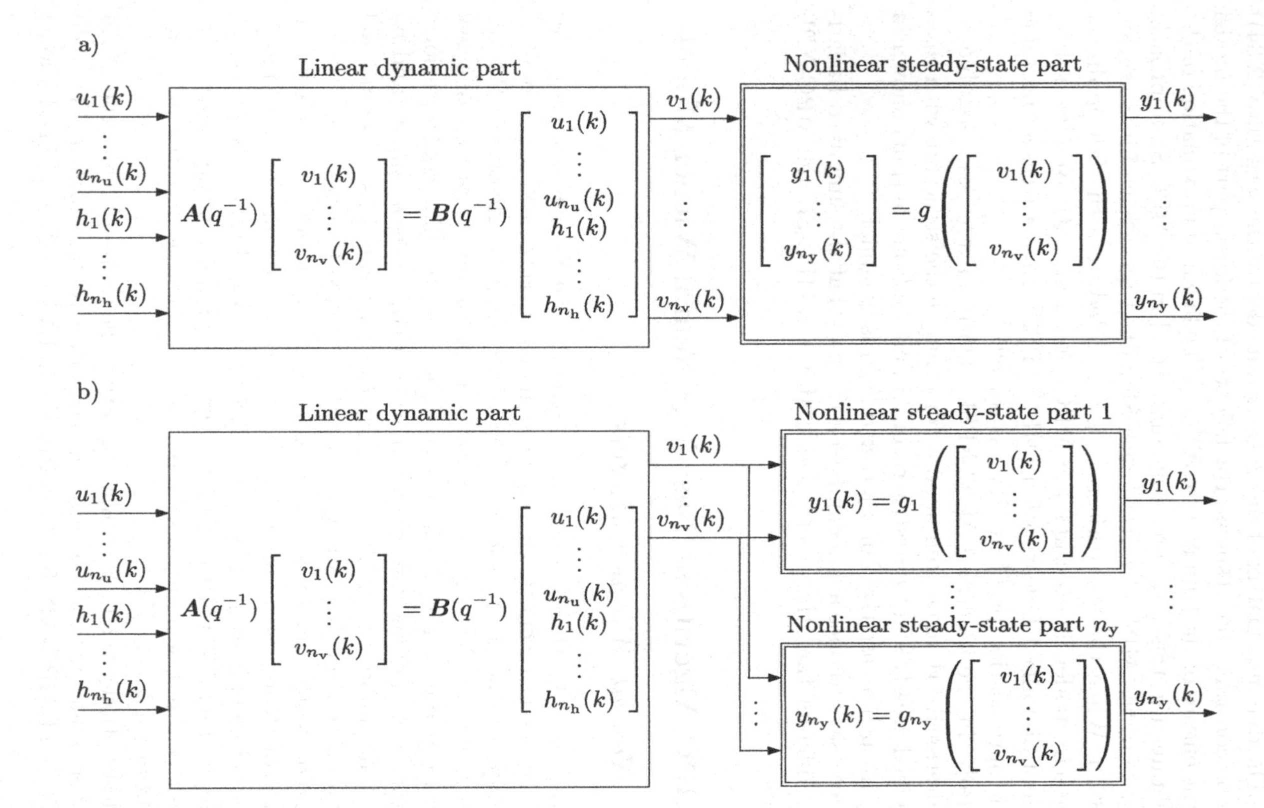

Fig. 3.4. The multi-input multi-output Wiener structures: a) the model with one steady-state part, b) the model with n_y steady-state parts

with n_{y} single-input single-output steady-state parts shown in Fig. 3.5 (MPC-NPL, MPC-NPSL and MPC-SSL algorithms). In both models the linear dynamic part is defined by the difference equation

$$\boldsymbol{A}(q^{-1})v(k) = \boldsymbol{B}(q^{-1})\begin{bmatrix} u(k) \\ h(k) \end{bmatrix} \tag{3.17}$$

where the matrices

$$\boldsymbol{A}(q^{-1}) = \begin{bmatrix} A_{1,1}(q^{-1}) & \dots & 0 \\ \vdots & \ddots & \vdots \\ 0 & \dots & A_{n_{\mathrm{v}},n_{\mathrm{v}}}(q^{-1}) \end{bmatrix}$$
$$\boldsymbol{B}(q^{-1}) = \begin{bmatrix} B_{1,1}(q^{-1}) & \dots & B_{1,n_{\mathrm{u}}+n_{\mathrm{h}}}(q^{-1}) \\ \vdots & \ddots & \vdots \\ B_{n_{\mathrm{v}},1}(q^{-1}) & \dots & B_{n_{\mathrm{v}},n_{\mathrm{u}}+n_{\mathrm{h}}}(q^{-1}) \end{bmatrix} \tag{3.18}$$

consist of polynomials of the variable q^{-1} defined by Eq. (3.7). The matrices $\boldsymbol{A}(q^{-1})$ and $\boldsymbol{B}(q^{-1})$ used in the Wiener model have different dimensionality than in the case of the Hammerstein model (Eq. (3.6)). It results from the reversed order of both parts of Hammerstein and Wiener model. From Eqs. (3.17) and (3.18), one can calculate the consecutive outputs of the dynamic part of the model ($s = 1, \dots, n_{\mathrm{v}}$)

$$v_s(k) = \sum_{n=1}^{n_{\mathrm{u}}}\sum_{l=1}^{n_{\mathrm{B}}} b_l^{s,n} u_n(k-l) + \sum_{n=1}^{n_{\mathrm{h}}}\sum_{l=1}^{n_{\mathrm{B}}} b_l^{s,n_{\mathrm{u}}+n} h_n(k-l) - \sum_{l=1}^{n_{\mathrm{A}}} a_l^s v_s(k-l) \tag{3.19}$$

The nonlinear steady-state parts are described by

$$y(k) = g(v(k))$$

where the steady-state parts are characterised by the function $g\colon \mathbb{R}^{n_{\mathrm{v}}} \to \mathbb{R}^{n_{\mathrm{y}}}$. In the Wiener model shown in Fig. 3.4b n_{y} DLP neural networks are used as the steady-state parts. The structure of such networks is depicted in

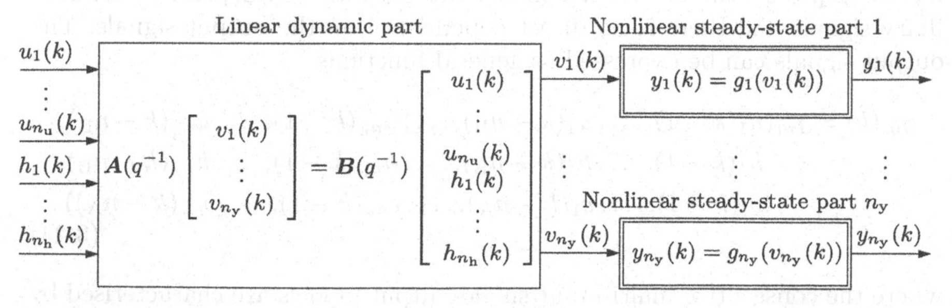

Fig. 3.5. The multi-input multi-output Wiener structure with n_{y} simplified steady-state parts

Fig. 3.6. Each network has n_{v} inputs $(v_1, \ldots, v_{n_{\mathrm{v}}})$, K^m hidden nodes with the nonlinear transfer function $\varphi\colon \mathbb{R} \rightarrow \mathbb{R}$ and one output $y_m(k)$. The output signals of the consecutive networks $(m = 1, \ldots, n_{\mathrm{y}})$ are

$$y_m(k) = w_0^{2,m} + \sum_{i=1}^{K^m} w_i^{2,m} \varphi(z_i^m(k)) \tag{3.20}$$

where

$$z_i^m(k) = w_{i,0}^{1,m} + \sum_{s=1}^{n_{\mathrm{v}}} w_{i,s}^{1,m} v_s(k) \tag{3.21}$$

is the sum of the input signals connected to the ith hidden node $(i = 1, \ldots, K^m)$. The weights of the first layer are denoted by $w_{i,j}^{1,m}$, where $i = 1, \ldots, K^m$, $j = 0, \ldots, n_{\mathrm{v}}$, $m = 1, \ldots, n_{\mathrm{y}}$, the weights of the second layer are denoted by $w_i^{2,m}$, where $i = 0, \ldots, K^m$, $m = 1, \ldots, n_{\mathrm{y}}$. From Eqs. (3.20) and (3.21) one obtains

$$y_m(k) = w_0^{2,m} + \sum_{i=1}^{K^m} w_i^{2,m} \varphi\left(w_{i,0}^{1,m} + \sum_{s=1}^{n_{\mathrm{v}}} w_{i,s}^{1,m} v_s(k) \right) \tag{3.22}$$

Using Eqs. (3.19) and (3.22), one has

$$\begin{aligned} y_m(k) = w_0^{2,m} + \sum_{i=1}^{K^m} w_i^{2,m} \varphi\Bigg(w_{i,0}^{1,m} + \sum_{s=1}^{n_{\mathrm{v}}} w_{i,s}^{1,m} \Bigg[& \sum_{n=1}^{n_{\mathrm{u}}} \sum_{l=1}^{n_{\mathrm{B}}} b_l^{s,n} u_n(k-l) \\ & + \sum_{n=1}^{n_{\mathrm{h}}} \sum_{l=1}^{n_{\mathrm{B}}} b_l^{s,n_{\mathrm{u}}+n} h_n(k-l) \\ & - \sum_{l=1}^{n_{\mathrm{A}}} a_l^s v_s(k-l) \Bigg] \Bigg) \end{aligned} \tag{3.23}$$

The current outputs of the Wiener model depend on the input and measured disturbance signals from previous sampling instants. Unlike the Hammerstein model (Eq.(3.9)), in the Wiener model the outputs also depend on all auxiliary signals and there is no direct dependence on the output signals. The output signals can be expressed as general functions

$$\begin{aligned} y_m(k) = f_m(& u_1(k-1), \ldots, u_1(k-n_{\mathrm{B}}), \ldots, u_{n_{\mathrm{u}}}(k-1), \ldots, u_{n_{\mathrm{u}}}(k-n_{\mathrm{B}}), \\ & h_1(k-1), \ldots, h_1(k-n_{\mathrm{B}}), \ldots, h_{n_{\mathrm{h}}}(k-1), \ldots, h_{n_{\mathrm{h}}}(k-n_{\mathrm{B}}), \\ & v_1(k-1), \ldots, v_1(k-n_{\mathrm{A}}), \ldots, v_{n_{\mathrm{v}}}(k-1), \ldots, v_{n_{\mathrm{v}}}(k-n_{\mathrm{A}})) \end{aligned} \tag{3.24}$$

where the consecutive multi-input single-output models are characterised by the functions $f_m\colon \mathbb{R}^{n_{\mathrm{v}} n_{\mathrm{A}} + (n_{\mathrm{u}} + n_{\mathrm{h}}) n_{\mathrm{B}}} \rightarrow \mathbb{R}$, $m = 1, \ldots, n_{\mathrm{y}}$.

For the neural Wiener model with n_{y} simplified steady-state parts shown in Fig. 3.5, from Eq. (3.23), one has

$$y_m(k) = w_0^{2,m} + \sum_{i=1}^{K^m} w_i^{2,m} \varphi\left(w_{i,0}^{1,m} + w_{i,1}^{1,m} \left[\sum_{n=1}^{n_{\mathrm{u}}} \sum_{l=1}^{n_{\mathrm{B}}} b_l^{m,n} u_n(k-l) + \sum_{n=1}^{n_{\mathrm{h}}} \sum_{l=1}^{n_{\mathrm{B}}} b_l^{m,n_{\mathrm{u}}+n} h_n(k-l) - \sum_{l=1}^{n_{\mathrm{A}}} a_l^m v_m(k-l) \right] \right) \quad (3.25)$$

Similarly to Eq. (3.24), the consecutive output signals are functions of input and measured disturbance signals from previous sampling instants and they depend in a simplified way on some auxiliary variables

$$\begin{aligned} y_m(k) = f_m(&u_1(k-1), \ldots, u_1(k-n_{\mathrm{B}}), \ldots, u_{n_{\mathrm{u}}}(k-1), \ldots, u_{n_{\mathrm{u}}}(k-n_{\mathrm{B}}), \\ &h_1(k-1), \ldots, h_1(k-n_{\mathrm{B}}), \ldots, h_{n_{\mathrm{h}}}(k-1), \ldots, h_{n_{\mathrm{h}}}(k-n_{\mathrm{B}}), \\ &v_m(k-1), \ldots, v_m(k-n_{\mathrm{A}})) \end{aligned} \quad (3.26)$$

where $f_m \colon \mathbb{R}^{n_{\mathrm{A}}+(n_{\mathrm{u}}+n_{\mathrm{h}})n_{\mathrm{B}}} \to \mathbb{R}$, $m = 1, \ldots, n_{\mathrm{y}}$.

Considering Hammerstein and Wiener models (Eqs. (3.9) and (3.23) or (3.25)), the impact of the order in which two models parts are connected is clear. In the Hammerstein structure nonlinearity is concerned only with the inputs and measured disturbances, there is a linear relation between the current outputs and the output signals from some previous sampling instants. In the Wiener structure the current outputs are nonlinear functions of the

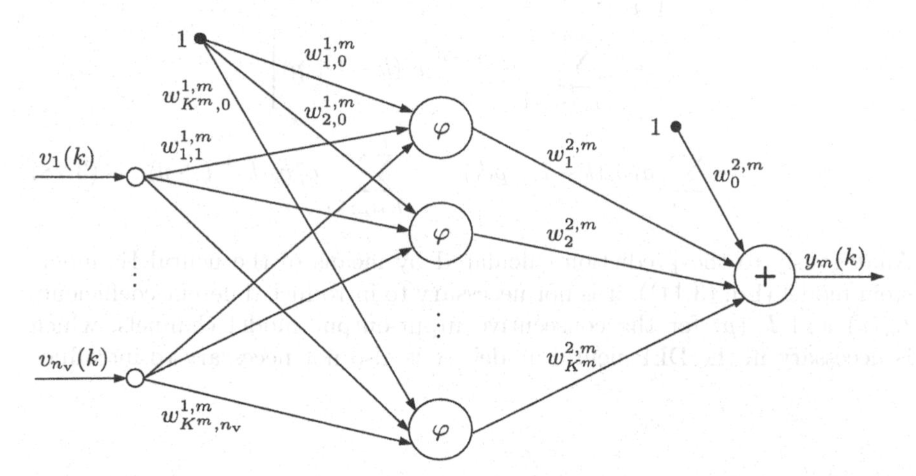

Fig. 3.6. The structure of the neural network used in the nonlinear part of the multi-input multi-output Wiener model with n_{y} steady-state parts, $m = 1, \ldots, n_{\mathrm{y}}$

inputs and the auxiliary signals, no dependence on previous output values exists. Intuitively, the Wiener model is "more nonlinear", it is expected to approximate behaviour of nonlinear processes in a better way. That is why the Wiener model is much more frequently used in practice, including its applications in MPC algorithms.

3.3.2 MPC-NO Algorithm

The general formulation of the MPC-NO algorithm discussed in Chapter 2.2 is universal. For the particular structure of the Wiener model, the predicted output trajectory and the derivatives of that trajectory with respect to the future control sequence must be derived.

The Wiener model with n_{y} separate steady-state nonlinear parts shown in Fig. 3.4b is used in the MPC-NO algorithm. From the general prediction equation (1.14) and Eq. (3.22), the predicted output trajectory for consecutive outputs ($m = 1, \ldots, n_{\mathrm{y}}$) is

$$\hat{y}_m(k+p|k) = w_0^{2,m} + \sum_{i=1}^{K^m} w_i^{2,m} \varphi\left(w_{i,0}^{1,m} + \sum_{s=1}^{n_{\mathrm{v}}} w_{i,s}^{1,m} v_s(k+p|k)\right) + d_m(k) \tag{3.27}$$

where, from Eq. (3.19), one has

$$\begin{aligned}
v_s(k+p|k) = &\sum_{n=1}^{n_{\mathrm{u}}} \left[\sum_{l=1}^{I_{\mathrm{uf}}(p)} b_l^{s,n} u_n(k-l+p|k) + \sum_{l=I_{\mathrm{uf}}(p)+1}^{n_{\mathrm{B}}} b_l^{s,n} u_n(k-l+p) \right] \\
&+ \sum_{n=1}^{n_{\mathrm{h}}} \left[\sum_{l=1}^{I_{\mathrm{uf}}(p)} b_l^{s,n_{\mathrm{u}}+n} h_n(k-l+p|k) \right. \\
&\qquad \left. + \sum_{l=I_{\mathrm{uf}}(p)+1}^{n_{\mathrm{B}}} b_l^{s,n_{\mathrm{u}}+n} h_n(k-l+p) \right] \\
&- \sum_{l=1}^{I_{\mathrm{yf}}(p)} a_l^s v_s(k-l+p|k) - \sum_{l=I_{\mathrm{yf}}(p)+1}^{n_{\mathrm{A}}} a_l^s v_s(k-l+p)
\end{aligned} \tag{3.28}$$

Analogously to the prediction calculated by means of the neural Hammerstein model (Eq. (3.11)), it is not necessary to introduce different coefficients $I_{\mathrm{uf}}(p)$ and $I_{\mathrm{yf}}(p)$ for the consecutive input-output model channels, which is necessary in the DLP neural model. It is also not necessary to introduce the additional symbol $I_{\mathrm{hf}}(p)$. Using Eqs. (1.15) and (3.23), the unmeasured disturbances are estimated from

$$d_m(k) = y_m(k) - w_0^{2,m} - \sum_{i=1}^{K^m} w_i^{2,m} \varphi\left(w_{i,0}^{1,m} + \sum_{s=1}^{n_{\mathrm{v}}} w_{i,s}^{1,m} \left[\sum_{n=1}^{n_{\mathrm{u}}} \sum_{l=1}^{n_{\mathrm{B}}} b_l^{s,n} u_n(k-l) + \sum_{n=1}^{n_{\mathrm{h}}} \sum_{l=1}^{n_{\mathrm{B}}} b_l^{s,n_{\mathrm{u}}+n} h_n(k-l) - \sum_{l=1}^{n_{\mathrm{A}}} a_l^s v_s(k-l) \right] \right)$$

Differentiating Eq. (3.27), one obtains the entries of the matrix (2.16)

$$\frac{\partial \hat{y}_m(k+p|k)}{\partial u_n(k+r|k)} = \sum_{i=1}^{K^m} w_i^{2,m} \frac{\mathrm{d}\varphi(z_i^m(k+p|k))}{\mathrm{d}z_i^m(k+p|k)} \sum_{s=1}^{n_{\mathrm{v}}} w_{i,s}^{1,m} \frac{\partial v_s(k+p|k)}{\partial u_n(k+r|k)}$$

where, from Eq. (3.21), one has $z_i^m(k+p|k) = w_{i,0}^{1,m} + \sum_{s=1}^{n_{\mathrm{v}}} w_{i,s}^{1,m} v_s(k+p|k)$, and differentiating Eq. (3.28), one obtains

$$\frac{\partial v_s(k+p|k)}{\partial u_n(k+r|k)} = \sum_{l=1}^{I_{\mathrm{uf}}(p)} b_l^{s,n} \frac{\partial u_n(k-l+p|k))}{\partial u_n(k+r|k)} - \sum_{l=1}^{I_{\mathrm{yf}}(p)} a_l^s \frac{\partial v_s(k-l+p|k)}{\partial u_n(k+r|k)} \tag{3.29}$$

3.3.3 *Suboptimal MPC Algorithms with On-Line Linearisation*

The MPC-NPLPT algorithm with linearisation of the predicted output trajectory along the future input trajectory can be easily developed for the Wiener model. Unfortunately, MPC-SL and MPC-NPL algorithms with model linearisation require the inverse of the steady-state part. That is why the general neural Wiener model with n_{y} steady-state multi-input single-output parts shown in Fig. 3.4b is used in the MPC-NPLPT algorithm whereas the Wiener model with n_{y} simplified single-input single-output steady-state parts shown in Fig. 3.5 is used in the MPC-NPL algorithm. The same neural Wiener model is also used in the MPC algorithms with simplified linearisation (MPC-SSL and MPC-NPSL).

MPC-NPLPT Algorithm

The general formulation of the MPC-NPLPT algorithm discussed in Chapter 2.3.4 holds true, but the specific structure of the neural Wiener model must be taken into account during calculation of the predicted output trajectory and its derivatives with respect to the future control signals.

For the neural Wiener model with n_{y} steady-state parts shown in Fig. 3.4b, using Eq. (3.27), the predicted output trajectory for the tth internal iteration is

$$\hat{y}_m^t(k+p|k) = w_0^{2,m} + \sum_{i=1}^{K^m} w_i^{2,m} \varphi\left(w_{i,0}^{1,m} + \sum_{s=1}^{n_{\mathrm{v}}} w_{i,s}^{1,m} v_s^t(k+p|k)\right) + d_m(k) \tag{3.30}$$

where, from Eq. (3.28), one has

$$\begin{aligned} v_s^t(k+p|k) = &\sum_{n=1}^{n_{\mathrm{u}}} \left[\sum_{l=1}^{I_{\mathrm{uf}}(p)} b_l^{s,n} u_n^t(k-l+p|k) + \sum_{l=I_{\mathrm{uf}}(p)+1}^{n_{\mathrm{B}}} b_l^{s,n} u_n(k-l+p)\right] \\ &+ \sum_{n=1}^{n_{\mathrm{h}}} \left[\sum_{l=1}^{I_{\mathrm{uf}}(p)} b_l^{s,n_{\mathrm{u}}+n} h_n(k-l+p|k) \right. \\ &\left. + \sum_{l=I_{\mathrm{uf}}(p)+1}^{n_{\mathrm{B}}} b_l^{s,n_{\mathrm{u}}+n} h_n(k-l+p)\right] \\ &- \sum_{l=1}^{I_{\mathrm{yf}}(p)} a_l^s v_s^t(k-l+p|k) - \sum_{l=I_{\mathrm{yf}}(p)+1}^{n_{\mathrm{A}}} a_l^s v_s(k-l+p) \end{aligned} \tag{3.31}$$

The entries of the matrix (2.53), i.e. the derivatives of the predicted output trajectory calculated for the previous internal iteration $(t-1)$ with respect to the future control signals calculated in the same internal iteration, are obtained by differentiating Eq. (3.30) and by replacing the index t by $t-1$

$$\frac{\partial \hat{y}_m^{t-1}(k+p|k)}{\partial u_n^{t-1}(k+r|k)} = \sum_{i=1}^{K^m} w_i^{2,m} \frac{\mathrm{d}\varphi(z_i^{t-1}(k+p|k))}{\mathrm{d}z_i^{t-1}(k+p|k)} \sum_{s=1}^{n_{\mathrm{v}}} w_{i,s}^{1,m} \frac{\partial v_s^{t-1}(k+p|k)}{\partial u_n^{t-1}(k+r|k)}$$

Differentiating Eq. (3.31), one obtains

$$\frac{\partial v_s^{t-1}(k+p|k)}{\partial u_n^{t-1}(k+r|k)} = \sum_{l=1}^{I_{\mathrm{uf}}(p)} b_l^{s,n} \frac{\partial u_n^{t-1}(k-l+p|k)}{\partial u_n^{t-1}(k+r|k)} - \sum_{l=1}^{I_{\mathrm{yf}}(p)} a_l^m \frac{\partial v_s^{t-1}(k-l+p|k)}{\partial u_n^{t-1}(k+r|k)}$$

MPC-NPL Algorithm

For the DLP neural models and for the Hammerstein systems it is possible to express the output signals for the current sampling instant k as some nonlinear functions of previous input, disturbance and output signals (Eqs. (2.1) and (3.10)). For the single-input single-output process (without any measured disturbance) the mentioned models have the form

$$y(k) = f(\boldsymbol{x}(k)) = f(u(k-\tau), \ldots, u(k-n_{\mathrm{B}}), y(k-1), \ldots, y(k-n_{\mathrm{A}}))$$

A linear approximation of such a model for the current operating point is

$$y(k) = f(\bar{\boldsymbol{x}}(k)) + \sum_{l=\tau}^{n_{\mathrm{B}}} b_l(k)(u(k-l) - \bar{u}(k-l)) - \sum_{l=1}^{n_{\mathrm{A}}} a_l(k)(y(k-l) - \bar{y}(k-l))$$

where the signals $\bar{u}(k-l)$ and $\bar{y}(k-l)$ are recorded in previous sampling instants and

$$a_l(k) = -\left.\frac{\partial f(\boldsymbol{x}(k))}{\partial y(k-l)}\right|_{\boldsymbol{x}(k)=\bar{\boldsymbol{x}}(k)}, \quad b_l(k) = \left.\frac{\partial f(\boldsymbol{x}(k))}{\partial u(k-l)}\right|_{\boldsymbol{x}(k)=\bar{\boldsymbol{x}}(k)}$$

In the Wiener model, because the steady-state nonlinear part follows the linear dynamic one, the auxiliary signals, not the output ones, are argument of the model. From Eq. (3.24), one has

$$y(k) = f(u(k-\tau), \ldots, u(k-n_{\mathrm{B}}), v(k-1), \ldots, v(k-n_{\mathrm{A}})) \tag{3.32}$$

In order to develop the MPC-SL or MPC-NPL algorithms for the Wiener model it is necessary to use the inverse model of the steady-state part

$$v(k) = \tilde{g}(y(k))$$

Using the Wiener model (3.32) and the inverse model, one has

$$y(k) = f(\boldsymbol{x}(k)) = f(u(k-\tau), \ldots, u(k-n_{\mathrm{B}}), \tilde{g}(y(k-1)), \ldots, \tilde{g}(y(k-n_{\mathrm{A}}))$$

The obtained relation is used for linearisation. Although the coefficients $b_l(k)$ of the linearised model are calculated in the same way it is done for the DLP neural model and the Hammerstein one, but coefficients

$$\begin{aligned} a_l(k) &= -\left.\frac{\partial f(\boldsymbol{x}(k))}{\partial y(k-l)}\right|_{\boldsymbol{x}(k)=\bar{\boldsymbol{x}}(k)} \\ &= -\left.\frac{\partial f(\boldsymbol{x}(k))}{\partial \tilde{g}(y(k-l))}\right|_{\boldsymbol{x}(k)=\bar{\boldsymbol{x}}(k)} \left.\frac{\partial \tilde{g}(y(k-l))}{\partial y(k-l)}\right|_{y(k-l)=\bar{y}(k-l)} \end{aligned}$$

depend not only on parameters of the Wiener model, but also on parameters of the inverse of its steady-state part.

In the MPC-SL and MPC-NPL algorithms for a multi-input multi-output process it is convenient to use the Wiener structure with n_{y} simplified steady-state parts with one input and one output shown in Fig. 3.5. From Eq. (3.26) and using n_{y} inverse models of the steady-state parts

$$\begin{aligned} v_1(k) &= \tilde{g}_1(y_1(k)) \\ &\vdots \\ v_{n_{\mathrm{y}}}(k) &= \tilde{g}_{n_{\mathrm{y}}}(y_{n_{\mathrm{y}}}(k)) \end{aligned}$$

the consecutive outputs ($m = 1, \ldots, n_{\mathrm{y}}$) of the model can be expressed as the functions

$$y_m(k) = f_m(u_1(k-1), \ldots, u_1(k-n_{\mathrm{B}}), \ldots, u_{n_{\mathrm{u}}}(k-1), \ldots, u_{n_{\mathrm{u}}}(k-n_{\mathrm{B}}), \\ h_1(k-1), \ldots, h_1(k-n_{\mathrm{B}}), \ldots, h_{n_{\mathrm{h}}}(k-1), \ldots, h_{n_{\mathrm{h}}}(k-n_{\mathrm{B}}), \\ \tilde{g}_m(y_m(k-1)), \ldots, \tilde{g}_m(y_m(k-n_{\mathrm{A}})))$$

As the inverse models n_{y} neural networks of the structure shown in Fig. 3.7 are used. Each network has one input ($y_m(k)$), one output ($v_m(k)$) and $\widetilde{K}^m$ hidden nodes, where $m = 1, \ldots, n_{\mathrm{y}}$. The inverse models are described by

$$v_m(k) = \tilde{g}_m(y_m(k)) = \widetilde{w}_0^{2,m} + \sum_{i=1}^{\widetilde{K}^m} \widetilde{w}_i^{2,m} \varphi(\widetilde{w}_{i,0}^{1,m} + \widetilde{w}_{i,1}^{1,m} y_m(k)) \tag{3.33}$$

The weights of the first layer are denoted by $\widetilde{w}_{i,j}^{1,m}$, where $m = 1, \ldots, n_{\mathrm{y}}$, $i = 1, \ldots, \widetilde{K}^m$, $j = 0, 1$, the weights of the second layer are denoted by $\widetilde{w}_i^{2,m}$, where $m = 1, \ldots, n_{\mathrm{y}}$, $i = 0, \ldots, \widetilde{K}^m$. Using the inverse models and Eq. (3.25), one obtains

$$\begin{aligned} y_m(k) = {} & w_0^{2,m} \\ & + \sum_{i=1}^{K^m} w_i^{2,m} \varphi\Bigg(w_{i,0}^{1,m} + w_{i,1}^{1,m} \Bigg[\sum_{n=1}^{n_{\mathrm{u}}} \sum_{l=1}^{n_{\mathrm{B}}} b_l^{m,n} u_n(k-l) \\ & \qquad + \sum_{n=1}^{n_{\mathrm{h}}} \sum_{l=1}^{n_{\mathrm{B}}} b_l^{m,n_{\mathrm{u}}+n} h_n(k-l) \\ & \qquad - \sum_{l=1}^{n_{\mathrm{A}}} a_l^m \left\{ \widetilde{w}_0^{2,m} + \sum_{i=1}^{\widetilde{K}^m} \widetilde{w}_i^{2,m} \varphi(\widetilde{w}_{i,0}^{1,m} + \widetilde{w}_{i,1}^{1,m} y_m(k-l)) \right\} \Bigg] \Bigg) \end{aligned} \tag{3.34}$$

Using Eqs. (2.27) and (2.28) and differentiating Eq. (3.34), it is possible to find analytically the coefficients of the linearised model

$$a_l^m(k) = \sum_{i=1}^{K^m} w_i^{2,m} \frac{\mathrm{d}\varphi(z_i^m(k-l))}{\mathrm{d}z_i^m(k-l)} w_{i,1}^{1,m} a_l^m \sum_{i=1}^{\widetilde{K}^m} \widetilde{w}_i^{2,m} \frac{\mathrm{d}\varphi(\tilde{z}_i^m(k-l))}{\mathrm{d}\tilde{z}_i^m(k-l)} \widetilde{w}_{i,1}^{1,m}$$

for $m = 1, \ldots, n_{\mathrm{y}}$ and

$$b_l^{m,n}(k) = \sum_{i=1}^{K^m} w_i^{2,m} \frac{\mathrm{d}\varphi(z_i^m(k-l))}{\mathrm{d}z_i^m(k-l)} w_{i,1}^{1,m} b_l^{m,n}$$

for all $l = 1, \ldots, n_{\mathrm{B}}$, $m = 1, \ldots, n_{\mathrm{y}}$, $n = 1, \ldots, n_{\mathrm{u}}$. Using Eq. (3.21) and remembering that the steady-state parts have one input and one output, one has $z_i^m(k-l) = w_{i,0}^{1,m} + w_{i,1}^{1,m} v_m(k-l)$, whereas using Eq. (3.33), one obtains $\tilde{z}_i^m(k-l) = \widetilde{w}_{i,0}^{1,m} + \widetilde{w}_{i,1}^{1,m} y_m(k-l)$.

The nonlinear free trajectory is calculated directly from the neural Wiener model, in a very similar way it is done in the MPC-NO algorithm, but it is necessary to remember that the steady-state parts of the model have only one input and one output. From Eq. (3.27) one has

$$y_m^0(k+p|k) = w_0^{2,m} + \sum_{i=1}^{K^m} w_i^{2,m} \varphi\left(w_{i,0}^{1,m} + w_{i,1}^{1,m} v_m^0(k+p|k)\right) + d_m(k) \quad (3.35)$$

where, using Eq. (3.28), one obtains

$$\begin{aligned} v_m^0(k+p|k) = &\sum_{n=1}^{n_u}\left[\sum_{l=1}^{I_{uf}(p)} b_l^{m,n} u_n(k-1) + \sum_{l=I_{uf}(p)+1}^{n_B} b_l^{m,n} u_n(k-l+p)\right] \\ &+ \sum_{n=1}^{n_h}\left[\sum_{l=1}^{I_{uf}(p)} b_l^{m,n_u+n} h_n(k-l+p|k) \right.\\ &\left. + \sum_{l=I_{uf}(p)+1}^{n_B} b_l^{m,n_u+n} h_n(k-l+p)\right] \\ &- \sum_{l=1}^{I_{yf}(p)} a_l^m v_m^0(k-l+p|k) - \sum_{l=I_{yf}(p)+1}^{n_A} a_l^m v_m(k-l+p) \end{aligned} \quad (3.36)$$

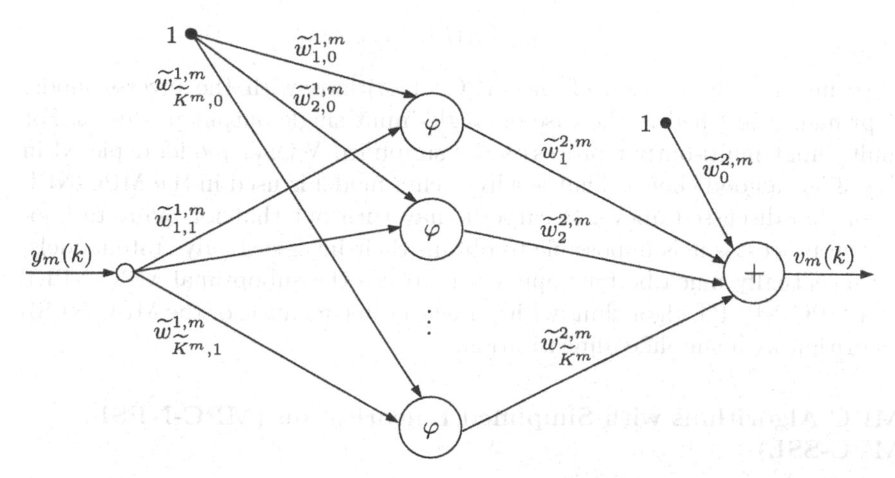

Fig. 3.7. The structure of the neural network used in the inverse model of the steady-state nonlinear part of the multi-input multi-output Wiener model with n_y simplified steady-state parts, $m = 1, \ldots, n_y$

From Eqs. (1.15) and (3.25), the unmeasured disturbances are estimated as

$$\begin{aligned} d_m(k) = y_m(k) \\ & - w_0^{2,m} - \sum_{i=1}^{K^m} w_i^{2,m} \varphi \Bigg(w_{i,0}^{1,m} + w_{i,1}^{1,m} \Bigg[\sum_{n=1}^{n_\mathrm{u}} \sum_{l=1}^{n_\mathrm{B}} b_l^{m,n} u_n(k-l) \\ & + \sum_{n=1}^{n_\mathrm{h}} \sum_{l=1}^{n_\mathrm{B}} b_l^{m,n_\mathrm{u}+n} h_n(k-l) \\ & - \sum_{l=1}^{n_\mathrm{A}} a_l^m v_m(k-l) \Bigg] \Bigg) \end{aligned}$$

There are three important disadvantages of using the inverse model in the MPC-NPL algorithm. Firstly, not all processes have the inverse model. Typically, technological processes are characterised by saturations. Secondly, it is necessary to point out that the Wiener system together with the inverse steady-state model is linearised. In consequence, control accuracy depends not only on the quality of the model, but also on the quality of the inverse one. Thirdly, the discussed approach has some important limitations when the Wiener system has a more complicated steady-state part. For the Wiener models depicted in Fig. 3.4, the inverse models must be

$$\begin{aligned} v_1(k) &= \tilde{g}_1(y_1(k), \ldots, y_{n_\mathrm{y}}(k)) \\ &\vdots \\ v_{n_\mathrm{y}}(k) &= \tilde{g}_{n_\mathrm{y}}(y_1(k), \ldots, y_{n_\mathrm{y}}(k)) \end{aligned}$$

It seems that application of the MPC algorithms with the inverse model is primarily justified in the case of single-input single-output processes. For multi-input multi-output processes the simplified Wiener model depicted in Fig. 3.5 is a good choice. That is why such a model is used in the MPC-NPL algorithm discussed above. Because it may turn out that for many technological processes it is impossible to obtain their inverse steady-state models, a conceptually much better approach is to use the suboptimal MPC-NPLT and MPC-NPLPT algorithms with trajectory linearisation or the MPC-NPSL algorithm with simplified linearisation.

MPC Algorithms with Simplified Linearisation (MPC-NPSL, MPC-SSL)

The MPC algorithms with simplified linearisation can be easily derived for the Wiener model. In the following text the neural Wiener model with n_y simplified steady-state parts shown in Fig. 3.5 is used. For the current operating conditions the steady-state part may be described by

$$y(k) = \boldsymbol{K}^\mathrm{W}(k) v(k) \tag{3.37}$$

Taking into account the structure of the model, the gain matrix is diagonal

$$\boldsymbol{K}^{\mathrm{W}}(k)=\begin{bmatrix}k_{1,1}^{\mathrm{W}}(k) & \ldots & 0\\ \vdots & \ddots & \vdots\\ 0 & \ldots & k_{n_{\mathrm{y}},n_{\mathrm{y}}}^{\mathrm{W}}(k)\end{bmatrix}=\begin{bmatrix}\frac{\partial y_1(k)}{\partial v_1(k)} & \ldots & 0\\ \vdots & \ddots & \vdots\\ 0 & \ldots & \frac{\partial y_{n_{\mathrm{y}}}(k)}{\partial v_{n_{\mathrm{y}}}(k)}\end{bmatrix}$$

From Eq. (3.37) it follows that $v(k)=(\boldsymbol{K}^{\mathrm{W}})^{-1}(k)y(k)$. Using the description of the linear part of the model given by Eq. (3.17), one has

$$\boldsymbol{A}(q^{-1})(\boldsymbol{K}^{\mathrm{W}})^{-1}(k)y(k)=\boldsymbol{B}(q^{-1})\begin{bmatrix}u(k)\\ h(k)\end{bmatrix} \tag{3.38}$$

Because the gain matrix $\boldsymbol{K}^{\mathrm{W}}(k)$ is diagonal, one obtains

$$\boldsymbol{A}(q^{-1})y(k)=\boldsymbol{K}^{\mathrm{W}}(k)\boldsymbol{B}(q^{-1})\begin{bmatrix}u(k)\\ h(k)\end{bmatrix} \tag{3.39}$$

Taking into account the structure of the polynomial matrices $\boldsymbol{A}(q^{-1})$ and $\boldsymbol{B}(q^{-1})$ defined by Eqs. (3.18), it follows that

$$\begin{bmatrix}A_{1,1}(q^{-1})y_1(k)\\ \vdots\\ A_{n_{\mathrm{y}},n_{\mathrm{y}}}(q^{-1})y_{n_{\mathrm{y}}}(k)\end{bmatrix}=$$

$$\begin{bmatrix}B_{1,1}(q^{-1})k_{1,1}^{\mathrm{W}}(k) & \ldots & B_{1,n_{\mathrm{u}}+n_{\mathrm{h}}}(q^{-1})k_{1,1}^{\mathrm{W}}(k)\\ \vdots & \ddots & \vdots\\ B_{n_{\mathrm{y}},1}(q^{-1})k_{n_{\mathrm{y}},n_{\mathrm{y}}}^{\mathrm{W}}(k) & \ldots & B_{n_{\mathrm{y}},n_{\mathrm{u}}+n_{\mathrm{h}}}(q^{-1})k_{n_{\mathrm{y}},n_{\mathrm{y}}}^{\mathrm{W}}(k)\end{bmatrix}\begin{bmatrix}u_1(k)\\ \vdots\\ u_{n_{\mathrm{u}}}(k)\\ h_1(k)\\ \vdots\\ h_{n_{\mathrm{h}}}(k)\end{bmatrix}$$

which leads to the coefficients of the linearised model

$$a_l^m(k)=a_l^m,\ b_l^{m,n}(k)=b_l^{m,n}k_{m,m}^{\mathrm{W}}(k)$$

Taking into account the structure of the nonlinear neural Wiener model defined by Eqs. (3.20) and (3.21) and remembering that the steady-state parts of the models have one input and one output, the entries of the diagonal gain matrix $\boldsymbol{K}^{\mathrm{W}}(k)$ are

$$k_{m,m}^{\mathrm{W}}(k)=\sum_{i=1}^{K^m}w_i^{2,m}\frac{\mathrm{d}\varphi(z_i^m(k))}{\mathrm{d}z_i^m(k)}w_{i,1}^{1,m}$$

for $m=1,\ldots,n_{\mathrm{y}}$.

The step-response matrices (2.31) are calculated as the products of the gain matrix and of the constant step-response matrices of the linear part of the model (they are of dimensionality $n_{\mathrm{y}} \times n_{\mathrm{u}}$), in a similar way it is done in the case of the Hammerstein model (Eq. (3.16)

$$\boldsymbol{S}_p(k) = \boldsymbol{K}^{\mathrm{W}}(k)\boldsymbol{S}_p = \begin{bmatrix} s_p^{1,1}k_{1,1}^{\mathrm{W}}(k) & \dots & s_p^{1,n_{\mathrm{u}}}k_{1,1}^{\mathrm{W}}(k) \\ \vdots & \ddots & \vdots \\ s_p^{n_{\mathrm{y}},1}k_{n_{\mathrm{y}},n_{\mathrm{y}}}^{\mathrm{W}}(k) & \dots & s_p^{n_{\mathrm{y}},n_{\mathrm{y}}}k_{n_{\mathrm{y}},n_{\mathrm{y}}}^{\mathrm{W}}(k) \end{bmatrix}$$

The dynamic matrix of the linearised model of the structure (2.34) can be used in the MPC-SL and MPC-NPL algorithms. The nonlinear free trajectory is calculated from Eqs. (3.35) and (3.36), in the same way it is done in the MPC-NPL approach.

Although simplified linearisation of the Hammerstein model with different steady-state parts is quite straightforward, it is not true for the Wiener structure. Taking into account the general Wiener structures with one multi-input multi-output steady-state part or with n_{y} multi-input single output parts shown in Fig. 3.4 and the simplifying condition $n_{\mathrm{v}} = n_{\mathrm{y}}$, the gain matrix $\boldsymbol{K}^{\mathrm{W}}(k)$ is square, but it is not diagonal. In consequence, it is impossible to eliminate the inverse matrix $(\boldsymbol{K}^{\mathrm{W}})^{-1}(k)$, the linearised model cannot be transformed into the simple form (3.39), it is necessary to use Eq. (3.38). If $n_{\mathrm{v}} \neq n_{\mathrm{y}}$, the gain matrix $\boldsymbol{K}^{\mathrm{W}}(k)$ is not square. In such a case $v(k) = (\boldsymbol{K}^{\mathrm{W}})^{+}(k)y(k)$ (where $(\boldsymbol{K}^{\mathrm{W}})^{+}(k)$ denotes a pseudoinverse matrix), the coefficients of the linearised model and the dynamic matrix are found from the relation

$$\boldsymbol{A}(q^{-1})(\boldsymbol{K}^{\mathrm{W}})^{+}(k)y(k) = \boldsymbol{B}(q^{-1})\begin{bmatrix} u(k) \\ h(k) \end{bmatrix}$$

To summarise, there are some restrictions concerned with the application of Wiener models in the MPC algorithms with simplified linearisation when compared with the Hammerstein model case. It is straightforward to use such algorithms for single-input single-output processes and for multivariable processes described by simplified single-input single-output steady-state parts. For more complicated Wiener models it is necessary to use the algorithms with trajectory linearisation (the MPC-NPLT and MPC-NPLPT approaches).

3.4 Example 3.1

The process under consideration is a polymerisation reaction taking place in a jacketed continuous stirred tank reactor shown in Fig. 3.8. The reaction is the free-radical polymerisation of methyl methacrylate with azo-bis-isobutyronitrile as initiator and toluene as solvent. The Number Average Molecular Weight (NAMW) (kg/kmol) is controlled by adjusting the initiator flow rate F_{I} (m^3/h), changes of the monomer flow rate F (m^3/h) may be treated as disturbances of the process.

The continuous-time first-principle model of the reactor [65] is comprised of four nonlinear ordinary differential equations

$$\begin{aligned}
\frac{\mathrm{d}C_{\mathrm{m}}(t)}{\mathrm{d}t} &= -\left[Z_{\mathrm{P}} \exp\left(\frac{-E_{\mathrm{P}}}{RT}\right) + Z_{\mathrm{f_m}} \exp\left(\frac{-E_{\mathrm{f_m}}}{RT}\right)\right] C_{\mathrm{m}}(t) P_0(t) \\
&\quad - \frac{F(t)C_{\mathrm{m}}(t)}{V} + \frac{F(t)C_{\mathrm{m_{in}}}}{V} \\
\frac{\mathrm{d}C_{\mathrm{I}}(t)}{\mathrm{d}t} &= -Z_{\mathrm{I}} \exp\left(\frac{-E_{\mathrm{I}}}{RT}\right) C_{\mathrm{I}}(t) - \frac{F(t)C_{\mathrm{I}}(t)}{V} + \frac{F_{\mathrm{I}}(t)C_{\mathrm{I_{in}}}}{V} \\
\frac{\mathrm{d}D_0(t)}{\mathrm{d}t} &= \left[0.5Z_{\mathrm{T_c}} \exp\left(\frac{-E_{\mathrm{T_c}}}{RT}\right) + Z_{\mathrm{T_d}} \exp\left(\frac{-E_{\mathrm{T_d}}}{RT}\right)\right] P_0^2(t) \\
&\quad + Z_{\mathrm{f_m}} \exp\left(\frac{-E_{\mathrm{f_m}}}{RT}\right) C_{\mathrm{m}}(t) P_0(t) - \frac{F(t)D_0(t)}{V} \\
\frac{\mathrm{d}D_{\mathrm{I}}(t)}{\mathrm{d}t} &= M_{\mathrm{m}}\left[Z_{\mathrm{P}} \exp\left(\frac{-E_{\mathrm{P}}}{RT}\right) + Z_{\mathrm{f_m}} \exp\left(\frac{-E_{\mathrm{f_m}}}{RT}\right)\right] C_{\mathrm{m}}(t) P_0(t) - \frac{F(t)D_{\mathrm{I}}(t)}{V}
\end{aligned}$$

where

$$P_0(t) = \sqrt{\frac{2f^* C_{\mathrm{I}}(t) Z_{\mathrm{I}} \exp\left(\frac{-E_{\mathrm{I}}}{RT}\right)}{Z_{\mathrm{T_d}} \exp\left(\frac{-E_{\mathrm{T_d}}}{RT}\right) + Z_{\mathrm{T_c}} \exp\left(\frac{-E_{\mathrm{T_c}}}{RT}\right)}}$$

and one output equation

$$\mathrm{NAMW}(t) = \frac{D_{\mathrm{I}}(t)}{D_0(t)} \tag{3.40}$$

The parameters of the first-principle model are given in Table 3.1, the nominal operating point is given in Table. 3.2.

Both steady-state and dynamic properties of the process are nonlinear. Fig. 3.9 shows the steady-state process characteristics $\mathrm{NAMW}(F_{\mathrm{I}})$ and the dependence of the steady-state gain on the operating point. For the assumed

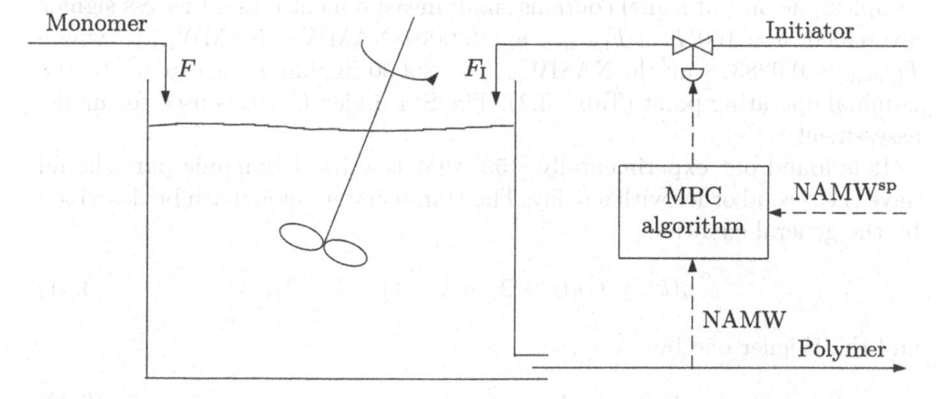

Fig. 3.8. The polymerisation reactor control system structure

Table 3.1. The parameters of the first-principle model of the polymerisation reactor

$C_{\mathrm{I_{in}}} = 8$ kmol/m^3	$R = 8.314$ kJ/(kmol K)
$C_{\mathrm{m_{in}}} = 6$ kmol/m^3	$T = 335$ K
$E_{\mathrm{T_c}} = 2.9442 \times 10^3$ kJ/kmol	$Z_{\mathrm{T_c}} = 3.8223 \times 10^{10}$ m^3/(kmol h)
$E_{\mathrm{T_d}} = 2.9442 \times 10^3$ kJ/kmol	$Z_{\mathrm{T_d}} = 3.1457 \times 10^{11}$ m^3/(kmol h)
$E_{\mathrm{f_m}} = 7.4478 \times 10^4$ kJ/kmol	$Z_{\mathrm{f_m}} = 1.0067 \times 10^{15}$ m^3/(kmol h)
$E_{\mathrm{I}} = 1.2550 \times 10^5$ kJ/kmol	$Z_{\mathrm{I}} = 3.7920 \times 10^{18}$ 1/h
$E_{\mathrm{P}} = 1.8283 \times 10^4$ kJ/kmol	$Z_{\mathrm{P}} = 1.7700 \times 10^9$ m^3/(kmol h)
$f^* = 0.58$	$V = 0.1$ m^3
$M_{\mathrm{m}} = 100.12$ kg/kmol	

Table 3.2. The nominal operating point of the polymerisation reactor

$C_{\mathrm{I}} = 2.2433 \times 10^{-1}$ kmol/m^3	$F = 1$ m^3/h
$C_{\mathrm{m}} = 5.3745$ kmol/m^3	$F_{\mathrm{I}} = 0.028328$ m^3/h
$D_0 = 3.1308 \times 10^{-3}$ kmol/m^3	NAMW = 20000 kg/kmol
$D_{\mathrm{I}} = 6.2616 \times 10^{-1}$ kmol/m^3	

range of the input signal the gain changes more than 37 times: from -4.31×10^6 to -1.16×10^5. The reactor is a popular benchmark process frequently used for comparing different nonlinear models and control algorithms [64, 65, 149, 177, 191, 316].

Polymerisation Reactor Modelling

The first-principle model is treated as the "real" process. The differential equations are solved by means of the Euler method (for the sampling period 108 seconds it practically gives the same results as the Runge-Kutta method of order 45). From open-loop simulations of the first-principle model the training, validation and test data sets are obtained. Each set has 2000 samples, the output signal contains small measurement noise. Process signals are scaled: $u = 100(F_{\mathrm{I}} - F_{\mathrm{I,nom}})$, $y = 0.0001(\mathrm{NAMW} - \mathrm{NAMW_{nom}})$, where $F_{\mathrm{I,nom}} = 0.028328$ m^3/h, $\mathrm{NAMW_{nom}} = 20000$ kg/kmol correspond to the nominal operating point (Table 3.2). The SSE index (1.20) is used for model assessment.

It is found out experimentally [153] that the linear dynamic part should have the second order with a delay. The Hammerstein model can be described by the general equation

$$y(k) = f(u(k-2), y(k-1), y(k-2)) \tag{3.41}$$

and the Wiener one by

$$y(k) = f(u(k-2), v(k-1), v(k-2)) \tag{3.42}$$

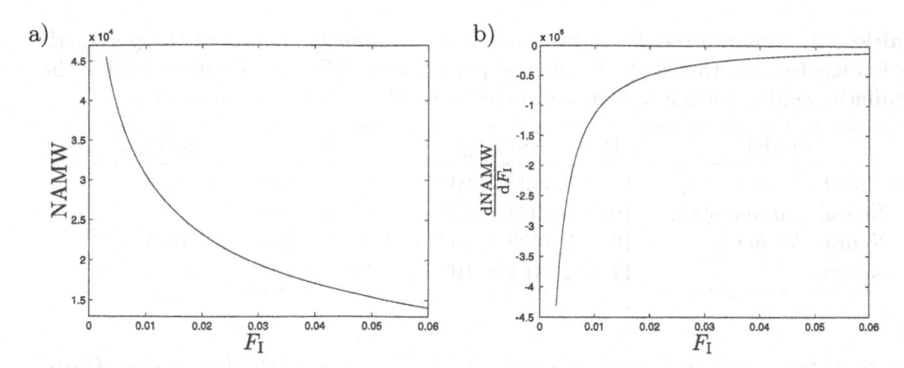

Fig. 3.9. a) The steady-state characteristics NAMW(F_{I}) of the polymerisation reactor, b) the dependence of the steady-state gain on the operating point

Table 3.3 shows the influence of the number of hidden nodes used in the nonlinear steady-state part on the number of parameters and accuracy of both model types. As a reasonable compromise between accuracy and complexity the models with the networks containing 2 hidden nodes are chosen. Although both models have the same number of parameters, the neural Wiener model is significantly better than the Hammerstein one (the validation error is approximately 76 times lower).

The linear model of the same order of dynamics

$$y(k) = b_2u(k-2) - a_1y(k-1) - a_2y(k-2) \tag{3.43}$$

is next found. A classical DLP (black-box) model is also found, it can be described by Eq. (3.41). The neural model has 2 hidden nodes. Table 3.4 compares the four discussed classes of models in terms of complexity and accuracy. It is interesting to notice that all three nonlinear models have a similar number of parameter. The linear model is the worst one, for the validation data set accuracy of the Hammerstein one is approximately 18 times worse than that of the DLP network and the neural Wiener model is more than 4 times better than the DLP network. The validation error for the Wiener

Table 3.3. Comparison of the neural Hammerstein and Wiener empirical models of the polymerisation reactor in terms of the number of parameters (NP) and accuracy (SSE)

K	NP	Hammerstein models		Wiener models	
		$\mathrm{SSE_{train}}$	$\mathrm{SSE_{val}}$	$\mathrm{SSE_{train}}$	$\mathrm{SSE_{val}}$
1	7	5.1852×10^1	8.8422×10^1	8.6410×10^0	1.4008×10^1
2	10	5.0033×10^1	8.1773×10^1	7.4406×10^{-1}	1.0652×10^0
3	13	5.0405×10^1	8.2356×10^1	6.6612×10^{-1}	1.0056×10^0
4	16	5.0414×10^1	8.2396×10^1	6.6628×10^{-1}	1.0095×10^0
5	19	5.0452×10^1	8.2433×10^1	6.6568×10^{-1}	1.0124×10^0

Table 3.4. Comparison of empirical linear and nonlinear models of the polymerisation reactor in terms of the number of parameters (NP) and accuracy (SSE); the nonlinear models have 2 hidden nodes and a similar number of parameters

Model	NP	SSE_{train}	SSE_{val}	SSE_{test}
Linear	3	3.4023×10^{2}	6.2558×10^{2}	–
Neural Hammerstein	10	5.0033×10^{1}	8.1773×10^{1}	–
Neural Wiener	10	7.4406×10^{-1}	1.0652×10^{0}	8.2166×10^{-1}
Neural	11	2.3161×10^{0}	4.5390×10^{0}	–

model is approximately 76 times lower in comparison with that for the Hammerstein structure. The straightforward choice is the neural Wiener model, for which the test error is also calculated. Its value is relatively low, comparable with those for the training and validation sets. Fig. 3.10 compares the validation data set and the outputs of the considered models, the differences between the data and model outputs are also shown.

Polymerisation Reactor Predictive Control

The following nonlinear MPC algorithms are used in the control system of the polymerisation reactor:

a) the linear MPC algorithm based on the linear model (3.43),
b) the nonlinear MPC-NO algorithm based on the classical DLP neural model (3.41) with 2 hidden nodes,
c) the nonlinear MPC-NO algorithm based on the neural Hammerstein model (3.41) with 2 hidden nodes in its steady-state part,
d) the nonlinear MPC-NO algorithm based on the neural Wiener model (3.42) with 2 hidden nodes in its steady-state part,
e) the nonlinear MPC-NPSL algorithms based on the same neural Wiener model,
f) the nonlinear MPC-NPL algorithm based on the same neural Wiener model, an additional DLP neural network with 2 hidden nodes is used as an inverse model of its steady-state part.

All algorithms have the same parameters: $N = 10$, $N_{\mathrm{u}} = 3$, $\lambda_p = 0.2$ for $p = 1, \ldots, N$. The same constraints are imposed on the value of the control signal: $F_{\mathrm{I}}^{\min} = 0.003\ \mathrm{m}^3/\mathrm{h}$, $F_{\mathrm{I}}^{\max} = 0.06\ \mathrm{m}^3/\mathrm{h}$.

Because the linear model is very inaccurate, the linear MPC algorithm does not work properly which is illustrated in Fig. 3.11. For the assumed set-point trajectory the algorithm works only for the first set-point change, for the remaining ones oscillations occur.

Table 3.5 compares control accuracy of all five considered nonlinear MPC algorithms in terms of the sum of squared errors over the simulation horizon (SSE_{sim}). Unlike the linear MPC algorithm, all nonlinear MPC ones work much better, there are no oscillations. The numerical results given in

Table 3.5 should be interpreted in two respects. Firstly, it is interesting to compare accuracy of the "ideal" MPC-NO algorithms based on three different nonlinear models (the DLP neural model, the neural Hammerstein model and the neural Wiener model are used). Because the neural Hammerstein structure is much worse than the two others, the resulting MPC-NO algorithms is characterised by the highest control error. Thanks to the negative

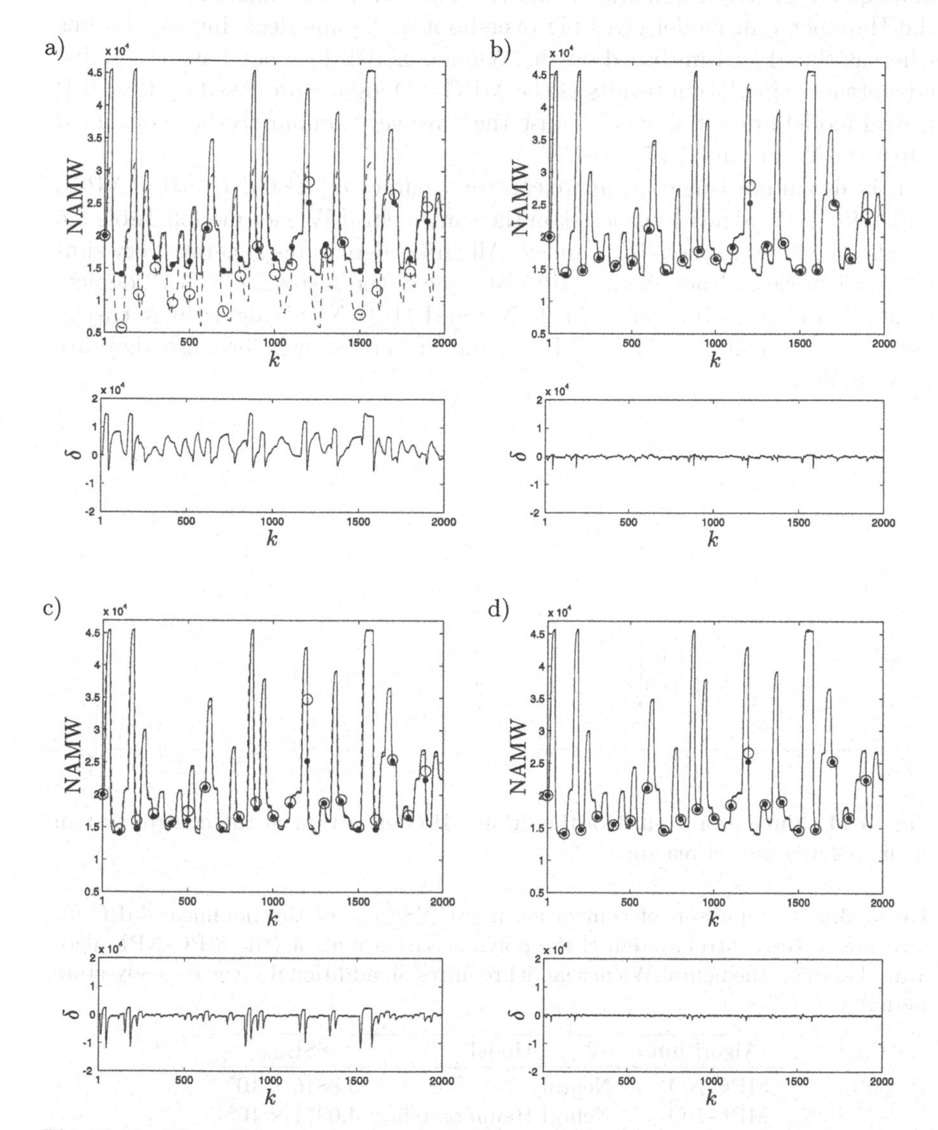

Fig. 3.10. The validation data set (*solid line with dots*) vs. the output of the model of the polymerisation reactor (*dashed line with circles*): a) the linear model, b) the classical black-box neural model, c) the neural Hammerstein model, d) the neural Wiener model; δ – differences between the data and model outputs

feedback the SSE_{sim} index is only by some 4% worse than in the case of the MPC-NO algorithm based on the neural Wiener model whereas the neural Hammerstein model itself is approximately 76 times worse than the neural Wiener one (Table 3.4). Fig. 3.12 shows simulation results of the MPC-NO algorithms based on the neural Hammerstein and Wiener models. The obtained trajectories, much better than the value of the errors SSE_{sim}, show inadequacy of the Hammerstein model. The MPC-NO algorithm based on the Hammerstein model gives big overshoot and some decaying oscillations whereas the algorithm based on the Wiener model does not have those disadvantages. Simulation results of the MPC-NO algorithm based on the DLP neural model are not shown because they are very similar to those obtained when the Wiener model is used.

It is also interesting to compare control accuracy of MPC-NO, MPC-NPSL and MPC-NPL algorithms based on the same neural Wiener model. Table 3.5 shows the values of the SSE_{sim} index. All three algorithms generate very similar trajectories, although the MPC-NO one is the fastest. Fig. 3.13 depicts simulation results obtained in MPC-NO and MPC-NPSL algorithms, trajectories obtained in the MPC-NPL scheme are not shown, because they are very similar.

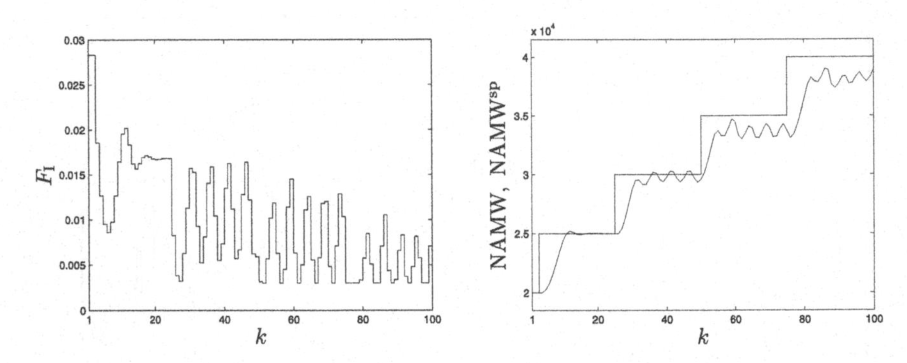

Fig. 3.11. Simulation results of the linear MPC algorithm in the control system of the polymerisation reactor

Table 3.5. Comparison of control accuracy (SSE_{sim}) of the nonlinear MPC algorithms in the control system of the polymerisation reactor (the MPC-NPL algorithm based on the neural Wiener model requires an additional inverse steady-state model)

Algorithm	Model	SSE_{sim}
MPC-NO	Neural	3.8816×10^8
MPC-NO	Neural Hammerstein	4.0871×10^8
MPC-NO	Neural Wiener	3.9334×10^8
MPC-NPL	Neural Wiener	3.9525×10^8
MPC-NPSL	Neural Wiener	3.9501×10^8

Table 3.6 shows computational burden of MPC-NO, MPC-NPL and MPC-NPSL algorithms based on the same neural Wiener model. Different control horizons are considered, the prediction horizon is $N = 10$. The MPC-NPSL algorithm with simplified linearisation is the best, when compared with the MPC-NO one, the computational complexity reduction factor is in the range 4.05–9.46. In the MPC-NPL algorithm not only the Wiener model, but also the inverse steady-state model is used, and linearisation is not performed in a simplified manner. As a result computational burden of the MPC-NPL algorithm is slightly higher than that of the MPC-NPSL approach, which is particularly noticeable for short control horizons (for longer horizons quadratic optimisation has a predominant influence on computational burden).

Because for the considered polymerisation reactor the suboptimal MPC-NPSL algorithm gives control accuracy very close to that obtained in the

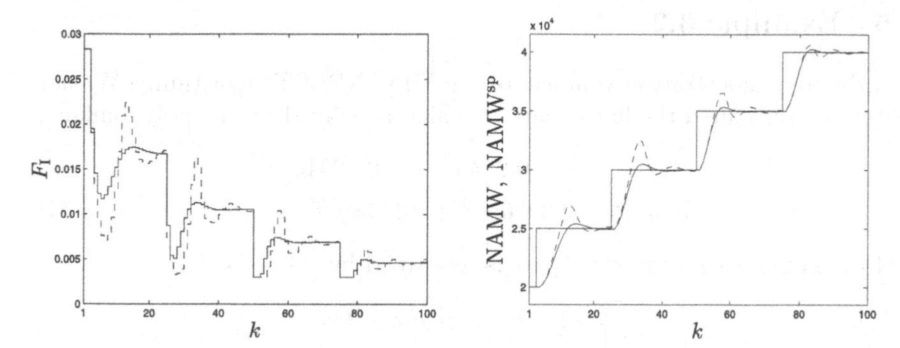

Fig. 3.12. Simulation results: the MPC-NO algorithm based on the neural Wiener model (*solid line*) and the MPC-NO algorithm based on the neural Hammerstein model (*dashed line*) in the control system of the polymerisation reactor

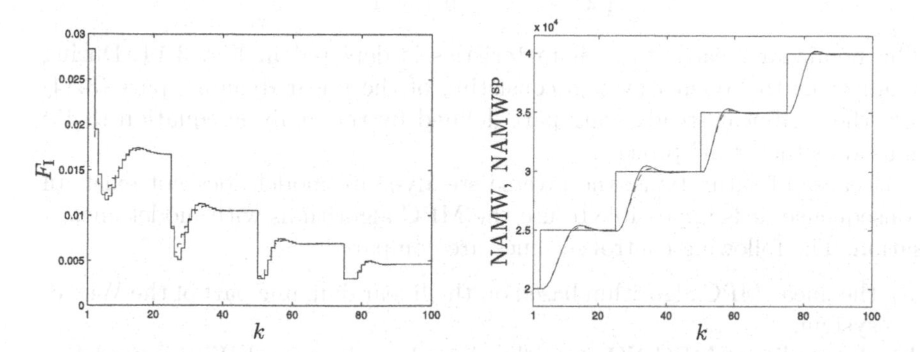

Fig. 3.13. Simulation results: the MPC-NO algorithm based on the neural Wiener model (*solid line*) and the MPC-NPSL algorithm based on the same model (*dashed line*) in the control system of the polymerisation reactor

Table 3.6. Comparison of computational complexity (in MFLOPS) of nonlinear MPC-NO, MPC-NPL and MPC-NPSL algorithms based on the neural Wiener model in the control system of the polymerisation reactor for different control horizons, $N = 10$ (the MPC-NPL algorithm requires an additional inverse steady-state model)

Algorithm	$N_\mathrm{u} = 1$	$N_\mathrm{u} = 2$	$N_\mathrm{u} = 3$	$N_\mathrm{u} = 4$	$N_\mathrm{u} = 5$	$N_\mathrm{u} = 10$
MPC-NO	0.66	1.10	1.52	2.04	2.54	7.85
MPC-NPL	0.08	0.12	0.19	0.28	0.42	1.76
MPC-NPSL	0.06	0.09	0.16	0.26	0.39	1.73

MPC-NO one, it is not necessary to use much more complicated MPC algorithm with trajectory linearisation.

3.5 Example 3.2

In order to demonstrate advantages of the MPC-NPLPT algorithm, a Wiener system is considered the linear part of which is defined by the polynomials

$$\begin{aligned} \boldsymbol{A}(q^{-1}) &= 1 - 0.9744q^{-1} + 0.2231q^{-2} \\ \boldsymbol{B}(q^{-1}) &= 0.3096q^{-1} + 0.1878q^{-2} \end{aligned} \tag{3.44}$$

and the nonlinear steady-state part is described by

$$y(v) = \begin{cases} -2 & \text{if } v < -11 \\ 0.1v - 0.9 & \text{if } -11 \leq v < -1 \\ v & \text{if } -1 \leq v < 1 \\ 0.1v + 0.9 & \text{if } 1 \leq v < 11 \\ 2 & \text{if } v \geq 11 \end{cases} \tag{3.45}$$

The nonlinear steady-state characteristics is depicted in Fig. 3.14. During simulations the Wiener system consisting of the linear dynamic part (3.44) and the nonlinear steady-state part defined by the analytic equation (3.45) is used as the "real" process.

Because of saturations, the inverse steady-state model does not exist. In consequence, it is impossible to use the MPC algorithms with model linearisation. The following control schemes are compared:

a) the linear MPC algorithm based on the linear dynamic part of the Wiener system,
b) the nonlinear MPC-NO algorithm based on the neural Wiener model,
c) the nonlinear MPC-NPSL algorithm based on the same model,
d) the nonlinear MPC-NPLPT algorithm based on the same model.

All three nonlinear MPC algorithms use the neural Wiener model, the linear dynamic part of which is the same that of the Wiener system whereas the steady-state nonlinear part is approximated by means of a neural network of the DLP type containing $\widetilde{K} = 5$ hidden nodes. The obtained approximation accuracy is very high, as shown in Fig. 3.14. All MPC algorithms have the same parameters: $N = 10$, $N_{\mathrm{u}} = 3$, $\lambda_p = 0.25$, the additional parameters of the MPC-NPLPT scheme are: $\delta_{\mathrm{u}} = \delta_{\mathrm{y}} = 10^{-1}$, $N_0 = 2$, $n_{\mathrm{max}} = 5$. In all algorithms the same magnitude and rate constraints are taken into account: $u^{\mathrm{min}} = -5$, $u^{\mathrm{max}} = 5$, $\triangle u^{\mathrm{max}} = 2.5$.

Because the considered process is significantly nonlinear, control quality of the linear MPC algorithm is unsatisfactory which is illustrated in Fig. 3.15. For the first set-point change the algorithm is slow, for the second change a big overshoot occurs. Fig. 3.16 compares the trajectories obtained in the "ideal" MPC-NO algorithm and in the MPC-NPSL one. Unfortunately, although for the polymerisation reactor the MPC-NPSL algorithm gives very similar results as the MPC-NO one, it is not true for the considered process. In particular, for the second set-point change the suboptimal algorithm

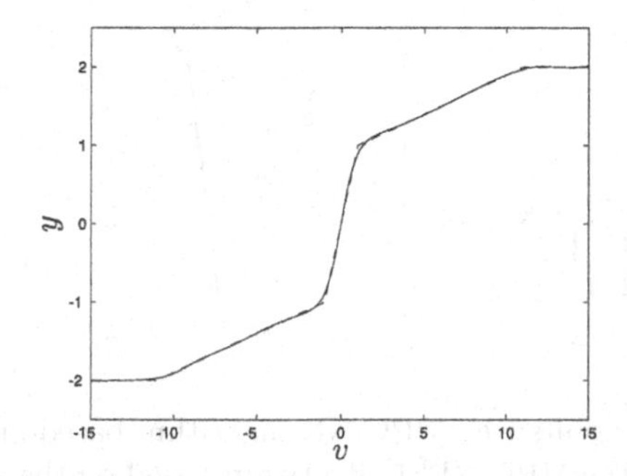

Fig. 3.14. The nonlinear steady-state characteristics $y(v)$ of the Wiener system (*solid line*) and its neural approximation (*dashed line*)

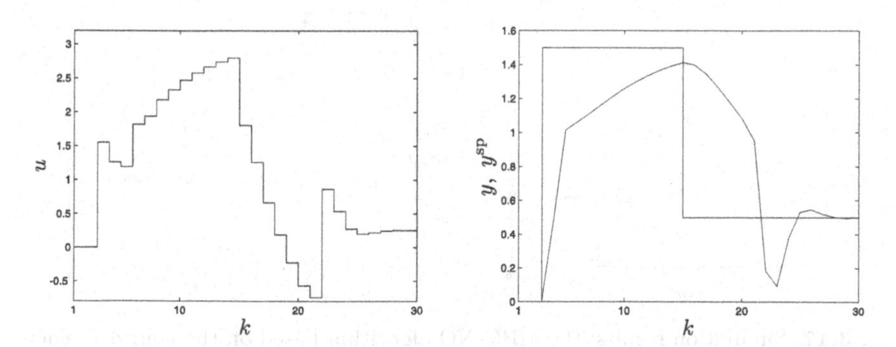

Fig. 3.15. Simulation results of the linear MPC algorithm in the control system of the Wiener system

is much worse. Fig. 3.17 shows simulation results of the MPC-NPLPT algorithm and of the MPC-NO one. Unlike the MPC-NPSL approach with simplified linearisation, the trajectory linearisation approach gives very good control accuracy, practically the same as in the MPC-NO one. The squared sum of control errors for the whole simulation horizon (SSE_{sim}) for MPC-NPLPT and MPC-NO algorithms is very similar: 7.0608 and 7.0411, respectively, whereas for the MPC-NPSL algorithm its value increases to 8.0012. Fig. 3.18 shows the number of internal iterations of the MPC-NPLPT algorithm. When the difference between the current value of the output and the set-point is significant, the algorithm needs four or two internal iterations, but for some 50% of the simulations one internal iteration is sufficient.

Table 3.7 compares computational burden of nonlinear MPC-NO and MPC-NPLPT algorithms based on the same neural Wiener model. Different

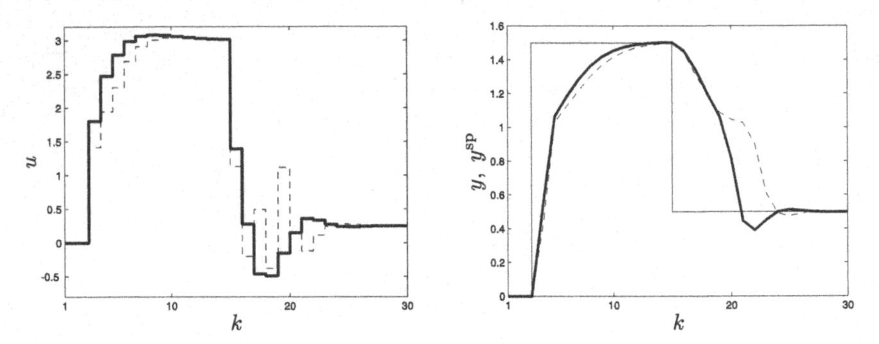

Fig. 3.16. Simulation results: the MPC-NO algorithm based on the neural Wiener model (*solid line*) and the MPC-NPSL algorithm based on the same model (*dashed line*) in the control system of the Wiener system

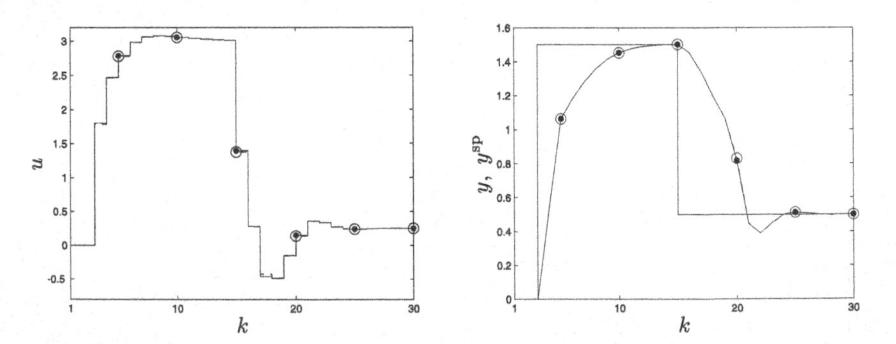

Fig. 3.17. Simulation results: the MPC-NO algorithm based on the neural Wiener model (*solid line with dots*) and the MPC-NPLPT algorithm based on the same model (*dashed line with circles*) in the control system of the Wiener system

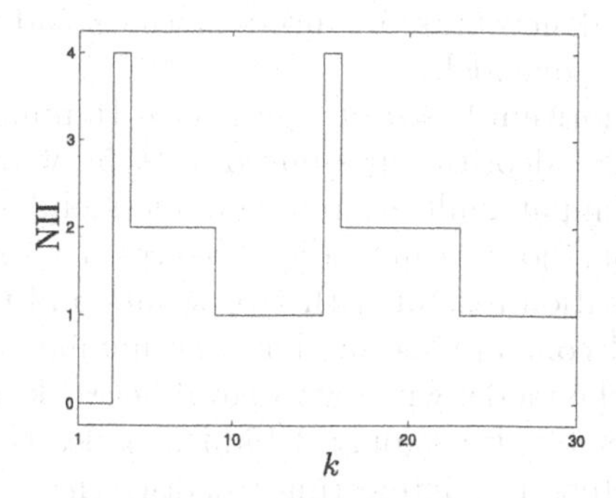

Fig. 3.18. The number of internal iterations (NII) in the MPC-NPLPT algorithm based on the neural Wiener model in the control system of the Wiener system

Table 3.7. Comparison of computational complexity (in MFLOPS) of nonlinear MPC algorithms based on the neural Wiener model in the control system of the Wiener system for different control horizons, $N = 10$

Algorithm	$N_u = 1$	$N_u = 2$	$N_u = 3$	$N_u = 4$	$N_u = 5$	$N_u = 10$
MPC-NO	0.43	0.67	1.06	1.54	1.90	4.39
MPC-NPLPT	0.10	0.17	0.25	0.35	0.47	1.51

control horizons are considered, the prediction horizon is $N = 10$. The suboptimal MPC-NPLPT algorithm is noticeable less computationally demanding than the MPC-NO approach, the computational complexity reduction factor is in the range 2.91–4.42.

3.6 Literature Review

The literature concerned with applications of cascade models is very rich. For example, the Hammerstein structure can be used as models of the following processes: neutralisation reactors (pH reactors) [73, 247, 302], distillation columns [69, 73, 136], heat exchangers [69] and diesel engines [11]. For the Wiener structure some example applications are: neutralisation reactors [82, 230, 247], distillation columns [32, 229, 136], polymerisation reactors [44] and separators [13]. The Wiener structure can be also used for modelling biological phenomena, e.g. an insulin administration process during operations [190]. An excellent review of identification algorithms for cascade Hammerstein and Wiener structures and a list of applications is given is [103].

Because the cascade structures are frequently used for modelling of numerous processes, they are also used in MPC. Typically, the inverse of the steady-state part of the model is used to compensate for the nonlinear nature of the process. For the Hammerstein structure, such an approach is discussed in [73], for the Wiener structure in [6, 44, 229, 230, 296]. In some works, e.g. in [6, 153], neural networks are used in the steady-state part of the model and in the inverse model.

The MPC-NPL algorithm based on the neural Hammerstein model is an extended version of the algorithm presented in [155], where implementation details for two multi-input multi-output Hammerstein structures shown in Fig. 3.1 are given (but no measured disturbances are taken into account). A simulated neutralisation reactor with two inputs and two outputs is consider to evaluate and compare the models, the number of auxiliary signals (n_v) is two. Finally, the model with two neural networks in the steady-state part is chosen, each network has only 3 hidden nodes ($K^1 = K^2 = 3$), the model has 38 parameters. It is interesting to notice that accuracy of the chosen model is similar to that of the Hammerstein structure with one neural network in the steady-state part which has as many as $K = 7$ hidden nodes (the model has 49 parameters). Next, for the chosen model the MPC-NPL and MPC-NO algorithms are developed. The suboptimal algorithm turns out to give trajectories practically the same as those obtained in the MPC-NO strategy. Efficiency of the MPC-NPL algorithm based on the neural Hammerstein structure is also shown in [170], in which a simulated methanol-water distillation column with two inputs and two outputs is considered.

An application of the neural Hammerstein model in set-pointy optimisation cooperating with MPC to a simulated multi-input multi-output neutralisation reactor is presented in [148] (it is also discussed in the eighth chapter of this book). The model has three inputs (including one measured disturbance) and two outputs, the number of auxiliary signals (n_v) is 2. Accuracy of Hammerstein systems with polynomial and neural steady-state parts are compared. Although the polynomial Hammerstein model has a lot of parameters, it is significantly less precise. In particular, its steady-state characteristics is unsatisfactory. The chosen Hammerstein model contains two separate neural networks which have $K^1 = K^2 = 5$ hidden nodes (the model has 64 parameters).

The MPC-NPL algorithm based on the neural Wiener model with one input and one output is discussed in [153]. Its advantages (good control accuracy, computational efficiency) are demonstrated in the control system of the simulated polymerisation reactor discussed in this chapter, the inverse characteristics of which exists (a neural network is used for approximation). The MPC-NPSL algorithm based on the simplified multi-input multi-output neural Wiener model shown in Fig. 3.5 is discussed in [150]. The MPC-NPSL, MPC-NPLT and MPC-NPLPT algorithms for the single-input single-output neural Wiener model are discussed in [143], their properties are compared in the control system of the polymerisation reactor discussed in this chapter and a simulated single-input single-output neutralisation reactor is considered. For the neutralisation reactor the simplified linearisation method used in the MPC-NPSL algorithm is not sufficient, the single trajectory linearisation used in the MPC-NPLT approach is much better, but the MPC-NPLPT algorithm is only able to give control quality comparable with that obtained in the MPC-NO one. The MPC-NPSL algorithm for the neural Wiener model is discussed in [9].

The suboptimal MPC algorithms discussed in this chapter may use cascade models of some alternative structures. For example, the MPC-NPSL algorithm based on the single-input single-output Hammerstein model with a fuzzy steady-state part is discussed in [197], a similar approach based on a fuzzy Wiener model is presented in [198]. A family of suboptimal MPC algorithms based on the LS SVM model, namely MPC-NPL, MPC-NPLT and MPC-NPLPT ones are detailed in [144].

3.7 Chapter Summary

The Hammerstein and Wiener cascade structures are viable alternatives to the classical black-box neural models. They can be used for modelling of numerous processes [103], in particular when nonlinearity is concerned with the sensors or actuators. Furthermore, they may turn out to be better than the black-box structures. For example, the neural Wiener model of the polymerisation reactor is more than 4 times more precise than the DLP neural structure (both models have a similar number of parameters). Although different steady-state parts may be used, neural networks seem to be a good choice. When compared to polynomials frequently used in the steady-state part of cascade models, advantages of the neural approach are apparent: they are universal approximators and they have good numerical properties. The neural approach is very advantageous in the case of multi-input multi-output processes: the neural cascade models typically have a significantly lower number of parameters than the polynomial ones [148, 155].

This chapter gives implementation details of some prototype MPC algorithms discussed in the previous chapter (MPC-NO, MPC-NPL and MPC-NPLPT approaches are considered) for neural Hammerstein and Wiener models. Additionally, MPC-SSL and MPC-NPSL algorithms with simplified linearisation are discussed. The simplified approach to on-line linearisation is possible due to the specific cascade structure of the models considered. Although many control algorithms for the cascade models discussed elsewhere (not only MPC algorithms) need inverse steady-state models, almost all the algorithms discussed in this chapter do not need inverse models (the MPC-NPL algorithm for the neural Wiener model is the only exception). In place of the MPC-NPL algorithm it is also possible to use the MPC-SL approach, the only difference is the fact that the linearised model, not the nonlinear one, is used for free trajectory calculation. Using the derivations presented in this chapter, it is also possible to easily develop the MPC-NPLT algorithm (it is very similar to the MPC-NPLPT structure, but linearisation is carried out once at each sampling instant) and the Newton-like MPC-NNPAPT algorithm. All discussed MPC algorithms can be also formulated in their explicit versions (a simple matrix decomposition is carried out and linear equations are solved in place of quadratic optimisation). Finally, the presented algorithms can be derived for more complicated cascade models, e.g. for Hammerstein-Wiener and Wiener-Hammerstein systems consisting of three parts [103].

Two processes are considered to present advantages of the cascade models and of the discussed MPC algorithms. For the polymerisation reactor, the Wiener structure turns out to be better than the Hammerstein one. The simple MPC-NPSL algorithm based on such a model with simplified on-line linearisation is sufficient (it gives control accuracy comparable with that of the MPC-NO approach), it is not necessary to use more advanced trajectory linearisation. At the same time, it is a few times (4.05–9.46, which depends on the horizons) less computationally demanding. For the polymerisation reactor it is also possible to use the MPC-NPL algorithm with precise linearisation because the inverse characteristics of the process exists. Three MPC algorithms based on the neural Wiener model are used for the second example process: MPC-NO, MPC-NPSL and MPC-NPLPT approaches. As the inverse characteristics does not exist, it is not possible to use the MPC-NPL strategy. Unlike the polymerisation reactor case, in the second example the MPC-NPSL algorithms gives control quality far from that of the MPC-NO algorithm. That is why the MPC-NPLPT approach is used. It gives very high control accuracy, practically the same as the MPC-NO approach, but at the same time it is a few times less computationally demanding.

4

MPC Algorithms Based on Neural State-Space Models

The objective of this chapter is to discuss nonlinear MPC algorithms based on neural state-space models. Implementation details of the MPC-NO scheme as well as of two suboptimal MPC-NPL and MPL-NPLPT algorithms are presented. All the algorithms are considered in two versions: with the state set-point trajectory and with the output set-point trajectory. Simulation results are concerned with the polymerisation reactor introduced in the previous chapter. It is assumed that all state variables can be measured, but in practice some of them may be unavailable and an observer must be used.

4.1 Neural State-Space Models

The first implemented MPC algorithms used non-parametric models: the impulse-response model was used in the MAC approach and the step-response one was used in the DMC algorithm. The GPC algorithm, which was developed some decade later, used the parametric input-output model (the difference equation). The mentioned models, in particular the step-response one, can be obtained in a relatively simple manner, which is very important in the industrial environment. Furthermore, the mentioned models can be efficiently used to describe properties of typical technological processes, e.g. chemical reactors and distillation columns. The development of hardware systems used for control (the last part of the XX century and the XXI century) made it possible to use in practice MPC algorithms based on more complicated state-space models. It is quite straightforward to use such MPC algorithms when the state-space model is available, which is true for many processes, e.g. for DC motors, electro-mechanic systems or robots.

In spite of the fact that the MPC algorithms based on the input-output models predominate in industrial applications, the algorithms based on the state-space models are also used. They are available in commercial MPC software packages for MPC [262]. It is necessary to emphasise the fact that the state-space description is much more general than the input-output one,

M. Ławryńczuk, *Computationally Efficient Model Predictive Control Algorithms*, Studies in Systems, Decision and Control 3,
DOI: 10.1007/978-3-319-04229-9_4, © Springer International Publishing Switzerland 2014

it is convenient for theoretical analysis (the state-space description is used for stability and robustness investigation of MPC [203]).

The state-space model of a multi-input multi-output process has the following general form

$$\begin{aligned} x(k+1) &= f(x(k), u(k), h(k)) \\ y(k) &= g(x(k)) \end{aligned} \tag{4.1}$$

where $x(k) \in \mathbb{R}^{n_{\mathrm{x}}}$ denotes the state vector, $u \in \mathbb{R}^{n_{\mathrm{u}}}$, $h \in \mathbb{R}^{n_{\mathrm{h}}}$, $y \in \mathbb{R}^{n_{\mathrm{y}}}$. The model is characterised by the nonlinear functions $f\colon \mathbb{R}^{n_{\mathrm{x}}+n_{\mathrm{u}}+n_{\mathrm{h}}} \to \mathbb{R}^{n_{\mathrm{x}}}$ and $g\colon \mathbb{R}^{n_{\mathrm{x}}} \to \mathbb{R}^{n_{\mathrm{y}}}$. The state equation may be rewritten for the sampling instant k. The model becomes

$$\begin{aligned} x(k) &= f(x(k-1), u(k-1), h(k-1)) \\ y(k) &= g(x(k)) \end{aligned} \tag{4.2}$$

Fig. 4.1 depicts the general structure of the neural state-space models. It consists of two DLP neural networks: the first one is responsible for the state equation, the second one – for the output equation. The structures of both neural networks are shown in Fig. 4.2. The first network has $n_{\mathrm{x}} + n_{\mathrm{u}} + n_{\mathrm{h}}$ inputs, K^{I} hidden nodes and n_{x} outputs. The weights of the first layer of the first network are denoted by $w_{i,j}^{1,\mathrm{I}}$, where $i = 1, \ldots, K^{\mathrm{I}}$, $j = 0, \ldots, n_{\mathrm{x}}+n_{\mathrm{u}}+n_{\mathrm{h}}$, the weights of the second layer are denoted by $w_{i,j}^{2,\mathrm{I}}$, where $i = 1, \ldots, n_{\mathrm{x}}$, $j = 0, \ldots, K^{\mathrm{I}}$. The second network has n_{x} inputs, K^{II} hidden nodes and n_{y} outputs. The weights of the first layer of the second network are denoted by $w_{i,j}^{1,\mathrm{II}}$, where $i = 1, \ldots, K^{\mathrm{II}}$, $j = 0, \ldots, n_{\mathrm{x}}$, the weights of the second layer are denoted by $w_{i,j}^{2,\mathrm{I}}$, where $i = 1, \ldots, n_{\mathrm{y}}$, $j = 0, \ldots, K^{\mathrm{II}}$. The transfer function of both networks is $\varphi\colon \mathbb{R} \to \mathbb{R}$.

The consecutive output signals of the first network (the states) are

$$x_i(k) = w_{i,0}^{2,\mathrm{I}} + \sum_{j=1}^{K^{\mathrm{I}}} w_{i,j}^{2,\mathrm{I}} \varphi(z_j^{\mathrm{I}}(k)) \tag{4.3}$$

where $i = 1, \ldots, n_{\mathrm{x}}$, and the sum of the input signals connected to the jth hidden node ($j = 1, \ldots, K^{\mathrm{I}}$) is

$$\begin{aligned} z_j^{\mathrm{I}}(k) = w_{j,0}^{1,\mathrm{I}} &+ \sum_{s=1}^{n_{\mathrm{x}}} w_{j,s}^{1,\mathrm{I}} x_s(k-1) + \sum_{n=1}^{n_{\mathrm{u}}} w_{j,n_{\mathrm{x}}+n}^{1,\mathrm{I}} u_n(k-1) \\ &+ \sum_{n=1}^{n_{\mathrm{h}}} w_{j,n_{\mathrm{x}}+n_{\mathrm{u}}+n}^{1,\mathrm{I}} h_n(k-1) \end{aligned} \tag{4.4}$$

From Eqs. (4.3) and (4.4), one has

$$x_i(k) = w_{i,0}^{2,\mathrm{I}} + \sum_{j=1}^{K^{\mathrm{I}}} w_{i,j}^{2,\mathrm{I}} \varphi\Bigg(w_{j,0}^{1,\mathrm{I}} + \sum_{s=1}^{n_{\mathrm{x}}} w_{j,s}^{1,\mathrm{I}} x_s(k-1) + \sum_{n=1}^{n_{\mathrm{u}}} w_{j,n_{\mathrm{x}}+n}^{1,\mathrm{I}} u_n(k-1) + \sum_{n=1}^{n_{\mathrm{h}}} w_{j,n_{\mathrm{x}}+n_{\mathrm{u}}+n}^{1,\mathrm{I}} h_n(k-1) \Bigg) \tag{4.5}$$

The consecutive output signals of the second network (the outputs of the model) are

$$y_m(k) = w_{m,0}^{2,\mathrm{II}} + \sum_{l=1}^{K^{\mathrm{II}}} w_{m,l}^{2,\mathrm{II}} \varphi(z_l^{\mathrm{II}}(k)) \tag{4.6}$$

where $m = 1, \ldots, n_{\mathrm{y}}$, and the sum of the input signals connected to the lth hidden node ($l = 1, \ldots, K^{\mathrm{II}}$) is

$$z_l^{\mathrm{II}}(k) = w_{l,0}^{1,\mathrm{II}} + \sum_{i=1}^{n_{\mathrm{x}}} w_{l,i}^{1,\mathrm{II}} x_i(k) \tag{4.7}$$

Using Eqs. (4.6) and (4.7), it is possible to obtain the model outputs

$$y_m(k) = w_{m,0}^{2,\mathrm{II}} + \sum_{l=1}^{K^{\mathrm{II}}} w_{m,l}^{2,\mathrm{II}} \varphi\left(w_{l,0}^{1,\mathrm{II}} + \sum_{i=1}^{n_{\mathrm{x}}} w_{l,i}^{1,\mathrm{II}} x_i(k) \right) \tag{4.8}$$

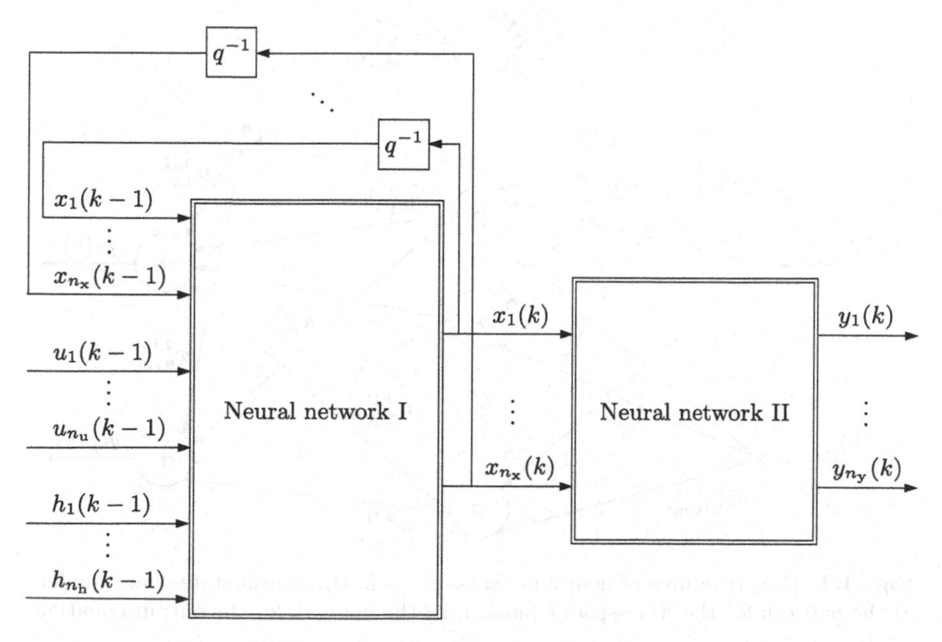

Fig. 4.1. The structure of the neural state-space model

Of course, it is possible to use some alternative neural state-space models. For the output equation not one, but a few networks can be used (a straightforward choice is n_y separate networks). Similarly, for the state equations as many as n_x separate networks can be used. The choice usually depends on properties of the process and availability of training algorithms. The MPC algorithms presented next can be easily modified to deal with alternative models.

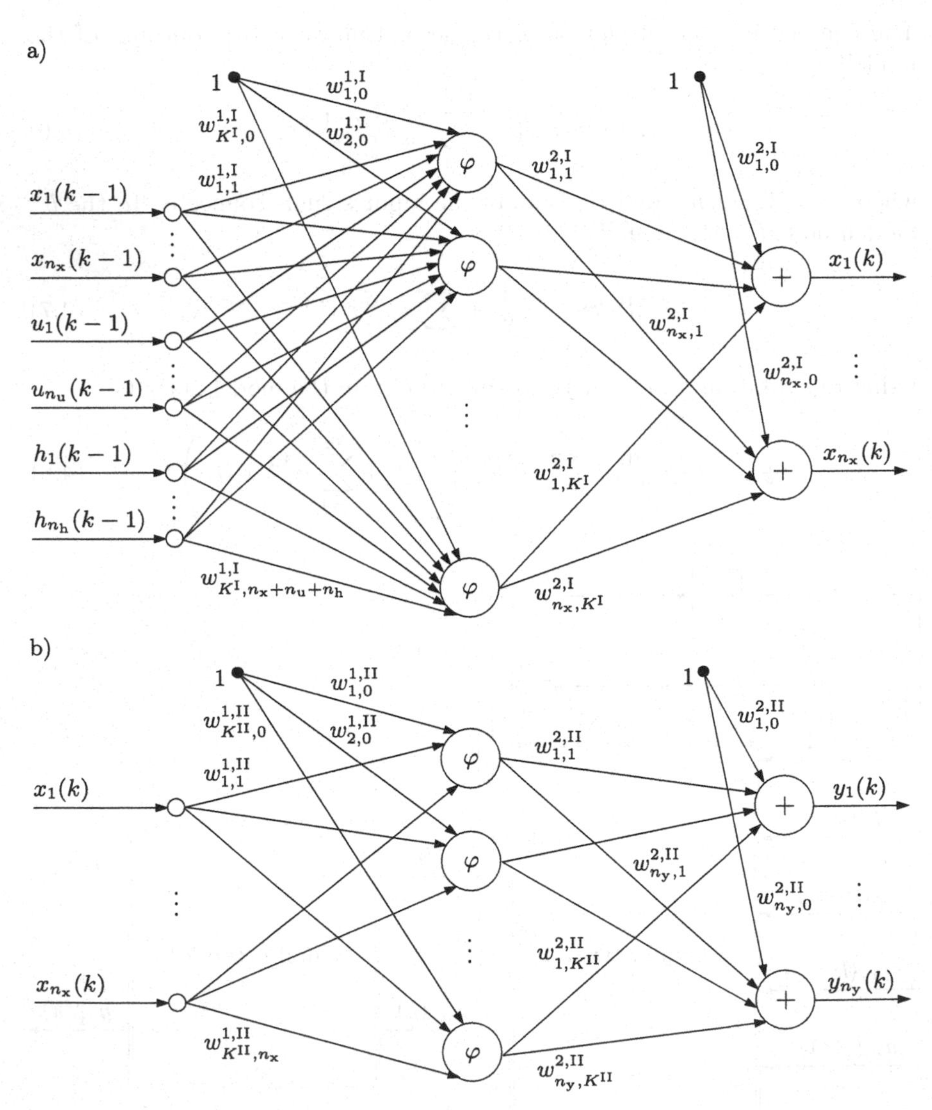

Fig. 4.2. The structures of neural networks used in the neural state-space model: a) the network for the state-space equation, b) the network for the output equation

4.2 Problem Formulation

The general formulation of MPC given in Chapter 1.1, in which the output set-point trajectory is considered, can be naturally used when the state-space model is used for prediction. Furthermore, for the state-space model an alternative formulation is possible in which the state-space set-point trajectory is considered. In such a case only the state model is necessary, it is described by the first equation (4.1), i.e. the state equation. The model shown in Fig. 4.1 consists of only the first neural network. Similarly to the optimisation task (1.6), the MPC optimisation problem with the state set-point trajectory is

$$\min_{\substack{\boldsymbol{u}(k) \text{ or } \triangle\boldsymbol{u}(k) \\ \varepsilon^{\min}(k+p),\ \varepsilon^{\max}(k+p)}} \Bigg\{ J(k) = \sum_{p=1}^{N} \left\| x^{\mathrm{sp}}(k+p|k) - \hat{x}(k+p|k) \right\|^2_{\boldsymbol{\Phi}_p} + \sum_{p=0}^{N_{\mathrm{u}}-1} \left\| \triangle u(k+p|k) \right\|^2_{\boldsymbol{\Lambda}_p} + \rho^{\min} \sum_{p=1}^{N} \left\| \varepsilon^{\min}(k+p) \right\|^2 + \rho^{\max} \sum_{p=1}^{N} \left\| \varepsilon^{\max}(k+p) \right\|^2 \Bigg\}$$

subject to (4.9)

$$\begin{aligned}
& u^{\min} \leq u(k+p|k) \leq u^{\max},\ p=0,\ldots,N_{\mathrm{u}}-1 \\
& -\triangle u^{\max} \leq \triangle u(k+p|k) \leq \triangle u^{\max},\ p=0,\ldots,N_{\mathrm{u}}-1 \\
& x^{\min} - \varepsilon^{\min}(k+p) \leq \hat{x}(k+p|k) \leq x^{\max} + \varepsilon^{\max}(k+p),\ p=1,\ldots,N \\
& \varepsilon^{\min}(k+p) \geq 0,\ \varepsilon^{\max}(k+p) \geq 0,\ p=1,\ldots,N
\end{aligned}$$

There are two differences when the above problem is compared with the general MPC optimisation problem (1.6) used if the output set-point trajectory is used: the predicted state (not output) errors are taken into account in the first part of the minimised cost-function, the output constraints are replaced by the state ones. Both output and state set-point trajectories are not used at the same time because the outputs are some functions of the state. If an output set-point trajectory inconsistent with the state one were used, it would lead to some stiffness. The weighting matrix $\boldsymbol{\Phi}_p = \mathrm{diag}(\phi_{p,1}, \ldots, \phi_{p,n_{\mathrm{x}}}) \geq 0$ is of dimensionality $n_{\mathrm{x}} \times n_{\mathrm{x}}$, where the coefficients $\phi_{p,m} \geq 0$ have a similar role as the coefficients $\mu_{p,m}$. Defining the state set-point trajectory and the predicted state trajectory vectors of length $n_{\mathrm{x}}N$

$$\boldsymbol{x}^{\mathrm{sp}}(k) = \begin{bmatrix} x^{\mathrm{sp}}(k+1|k) \\ \vdots \\ x^{\mathrm{sp}}(k+N|k) \end{bmatrix},\ \hat{\boldsymbol{x}}(k) = \begin{bmatrix} \hat{x}(k+1|k) \\ \vdots \\ \hat{x}(k+N|k) \end{bmatrix}$$

and the weighting matrix $\boldsymbol{\Phi} = \mathrm{diag}(\boldsymbol{\Phi}_1, \ldots, \boldsymbol{\Phi}_N) \geq 0$, of dimensionality $n_{\mathrm{x}}N \times n_{\mathrm{x}}N$, the optimisation problem (4.9) may be rewritten in the following compact form

$$\begin{aligned}
&\min_{\substack{\boldsymbol{u}(k) \text{ or } \triangle\boldsymbol{u}(k) \\ \boldsymbol{\varepsilon}^{\min}(k),\ \boldsymbol{\varepsilon}^{\max}(k)}} \Big\{ J(k) = \left\|\boldsymbol{x}^{\mathrm{sp}}(k) - \hat{\boldsymbol{x}}(k)\right\|^2_{\boldsymbol{\Phi}} + \left\|\triangle\boldsymbol{u}(k)\right\|^2_{\boldsymbol{\Lambda}} \\
&\qquad\qquad + \rho^{\min}\left\|\boldsymbol{\varepsilon}^{\min}(k)\right\|^2 + \rho^{\max}\left\|\boldsymbol{\varepsilon}^{\max}(k)\right\|^2 \Big\} \\
&\text{subject to} \\
&\boldsymbol{u}^{\min} \leq \boldsymbol{u}(k) \leq \boldsymbol{u}^{\max} \\
&-\triangle\boldsymbol{u}^{\max} \leq \triangle\boldsymbol{u}(k) \leq \triangle\boldsymbol{u}^{\max} \\
&\boldsymbol{x}^{\min} - \boldsymbol{\varepsilon}^{\min}(k) \leq \hat{\boldsymbol{x}}(k) \leq \boldsymbol{x}^{\max} + \boldsymbol{\varepsilon}^{\max}(k) \\
&\boldsymbol{\varepsilon}^{\min}(k) \geq 0,\ \boldsymbol{\varepsilon}^{\max}(k) \geq 0
\end{aligned} \tag{4.10}$$

The vectors $\boldsymbol{\varepsilon}^{\min}(k)$ and $\boldsymbol{\varepsilon}^{\max}(k)$, which are the additional decision variables of the optimisation problem defining violation of the original state constraints, have the form defined by Eqs. (1.8), but length $n_{\mathrm{x}}N$. The number of decision variables is $n_{\mathrm{u}}N_{\mathrm{u}}+2n_{\mathrm{x}}N$, the number of constraints is $4n_{\mathrm{u}}N_{\mathrm{u}}+4n_{\mathrm{x}}N$. Of course, it is possible to use the same vectors $\varepsilon^{\min}(k)$ and $\varepsilon^{\max}(k)$ of length n_{x} for the whole prediction horizon (similarly to the formulation (1.10)). In such a case the MPC optimisation task is

$$\begin{aligned}
&\min_{\substack{\boldsymbol{u}(k) \text{ or } \triangle\boldsymbol{u}(k) \\ \varepsilon^{\min}(k),\ \varepsilon^{\max}(k)}} \Bigg\{ J(k) = \sum_{p=1}^{N} \left\|x^{\mathrm{sp}}(k+p|k) - \hat{x}(k+p|k)\right\|^2_{\boldsymbol{\Phi}_p} \\
&\qquad\qquad + \sum_{p=0}^{N_{\mathrm{u}}-1} \left\|\triangle u(k+p|k)\right\|^2_{\boldsymbol{\Lambda}_p} \\
&\qquad\qquad + \rho^{\min}\left\|\varepsilon^{\min}(k)\right\|^2 + \rho^{\max}\left\|\varepsilon^{\max}(k)\right\|^2 \Bigg\} \\
&\text{subject to} \\
&u^{\min} \leq u(k+p|k) \leq u^{\max},\ p = 0, \ldots, N_{\mathrm{u}}-1 \\
&-\triangle u^{\max} \leq \triangle u(k+p|k) \leq \triangle u^{\max},\ p = 0, \ldots, N_{\mathrm{u}}-1 \\
&x^{\min} - \varepsilon^{\min}(k) \leq \hat{x}(k+p|k) \leq x^{\max} + \varepsilon^{\max}(k),\ p = 1, \ldots, N \\
&\varepsilon^{\min}(k) \geq 0,\ \varepsilon^{\max}(k) \geq 0
\end{aligned}$$

which can be transformed to the following compact form

$$\min_{\substack{\boldsymbol{u}(k) \text{ or } \triangle\boldsymbol{u}(k)\\ \varepsilon^{\min}(k),\ \varepsilon^{\max}(k)}} \Big\{ J(k) = \|\boldsymbol{x}^{\mathrm{sp}}(k) - \hat{\boldsymbol{x}}(k)\|^2_{\boldsymbol{\Phi}} + \|\triangle\boldsymbol{u}(k)\|^2_{\boldsymbol{\Lambda}} + \rho^{\min}\left\|\varepsilon^{\min}(k)\right\|^2 + \rho^{\max}\left\|\varepsilon^{\max}(k)\right\|^2 \Big\}$$

subject to

$$\boldsymbol{u}^{\min} \leq \boldsymbol{u}(k) \leq \boldsymbol{u}^{\max}$$
$$-\triangle\boldsymbol{u}^{\max} \leq \triangle\boldsymbol{u}(k) \leq \triangle\boldsymbol{u}^{\max}$$
$$\boldsymbol{x}^{\min} - \boldsymbol{\varepsilon}^{\min}(k) \leq \hat{\boldsymbol{x}}(k) \leq \boldsymbol{x}^{\max} + \boldsymbol{\varepsilon}^{\max}(k)$$
$$\varepsilon^{\min}(k) \geq 0,\ \varepsilon^{\max}(k) \geq 0$$

The vectors $\boldsymbol{\varepsilon}^{\min}(k)$ and $\boldsymbol{\varepsilon}^{\max}(k)$ have the structure defined by Eqs. (1.12), but length $n_{\mathrm{x}}N$. The number of decision variables drops to $n_{\mathrm{u}}N_{\mathrm{u}} + 2n_{\mathrm{x}}$, the number of constraints decreases to $4n_{\mathrm{u}}N_{\mathrm{u}} + 2n_{\mathrm{x}}N + 2n_{\mathrm{x}}$.

4.3 MPC-NO Algorithm

Two versions of the MPC-NO algorithm are discussed next: the output and the state set-point trajectories are considered. The general algorithm formulation discussed in Chapter 2.2, in which the input-output neural model is used for prediction, of course holds true for the state-space model provided that the minimised cost-function takes into account the output set-point trajectory. Replacing the input-output model by the state-space one entails deriving the predicted trajectory and its derivatives with respect to the future control policy for a new model type. When the state set-point trajectory is taken into account, the predicted output trajectory $\hat{\boldsymbol{y}}(k)$ must be replaced by the state trajectory $\hat{\boldsymbol{x}}(k)$.

The predicted state vector for the sampling instant $k+p$ calculated at the current instant k is calculated from

$$\hat{x}(k+p|k) = x(k+p|k) + \nu(k) \tag{4.11}$$

where $x(k+p|k)$ denotes the state vector obtained from the state equation (4.2), $\nu(k)$ denotes the estimation of the unmeasured state disturbance which can also represent state modelling errors. The state prediction equation (4.11) is very similar to the output prediction formula (1.13): in both approaches the predicted vector is a sum of a model output and of a disturbance estimation. The DMC prediction model is used, in which it is assumed that the same disturbance acts on the process over the whole prediction horizon. Similarly to Eq. (1.15), the unmeasured state disturbance is calculated as a difference between the measured (observed) state and the state calculated from the model using signals up to the sampling instant $k-1$

$$\begin{aligned}\nu(k) &= x(k) - x(k|k-1)\\ &= x(k) - f(x(k-1), u(k-1), h(k-1))\end{aligned} \tag{4.12}$$

Using the state equation (4.2) recurrently, from the general state prediction equation (4.11), for $p = 1$ one has

$$\hat{x}(k+1|k) = f(x(k), u(k|k), h(k)) + \nu(k)$$

and for $p = 2, \ldots, N$ the recurrent formula can be used

$$\hat{x}(k+p|k) = f(\hat{x}(k+p-1|k), u(k+p-1|k), h(k+p-1|k)) + \nu(k)$$

Using Eq. (4.3), it is possible to calculate the predicted states

$$\hat{x}_i(k+p|k) = w_{i,0}^{2,\mathrm{I}} + \sum_{j=1}^{K^{\mathrm{I}}} w_{i,j}^{2,\mathrm{I}} \varphi(z_j^{\mathrm{I}}(k+p|k)) + \nu_i(k) \tag{4.13}$$

where, using Eq. (4.4), for the sampling instant $k+1$ one has

$$z_j^{\mathrm{I}}(k+1|k) = w_{j,0}^{1,\mathrm{I}} + \sum_{s=1}^{n_{\mathrm{x}}} w_{j,s}^{1,\mathrm{I}} x_s(k) + \sum_{n=1}^{n_{\mathrm{u}}} w_{j,n_{\mathrm{x}}+n}^{1,\mathrm{I}} u_n(k|k) + \sum_{n=1}^{n_{\mathrm{h}}} w_{j,n_{\mathrm{x}}+n_{\mathrm{u}}+n}^{1,\mathrm{I}} h_n(k) \tag{4.14}$$

and for any sampling instant $k+p$, where $p = 2, \ldots, N$, one obtains

$$z_j^{\mathrm{I}}(k+p|k) = w_{j,0}^{1,\mathrm{I}} + \sum_{s=1}^{n_{\mathrm{x}}} w_{j,s}^{1,\mathrm{I}} \hat{x}_s(k+p-1|k) + \sum_{n=1}^{n_{\mathrm{u}}} w_{j,n_{\mathrm{x}}+n}^{1,\mathrm{I}} u_n(k+p-1|k) + \sum_{n=1}^{n_{\mathrm{h}}} w_{j,n_{\mathrm{x}}+n_{\mathrm{u}}+n}^{1,\mathrm{I}} h_n(k+p-1|k) \tag{4.15}$$

Using Eqs. (4.5) and (4.12), the current estimation of the unmeasured state disturbance is

$$\nu_i(k) = x_i(k) - w_{i,0}^{2,\mathrm{I}} - \sum_{j=1}^{K^{\mathrm{I}}} w_{i,j}^{2,\mathrm{I}} \varphi\left(w_{j,0}^{1,\mathrm{I}} + \sum_{s=1}^{n_{\mathrm{x}}} w_{j,s}^{1,\mathrm{I}} x_s(k-1) + \sum_{n=1}^{n_{\mathrm{u}}} w_{j,n_{\mathrm{x}}+n}^{1,\mathrm{I}} u_n(k-1) + \sum_{n=1}^{n_{\mathrm{h}}} w_{j,n_{\mathrm{x}}+n_{\mathrm{u}}+n}^{1,\mathrm{I}} h_n(k-1) \right) \tag{4.16}$$

Using the general output prediction formula (1.14) and Eq. (4.8), the predicted output trajectory for consecutive outputs ($m = 1, \ldots, n_{\mathrm{y}}$) is

$$\hat{y}_m(k+p|k) = w_{m,0}^{2,\mathrm{II}} + \sum_{l=1}^{K^{\mathrm{II}}} w_{m,l}^{2,\mathrm{II}} \varphi\left(w_{l,0}^{1,\mathrm{II}} + \sum_{i=1}^{n_{\mathrm{x}}} w_{l,i}^{1,\mathrm{II}} \hat{x}_i(k+p|k) \right) + d_m(k) \tag{4.17}$$

The estimation of the state disturbances $\nu(k)$ must be taken into account when the estimation of the unmeasured output disturbance is calculated [316]

$$\begin{aligned} d_m(k) &= y_m(k) - g_m(x(k|k-1) + \nu(k)) \\ &= y_m(k) - g_m(f(x(k-1), u(k-1), h(k-1)) + \nu(k)) \end{aligned}$$

From Eqs. (1.15) and (4.8), the unmeasured output disturbances can be found in the following way

$$d_m(k) = y_m(k) - w_{m,0}^{2,\mathrm{II}} - \sum_{l=1}^{K^{\mathrm{II}}} w_{m,l}^{2,\mathrm{II}} \varphi\left(w_{l,0}^{1,\mathrm{II}} + \sum_{i=1}^{n_{\mathrm{x}}} w_{l,i}^{1,\mathrm{II}} (x_i(k) + \nu_i(k)) \right) \tag{4.18}$$

The entries of the matrix (2.16), i.e. the derivatives of the predicted output trajectory with respect to the future control signals, are calculated differentiating Eq. (4.17)

$$\frac{\partial \hat{y}_m(k+p|k)}{\partial u_n(k+r|k)} = \sum_{l=1}^{K^{\mathrm{II}}} w_{m,l}^{2,\mathrm{II}} \frac{\mathrm{d}\varphi(z_l^{\mathrm{II}}(k+p|k))}{\mathrm{d}z_l^{\mathrm{II}}(k+p|k)} \sum_{i=1}^{n_{\mathrm{x}}} w_{l,i}^{1,\mathrm{II}} \frac{\partial \hat{x}_i(k+p|k)}{\partial u_n(k+r|k)}$$

where

$$z_l^{\mathrm{II}}(k+p|k) = w_{l,0}^{1,\mathrm{II}} + \sum_{i=1}^{n_{\mathrm{x}}} w_{l,i}^{1,\mathrm{II}} \hat{x}_i(k+p|k)$$

Differentiating Eq. (4.13), one has

$$\frac{\partial \hat{x}_i(k+p|k)}{\partial u_n(k+r|k)} = \sum_{j=1}^{K^{\mathrm{I}}} w_{i,j}^{2,\mathrm{I}} \frac{\mathrm{d}\varphi(z_j^{\mathrm{I}}(k+p|k))}{\mathrm{d}z_j^{\mathrm{I}}(k+p|k)} \frac{\partial z_j^{\mathrm{I}}(k+p|k)}{\partial u_n(k+r|k)}$$

where, using Eq. (4.14), it can be noticed that for $p = 1$ one has

$$\frac{\partial z_j^{\mathrm{I}}(k+1|k)}{\partial u_n(k+r|k)} = \begin{cases} w_{j,n_{\mathrm{x}}+n}^{1,\mathrm{I}} & \text{if } r = 0 \\ 0 & \text{if } r = 1, \ldots, N_{\mathrm{u}} - 1 \end{cases}$$

and from Eq. (4.15), for $p = 2, \ldots, N$, one obtains

$$\frac{\partial z_j^{\mathrm{I}}(k+p|k)}{\partial u_n(k+r|k)} = \sum_{s=1}^{n_{\mathrm{x}}} w_{j,s}^{1,\mathrm{I}} \frac{\partial \hat{x}_s(k+p-1|k)}{\partial u_n(k+r|k)} + w_{j,n_{\mathrm{x}}+n}^{1,\mathrm{I}} \frac{\partial u_n(k+p-1|k)}{\partial u_n(k+r|k)} \tag{4.19}$$

The equations given above make it possible to implement the MPC-NO algorithm based on the neural state-space model provided that the output trajectory is used. The optimisation problem (2.5) is solved at each sampling instant, it can be transformed into the standard formulation (2.8), typical of nonlinear optimisation.

If the state set-point trajectory is considered, the state-space model consists of only the first neural network. In such a case it is necessary to calculate only the predicted state trajectory and its derivatives with respect to the future control policy. Similarly to the optimisation problem (2.5), using the formulation (4.10), the following optimisation problem is solved at each sampling instant of the discussed algorithm

$$\min_{\substack{\boldsymbol{u}(k)\\ \boldsymbol{\varepsilon}^{\min}(k),\ \boldsymbol{\varepsilon}^{\max}(k)}} \left\{ J(k) = \left\| \boldsymbol{x}^{\mathrm{sp}}(k) - \hat{\boldsymbol{x}}(k) \right\|^2_{\boldsymbol{\Phi}} + \left\| \boldsymbol{J}^{\mathrm{NO}} \boldsymbol{u}(k) + \boldsymbol{u}^{\mathrm{NO}}(k) \right\|^2_{\boldsymbol{\Lambda}} + \rho^{\min} \left\| \boldsymbol{\varepsilon}^{\min}(k) \right\|^2 + \rho^{\max} \left\| \boldsymbol{\varepsilon}^{\max}(k) \right\|^2 \right\}$$

subject to

$$\boldsymbol{u}^{\min} \leq \boldsymbol{u}(k) \leq \boldsymbol{u}^{\max}$$
$$-\triangle \boldsymbol{u}^{\max} \leq \boldsymbol{J}^{\mathrm{NO}} \boldsymbol{u}(k) + \boldsymbol{u}^{\mathrm{NO}}(k) \leq \triangle \boldsymbol{u}^{\max}$$
$$\boldsymbol{x}^{\min} - \boldsymbol{\varepsilon}^{\min}(k) \leq \hat{\boldsymbol{x}}(k) \leq \boldsymbol{x}^{\max} + \boldsymbol{\varepsilon}^{\max}(k)$$
$$\boldsymbol{\varepsilon}^{\min}(k) \geq 0,\ \boldsymbol{\varepsilon}^{\max}(k) \geq 0$$

The above task can be easily transformed to the standard form (2.8), but the predicted output trajectory $\hat{\boldsymbol{y}}(k)$ must be replaced by the predicted state trajectory $\hat{\boldsymbol{x}}(k)$, the output set-point trajectory $\hat{\boldsymbol{y}}^{\mathrm{sp}}(k)$ must be replaced by the state set-point trajectory $\hat{\boldsymbol{x}}^{\mathrm{sp}}(k)$, the output constrains must be replaced by the state constraints, the number of outputs (n_{y}) must be replaced by the number of states (n_{x}). The decision variables are defined by the vector (2.6), but its length is $n_{\mathrm{u}} N_{\mathrm{u}} + 2 n_{\mathrm{x}} N$. Analogously to Eq. (2.9), the minimised cost-function is

$$f_{\mathrm{opt}}(\boldsymbol{x}_{\mathrm{opt}}(k)) = \left\| \boldsymbol{x}^{\mathrm{sp}}(k) - \hat{\boldsymbol{x}}(k) \right\|^2_{\boldsymbol{\Phi}} + \left\| \boldsymbol{J}^{\mathrm{NO}} \boldsymbol{N}_1 \boldsymbol{x}_{\mathrm{opt}}(k) + \boldsymbol{u}^{\mathrm{NO}}(k) \right\|^2_{\boldsymbol{\Lambda}} + \rho^{\min} \left\| \boldsymbol{N}_2 \boldsymbol{x}_{\mathrm{opt}}(k) \right\|^2 + \rho^{\max} \left\| \boldsymbol{N}_3 \boldsymbol{x}_{\mathrm{opt}}(k) \right\|^2 \tag{4.20}$$

Similarly to Eq. (2.10), the constraints are defined by

$$\boldsymbol{g}_{\mathrm{opt}}(\boldsymbol{x}_{\mathrm{opt}}(k)) = \begin{bmatrix} -\boldsymbol{N}_1 \boldsymbol{x}_{\mathrm{opt}}(k) + \boldsymbol{u}^{\min} \\ \boldsymbol{N}_1 \boldsymbol{x}_{\mathrm{opt}}(k) - \boldsymbol{u}^{\max} \\ -\boldsymbol{J}^{\mathrm{NO}} \boldsymbol{N}_1 \boldsymbol{x}_{\mathrm{opt}}(k) - \boldsymbol{u}^{\mathrm{NO}}(k) - \triangle \boldsymbol{u}^{\max} \\ \boldsymbol{J}^{\mathrm{NO}} \boldsymbol{N}_1 \boldsymbol{x}_{\mathrm{opt}}(k) + \boldsymbol{u}^{\mathrm{NO}}(k) - \triangle \boldsymbol{u}^{\max} \\ -\hat{\boldsymbol{x}}(k) - \boldsymbol{N}_2 \boldsymbol{x}_{\mathrm{opt}}(k) + \boldsymbol{x}^{\min} \\ \hat{\boldsymbol{x}}(k) - \boldsymbol{N}_3 \boldsymbol{x}_{\mathrm{opt}}(k) - \boldsymbol{x}^{\max} \\ -\boldsymbol{N}_2 \boldsymbol{x}_{\mathrm{opt}}(k) \\ -\boldsymbol{N}_3 \boldsymbol{x}_{\mathrm{opt}}(k) \end{bmatrix} \tag{4.21}$$

where, in place of the auxiliary matrices (2.7), one must use

$$\boldsymbol{N}_1 = \left[\boldsymbol{I}_{n_{\mathrm{u}} N_{\mathrm{u}} \times n_{\mathrm{u}} N_{\mathrm{u}}}\ \boldsymbol{0}_{n_{\mathrm{u}} N_{\mathrm{u}} \times 2 n_{\mathrm{x}} N} \right]$$
$$\boldsymbol{N}_2 = \left[\boldsymbol{0}_{n_{\mathrm{x}} N \times n_{\mathrm{u}} N_{\mathrm{u}}}\ \boldsymbol{I}_{n_{\mathrm{x}} N \times n_{\mathrm{x}} N}\ \boldsymbol{0}_{n_{\mathrm{x}} N \times n_{\mathrm{x}} N} \right]$$
$$\boldsymbol{N}_3 = \left[\boldsymbol{0}_{n_{\mathrm{x}} N \times (n_{\mathrm{u}} N_{\mathrm{u}} + n_{\mathrm{x}} N)}\ \boldsymbol{I}_{n_{\mathrm{x}} N \times n_{\mathrm{x}} N} \right]$$

Differentiating the cost-function (4.20) with respect to the decision variables vector (2.6), similarly to Eq. (2.11), one obtains the gradient vector of length $n_{\mathrm{u}}N_{\mathrm{u}} + 2n_{\mathrm{x}}N$ used in on-line nonlinear optimisation

$$\begin{aligned}\frac{\mathrm{d}f_{\mathrm{opt}}(\boldsymbol{x}_{\mathrm{opt}}(k))}{\mathrm{d}\boldsymbol{x}_{\mathrm{opt}}(k)} &= 2\left(\frac{\mathrm{d}\hat{\boldsymbol{x}}(k)}{\mathrm{d}\boldsymbol{x}_{\mathrm{opt}}(k)}\right)^{\mathrm{T}}\boldsymbol{\Phi}(\hat{\boldsymbol{x}}(k)-\boldsymbol{x}^{\mathrm{sp}}(k))\\ &+2\boldsymbol{N}_1^{\mathrm{T}}(\boldsymbol{J}^{\mathrm{NO}})^{\mathrm{T}}\boldsymbol{\Lambda}(\boldsymbol{J}^{\mathrm{NO}}\boldsymbol{N}_1\boldsymbol{x}_{\mathrm{opt}}(k)+\boldsymbol{u}^{\mathrm{NO}}(k))\\ &+2\rho^{\min}\boldsymbol{N}_2^{\mathrm{T}}\boldsymbol{N}_2\boldsymbol{x}_{\mathrm{opt}}(k)+2\rho^{\max}\boldsymbol{N}_3^{\mathrm{T}}\boldsymbol{N}_3\boldsymbol{x}_{\mathrm{opt}}(k)\end{aligned}$$

Similarly to Eq. (2.12), the matrix of derivatives of the predicted state trajectory, of dimensionality $n_{\mathrm{x}}N \times (n_{\mathrm{u}}N_{\mathrm{u}} + 2n_{\mathrm{x}}N)$, is

$$\frac{\mathrm{d}\hat{\boldsymbol{x}}(k)}{\mathrm{d}\boldsymbol{x}_{\mathrm{opt}}(k)} = \left[\frac{\mathrm{d}\hat{\boldsymbol{x}}(k)}{\mathrm{d}\boldsymbol{u}(k)}\ \frac{\mathrm{d}\hat{\boldsymbol{x}}(k)}{\mathrm{d}\boldsymbol{\varepsilon}^{\min}(k)}\ \frac{\mathrm{d}\hat{\boldsymbol{x}}(k)}{\mathrm{d}\boldsymbol{\varepsilon}^{\max}(k)}\right]$$

where the derivatives of the predicted state trajectory with respect to the vectors $\boldsymbol{\varepsilon}^{\min}(k)$ and $\boldsymbol{\varepsilon}^{\max}(k)$ are all zeros matrices of dimensionality $n_{\mathrm{x}}N \times n_{\mathrm{x}}N$, the matrix of derivatives of the constraints (4.21), of dimensionality $(4n_{\mathrm{u}}N_{\mathrm{u}} + 4n_{\mathrm{x}}N) \times (n_{\mathrm{u}}N_{\mathrm{u}} + 2n_{\mathrm{x}}N)$, similarly to Eq. (2.13), has the structure

$$\frac{\mathrm{d}\boldsymbol{g}_{\mathrm{opt}}(\boldsymbol{x}_{\mathrm{opt}}(k))}{\mathrm{d}\boldsymbol{x}_{\mathrm{opt}}(k)} = \begin{bmatrix} -\boldsymbol{N}_1\\ \boldsymbol{N}_1\\ -\boldsymbol{J}^{\mathrm{NO}}\boldsymbol{N}_1\\ \boldsymbol{J}^{\mathrm{NO}}\boldsymbol{N}_1\\ -\dfrac{\mathrm{d}\hat{\boldsymbol{x}}(k)}{\mathrm{d}\boldsymbol{x}_{\mathrm{opt}}(k)}-\boldsymbol{N}_2\\ \dfrac{\mathrm{d}\hat{\boldsymbol{x}}(k)}{\mathrm{d}\boldsymbol{x}_{\mathrm{opt}}(k)}-\boldsymbol{N}_3\\ -\boldsymbol{N}_2\\ -\boldsymbol{N}_3 \end{bmatrix} \tag{4.22}$$

The matrix of derivatives of the predicted state trajectory with respect to the future control sequence, of dimensionality $n_{\mathrm{x}}N \times n_{\mathrm{u}}N_{\mathrm{u}}$, is

$$\frac{\mathrm{d}\hat{\boldsymbol{x}}(k)}{\mathrm{d}\boldsymbol{u}(k)} = \begin{bmatrix} \dfrac{\partial\hat{x}(k+1|k)}{\partial u(k|k)} & \cdots & \dfrac{\partial\hat{x}(k+1|k)}{\partial u(k+N_{\mathrm{u}}-1|k)}\\ \vdots & \ddots & \vdots\\ \dfrac{\partial\hat{x}(k+N|k)}{\partial u(k|k)} & \cdots & \dfrac{\partial\hat{x}(k+N|k)}{\partial u(k+N_{\mathrm{u}}-1|k)} \end{bmatrix}$$

where the matrices

$$\frac{\partial\hat{x}(k+p|k)}{\partial u(k+r|k)} = \begin{bmatrix} \dfrac{\partial\hat{x}_1(k+p|k)}{\partial u_1(k+r|k)} & \cdots & \dfrac{\partial\hat{x}_1(k+p|k)}{\partial u_{n_{\mathrm{u}}}(k+r|k)}\\ \vdots & \ddots & \vdots\\ \dfrac{\partial\hat{x}_{n_{\mathrm{x}}}(k+p|k)}{\partial u_1(k+r|k)} & \cdots & \dfrac{\partial\hat{x}_{n_{\mathrm{x}}}(k+p|k)}{\partial u_{n_{\mathrm{u}}}(k+r|k)} \end{bmatrix}$$

of dimensionality $n_x \times n_u$ are calculated for all $p = 1, \ldots, N$ and $r = 0, \ldots, N_u - 1$.

If the additional decision variables concerned with the soft constraints are of the type given by Eq. (1.12), i.e. they are constant for the whole prediction horizon, the implementation details of the MPC-NO algorithm with the state set-point trajectory discussed so far hold true, but the decision variables vector is of length $n_u N_u + 2n_x$ and is given by Eq. (2.14), the number of constraints is $4n_u N_u + 2n_x N + 2n_x$ and the auxiliary matrices are

$$\begin{aligned} \boldsymbol{N}_1 &= \left[\boldsymbol{I}_{n_u N_u \times n_u N_u} \ \boldsymbol{0}_{n_u N_u \times 2n_x} \right] \\ \boldsymbol{N}_2 &= \left[\boldsymbol{0}_{n_x \times n_u N_u} \ \boldsymbol{I}_{n_x \times n_x} \ \boldsymbol{0}_{n_x \times n_x} \right] \\ \boldsymbol{N}_3 &= \left[\boldsymbol{0}_{n_x \times (n_u N_u + n_x)} \ \boldsymbol{I}_{n_x \times n_x} \right] \end{aligned}$$

4.4 Suboptimal MPC Algorithms with On-Line Linearisation

Two suboptimal MPC algorithms based on the neural state-space model are discussed: MPC-NPL and MPC-NPLPT ones, in two versions: with the output and state set-point trajectories. The general description of MPC presented in Chapter 2.3 holds true, but when the output set-point trajectory is used, some changes are necessary (linearisation, calculation of the predicted trajectory). More serious modifications are necessary when the state set-point trajectory is taken into account.

MPC-NPL Algorithm

Using the Taylor series expansion formula (2.25), the linear approximation of the nonlinear state-space model (4.2) is

$$\begin{aligned} x(k) &= f(\bar{x}(k-1), \bar{u}(k-1), \bar{h}(k-1)) + \boldsymbol{A}(k)(x(k-1) - \bar{x}(k-1)) \\ &\quad + \boldsymbol{B}(k)(u(k-1) - \bar{u}(k-1)) \\ y(k) &= g(\bar{x}(k)) + \boldsymbol{C}(k)(x(k) - \bar{x}(k)) \end{aligned} \tag{4.23}$$

where the measured signals $\bar{x}(k-1)$, $\bar{u}(k-1)$, $\bar{h}(k-1)$ define the current operating point. The linearised model (4.23) may be transform to its incremental form

$$\begin{aligned} \delta\tilde{x}(k) &= \boldsymbol{A}(k)\delta x(k-1) + \boldsymbol{B}(k)\delta u(k-1) \\ \delta\tilde{y}(k) &= \boldsymbol{C}(k)\delta x(k) \end{aligned}$$

where $\delta\tilde{x}(k) = x(k) - f(\bar{x}(k-1), \bar{u}(k-1), \bar{h}(k-1))$, $\delta x(k-1) = x(k-1) - \bar{x}(k-1)$, $\delta u(k-1) = u(k-1) - \bar{u}(k-1)$, $\delta\tilde{y}(k) = y(k) - g(\bar{x}(k))$ and

$\delta x(k) = x(k) - \bar{x}(k)$. Neglecting the notation of incremental variables, the linearised model becomes

$$\begin{aligned} x(k) &= \boldsymbol{A}(k)x(k-1) + \boldsymbol{B}(k)u(k-1) \\ y(k) &= \boldsymbol{C}(k)x(k) \end{aligned} \tag{4.24}$$

The matrices of the linearised model, of dimensionality $n_{\mathrm{x}} \times n_{\mathrm{x}}$, $n_{\mathrm{x}} \times n_{\mathrm{u}}$ and $n_{\mathrm{y}} \times n_{\mathrm{x}}$, respectively, are calculated analytically on-line from the general equations

$$\begin{aligned} \boldsymbol{A}(k) &= \left.\frac{\mathrm{d}f(x(k-1), u(k-1), h(k-1))}{\mathrm{d}x(k-1)}\right|_{\substack{x(k-1)=\bar{x}(k-1)\\ u(k-1)=\bar{u}(k-1)\\ h(k-1)=\bar{h}(k-1)}} \\ \boldsymbol{B}(k) &= \left.\frac{\mathrm{d}f(x(k-1), u(k-1), h(k-1))}{\mathrm{d}u(k-1)}\right|_{\substack{x(k-1)=\bar{x}(k-1)\\ u(k-1)=\bar{u}(k-1)\\ h(k-1)=\bar{h}(k-1)}} \\ \boldsymbol{C}(k) &= \left.\frac{\mathrm{d}g(x(k))}{\mathrm{d}x(k)}\right|_{x(k)=\bar{x}(k)} \end{aligned} \tag{4.25}$$

The state and input signals from the previous iteration are arguments of the linearised model. The measured disturbances, which define the operating point, of course influence the model matrices $\boldsymbol{A}(k)$, $\boldsymbol{B}(k)$ and $\boldsymbol{C}(k)$.

The current state vector $x(k)$ may be measured or calculated by an observer, the current values of disturbances $h(k)$ may be also measured. They can be used for linearisation. For that purpose it would be necessary to take into account the manipulated variables for the current sampling instant, but they are calculated from the MPC optimisation problem solved after linearisation. That is why for definition of the current operating point, which is used for linearisation, the signals $u(k-1)$ calculated and used for control at the previous sampling instant or the signals $u(k|k-1)$ calculated for the current iteration at the previous iteration may be used in place of the unavailable current control signals $u(k|k)$.

Using recurrently the linearised state equation (4.24), from the general prediction equation (4.11) it is possible to calculate the predicted state vector for the whole prediction horizon ($p = 1, \ldots, N$)

$$\begin{aligned} \hat{x}(k+1|k) &= \boldsymbol{A}(k)x(k) + \boldsymbol{B}(k)u(k|k) + \nu(k) \\ \hat{x}(k+2|k) &= \boldsymbol{A}(k)\hat{x}(k+1|k) + \boldsymbol{B}(k)u(k+1|k) + \nu(k) \\ \hat{x}(k+3|k) &= \boldsymbol{A}(k)\hat{x}(k+2|k) + \boldsymbol{B}(k)u(k+2|k) + \nu(k) \\ &\vdots \end{aligned}$$

The state predictions can be expressed as functions of the increments of the future control increments (similarly to Eq. (2.30) the influence of the past is not taken into account)

$$\begin{aligned}\hat{x}(k+1|k) &= \boldsymbol{B}(k)\triangle u(k|k) + \ldots \\ \hat{x}(k+2|k) &= (\boldsymbol{A}(k)+\boldsymbol{I})\boldsymbol{B}(k)\triangle u(k|k) + \boldsymbol{B}(k)\triangle u(k+1|k) + \ldots \\ \hat{x}(k+3|k) &= (\boldsymbol{A}^2(k)+\boldsymbol{A}(k)+\boldsymbol{I})\boldsymbol{B}(k)\triangle u(k|k) \\ &\quad + (\boldsymbol{A}(k)+\boldsymbol{I})\boldsymbol{B}(k)\triangle u(k+1|k) + \boldsymbol{B}(k)\triangle u(k+2|k) + \ldots \\ &\vdots\end{aligned} \tag{4.26}$$

From Eq. (4.26), the state vector predicted over the prediction horizon, of length $n_{\mathrm{x}}N$, can be expressed as

$$\hat{\boldsymbol{x}}(k) = \underbrace{\boldsymbol{P}(k)\triangle\boldsymbol{u}(k)}_{\text{future}} + \underbrace{\boldsymbol{x}^0(k)}_{\text{past}} \tag{4.27}$$

where the matrix

$$\boldsymbol{P}(k) = \begin{bmatrix} \boldsymbol{B}(k) & \ldots & \boldsymbol{0}_{n_{\mathrm{x}}\times n_{\mathrm{u}}} \\ (\boldsymbol{A}(k)+\boldsymbol{I})\boldsymbol{B}(k) & \ldots & \boldsymbol{0}_{n_{\mathrm{x}}\times n_{\mathrm{u}}} \\ \vdots & \ddots & \vdots \\ (\sum_{i=1}^{N_{\mathrm{u}}-1}\boldsymbol{A}^i(k)+\boldsymbol{I})\boldsymbol{B}(k) & \ldots & \boldsymbol{B}(k) \\ (\sum_{i=1}^{N_{\mathrm{u}}}\boldsymbol{A}^i(k)+\boldsymbol{I})\boldsymbol{B}(k) & \ldots & (\boldsymbol{A}(k)+\boldsymbol{I})\boldsymbol{B}(k) \\ \vdots & \ddots & \vdots \\ (\sum_{i=1}^{N-1}\boldsymbol{A}^i(k)+\boldsymbol{I})\boldsymbol{B}(k) & \ldots & (\sum_{i=1}^{N-N_{\mathrm{u}}}\boldsymbol{A}^i(k)+\boldsymbol{I})\boldsymbol{B}(k) \end{bmatrix} \tag{4.28}$$

is of dimensionality $n_{\mathrm{x}}N \times n_{\mathrm{u}}N_{\mathrm{u}}$ and the state free trajectory vector

$$\boldsymbol{x}^0(k) = \begin{bmatrix} x^0(k+1|k) \\ \vdots \\ x^0(k+N|k) \end{bmatrix}$$

is of length $n_{\mathrm{x}}N$. From the linearised output equation (4.24), one obtains the predicted output trajectory

$$\hat{\boldsymbol{y}}(k) = \underbrace{\widetilde{\boldsymbol{C}}(k)\boldsymbol{P}(k)\triangle\boldsymbol{u}(k)}_{\text{future}} + \underbrace{\boldsymbol{y}^0(k)}_{\text{past}} \tag{4.29}$$

where the matrix $\widetilde{\boldsymbol{C}}(k) = \mathrm{diag}(\boldsymbol{C}(k), \ldots, \boldsymbol{C}(k))$ is of dimensionality $n_{\mathrm{y}}N \times n_{\mathrm{x}}N$. The matrices $\boldsymbol{P}(k)$ and $\widetilde{\boldsymbol{C}}(k)$ are calculated at each sampling instant using the current linear approximation (4.24) of the nonlinear model.

It is interesting to notice that the equations derived above are quite similar to the prediction equation (2.33) used in the MPC-NPL algorithm based on the input-output model. The first part of the right side of Eq. (4.27) depends only on the future (it is a function of the matrix $\boldsymbol{P}(k)$ calculated from the current linear approximation of the nonlinear model and of the future control increments $\triangle\boldsymbol{u}(k)$). The state free trajectory $\boldsymbol{x}^0(k)$ depends only on the past. Similarly, the first part of the right side of Eq. (4.29) depends only on the future, the output free trajectory $\boldsymbol{y}^0(k)$ depends only on the past. The state free trajectory or the output free trajectory is calculated from the full nonlinear model. Of course, the free trajectories can be calculated from the linearised model, which leads to the MPC-SL algorithm.

Using the suboptimal output prediction equation (4.29), the general MPC optimisation problem with the output set-point trajectory (1.9) can be transformed to the following quadratic optimisation task

$$\min_{\substack{\triangle\boldsymbol{u}(k)\\ \boldsymbol{\varepsilon}^{\min}(k),\ \boldsymbol{\varepsilon}^{\max}(k)}} \Big\{J(k) = \left\|\boldsymbol{y}^{\mathrm{sp}}(k) - \widetilde{\boldsymbol{C}}(k)\boldsymbol{P}(k)\triangle\boldsymbol{u}(k) - \boldsymbol{y}^0(k)\right\|^2_{\boldsymbol{M}} + \left\|\triangle\boldsymbol{u}(k)\right\|^2_{\boldsymbol{\Lambda}} + \rho^{\min}\left\|\boldsymbol{\varepsilon}^{\min}(k)\right\|^2 + \rho^{\max}\left\|\boldsymbol{\varepsilon}^{\max}(k)\right\|^2\Big\}$$

subject to

$$\boldsymbol{u}^{\min} \leq \boldsymbol{J}\triangle\boldsymbol{u}(k) + \boldsymbol{u}(k-1) \leq \boldsymbol{u}^{\max}$$
$$-\triangle\boldsymbol{u}^{\max} \leq \triangle\boldsymbol{u}(k) \leq \triangle\boldsymbol{u}^{\max}$$
$$\boldsymbol{y}^{\min} - \boldsymbol{\varepsilon}^{\min}(k) \leq \widetilde{\boldsymbol{C}}(k)\boldsymbol{P}(k)\triangle\boldsymbol{u}(k) + \boldsymbol{y}^0(k) \leq \boldsymbol{y}^{\max} + \boldsymbol{\varepsilon}^{\max}(k)$$
$$\boldsymbol{\varepsilon}^{\min}(k) \geq \boldsymbol{0},\ \boldsymbol{\varepsilon}^{\max}(k) \geq \boldsymbol{0}$$

The obtained optimisation problem is very similar to the task (2.36), which is used in the MPC-NPL algorithm based on the input-output model. The only difference is the prediction equation, which results from replacing the input-output model by the state-space one.

Alternatively, the MPC-NPL algorithm with the state set-point trajectory can be used. In such a case the model consists of only the first neural network. From the suboptimal state prediction equation (4.27), the general MPC optimisation problem with the state set-point trajectory (4.10) can be transformed to the following quadratic optimisation task

$$\min_{\substack{\triangle\boldsymbol{u}(k)\\ \boldsymbol{\varepsilon}^{\min}(k),\ \boldsymbol{\varepsilon}^{\max}(k)}} \Big\{J(k) = \left\|\boldsymbol{x}^{\mathrm{sp}}(k) - \boldsymbol{P}(k)\triangle\boldsymbol{u}(k) - \boldsymbol{x}^0(k)\right\|^2_{\boldsymbol{\Phi}} + \left\|\triangle\boldsymbol{u}(k)\right\|^2_{\boldsymbol{\Lambda}} + \rho^{\min}\left\|\boldsymbol{\varepsilon}^{\min}(k)\right\|^2 + \rho^{\max}\left\|\boldsymbol{\varepsilon}^{\max}(k)\right\|^2\Big\}$$

subject to

$$\boldsymbol{u}^{\min} \leq \boldsymbol{J}\triangle\boldsymbol{u}(k) + \boldsymbol{u}(k-1) \leq \boldsymbol{u}^{\max}$$
$$-\triangle\boldsymbol{u}^{\max} \leq \triangle\boldsymbol{u}(k) \leq \triangle\boldsymbol{u}^{\max}$$
$$\boldsymbol{x}^{\min} - \boldsymbol{\varepsilon}^{\min}(k) \leq \boldsymbol{P}(k)\triangle\boldsymbol{u}(k) + \boldsymbol{x}^0(k) \leq \boldsymbol{x}^{\max} + \boldsymbol{\varepsilon}^{\max}(k)$$
$$\boldsymbol{\varepsilon}^{\min}(k) \geq \boldsymbol{0},\ \boldsymbol{\varepsilon}^{\max}(k) \geq \boldsymbol{0}$$

When the state set-point trajectory is used, the vectors $\boldsymbol{\varepsilon}^{\min}(k)$ and $\boldsymbol{\varepsilon}^{\max}(k)$ are of length $n_{\mathrm{x}}N$.

The linear approximation of the nonlinear state-space model (4.2) is defined by the general equations (4.24). At each iteration of the algorithm the parameters of the linearised model are calculated for the neural state-space model characterised by Eqs. (4.3), (4.4), (4.5), (4.6), (4.7) and (4.8). Taking into account Eqs. (4.25), the matrices of the linearised model are

$$\boldsymbol{A}(k) = \begin{bmatrix} \sum_{j=1}^{K^{\mathrm{I}}} w_{1,j}^{2,\mathrm{I}} \frac{\mathrm{d}\varphi(z_j^{\mathrm{I}}(k))}{\mathrm{d}z_j^{\mathrm{I}}(k)} w_{j,1}^{1,\mathrm{I}} & \cdots & \sum_{j=1}^{K^{\mathrm{I}}} w_{1,j}^{2,\mathrm{I}} \frac{\mathrm{d}\varphi(z_j^{\mathrm{I}}(k))}{\mathrm{d}z_j^{\mathrm{I}}(k)} w_{j,n_{\mathrm{x}}}^{1,\mathrm{I}} \\ \vdots & \ddots & \vdots \\ \sum_{j=1}^{K^{\mathrm{I}}} w_{n_{\mathrm{x}},j}^{2,\mathrm{I}} \frac{\mathrm{d}\varphi(z_j^{\mathrm{I}}(k))}{\mathrm{d}z_j^{\mathrm{I}}(k)} w_{j,1}^{1,\mathrm{I}} & \cdots & \sum_{j=1}^{K^{\mathrm{I}}} w_{n_{\mathrm{x}},j}^{2,\mathrm{I}} \frac{\mathrm{d}\varphi(z_j^{\mathrm{I}}(k))}{\mathrm{d}z_j^{\mathrm{I}}(k)} w_{j,n_{\mathrm{x}}}^{1,\mathrm{I}} \end{bmatrix}$$

$$\boldsymbol{B}(k) = \begin{bmatrix} \sum_{j=1}^{K^{\mathrm{I}}} w_{1,j}^{2,\mathrm{I}} \frac{\mathrm{d}\varphi(z_j^{\mathrm{I}}(k))}{\mathrm{d}z_j^{\mathrm{I}}(k)} w_{j,n_{\mathrm{x}}+1}^{1,\mathrm{I}} & \cdots & \sum_{j=1}^{K^{\mathrm{I}}} w_{1,j}^{2,\mathrm{I}} \frac{\mathrm{d}\varphi(z_j^{\mathrm{I}}(k))}{\mathrm{d}z_j^{\mathrm{I}}(k)} w_{j,n_{\mathrm{x}}+n_{\mathrm{u}}}^{1,\mathrm{I}} \\ \vdots & \ddots & \vdots \\ \sum_{j=1}^{K^{\mathrm{I}}} w_{n_{\mathrm{x}},j}^{2,\mathrm{I}} \frac{\mathrm{d}\varphi(z_j^{\mathrm{I}}(k))}{\mathrm{d}z_j^{\mathrm{I}}(k)} w_{j,n_{\mathrm{x}}+1}^{1,\mathrm{I}} & \cdots & \sum_{j=1}^{K^{\mathrm{I}}} w_{n_{\mathrm{x}},j}^{2,\mathrm{I}} \frac{\mathrm{d}\varphi(z_j^{\mathrm{I}}(k))}{\mathrm{d}z_j^{\mathrm{I}}(k)} w_{j,n_{\mathrm{x}}+n_{\mathrm{u}}}^{1,\mathrm{I}} \end{bmatrix}$$

$$\boldsymbol{C}(k) = \begin{bmatrix} \sum_{l=1}^{K^{\mathrm{II}}} w_{1,l}^{2,\mathrm{II}} \frac{\mathrm{d}\varphi(z_l^{\mathrm{II}}(k))}{\mathrm{d}z_l^{\mathrm{II}}(k)} w_{l,1}^{1,\mathrm{II}} & \cdots & \sum_{l=1}^{K^{\mathrm{II}}} w_{1,l}^{2,\mathrm{II}} \frac{\mathrm{d}\varphi(z_l^{\mathrm{II}}(k))}{\mathrm{d}z_l^{\mathrm{II}}(k)} w_{l,n_{\mathrm{x}}}^{1,\mathrm{II}} \\ \vdots & \ddots & \vdots \\ \sum_{l=1}^{K^{\mathrm{II}}} w_{n_{\mathrm{y}},l}^{2,\mathrm{II}} \frac{\mathrm{d}\varphi(z_l^{\mathrm{II}}(k))}{\mathrm{d}z_l^{\mathrm{II}}(k)} w_{l,1}^{1,\mathrm{II}} & \cdots & \sum_{l=1}^{K^{\mathrm{II}}} w_{n_{\mathrm{y}},l}^{2,\mathrm{II}} \frac{\mathrm{d}\varphi(z_l^{\mathrm{II}}(k))}{\mathrm{d}z_l^{\mathrm{II}}(k)} w_{l,n_{\mathrm{x}}}^{1,\mathrm{II}} \end{bmatrix}$$

The elements of the nonlinear output free trajectory $\boldsymbol{y}^0(k)$ are calculated recurrently for all $m = 1, \ldots, n_{\mathrm{y}}$ and $p = 1, \ldots, N$ using the nonlinear model of the process (in the same way it is done in the MPC-NO algorithm). From the nonlinear output prediction equation (4.17) and considering only the past, one has

$$y_m^0(k+p|k) = w_{m,0}^{2,\mathrm{II}} + \sum_{l=1}^{K^{\mathrm{II}}} w_{m,l}^{2,\mathrm{II}} \varphi\left(w_{l,0}^{1,\mathrm{II}} + \sum_{i=1}^{n_{\mathrm{x}}} w_{l,i}^{1,\mathrm{II}} x_i^0(k+p|k)\right) + d_m(k)$$

The elements of the nonlinear state trajectory $\boldsymbol{x}^0(k)$ are calculated recurrently for all $i = 1, \ldots, n_{\mathrm{x}}$ and $p = 1, \ldots, N$, also using the nonlinear model of the process. From the nonlinear state prediction equation (4.13), one obtains

$$x_i^0(k+p|k) = w_{i,0}^{2,\mathrm{I}} + \sum_{j=1}^{K^{\mathrm{I}}} w_{i,j}^{2,\mathrm{I}} \varphi(z_j^{\mathrm{I},0}(k+p|k)) + \nu_i(k)$$

where, considering only the past, from Eq. (4.14), it follows that

$$z_j^{\mathrm{I},0}(k+1|k) = w_{j,0}^{1,\mathrm{I}} + \sum_{s=1}^{n_\mathrm{x}} w_{j,s}^{1,\mathrm{I}} x_s(k) + \sum_{n=1}^{n_\mathrm{u}} w_{j,n_\mathrm{x}+n}^{1,\mathrm{I}} u_n(k-1)$$
$$+ \sum_{n=1}^{n_\mathrm{h}} w_{j,n_\mathrm{x}+n_\mathrm{u}+n}^{1,\mathrm{I}} h_n(k+p-1|k)$$

From Eq. (4.15), one has

$$z_j^{\mathrm{I},0}(k+p|k) = w_{j,0}^{1,\mathrm{I}} + \sum_{s=1}^{n_\mathrm{x}} w_{j,s}^{1,\mathrm{I}} x_s^0(k+p-1|k) + \sum_{n=1}^{n_\mathrm{u}} w_{j,n_\mathrm{x}+n}^{1,\mathrm{I}} u_n(k-1)$$
$$+ \sum_{n=1}^{n_\mathrm{h}} w_{j,n_\mathrm{x}+n_\mathrm{u}+n}^{1,\mathrm{I}} h_n(k+p-1|k)$$

The unmeasured state disturbances $\nu_i(k)$ and the unmeasured output disturbances $d_m(k)$ are calculated in the same way it is done in the MPC-NO algorithm for the neural state-space model, i.e. from Eqs. (4.16) and (4.18).

MPC-NPLPT Algorithm

The formulae used in the MPC-NPLPT algorithm are quite similar to those used in the MPC-NO one. From Eq. (4.17), in the tth internal iteration the predicted output trajectory is

$$\hat{y}_m^t(k+p|k) = w_{m,0}^{2,\mathrm{II}} + \sum_{l=1}^{K^{\mathrm{II}}} w_{m,l}^{2,\mathrm{II}} \varphi\left(w_{l,0}^{1,\mathrm{II}} + \sum_{i=1}^{n_\mathrm{x}} w_{l,i}^{1,\mathrm{II}} \hat{x}_i^t(k+p|k) \right) + d_m(k) \tag{4.30}$$

where, using Eq. (4.13), the predicted state trajectory is

$$\hat{x}_i^t(k+p|k) = w_{i,0}^{2,\mathrm{I}} + \sum_{j=1}^{K^\mathrm{I}} w_{i,j}^{2,\mathrm{I}} \varphi(z_j^{\mathrm{I},t}(k+p|k)) + \nu_i(k) \tag{4.31}$$

From Eq. (4.14), one has

$$z_j^{\mathrm{I},t}(k+1|k) = w_{j,0}^{1,\mathrm{I}} + \sum_{s=1}^{n_\mathrm{x}} w_{j,s}^{1,\mathrm{I}} x_s(k) + \sum_{n=1}^{n_\mathrm{u}} w_{j,n_\mathrm{x}+n}^{1,\mathrm{I}} u_n^t(k|k)$$
$$+ \sum_{n=1}^{n_\mathrm{h}} w_{j,n_\mathrm{x}+n_\mathrm{u}+n}^{1,\mathrm{I}} h_n(k) \tag{4.32}$$

and from Eq. (4.15), one obtains

$$z_j^{\mathrm{I},t}(k+p|k) = w_{j,0}^{1,\mathrm{I}} + \sum_{s=1}^{n_\mathrm{x}} w_{j,s}^{1,\mathrm{I}} \hat{x}_s^t(k+p-1|k) + \sum_{n=1}^{n_\mathrm{u}} w_{j,n_\mathrm{x}+n}^{1,\mathrm{I}} u_n^t(k+p-1|k)$$
$$+ \sum_{n=1}^{n_\mathrm{h}} w_{j,n_\mathrm{x}+n_\mathrm{u}+n}^{1,\mathrm{I}} h_n(k+p-1|k) \tag{4.33}$$

In the MPC-NPLPT algorithm with the output set-point trajectory, according to Eq. (2.52), linearisation of the predicted output trajectory is carried out for the future input trajectory $\boldsymbol{u}^{t-1}(k)$ at each internal iteration t. The entries of the matrix $\boldsymbol{H}^t(k)$, of the general structure (2.53), are calculated differentiating Eq. (4.30) which leads to

$$\frac{\partial \hat{y}_m^{t-1}(k+p|k)}{\partial u_n^{t-1}(k+r|k)} = \sum_{l=1}^{K^{\mathrm{II}}} w_{m,l}^{2,\mathrm{II}} \frac{\mathrm{d}\varphi(z_l^{\mathrm{II},t-1}(k+p|k))}{\mathrm{d}z_l^{\mathrm{II},t-1}(k+p|k)} \sum_{i=1}^{n_\mathrm{x}} w_{l,i}^{1,\mathrm{II}} \frac{\partial \hat{x}_i^{t-1}(k+p|k)}{\partial u_n^{t-1}(k+r|k)}$$

where, from Eq. (4.7), one has

$$z_l^{\mathrm{II},t-1}(k+p|k) = w_{l,0}^{1,\mathrm{II}} + \sum_{i=1}^{n_\mathrm{x}} w_{l,i}^{1,\mathrm{II}} \hat{x}_i^{t-1}(k+p|k)$$

and from Eq. (4.31), one obtains

$$\frac{\partial \hat{x}_i^{t-1}(k+p|k)}{\partial u_n^{t-1}(k+r|k)} = \sum_{j=1}^{K^{\mathrm{I}}} w_{i,j}^{2,\mathrm{I}} \frac{\mathrm{d}\varphi(z_j^{\mathrm{I},t-1}(k+p|k))}{\mathrm{d}z_j^{\mathrm{I},t-1}(k+p|k)} \frac{\partial z_j^{\mathrm{I},t-1}(k+p|k)}{\partial u_n^{t-1}(k+r|k)}$$

Using Eq. (4.32), it follows that

$$\frac{\partial z_j^{\mathrm{I},t-1}(k+1|k)}{\partial u_n^{t-1}(k+r|k)} = \begin{cases} w_{j,n_\mathrm{x}+n}^{1,\mathrm{I}} & \text{if } r=0 \\ 0 & \text{if } r=1,\ldots,N_\mathrm{u}-1 \end{cases}$$

and from Eq. (4.33), one has

$$\frac{\partial z_j^{\mathrm{I},t-1}(k+p|k)}{\partial u_n^{t-1}(k+r|k)} = \sum_{s=1}^{n_\mathrm{x}} w_{j,s}^{1,\mathrm{I}} \frac{\partial \hat{x}_s^{t-1}(k+p-1|k)}{\partial u_n^{t-1}(k+r|k)}$$
$$+ w_{j,n_\mathrm{x}+n}^{1,\mathrm{I}} \frac{\partial u_n^{t-1}(k+p-1|k)}{\partial u_n^{t-1}(k+r|k)}$$

If the state set-point trajectory is taken into account in the MPC-NPLPT algorithm, it is necessary to find on-line a linear approximation of the nonlinear predicted state trajectory $\hat{\boldsymbol{x}}^t(k)$ along the future control trajectory $\boldsymbol{u}^{t-1}(k)$. Using the Taylor series expansion formula (2.25), a linear approximation of the nonlinear function $\hat{\boldsymbol{x}}^t(\boldsymbol{u}^t(k))\colon \mathbb{R}^{n_\mathrm{u}N_\mathrm{u}} \to \mathbb{R}^{n_\mathrm{x}N}$ can be expressed as

$$\hat{\boldsymbol{x}}^t(k) = \hat{\boldsymbol{x}}^{t-1}(k) + \boldsymbol{H}^t(k)(\boldsymbol{u}^t(k) - \boldsymbol{u}^{t-1}(k)) \tag{4.34}$$

where the matrix

$$\boldsymbol{H}^t(k) = \left.\frac{\mathrm{d}\hat{\boldsymbol{x}}(k)}{\mathrm{d}\boldsymbol{u}(k)}\right|_{\substack{\hat{\boldsymbol{x}}(k)=\hat{\boldsymbol{x}}^{t-1}(k)\\ \boldsymbol{u}(k)=\boldsymbol{u}^{t-1}(k)}} = \frac{\mathrm{d}\hat{\boldsymbol{x}}^{t-1}(k)}{\mathrm{d}\boldsymbol{u}^{t-1}(k)}$$
$$= \begin{bmatrix} \frac{\partial \hat{x}^{t-1}(k+1|k)}{\partial u^{t-1}(k|k)} & \cdots & \frac{\partial \hat{x}^{t-1}(k+1|k)}{\partial u^{t-1}(k+N_{\mathrm{u}}-1|k)} \\ \vdots & \ddots & \vdots \\ \frac{\partial \hat{x}^{t-1}(k+N|k)}{\partial u^{t-1}(k|k)} & \cdots & \frac{\partial \hat{x}^{t-1}(k+N|k)}{\partial u^{t-1}(k+N_{\mathrm{u}}-1|k)} \end{bmatrix} \tag{4.35}$$

is of dimensionality of $n_{\mathrm{x}}N \times n_{\mathrm{u}}N_{\mathrm{u}}$. Linearisation of the state trajectory is very similar to linearisation of the output trajectory: Eq. (4.34) is the counterpart of Eq. (2.52), Eq. (4.35) is the counterpart of Eq. (2.53).

Using Eqs. (2.54) and (4.34), it is possible to formulate the quadratic optimisation problem solved in the MPC-NPLPT algorithm with the state set-point trajectory, which is similar to the task (2.56) solved when the output set-point trajectory is used

$$\min_{\substack{\triangle \boldsymbol{u}^t(k)\\ \boldsymbol{\varepsilon}^{\min}(k),\ \boldsymbol{\varepsilon}^{\max}(k)}} \Big\{ J(k) = \big\|\boldsymbol{x}^{\mathrm{sp}}(k) - \boldsymbol{H}^t(k)\boldsymbol{J}\triangle\boldsymbol{u}^t(k) - \hat{\boldsymbol{x}}^{t-1}(k) - \boldsymbol{H}^t(k)(\boldsymbol{u}(k-1) - \boldsymbol{u}^{t-1}(k))\big\|_{\boldsymbol{\Phi}}^2 + \|\triangle\boldsymbol{u}^t(k)\|_{\boldsymbol{\Lambda}}^2 + \rho^{\min}\left\|\boldsymbol{\varepsilon}^{\min}(k)\right\|^2 + \rho^{\max}\left\|\boldsymbol{\varepsilon}^{\max}(k)\right\|^2 \Big\}$$

subject to

$$\boldsymbol{u}^{\min} \leq \boldsymbol{J}\triangle\boldsymbol{u}^t(k) + \boldsymbol{u}(k-1) \leq \boldsymbol{u}^{\max}$$
$$-\triangle\boldsymbol{u}^{\max} \leq \triangle\boldsymbol{u}^t(k) \leq \triangle\boldsymbol{u}^{\max}$$
$$\boldsymbol{x}^{\min} - \boldsymbol{\varepsilon}^{\min}(k) \leq \boldsymbol{H}^t(k)\boldsymbol{J}\triangle\boldsymbol{u}^t(k) + \hat{\boldsymbol{x}}^{t-1}(k) + \boldsymbol{H}^t(k)(\boldsymbol{u}(k-1) - \boldsymbol{u}^{t-1}(k)) \leq \boldsymbol{x}^{\max} + \boldsymbol{\varepsilon}^{\max}(k)$$
$$\boldsymbol{\varepsilon}^{\min}(k) \geq 0,\ \boldsymbol{\varepsilon}^{\max}(k) \geq 0$$

One may notice that when the state set-point trajectory is used calculations are simpler than in the case of the output trajectory. It is because in the first case the model consists of only one neural network.

In the case of the state set-point trajectory, the internal iteration continuation condition (given by Eq. (2.57) when the output set-point trajectory is used) has the following form

$$\sum_{p=0}^{N_0}(x^{\mathrm{sp}}(k-p) - x(k-p))^2 \geq \delta_{\mathrm{x}}$$

where the quantity $\delta_{\mathrm{x}} > 0$ is adjusted experimentally.

4.5 Example 4.1

The considered process is the polymerisation reactor described in the previous chapter (p. 124). The reactor can be controlled effectively by MPC algorithms based on different neural models (of DLP, Hammerstein and Wiener types). All the models considered so far are input-output structures, it is interesting to verify MPC algorithms based on the state-space model.

Polymerisation Reactor Modelling

In order to obtain data necessary for identification of the state-space models, open-loop simulations of the first-principle model are performed. As a result, the training, validation and test data sets are obtained. Not only input and output signals are recorded (in the same way it is done for identification of input-outputs models), but also four state variables. Each set has 2000 samples. The output and state signals contain small measurement noise. Scaling of input and output signals is the same as previously, i.e. $u = 100(F_{\mathrm{I}} - F_{\mathrm{I,nom}})$, $y = 0.0001(\mathrm{NAMW} - \mathrm{NAMW}_{\mathrm{nom}})$, the state variables are scaled as: $x_1 = 3(C_{\mathrm{m}} - C_{\mathrm{m,nom}})$, $x_2 = 4(C_{\mathrm{I}} - C_{\mathrm{I,nom}})$, $x_3 = 400(D_0 - D_{0,\mathrm{nom}})$, $x_4 = 0.025(D_{\mathrm{I}} - D_{\mathrm{I,nom}})$, where $F_{\mathrm{I,nom}} = 0.028328$, $\mathrm{NAMW}_{\mathrm{nom}} = 20000$, $C_{\mathrm{m,nom}} = 5.3745$, $C_{\mathrm{I,nom}} = 2.2433 \times 10^{-1}$, $D_{0,\mathrm{nom}} = 3.1308 \times 10^{-3}$, $D_{\mathrm{I,nom}} = 6.2616 \times 10^{-1}$ correspond to the nominal operating point (Table 3.2).

The neural state-space model consists of two networks. They are trained separately. At first, the state network is trained using the recorded input and state signals. Next, the output network is trained using the recorded states. Both neural networks are independently verified using validation and test data sets. Finally, the complete model comprised of two networks is verified.

Accuracy of the first network is assessed using the SSE index which takes into account the state errors

$$\mathrm{SSE} = \sum_{k=1}^{n_{\mathrm{p}}} \sum_{n=1}^{n_{\mathrm{x}}} (x_n^{\mathrm{mod}}(k) - x_n(k))^2$$

where $x_n^{\mathrm{mod}}(k)$ denote the signals calculated from the model, $x_n(k)$ are the real values of the recorded states of the process, n_{p} is the number of samples, $n_{\mathrm{x}} = 4$. Accuracy of the second network is assessed by means of the SSE index defined by Eq. (1.20), which is used in the case of input-output models.

Table 4.1 compares empirical state linear and neural models of the polymerisation process in terms of the number of parameters and accuracy. For the yeast fermentation reactor (Table 2.3) and for the high-pressure distillation column (Table 2.6), the models with too many parameters have poor generalisation abilities (the validation error increases). That phenomenon is not observed in the case of the considered polymerisation reactor for the compared state networks, but one may expect that the generalisation error increases for more complicated networks. As a reasonable compromise between accuracy and complexity, the model with $K^{\mathrm{I}} = 4$ hidden nodes is chosen. Its validation error is by some 5.8% higher than that of the model with 5 hidden nodes, but it has as many as 10 weights less (which means that its number of parameters is lower by approximately 22%). For the validation data set, the neural model is approximately 130 times better than the linear one. The test error is also calculated for the chosen neural model. Its value is low, comparable with that for training and validation data sets. Fig. 4.3 compares the validation data set and the output x_1 of neural and linear state models, the differences between the data and model outputs are also shown.

Because, according to Eq. (3.40), the output NAMW of the first-principle model of the polymerisation reactor depends only on the state variables x_3 and x_4, it is natural to only take them as the inputs of the second neural network.

Table 4.1. Comparison of empirical state models of the polymerisation reactor in terms of the number of parameters (NP) and accuracy (SSE)

Model	NP	$\mathrm{SSE_{train}}$	$\mathrm{SSE_{val}}$	$\mathrm{SSE_{test}}$
Linear	20	5.4341×10^{1}	1.0795×10^{2}	–
Neural, $K^{\mathrm{I}} = 1$	14	3.0115×10^{2}	4.7162×10^{2}	–
Neural, $K^{\mathrm{I}} = 2$	24	1.5463×10^{1}	2.8606×10^{1}	–
Neural, $K^{\mathrm{I}} = 3$	34	7.1437×10^{-1}	1.6116×10^{0}	–
Neural, $K^{\mathrm{I}} = 4$	44	3.2193×10^{-1}	8.2677×10^{-1}	8.0422×10^{-1}
Neural, $K^{\mathrm{I}} = 5$	54	2.9298×10^{-1}	7.7903×10^{-1}	–
Neural, $K^{\mathrm{I}} = 6$	64	3.0500×10^{-1}	7.4233×10^{-1}	–
Neural, $K^{\mathrm{I}} = 7$	74	2.8785×10^{-1}	7.4196×10^{-1}	–

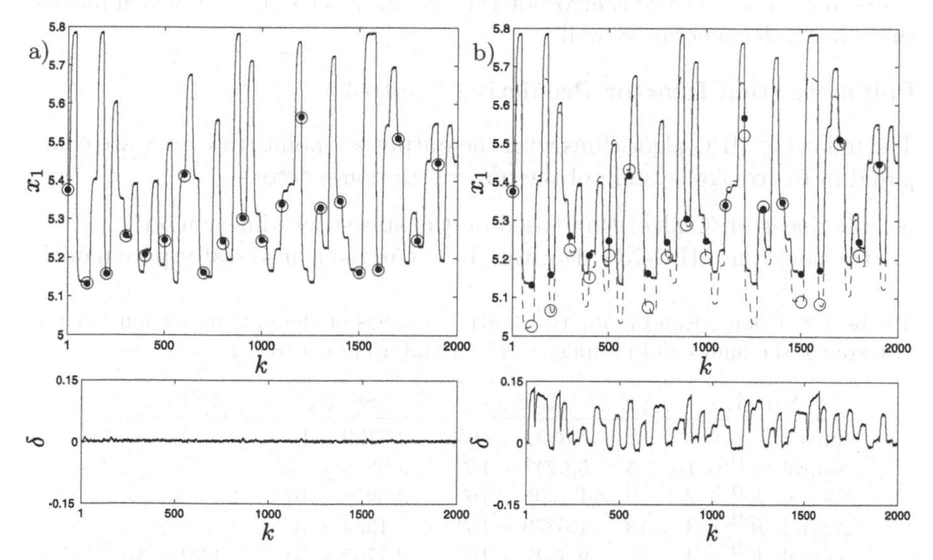

Fig. 4.3. Verification of the state model of the polymerisation reactor for the state variable x_1. The validation data set (*solid line with dots*) vs. the output x_1 of the model (*dashed line with circles*): a) the neural model, b) the linear model; δ – differences between the data and model outputs.

Table 4.2 compares empirical output linear and neural models of the process in terms of the number of parameters and accuracy. The model with $K^{\text{II}} = 4$ hidden nodes is chosen. Although its validation error is some 6.5% higher than the error for the network with 5 nodes, but it has 4 weights less (which means that its number of parameters is lower by approximately 19%). For the validation data set the neural model is more than 13 times more precise than the linear one. For the chosen neural model the test error is also calculated. Its value is low, comparable with that for training and validation data sets.

Because independent verification of the state and output networks gives good results, one may expect that they are likely to form a precise state-space model. Fig. 4.4 compares the validation data set and the outputs of the neural and linear state-space models, the differences between the data and model outputs are also shown. For the neural model the validation error is 1.1932, whereas for the linear one it soars to 5.4619×10^2, which means that the neural state-space model is approximately 458 times better. Fig. 4.4 is worth comparing with Fig. 3.10, in which similar data vs. model outputs comparisons are depicted, but for some chosen classes of input-output models of the polymerisation reactor. The obtained value of the SSE index should be also compared with the corresponding errors for some input-output models given in Table 3.4. High accuracy of the neural state-space model is very similar to that of the neural Wiener one, which is the best among other input-output structures. Nevertheless, it is necessary to point out that for identification of the state-space models not only to the output, but also to the state signals small measurement noise is deliberately added. For the neural state-space model the SSE error for the test data set is 1.3169, which means that the model generalises well.

Polymerisation Reactor Predictive Control

The following MPC algorithms with the output set-point trajectory are compared in the control system of the polymerisation reactor:

a) the linear MPC algorithm based on the state-space linear model,
b) the nonlinear MPC-NO algorithm based on the neural state-space model,

Table 4.2. Comparison of empirical output models of the polymerisation reactor in terms of the number of parameters (NP) and accuracy (SSE)

Model	NP	$\text{SSE}_{\text{train}}$	SSE_{val}	SSE_{test}
Linear	2	1.8832×10^1	3.7199×10^1	–
Neural, $K^{\text{II}} = 1$	5	5.6243×10^0	8.9816×10^0	–
Neural, $K^{\text{II}} = 2$	9	1.2495×10^0	3.6976×10^0	–
Neural, $K^{\text{II}} = 3$	13	1.0279×10^0	3.1334×10^0	–
Neural, $K^{\text{II}} = 4$	17	9.4247×10^{-1}	2.7709×10^0	2.3281×10^0
Neural, $K^{\text{II}} = 5$	21	8.2148×10^{-1}	2.9621×10^0	–
Neural, $K^{\text{II}} = 6$	25	7.6366×10^{-1}	2.8431×10^0	–
Neural, $K^{\text{II}} = 7$	29	6.8162×10^{-1}	2.7809×10^0	–

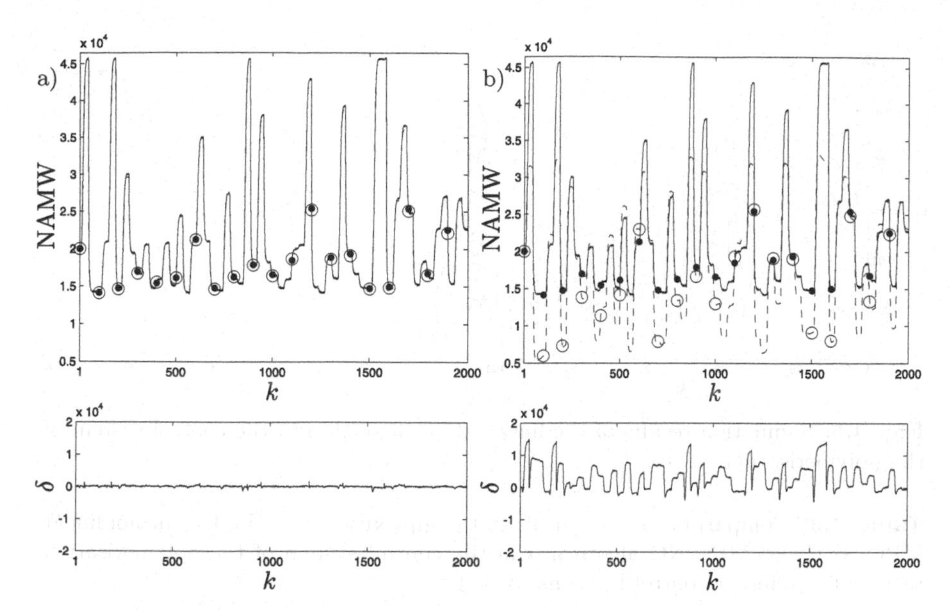

Fig. 4.4. Verification of the state and output models of the polymerisation reactor. The validation data set (*solid line with dots*) vs. the output y of the model (*dashed line with circles*): a) the neural model, b) the linear model; δ – differences between the data and model outputs.

c) the nonlinear MPC-NPL algorithm based on the same model.

The parameters of all algorithms are the same as in the case of the input-output MPC approaches discussed in the previous chapter (p. 128), namely: $N = 10$, $N_{\mathrm{u}} = 3$, $\lambda_p = 0.2$ for $p = 1, \ldots, N$, $F_{\mathrm{I}}^{\min} = 0.003$ m^3/h, $F_{\mathrm{I}}^{\max} = 0.06$ m^3/h. Because the objective of the following presentation is to compare the MPC algorithms themselves, it is assumed that all state variables are measured. In practice, however, it is rarely possible.

The MPC algorithm based on the linear state-space model gives unwanted oscillations, the process does not converge to the desired set-points, which is illustrated in Fig. 4.5. A similar result is obtained when for prediction a linear input-output model is used as shown in Fig. 3.11. Fig. 4.6 compares the simulation results obtained in two nonlinear MPC algorithms. The trajectories obtained in the suboptimal MPC-NPL algorithm are practically the same as in the MPC-NO one. Once again, one may notice that the results of two compared nonlinear algorithms are quite similar to those obtained when the neural input-output models are used (Figs. 3.12 and 3.13).

The value of the sum of squared errors over the simulation horizon ($\mathrm{SSE_{sim}}$) for the MPC-NO algorithm is 3.9550×10^8 whereas for the MPC-NLP one it is 3.9883×10^8. The obtained results are very similar to those for the MPC algorithms based on the neural input-output models (Table 3.5).

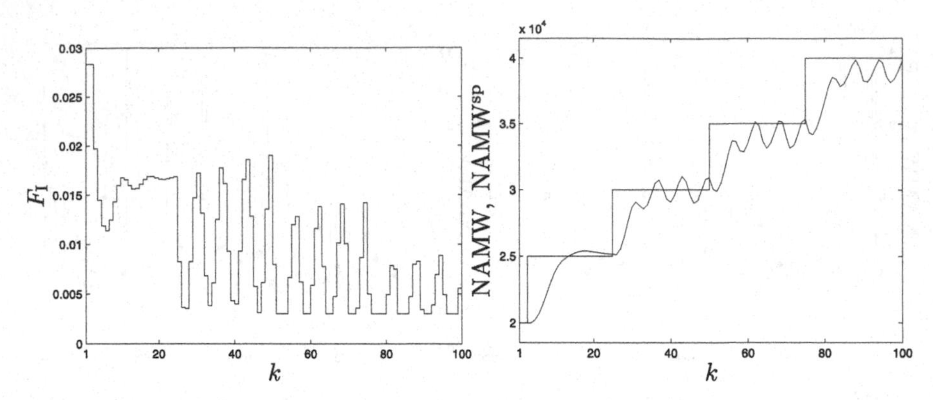

Fig. 4.5. Simulation results of the linear MPC algorithm in the control system of the polymerisation reactor

Table 4.3. Comparison of computational complexity (in MFLOPS) of nonlinear MPC-NPL and MPC-NO algorithms in the control system of the polymerisation reactor for different control horizons, $N = 10$

Algorithm	$N_u = 1$	$N_u = 2$	$N_u = 3$	$N_u = 4$	$N_u = 5$	$N_u = 10$
NO	4.07	7.43	10.01	13.75	16.37	46.02
NPL	0.65	0.77	0.91	1.09	1.31	3.02

Table 4.3 shows computational burden of compared nonlinear MPC algorithms for different control horizons. The MPC-NPL strategy is much more efficient than the MPC-NO one. Depending on the control horizon, the quantitative reduction of computational burden is in the range 6.2–15.2. It is worth comparing the obtained results with computational burden of the same MPC algorithms but based on the neural input-output Wiener model (Table 3.6). Unfortunately, computational burden of the MPC algorithms based on the state-space models is a few times higher. It is because the chosen neural state-space model (two networks) has as many as 61 parameters whereas the neural Wiener model has only 10 parameters (an inverse steady-state model must be used in the MPC-NPL scheme based on the neural input-output Wiener model, but it increases the number of parameters only by 7).

4.6 Literature Review

In spite of the fact that the state-space models are capable of describing a much wider class of nonlinear processes than the rudimentary input-output ones, their identification is significantly more demanding. In consequence, the MPC algorithms based on such models are less popular in the literature. Properties of the neural state-space models are discussed in [340], model stability issues are considered, a gradient training algorithm is detailed and

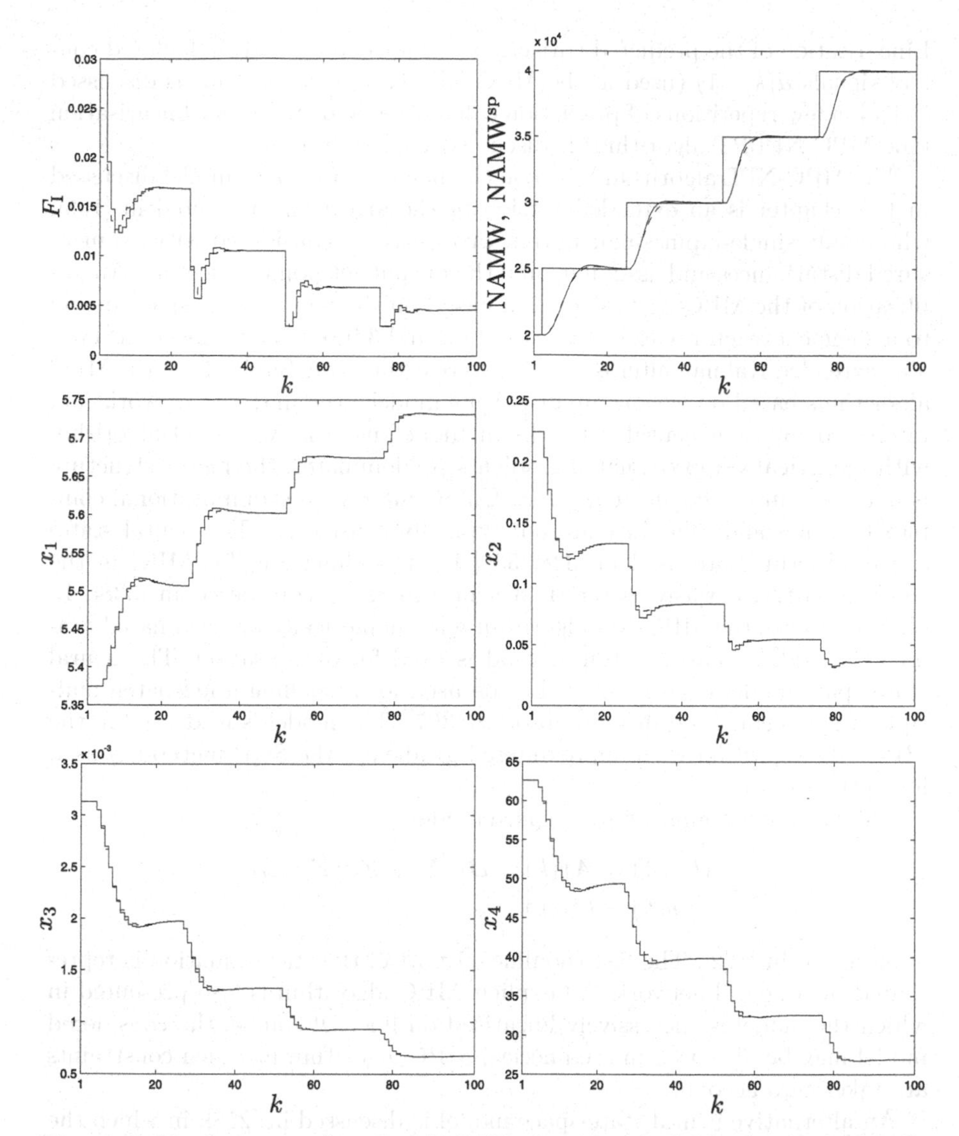

Fig. 4.6. Simulation results: the MPC-NO algorithm (*solid line*) and the MPC-NPL algorithm (*dashed line*) in the control system of the polymerisation reactor

modelling of a chemical reactor is reported. A similar model structure is presented in [241], but the prediction errors of the output signals are additional inputs of the first neural network, a gradient training algorithm is given.

There are relatively few works concerned with the suboptimal MPC algorithms based on the general, not only neural, state-space model. MPC-SL and MPC-NPL structures for such models are detailed in [58, 78, 129, 207].

Linearisation of the predicted trajectory along the previously calculated control signals $\boldsymbol{u}(k-1)$ (used in the MPC-NPLT$_{\boldsymbol{u}(k-1)}$ algorithm) is discussed in [58], a few repetitions of prediction calculation and trajectory linearisation (the MPC-NPLPT algorithm) is discussed in [132, 231].

The MPC-NPL algorithm based on the neural state-space model discussed in this chapter is an extended version of the algorithm presented in [158], where only single-input single-output processes are considered with no measured disturbances and assuming only the output set-point trajectory. An application of the MPC-NPL algorithm based on the neural state-space model to a chemical reactor with 4 state variables and 3 inputs is discussed in [113]. The extended Kalman filter is used for state estimation. Similarly to the MPC algorithms based on various input-output models, the majority of works are interested in using neural networks in modelling, the MPC-NO algorithm with numerically approximated gradients predominates, the model structure is not taken into account in any way. Unfortunately, the computational complexity issues and reliability are not typically considered. The neural state-space structures are used in [338, 339] for modelling and for MPC in the sugar industry, a waste water treatment process is considered in [338]. In all cited works the MPC-NO algorithm with numerically approximated gradient is used, a quasi-Newton method is used for optimisation. The neural state-space model is also reported to be used for modelling a fed-batch multiple stages sugar crystallisation process [307]. The model is next used in the MPC-NO algorithm with approximated gradients, the SQP method is used for optimisation.

The following nonlinear state-space model

$$\begin{aligned} x(k+1) &= \boldsymbol{A}x(k) + \boldsymbol{B}u(k) + \boldsymbol{E}\varphi(\boldsymbol{F}y(k)) \\ y(k) &= \boldsymbol{C}x(k) \end{aligned}$$

is discussed in [122]. The last (nonlinear) part of the state equation is represented by a neural network. An explicit MPC algorithm is also presented in which the model is successively linearised on-line. Of course, the considered model may be also used in a numerical MPC algorithm in which constraints are taken into account.

An alternative neural state-space model is discussed in [258], in which the feedback between the input and the output is not necessary. The signals $u(k-1), \ldots, u(k-n_{\mathrm{B}})$ and $y(k-1), \ldots, y(k-n_{\mathrm{A}})$ are the inputs of the first network, similarly to a typical input-output model. The network has two nonlinear hidden nodes. Its outputs (states) and the signals $u(k-1), \ldots, u(k-n_{\mathrm{B}})$ are inputs of the second network which has only one nonlinear hidden layer and a nonlinear output node. The output signal $y(k)$ is the output of the second network. A pruning algorithm based on the SVD decomposition is also discussed.

Efficient computational methods for solving nonlinear optimisation tasks used in the MPC-NO algorithm based on Differential Algebraic Equations (DAE) are developed in [33].

In practice estimation of unmeasured state variables is of fundamental importance. In order to apply the MPC algorithms discussed in this chapter an observer must usually be used. Observer development for MPC algorithms based on linear models is described in [186, 268, 312, 316]. In the nonlinear case it is necessary to use the Extended Kalman Filter (EKF) [10, 299] or the Moving Horizon Estimation (MHE) technique [8, 186, 208, 267]. An interesting comparison of both approaches is discussed in [92]. Some examples when the EKF technique may give wrong results, even when there are no modelling errors, are given in [92]. For the same processes the MHE approach correctly calculates state estimation and, at the same time, is less sensitive to the initial state and parameters. On the other hand, unfortunately, the estimator needs on-line optimisation, its computational burden is much higher than that of the EKF technique. In addition to the EKF and MHE approaches, some new state estimation methods are available. For example, the estimator described in [228], unlike the EKF one, does not require differentiation. Thanks to that property it can be used for processes for which the classical EKF approach does not work. It is also possible to use the neural EKF technique [119]. Two neural networks are used, the role of which is neural approximation of numerical linearisation and filtration: the first network calculates on-line the gain matrix, the second one estimates the state variables. The data sets necessary for training and validation are obtained from simulations (or application to the real process) of the classical (numerical) EKF procedure.

4.7 Chapter Summary

There are two fundamental reasons for which the state-space models and the MPC algorithms based on such models are worth considering. Firstly, the state-space representation is much more universal than the input-output one, it makes it possible to describe properties of much wider class of dynamic processes. In the case of some processes the state-space representation is natural. Secondly, the state-space representation is a straightforward choice when it comes to theoretical investigations of the model (e.g. stability analysis). Similarly, in stability and robustness analysis of the MPC algorithms state evolution is typically considered. Neural networks, thanks to good approximation abilities, a relatively low number of parameters and good numerical properties, can be efficiently used not only for input-output modelling, but also for the state-space representation.

This chapter details three MPC algorithms based on the neural state-space model: the MPC-NO as well as the suboptimal MPC-NPL and MPC-NPLPT schemes are described. For the considered polymerisation reactor the trajectories obtained in the MPC-NPL strategy are very similar to those given by the computationally demanding MPC-NO one. At the same time, the MPC-NPL algorithm is many times less computationally demanding.

Of course, for other processes it may be necessary to use the MPC-NPLPT algorithm with trajectory linearisation.

The MPC algorithms discussed in this chapter can be extended in two respects. Firstly, using the derived formulae, one may easily implement the simplest suboptimal MPC-SL algorithm, the MPC-NPLT one and the MPC-NNPAPT one. Of course, the suboptimal algorithms may be implemented in their explicit versions. Secondly, it is possible to derive MPC algorithms based on alternative state-space models, e.g. the neural structures discussed in [122, 258] or the neural state-space Hammerstein and Wiener models [103] can be considered. An interesting alternative is to use the LRGF neural network with dynamic neurons [244] (s. 28). Many processes may be described by nonlinear state equations and linear output relations. In such cases only one (state) neural network is sufficient for modelling. For such a state-space model structure the MPC algorithms discussed in this chapter can be significantly simplified. One may find in the literature an incremental state-space model in which the following extended state vector is used [186]

$$x_{\mathrm{e}}(k) = \begin{bmatrix} \triangle x(k) \\ y(k) \end{bmatrix} = \begin{bmatrix} x(k) - x(k-1) \\ y(k) \end{bmatrix}$$

Of course, it is possible to derive the MPC algorithms for the neural state-space model in its incremental form. Unfortunately, it is necessary to point out that an application of the incremental model leads to worsening the numerical properties of some matrices used in the suboptimal MPC optimisation problem (e.g. the matrix $\boldsymbol{P}(k)$ given by Eq. (4.28) in the case of the MPC-NPL algorithm).

5

MPC Algorithms Based on Neural Multi-Models

The objective of the next two chapters is to discuss MPC algorithms based on some neural structures inspired by the idea of predictive control. This chapter is concerned with MPC algorithms based on neural multi-models. The classical dynamic models, both input-output and state-space structures, are used recurrently in MPC algorithms as they calculate the predictions for the whole prediction horizon. In such a case the prediction error is propagated, which may badly affect control quality. In order to solve the problem, the multi-model is used in which independent submodels calculate predictions for the consecutive sampling instants of the prediction horizon. The structure of the neural multi-model is discussed in this chapter, implementation details of the MPC-NO algorithm and some suboptimal MPC schemes are given.

5.1 Neural Multi-Models

The role of the model in MPC algorithms is of fundamental importance: it is used on-line to calculate the predicted output or (and) state trajectory. When a sufficiently precise model is not available, it is practically impossible to design and apply any MPC algorithm. As emphasised in [278, 279], during model identification it is necessary to take into account its further role in MPC. The synergy between model identification and the model role in MPC is obvious. From the perspective of MPC algorithms, the model has to be able to make good predictions of future behaviour of the process not only one step ahead, but over the whole prediction horizon.

In MPC algorithms the classical input-output or state-space models are always used recurrently, because the output or state predictions depend recurrently on some previous predictions. In practice, the dynamic models, e.g. the neural ones, are trained in the simplest non-recurrent serial-parallel configuration. Although MPC algorithms based on such models frequently work satisfactorily (the negative feedback mechanism in MPC compensates modelling errors), one must be aware of the fact that the prediction error is

M. Ławryńczuk, *Computationally Efficient Model Predictive Control Algorithms*, Studies in Systems, Decision and Control 3,
DOI: 10.1007/978-3-319-04229-9_5, © Springer International Publishing Switzerland 2014

propagated over the prediction horizon. The prediction error, which is the discrepancy between the prediction and the real process output (or state), is usually quite low at the beginning of the prediction horizon but it grows for the consecutive sampling instants. In particular, the prediction error propagation phenomenon can be observed in the case of modelling errors and measurement errors. What concerns modelling errors, they are caused by using too low order of dynamics (assuming that the model is capable of approximating the nonlinear nature of the process and it is well trained). It may be done deliberately, because the real order of dynamic may be in practice very high or it is difficult to determine the actual order of dynamics.

To reduce the unwanted prediction error propagation phenomenon, one may use the model trained recurrently (in the parallel configuration). Alternatively, it is also possible to minimise during model identification all predictions over the whole prediction horizon, for all data samples (the minimised SSE cost-function is defined by Eq. (1.21)). Unfortunately, such models must be still used recurrently in MPC algorithms and the prediction error is still propagated, although usually it is expected to be lower than for the models trained non-recurrently. Moreover, recurrent training is much more computationally demanding than in the simplest serial-parallel configuration and the recurrent models may be sensitive to noise.

Elimination of the prediction error propagation phenomenon needs using models of a special structure, which takes into account their role in MPC algorithms (prediction calculation). The simplest, but also naive, approach is to assume that the current value of the output signal does not depend on any values of the output measured for some previous sampling instants. From the single-input single-output model (1.16), one obtains the simplified model

$$y(k) = f(u(k-\tau), \ldots, u(k-n_{\mathrm{B}})) \tag{5.1}$$

For the above model, using the general prediction equation (1.14), the output predictions over the predictions horizon can be expressed as the functions

$$\hat{y}(k+p|k) = f(\underbrace{u(k-\tau+p|k), \ldots, u(k|k)}_{I_{\mathrm{uf}}(p)}, \underbrace{u(k-1), \ldots, u(k-n_{\mathrm{B}}+p)}_{I_{\mathrm{u}}-I_{\mathrm{uf}}(p)}) + d(k)$$

Unlike the classical model (1.16), the model (5.1) is not used recurrently in MPC algorithms, the prediction errors are not propagated. Unfortunately, such models are likely to be unable to approximate the real nature of typical technological processes or it may turn out that to obtain descent models it is necessary to use a very high order of dynamics (defined by the parameter n_{B}). In consequence, such a model, e.g. of the Volterra series type, may have a huge number of parameters.

The general structure of the neural single-input single-output multi-model is depicted in Fig. 5.1. As many as N independent neural submodels are used for prediction, each of them calculates the prediction for only one sampling instant of the prediction horizon. Thanks to using separate submodels, the

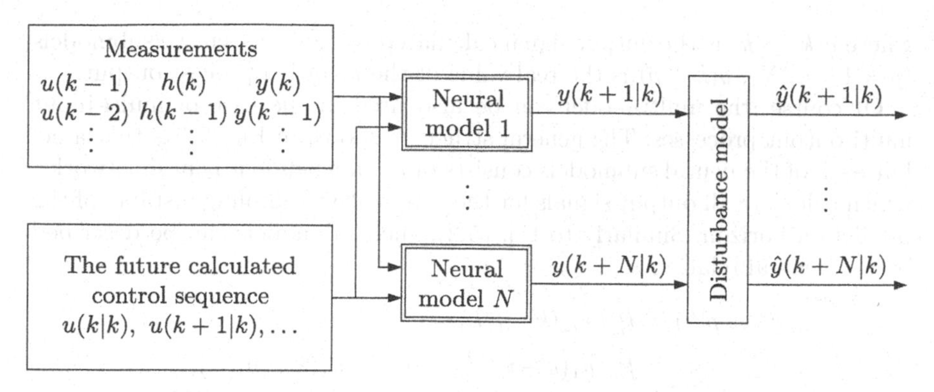

Fig. 5.1. The structure of the neural multi-model

multi-model is not used recurrently. Intuitively, one may expect that it is better to use for prediction N simpler models than one model, which must find the predictions for the whole prediction horizon N. For the first sampling instant of the prediction horizon ($p = 1$), the submodel is

$$y(k+1|k) = f^1(u(k-\tau+1), \ldots, u(k-n_{\mathrm{B}}), h(k-\tau_{\mathrm{h}}+1), \ldots, h(k-n_{\mathrm{C}})$$
$$y(k), \ldots, y(k-n_{\mathrm{A}}))$$

For the next sampling instant ($p = 2$), the submodel is

$$y(k+2|k) = f^2(u(k-\tau+2), \ldots, u(k-n_{\mathrm{B}}), h(k-\tau_{\mathrm{h}}+2), \ldots, h(k-n_{\mathrm{C}})$$
$$y(k), \ldots, y(k-n_{\mathrm{A}}))$$

All submodels for the consecutive sampling instants of the prediction horizon are defined in a similar way. The last submodel is

$$y(k+N|k) = f^N(u(k-\tau+N), \ldots, u(k-n_{\mathrm{B}}), h(k-\tau_{\mathrm{h}}+N), \ldots, h(k-n_{\mathrm{C}})$$
$$y(k), \ldots, y(k-n_{\mathrm{A}}))$$

In general, for any $p = 1, \ldots, N$, the submodels are described by the equation

$$\begin{aligned} y(k+p|k) &= f^p(\boldsymbol{x}(k+p|k)) \\ &= f^p(u(k-\tau+p), \ldots, u(k-n_{\mathrm{B}}), h(k-\tau_{\mathrm{h}}+p), \ldots, h(k-n_{\mathrm{C}}) \\ &\quad y(k), \ldots, y(k-n_{\mathrm{A}})) \end{aligned} \tag{5.2}$$

All submodels are trained independently in the serial-parallel configuration, which is a great advantage when compared with recurrent parallel training of the classical model. Accuracy of the submodels is assessed taking into account the SSE index (similarly to Eq. (1.20))

$$\mathrm{SSE}_p = \sum_{k=1}^{n_{\mathrm{p}}-N} (y(k+p|k) - y(k+p))^2 \tag{5.3}$$

where $y(k+p|k)$ is the output signal calculated by the consecutive submodels ($p = 1, \ldots, N$), $y(k+p)$ is the real value of the recorded process output.

Of course, the multi-model can be also used in the case of multi-input multi-output processes. The general structure shown in Fig. 5.1 is the same, but each of the neural submodels consists of n_{y} independent neural networks, which calculate all output signals for the consecutive sampling instants of the prediction horizon. Similarly to Eq. (5.2), such submodels can be described by the general relation

$$\begin{aligned} y_m(k+p|k) &= f_m^p(\boldsymbol{x}_m(k+p|k)) \\ &= f_m^p(u_1(k-\tau^{m,1}+p), \ldots, u_1(k-n_{\mathrm{B}}^{m,1}), \ldots, \\ &\quad u_{n_{\mathrm{u}}}(k-\tau^{m,n_{\mathrm{u}}}+p), \ldots, u_{n_{\mathrm{u}}}(k-n_{\mathrm{B}}^{m,n_{\mathrm{u}}}), \\ &\quad h_1(k-\tau_{\mathrm{h}}^{m,1}+p), \ldots, h_1(k-n_{\mathrm{C}}^{m,1}), \ldots, \\ &\quad h_{n_{\mathrm{h}}}(k-\tau_{\mathrm{h}}^{m,n_{\mathrm{h}}}+p), \ldots, h_{n_{\mathrm{h}}}(k-n_{\mathrm{C}}^{m,n_{\mathrm{h}}}), \\ &\quad y_m(k), \ldots, y_m(k-n_{\mathrm{A}}^m)) \end{aligned} \tag{5.4}$$

where $m = 1, \ldots, n_{\mathrm{y}}$, $p = 1, \ldots, N$. From Eq. (5.4), output predictions are the following functions

$$\begin{aligned} \hat{y}_m(k+p|k) = f_m^p(&\underbrace{u_n(k-\tau^{m,n}+p|k), \ldots, u_n(k|k)}_{I_{\mathrm{uf}}^{m,n}(p),\ n=1,\ldots,n_{\mathrm{u}}}, \\ &\underbrace{u_n(k-\max(\tau^{m,n}-p,1)), \ldots, u_n(k-n_{\mathrm{B}}^{m,n})}_{I_{\mathrm{up}}^{m,n}(p),\ n=1,\ldots,n_{\mathrm{u}}}, \\ &\underbrace{h_n(k-\tau_{\mathrm{h}}^{m,n}+p|k), \ldots, h_n(k+1|k)}_{I_{\mathrm{hf}}^{m,n}(p),\ n=1,\ldots,n_{\mathrm{h}}}, \\ &\underbrace{h_n(k-\max(\tau_{\mathrm{h}}^{m,n}-p,0)), \ldots, h_n(k-n_{\mathrm{C}}^{m,n})}_{I_{\mathrm{hp}}^{m,n}(p),\ n=1,\ldots,n_{\mathrm{h}}}, \\ &\underbrace{y_m(k), \ldots, y_m(k-n_{\mathrm{A}}^m)}_{n_{\mathrm{A}}^m+1}) + d_m(k+p|k) \end{aligned} \tag{5.5}$$

where $I_{\mathrm{uf}}^{m,n}(p) = \max(p-\tau^{m,n}+1, 0)$, $I_{\mathrm{up}}^{m,n}(p) = n_{\mathrm{B}}^{m,n} - \max(\tau^{m,n}-p, 1)+1$, $I_{\mathrm{hf}}^{m,n}(p) = \max(p-\tau_{\mathrm{h}}^{m,n}, 0)$, $I_{\mathrm{hp}}^{m,n}(p) = n_{\mathrm{C}}^{m,n} - \max(\tau_{\mathrm{h}}^{m,n} - p + 1, 1) + 2$. Because for the consecutive sampling instants an independent submodel is used, the classical prediction equation (1.14), in which the same model is used for prediction recurrently over the whole prediction horizon, is not correct. The prediction equation for the multi-model is

$$\hat{y}_m(k+p|k) = y_m(k+p|k) + d_m(k+p|k)$$

In place of Eq. (1.15), which makes it possible to estimate the unmeasured disturbance, one must use

$$d_m(k+p|k) = y_m(k) - f_m^p(\boldsymbol{x}_m(k|k-1))$$

where the signals $y_m(k)$ are measured and the signals $f_m^p(\boldsymbol{x}_m(k|k-1))$ are calculated from the multi-model, analogously to the signals $y_m(k|k-1)$ in Eq. (1.15). From Eq. (5.4), one has

$$\begin{aligned} d_m(k+p|k) = y_m(k) - f_m^p(&u_1(k-\tau^{m,1}),\ldots,u_1(k-n_{\mathrm{B}}^{m,1}-p),\ldots,\\ &u_{n_{\mathrm{u}}}(k-\tau^{m,n_{\mathrm{u}}}),\ldots,u_{n_{\mathrm{u}}}(k-n_{\mathrm{B}}^{m,n_{\mathrm{u}}}-p),\\ &h_1(k-\tau_{\mathrm{h}}^{m,1}),\ldots,h_1(k-n_{\mathrm{C}}^{m,1}-p),\ldots,\\ &h_{n_{\mathrm{h}}}(k-\tau_{\mathrm{h}}^{m,n_{\mathrm{h}}}),\ldots,h_{n_{\mathrm{h}}}(k-n_{\mathrm{C}}^{m,n_{\mathrm{h}}}-p),\\ &y_m(k-p),\ldots,y_m(k-n_{\mathrm{A}}^m-p)) \end{aligned}$$

Assuming that the DLP neural networks with one hidden layer are used in the multi-model, the predictions for consecutive outputs ($m=1,\ldots,n_{\mathrm{y}}$) and for consecutive sampling instants of the prediction horizon ($p=1,\ldots,N$) are calculated from

$$\begin{aligned} \hat{y}_m(k+p|k) &= f_m^p(\boldsymbol{x}_m(k+p|k)) + d_m(k+p|k)\\ &= w_0^{2,m,p} + \sum_{i=1}^{K^{m,p}} w_i^{2,m,p}\varphi(z_i^m(k+p|k)) + d_m(k+p|k) \end{aligned} \tag{5.6}$$

where the sum of the input signals connected to the ith hidden node ($i=1,\ldots,K^{m,p}$) in the submodel of the mth output used for prediction for the sampling instant $k+p$, according to Eq. (5.5), is

$$\begin{aligned} z_i^m(k+p|k) = w_{i,0}^{1,m,p} &+ \sum_{n=1}^{n_{\mathrm{u}}}\sum_{j=1}^{I_{\mathrm{uf}}^{m,n}(p)} w_{i,J_{\mathrm{u}}^{m,n}(p)+j}^{1,m,p} u_n(k-\tau^{m,n}+1-j+p|k)\\ &+ \sum_{n=1}^{n_{\mathrm{u}}}\sum_{j=1}^{I_{\mathrm{up}}^{m,n}(p)} w_{i,J_{\mathrm{u}}^{m,n}(p)+I_{\mathrm{uf}}^{m,n}(p)+j}^{1,m,p}\\ &\qquad \times u_n(k-\max(\tau^{m,n}-p,1)+1-j)\\ &+ \sum_{n=1}^{n_{\mathrm{h}}}\sum_{j=1}^{I_{\mathrm{hf}}^{m,n}(p)} w_{i,I_{\mathrm{u}}^m(p)+J_{\mathrm{h}}^{m,n}(p)+j}^{1,m,p} h_n(k-\tau_{\mathrm{h}}^{m,n}+1-j+p|k)\\ &+ \sum_{n=1}^{n_{\mathrm{h}}}\sum_{j=1}^{I_{\mathrm{hp}}^{m,n}(p)} w_{i,I_{\mathrm{u}}^m(p)+J_{\mathrm{h}}^{m,n}(p)+I_{\mathrm{hf}}^{m,n}(p)+j}^{1,m,p}\\ &\qquad \times h_n(k-\max(\tau_{\mathrm{h}}^{m,n}-p,0)+1-j)\\ &+ \sum_{j=1}^{n_{\mathrm{A}}^m+1} w_{i,I_{\mathrm{u}}^m(p)+I_{\mathrm{h}}^m(p)+j}^{1,m,p} y_m(k+1-j) \end{aligned} \tag{5.7}$$

Each network has $I^m(p)$ inputs, $K^{m,p}$ hidden nodes with the transfer function $\varphi\colon \mathbb{R}\to\mathbb{R}$ and one output $y_m(k+p|k)$. The weights of the first layer are denoted by $w_{i,j}^{1,m,p}$, where $i = 1, \ldots, K^{m,p}$, $j = 0, \ldots, I^m(p)$, $m = 1, \ldots, n_{\mathrm{y}}$, $p = 1, \ldots, N$, the weights of the second layer are denoted by $w_i^{2,m,p}$, where $i = 0, \ldots, K^{m,p}$, $m = 1, \ldots, n_{\mathrm{y}}$, $p = 1, \ldots, N$. Consecutive neural networks can be described by the functions $f_m^p \colon \mathbb{R}^{I^m(p)} \to \mathbb{R}$, where $m = 1, \ldots, n_{\mathrm{y}}$, $p = 1, \ldots, N$. For the mth output, for prediction for the sampling instant $k+p$, the number of network inputs which depend on the process input signal u_n $(n = 1, \ldots, n_{\mathrm{u}})$ is $I_{\mathrm{u}}^{m,n}(p) = I_{\mathrm{uf}}^{m,n}(p) + I_{\mathrm{up}}^{m,n}(p)$, the number of network inputs which depend on the process disturbance signal h_n $(n = 1, \ldots, n_{\mathrm{h}})$ is $I_{\mathrm{h}}^{m,n}(p) = I_{\mathrm{hf}}^{m,n}(p) + I_{\mathrm{hp}}^{m,n}(p)$, and the number of network inputs which depend on the process output y_m is always $n_{\mathrm{A}}^m + 1$. The number of network inputs which depend on all process input signals is $I_{\mathrm{u}}^m(p) = \sum_{n=1}^{n_{\mathrm{u}}} I_{\mathrm{u}}^{m,n}(p)$, the number of network inputs which depend on all process disturbance signals is $I_{\mathrm{h}}^m(p) = \sum_{n=1}^{n_{\mathrm{h}}} I_{\mathrm{h}}^{m,n}(p)$. The total number of inputs of the submodel used for prediction of the mth output for the sampling instant $k+p$ is then

$$
\begin{aligned}
I^m(p) &= I_{\mathrm{u}}^m(p) + I_{\mathrm{h}}^m(p) + n_{\mathrm{A}}^m + 1 = \sum_{n=1}^{n_{\mathrm{u}}} I_{\mathrm{u}}^{m,n}(p) + \sum_{n=1}^{n_{\mathrm{h}}} I_{\mathrm{h}}^{m,n}(p) + n_{\mathrm{A}}^m + 1 \\
&= \sum_{n=1}^{n_{\mathrm{u}}} (\max(p - \tau^{m,n} + 1, 0) + n_{\mathrm{B}}^{m,n} - \max(\tau^{m,n} - p, 1) + 1) \\
&\quad + \sum_{n=1}^{n_{\mathrm{h}}} (\max(p - \tau_{\mathrm{h}}^{m,n}, 0) + n_{\mathrm{C}}^{m,n} - \max(\tau_{\mathrm{h}}^{m,n} - p + 1, 1) + 2) + n_{\mathrm{A}}^m + 1
\end{aligned}
$$

where

$$
J_{\mathrm{u}}^{m,n}(p) = \begin{cases} 0 & \text{if } n = 1 \\ \sum_{i=1}^{n-1} I_{\mathrm{u}}^{m,i}(p) & \text{if } n = 2, \ldots, n_{\mathrm{u}} \end{cases}
$$

and

$$
J_{\mathrm{h}}^{m,n}(p) = \begin{cases} 0 & \text{if } n = 1 \\ \sum_{i=1}^{n-1} I_{\mathrm{h}}^{m,i}(p) & \text{if } n = 2, \ldots, n_{\mathrm{h}} \end{cases}
$$

The number of model inputs which are the decision variables of MPC algorithms grows for consecutive sampling instants of the prediction horizon. If the manipulated variables are constant for long periods, training may be numerically ill-conditioned. That is why it is desirable that all the input signals change over time.

In addition to the discussed multi-input multi-output multi-model structure ($n_{\mathrm{y}}N$ neural networks), for some specific processes one may use a lower number of networks. In the simplest case only N networks are used, each of which has n_{y} outputs and calculates predictions for all process outputs.

5.2 MPC-NO Algorithm

The general formulation of the MPC-NO algorithm, discussed in Chapter 2.2, and its structure, shown in Fig. 2.2, are universal. The vector of decision variables is calculated from the nonlinear optimisation problem (2.5), which can be transformed to the standard form (2.8). The MPC-NO algorithm can be also designed for the neural multi-model. In such a case the nonlinear predicted trajectory is determined from Eqs. (5.6) and (5.7). In order to calculate the entries of the matrix (2.16), i.e. the derivatives of that trajectory with respect to the future control policy, differentiating Eq. (5.6), one obtains

$$\frac{\partial \hat{y}_m(k+p|k)}{\partial u_n(k+r|k)} = \sum_{i=1}^{K^{m,p}} w_i^{2,m,p} \frac{\mathrm{d}\varphi(z_i^m(k+p|k))}{\mathrm{d}z_i^m(k+p|k)} \frac{\partial z_i^m(k+p|k)}{\partial u_n(k+r|k)}$$

where, differentiating Eq. (5.7), one has

$$\frac{\partial z_i^m(k+p|k)}{\partial u_n(k+r|k)} = \sum_{j=1}^{I_{\mathrm{uf}}^{m,n}(p)} w_{i,J_{\mathrm{u}}^{m,n}(p)+j}^{1,m,p} \frac{\partial u_n(k-\tau^{m,n}+1-j+p|k)}{\partial u_n(k+r|k)}$$

Because the multi-model is used for prediction, the derivatives of the predicted trajectory depend only on the calculated future control signals and on the model parameters. For other classes of input-output models, e.g. for the classical neural model or for the cascade Hammerstein and Wiener structures, the predictions are also functions of some previous predictions (Eqs. (2.23) and (3.13)). For the Wiener model it is additionally necessary to take into account the derivatives of the auxiliary signals between the dynamic and steady-state parts (Eq. (3.29)). For the state-space structure one must take into account the derivatives of the predicted state trajectory (Eq. (4.19)).

5.3 Suboptimal MPC Algorithms with On-Line Linearisation

The future control sequence is an argument of the multi-model (5.5). That is why the predicted output trajectory linearisation along the future control trajectory is straightforward rather than linearisation for the current operating point. Hence, the MPC-NPLT, MPC-NPLPT and MPC-NNPAPT algorithms, the general formulations of which and corresponding optimisation problems are detailed in Chapter 2.3, can be designed for the considered multi-model structure. Implementation details of the MPC-NPLPT algorithm based on the multi-model are given in the following text.

From (5.6), the predicted output trajectory for the tth internal iteration can be expressed as

$$\hat{y}_m^t(k+p|k) = w_0^{2,m,p} + \sum_{i=1}^{K^{m,p}} w_i^{2,m,p} \varphi(z_i^{m,t}(k+p|k)) + d_m(k+p|k) \quad (5.8)$$

where, from Eq. (5.7), one has

$$\begin{aligned} z_i^{m,t}(k+p|k) = w_{i,0}^{1,m,p} &+ \sum_{n=1}^{n_\mathrm{u}} \sum_{j=1}^{I_\mathrm{uf}^{m,n}(p)} w_{i,J_\mathrm{u}^{m,n}(p)+j}^{1,m,p} u_n^t(k-\tau^{m,n}+1-j+p|k) \\ &+ \sum_{n=1}^{n_\mathrm{u}} \sum_{j=1}^{I_\mathrm{up}^{m,n}(p)} w_{i,J_\mathrm{u}^{m,n}(p)+I_\mathrm{uf}^{m,n}(p)+j}^{1,m,p} \\ &\qquad \times u_n(k-\max(\tau^{m,n}-p,1)+1-j) \\ &+ \sum_{n=1}^{n_\mathrm{h}} \sum_{j=1}^{I_\mathrm{hf}^{m,n}(p)} w_{i,I_\mathrm{u}^m(p)+J_\mathrm{h}^{m,n}(p)+j}^{1,m,p} h_n(k-\tau_\mathrm{h}^{m,n}+1-j+p|k) \\ &+ \sum_{n=1}^{n_\mathrm{h}} \sum_{j=1}^{I_\mathrm{hp}^{m,n}(p)} w_{i,I_\mathrm{u}^m(p)+J_\mathrm{h}^{m,n}(p)+I_\mathrm{hf}^{m,n}(p)+j}^{1,m,p} \\ &\qquad \times h_n(k-\max(\tau_\mathrm{h}^{m,n}-p,0)+1-j) \\ &+ \sum_{j=1}^{n_\mathrm{A}^m+1} w_{i,I_\mathrm{u}^m(p)+I_\mathrm{h}^m(p)+j}^{1,m,p} y_m(k+1-j) \end{aligned} \tag{5.9}$$

The only difference between the above formula and Eq. (5.7) is the fact that in the latter the predicted trajectory takes into account the index of internal iterations t.

The entries of the matrix (2.53) are derivatives of the predicted output trajectory calculated in the previous internal iteration $(t-1)$ with respect to the future control sequence calculated in the same iteration. From Eq. (5.8), remembering that the index t must be replaced by the index $t-1$, one obtains

$$\frac{\partial \hat{y}_m^{t-1}(k+p|k)}{\partial u_n^{t-1}(k+r|k)} = \sum_{i=1}^{K^{m,p}} w_i^{2,m,p} \frac{\mathrm{d}\varphi(z_i^{m,t-1}(k+p|k))}{\mathrm{d}z_i^{m,t-1}(k+p|k)} \frac{\partial z_i^{m,t-1}(k+p|k)}{\partial u_n^{t-1}(k+r|k)}$$

where, from Eq. (5.9), one has

$$\frac{\partial z_i^{m,t-1}(k+p|k)}{\partial u_n^{t-1}(k+r|k)} = \sum_{j=1}^{I_\mathrm{uf}^{m,n}(p)} w_{i,J_\mathrm{u}^{m,n}(p)+j}^{1,m,p} \frac{\partial u_n^{t-1}(k-\tau^{m,n}+1-j+p|k)}{\partial u_n^{t-1}(k+r|k)}$$

The above formulae can be also used in the MPC-NPLT algorithm. The input trajectory, along which linearisation is carried out, is defined by the control signals used in the previous sampling instant or it is comprised by the last $n_\mathrm{u}N_\mathrm{u}$ elements of the optimal control trajectory determined at the previous sampling instant and not applied to the process. At each sampling instant linearisation is performed once, only one quadratic optimisation problem is solved.

5.4 Multi-Models in Adaptive MPC Algorithms

In all MPC algorithms discussed so far it is assumed that the model is trained and verified off-line, i.e. the model identification phase precedes the development of MPC structures. Some data sets are used for model identification. Such an approach is correct provided that properties of the process do not change significantly in time, some inevitable discrepancy between the process and its model are compensated by the negative feedback mechanism. Adaptive MPC algorithms based on the multi-models are worth mentioning because the submodels are trained in the simple serial-parallel configuration, parallel (recurrent) training is not necessary. In consequence, on-line identification of the multi-models seems to be less computationally difficult and more reliable than identification of alternative models (e.g. the input-output and state-space structures) in the parallel configuration.

Although the described DLP neural multi-models can be of course used in adaptive MPC algorithms, the application of the RBF multi-models seems to be an interesting option. Training of the DLP neural networks requires solving a nonlinear optimisation problem, but, provided that the parameters of the basis functions (the centres and the spreads) are fixed, determination of weights of the RBF networks needs solving an unconstrained quadratic optimisation problem. Such a problem can be efficiently solved by the least-squares method (the SVD decomposition may be used for this purpose), the global minimum is always found. The assumption that the RBF network has a constant number of basis functions and all their parameters are fixed is not restrictive, because, in practice, the general nature of the model (i.e. the order of dynamics, the type of nonlinearity) does not change in time, but the process gains and time constants are likely to change their values. An initial topology and parameters of the RBF neural network is chosen in the same way it is done in the case of the DLP structure, i.e. on the basis on the recorded data set. What is more, SVM or LS-SVM models are interesting alternatives to the RBF structure. In such cases training needs solving a quadratic optimisation problem or a least-squares one, which can be efficiently done on-line.

5.5 Example 5.1

The considered process is the polymerisation reactor (s. 124). It is demonstrated in the previous two chapters that the linear MPC algorithms, based on input-output or state-space models, do not work properly for the reactor. The MPC algorithms based on the neural models give good results, but they all base on the models which have a sufficient order of dynamics (chosen experimentally). Now, it is interesting to investigate modelling and predictive control when the order of dynamics is deliberately lowered.

Polymerisation Reactor Modelling

In order to demonstrate advantages of the neural multi-models, the following models of the polymerisation reactor are considered:

a) a high-order DLP neural model,
b) a low-order DLP neural model,
c) a low-order DLP neural multi-model,
d) a low-order linear multi-model.

The classical neural models are trained in the non-recurrent serial-parallel configuration and in the recurrent parallel configuration whereas the neural multi-models are always trained non-recurrently since the recurrent mode is not necessary for them. The SSE index (1.20) is minimised during training of the classical models, the SSE index (5.3) is used in the case of the multi-models. For identification the data sets used in Example 3.1 (p. 124) are used.

A high-order neural model. Provided that the order of dynamics is sufficiently high, the model has good approximation ability and the data sets are rich enough (i.e. the number of samples is sufficient and the operating point changes significantly), one may expect that accuracy of the classical neural model is good. The following fourth-order dynamic model is considered

$$\begin{aligned} y(k) = f(&u(k-1), u(k-2), u(k-3), u(k-4), \\ &y(k-1), y(k-2), y(k-3), y(k-4)) \end{aligned}$$

i.e. $\tau = 1$, $n_{\mathrm{A}} = n_{\mathrm{B}} = 4$. In order to obtain models of good accuracy, as many as 10 hidden nodes are used. Table 5.1 shows the influence of the training mode on accuracy of the classical high-order neural model. The model is trained in two configurations: non-recurrent (series-parallel) and recurrent (parallel). The values of the non-recurrent model errors ($\mathrm{SSE}_{\mathrm{nrec}}$) and of the recurrent errors ($\mathrm{SSE}_{\mathrm{rec}}$) are given for the models trained in two modes, i.e. the non-recurrent error is additionally calculated for the model trained recurrently and the recurrent error is found for the model trained non-recurrently. From the perspective of MPC algorithms, the model should have a relatively low value of the recurrent error. As one may expect, recurrent training gives better results than the non-recurrent approach. For the validation data set the recurrent model gives $\mathrm{SSE}_{\mathrm{rec}} = 0.2207$ whereas for the non-recurrent one the error increases to $\mathrm{SSE}_{\mathrm{rec}} = 0.6378$.

Taking into account the role of a model in MPC algorithms, the results presented in Table 5.1 clearly indicate superiority of the recurrent training configuration over the non-recurrent one. Figs. 5.2 and 5.3 depicts the step-responses of the high-order neural model trained in both configurations. For the step-response calculation, both models are used recurrently over the prediction horizon $N = 10$. Two input signal changes from the nominal operating point ($F_{\mathrm{I}} = 0.028328$ m^3/h, NAMW $= 20000$ kg/kmol) occurring at the sampling instant $k = 0$ are considered: to $F_{\mathrm{I}} = 0.0526$ m^3/h (which corresponds to the steady-state NAMW = 15000 kg/kmol) and to $F_{\mathrm{I}} = 0.0168$ m^3/h (which corresponds to the steady-state NAMW = 25000 kg/kmol). The step-responses of the model trained recurrently are very precise. It is necessary to point out, however, that the model trained non-recurrently also gives quite good predictions. Although the differences between the responses of the model and of the process are present, they are not significant. Such small prediction errors are likely to be compensated by the negative feedback mechanism present in MPC algorithms. To sum up, when the order of dynamics is sufficiently high and the number of model parameters is sufficient, one may use the significantly faster non-recurrent training mode (for the PC workstation with a microprocessor running at the frequency 3.1 MHz, 200 iterations of the BFGS algorithm in the non-recurrent configuration takes 35 seconds whereas in the recurrent one the calculation time soars to 125 seconds).

A low-order neural model. The real order of dynamics is in practice not known or models of a low-order are deliberately used. For the considered polymerisation reactor it has been found out experimentally that the second-order dynamics (3.41) is necessary, for which $\tau = n_{\mathrm{A}} = n_{\mathrm{B}} = 2$. Now it is deliberately assumed that the order of model dynamics is lower than the order of process dynamics. The considered neural model of a low order is

$$y(k) = f(u(k-2), y(k-1)) \tag{5.10}$$

i.e. $\tau = n_{\mathrm{B}} = 2$, $n_{\mathrm{A}} = 1$. 6 nodes are used in the hidden layer. Table 5.2 shows the influence of the training mode on accuracy of the classical low-order neural model. It is worth comparing the obtained results with Table 5.1, which is concerned with the classical high-order model. Similarly as previously, the low-order model is also trained in non-recurrent and recurrent configurations. Although the model trained recurrently has better prediction accuracy (for

Table 5.1. Accuracy of the classical high-order neural model of the polymerisation reactor trained non-recurrently and recurrently, the value of the SSE performance index actually minimised during training is given in bold

Training mode	Training data set		Validation data set	
	$\mathrm{SSE}_{\mathrm{nrec}}$	$\mathrm{SSE}_{\mathrm{rec}}$	$\mathrm{SSE}_{\mathrm{nrec}}$	$\mathrm{SSE}_{\mathrm{rec}}$
Non-recurrent	**0.2148**	0.2866	0.2412	0.6378
Recurrent	0.2523	**0.1636**	0.2574	0.2207

Table 5.2. Accuracy of the classical low-order neural model of the polymerisation reactor trained non-recurrently and recurrently, the value of the SSE performance index actually minimised during training is given in bold

Training mode	Training data set		Validation data set	
	$\mathrm{SSE}_{\mathrm{nrec}}$	$\mathrm{SSE}_{\mathrm{rec}}$	$\mathrm{SSE}_{\mathrm{nrec}}$	$\mathrm{SSE}_{\mathrm{rec}}$
Non-recurrent	**0.3671**	2.7605	0.5618	7.5103
Recurrent	0.4841	**1.1189**	0.7805	3.0351

the validation data set the recurrent model gives $\text{SSE}_{\text{rec}} = 3.0351$, whereas for the non-recurrent one the error increases to $\text{SSE}_{\text{rec}} = 7.5103$), it is necessary to point out that the low-order structure is much worse when used for prediction – the values of SSE_{rec} are a few times higher than in the case of the high-order structure. In consequence, one may expect that when the low-order model is used for prediction in MPC algorithm, it does not leads to good control.

Low-order neural and linear multi-models. It is assumed that the order of dynamics of the neural multi-model is the same as that of the low-order classical neural model (5.10), the prediction abilities of which are poor. From

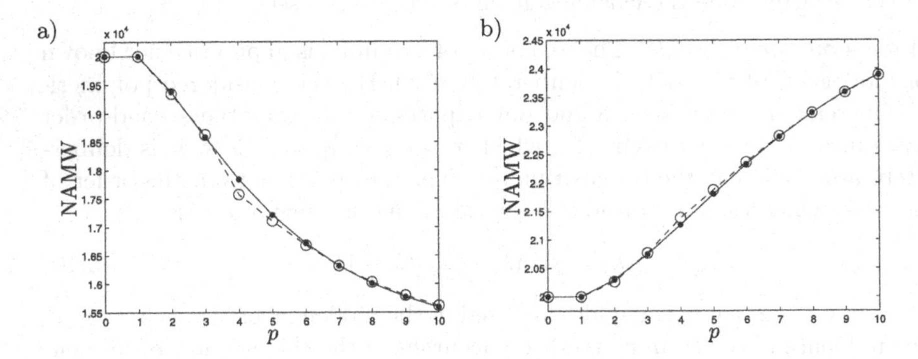

Fig. 5.2. The step-responses of the high-order neural model of the polymerisation reactor trained non-recurrently (*dashed line with circles*) vs. the outputs of the process (*solid line with dots*), the input signal changes from the nominal operating point to: a) $F_{\mathrm{I}} = 0.0526$ m^3/h, b) $F_{\mathrm{I}} = 0.0168$ m^3/h

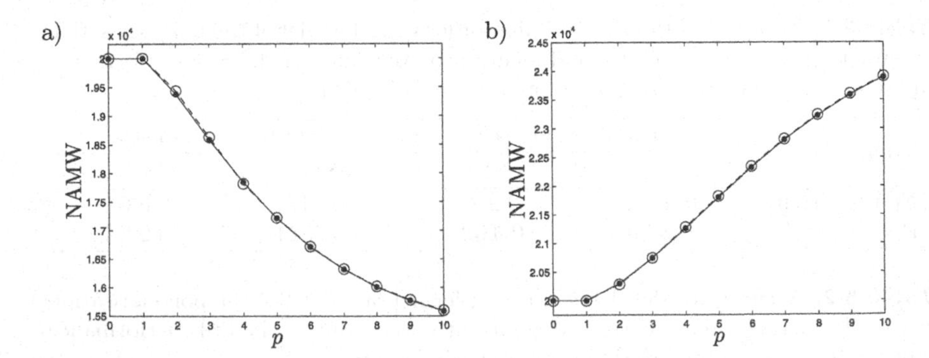

Fig. 5.3. The step-responses of the high-order neural model of the polymerisation reactor trained recurrently (*dashed line with circles*) vs. the outputs of the process (*solid line with dots*), the input signal changes from the nominal operating point to: a) $F_{\mathrm{I}} = 0.0526$ m^3/h, b) $F_{\mathrm{I}} = 0.0168$ m^3/h

Eq. (5.4), for $\tau = n_{\mathrm{B}} = 2$, $n_{\mathrm{A}} = 1$, the multi-model for the prediction horizon $N = 10$ can be described by the general equations

$$
\begin{aligned}
y(k+1|k) &= f_1(u(k-1), u(k-2), y(k), y(k-1)) \qquad (5.11)\\
y(k+2|k) &= f_2(u(k|k), u(k-1), u(k-2), y(k), y(k-1))\\
&\vdots\\
y(k+10|k) &= f_{10}(u(k+8|k), \ldots, u(k|k), u(k-1), u(k-2), y(k), y(k-1))
\end{aligned}
$$

In contrast to the classical neural model, the multi-model is always trained non-recurrently. The following low-order linear multi-model is also considered (it has the same arguments as the neural one (5.11))

$$
\begin{aligned}
y(k+1|k) &= b_1^1 u(k-1) + b_2^1 u(k-2) - a_1^1 y(k) - a_2^1 y(k-1)\\
y(k+2|k) &= b_1^2 u(k|k) + b_2^2 u(k-1) + b_3^2 u(k-2) - a_1^2 y(k) - a_2^2 y(k-1)\\
&\vdots\\
y(k+10|k) &= b_1^{10} u(k+8|k) + \ldots + b_9^{10} u(k|k) + b_{10}^{10} u(k-1) + b_{11}^{10} u(k-2)\\
&\quad - a_1^{10} y(k) - a_2^{10} y(k-1)
\end{aligned}
$$

Table 5.3 compares the low-order linear and neural multi-models in terms of the number of parameters and accuracy. Of course, each submodel (for $p = 1, \ldots, 10$) is trained independently, the optimal number of hidden nodes is chosen for each submodel individually, the test errors are calculated for the selected submodels. One may notice that for the neural submodels the errors are a few times lower than for the linear ones.

A graphical comparison of accuracy of all submodels which comprise the multi-model needs showing as many as 10 figures (i.e. the outputs of the consecutive submodels vs. the data sets). That is why Fig. 5.4 depicts correlation between the validation data set and the outputs of all low-order linear and neural submodels. In the case of ideal modelling, the correlation figure consists of a set of points which form a line of slope 45 degrees. From the correlation figure one may notice that the low-order neural multi-model is very precise whereas the linear multi-model of the same order is inaccurate.

The errors of the submodels which comprise the low-order neural multi-model are comparable with that of the classical low-order neural model trained non-recurrently (Table 5.2, the values of $\mathrm{SSE}_{\mathrm{nrek}}$). Hence, it is interesting to investigate prediction abilities of both low-order model classes. The step-responses of both models are shown in Fig. 5.5. The classical neural model must be used recurrently. Although the high-order classical neural model gives step-responses practically the same as the process (Figs. 5.2 and 5.3), the step-responses of the low-order one are significantly different. The low-order neural model works correctly only for the sampling instant $p = 1$, i.e. it correctly calculates the signal $y(k+1|k)$, but due to an inadequate order of dynamics, for the following sampling instants it does not work correctly, i.e. the prediction errors are significant. Conversely, the step-response of the neural multi-model, although its order of dynamics is

Table 5.3. Comparison of the low-order linear multi-models and of the low-order neural multi-models of the polymerisation reactor in terms of the number of parameters (NP) and accuracy (SSE)

p	Submodel	NP	SSE_{train}	SSE_{val}	SSE_{test}
1	Linear	4	3.9161×10^{0}	6.4341×10^{0}	–
	Neural, $K^1 = 5$	31	2.7933×10^{-1}	3.6070×10^{-1}	–
	Neural, $K^1 = 6$	37	2.6896×10^{-1}	3.2130×10^{-1}	3.7251×10^{-1}
	Neural, $K^1 = 7$	43	2.7740×10^{-1}	3.4560×10^{-1}	–
2	Linear	5	1.2295×10^{1}	2.0406×10^{1}	–
	Neural, $K^2 = 5$	36	2.7799×10^{-1}	4.1147×10^{-1}	–
	Neural, $K^2 = 6$	43	3.0015×10^{-1}	3.9809×10^{-1}	4.2891×10^{-1}
	Neural, $K^2 = 7$	50	2.5082×10^{-1}	4.1431×10^{-1}	–
3	Linear	6	2.3853×10^{1}	4.0351×10^{1}	–
	Neural, $K^3 = 3$	25	3.4853×10^{-1}	5.5523×10^{-1}	–
	Neural, $K^3 = 4$	33	3.0657×10^{-1}	4.4963×10^{-1}	4.5438×10^{-1}
	Neural, $K^3 = 5$	41	3.1647×10^{-1}	4.9580×10^{-1}	–
4	Linear	7	3.7943×10^{1}	6.3561×10^{1}	–
	Neural, $K^4 = 4$	37	3.0943×10^{-1}	4.6162×10^{-1}	–
	Neural, $K^4 = 5$	46	3.1106×10^{-1}	4.5482×10^{-1}	4.9032×10^{-1}
	Neural, $K^4 = 6$	55	2.8514×10^{-1}	4.7400×10^{-1}	–
5	Linear	8	5.3247×10^{1}	8.9335×10^{1}	–
	Neural, $K^5 = 5$	51	2.6908×10^{-1}	5.0170×10^{-1}	–
	Neural, $K^5 = 6$	61	2.4885×10^{-1}	4.6054×10^{-1}	5.6423×10^{-1}
	Neural, $K^5 = 7$	71	2.9899×10^{-1}	5.1909×10^{-1}	–
6	Linear	9	6.8990×10^{1}	1.1602×10^{2}	–
	Neural, $K^6 = 4$	45	2.8838×10^{-1}	5.1113×10^{-1}	–
	Neural, $K^6 = 5$	56	2.3441×10^{-1}	4.0483×10^{-1}	5.0712×10^{-1}
	Neural, $K^6 = 6$	67	2.1941×10^{-1}	4.1358×10^{-1}	–
7	Linear	10	8.4823×10^{1}	1.4364×10^{2}	–
	Neural, $K^7 = 4$	49	2.4951×10^{-1}	4.7613×10^{-1}	–
	Neural, $K^7 = 5$	61	2.4818×10^{-1}	4.2334×10^{-1}	5.8667×10^{-1}
	Neural, $K^7 = 6$	73	2.3003×10^{-1}	4.4202×10^{-1}	–
8	Linear	11	1.0050×10^{2}	1.7078×10^{2}	–
	Neural, $K^8 = 4$	53	2.4698×10^{-1}	4.5060×10^{-1}	–
	Neural, $K^8 = 5$	66	2.1081×10^{-1}	3.5349×10^{-1}	5.0166×10^{-1}
	Neural, $K^8 = 6$	79	2.1686×10^{-1}	3.9282×10^{-1}	–
9	Linear	12	1.1502×10^{2}	1.9776×10^{2}	–
	Neural, $K^9 = 5$	71	2.4359×10^{-1}	4.7499×10^{-1}	–
	Neural, $K^9 = 6$	85	2.2914×10^{-1}	3.9618×10^{-1}	6.5027×10^{-1}
	Neural, $K^9 = 7$	99	2.1402×10^{-1}	4.1273×10^{-1}	–
10	Linear	13	1.2895×10^{2}	2.2334×10^{2}	–
	Neural, $K^{10} = 5$	76	1.9755×10^{-1}	3.2733×10^{-1}	–
	Neural, $K^{10} = 6$	91	1.9999×10^{-1}	3.3268×10^{-1}	4.9387×10^{-1}
	Neural, $K^{10} = 7$	106	2.0371×10^{-1}	3.4394×10^{-1}	–

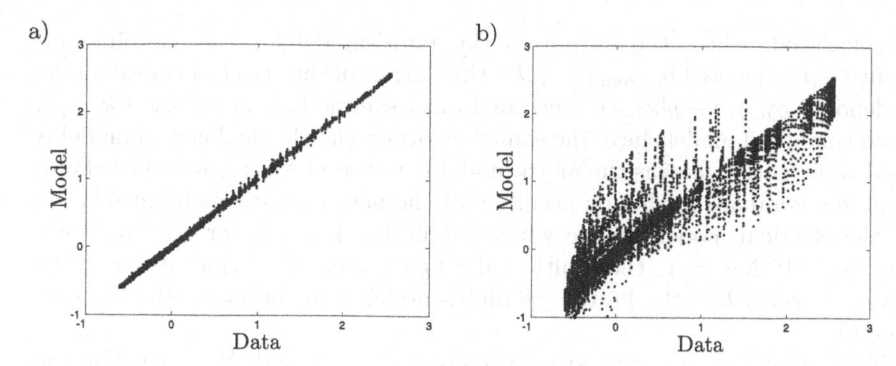

Fig. 5.4. The correlation between the validation data set and the output of the low-order multi-models of the polymerisation reactor: a) the neural multi-model, b) the linear multi-model

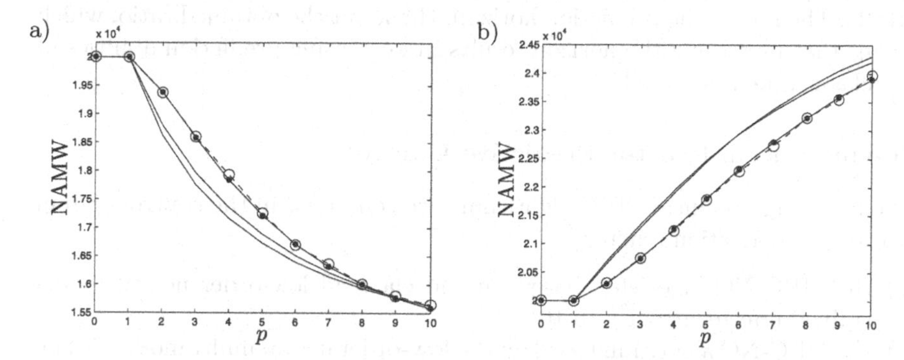

Fig. 5.5. The step-responses of the low-order neural model of the polymerisation reactor trained non-recurrently and recurrently (*solid lines, the model trained recurrently is slightly better*) and the step-responses of the low-order multi-model (*dashed line with circles*) vs. the outputs of the process (*solid line with dots*), the input signal changes from the nominal operating point to: a) $F_{\mathrm{I}} = 0.0526\ \mathrm{m}^3/\mathrm{h}$, b) $F_{\mathrm{I}} = 0.0168\ \mathrm{m}^3/\mathrm{h}$

also low (the same as that of the classical neural model), is very similar to that of the process.

Although the step-responses depicted in Fig. 5.5 show superiority of the neural multi-model over the classical neural model of the same low order, it would be beneficial to compare the models for a whole set of excitation signals. That is why the following ratio

$$R_N = \frac{1}{N}\sum_{k=1}^{n_{\mathrm{p}}} \frac{\sum_{p=1}^{N}(y_{\mathrm{mm}}(k+p|k) - y(k+p))^2}{\sum_{p=1}^{N}(y(k+p|k) - y(k+p))^2}$$

is calculated, where the output of the multi-model for the sampling instant $k+p$ is denoted by $y_{\mathrm{mm}}(k+p|k)$, the output of the classical neural model is denoted by $y(k+p|k)$, the recorded data sample is denoted by $y(k+p)$. Both compared models have the same low order, the classical neural model is used recurrently whereas the multi-model is not used recurrently. The coefficient R_N compare prediction accuracy of the neural multi-model in relation to the classical model for the whole prediction horizon, for all data samples (n_{p}). If $R_N < 1$, the multi-model has better prediction abilities; the lower the ratio R_N, the better the multi-model (and the worse the classical model).

Two prediction horizons are considered: $N = 5$ and $N = 10$. For the ratio R_N calculation, the classical low-order neural model trained recurrently is chosen, because it is better than the model trained non-recurrently). For the validation data set the following values are obtained: $R_5 = 0.2598$ and $R_{10} = 0.1823$. The longer the prediction horizon, the lower the obtained ratio, which means that increasing the horizon results in decreasing prediction abilities of the classical neural model.

Polymerisation Reactor Predictive Control

The following nonlinear MPC algorithms are compared in the control system of the polymerisation reactor:

a) the MPC-NO algorithm based on the classical low-order neural model trained non-recurrently (5.10),
b) the MPC-NO algorithm based on the low-order neural multi-model (5.11),
c) the suboptimal MPC-NPLT$_{\boldsymbol{u}(k|k-1)}$ algorithm based on the same low-order neural multi-model.

In all algorithms the same parameters are used (the same as in two previous chapters): $N = 10$, $N_{\mathrm{u}} = 3$, $\lambda_p = 0.2$, $F_{\mathrm{I}}^{\mathrm{min}} = 0.003\,\mathrm{m}^3/\mathrm{h}$, $F_{\mathrm{I}}^{\mathrm{max}} = 0.06\,\mathrm{m}^3/\mathrm{h}$. Because the process is characterised by a time-delay, the output $y(k+1|k)$ is independent of the currently calculated future control sequence. That is why the first submodel is not used in MPC algorithms.

The classical low-order neural model has rather poor prediction abilities. When it is used in the "ideal" MPC-NO algorithm, obtained control accuracy is not satisfactory which is demonstrated in Fig. 5.6. Fig. 5.7 compares the trajectories obtained in the MPC-NO and MPC-NPLT$_{\boldsymbol{u}(k|k-1)}$ structures based on the same low-order neural multi-model. Both MPC algorithms work correctly, the suboptimal approach gives results very similar to those obtained in the MPC-NO one. Thanks to on-line trajectory linearisation the results obtained in the suboptimal MPC-NPLT$_{\boldsymbol{u}(k|k-1)}$ algorithm are much closer to those obtained in the MPC-NO one than in the case of the suboptimal MPC algorithms with model linearisation for the current operating point (the differences between the MPC-NO and MPC-NPSL algorithms based on the neural Winer model (Fig. 3.13) or between the MPC-NO and MPC-NPL algorithms based on the neural state-space model (Fig. 4.6) are much bigger). The sum of squared errors over the simulation ($\mathrm{SSE}_{\mathrm{sim}}$) for the MPC-NO algorithm is 4.0297×10^8, whereas for the MPC-NPLT$_{\boldsymbol{u}(k|k-1)}$ one it is 3.9981×10^8. The obtained values are slightly higher when compared to the results for the MPC algorithms based on input-output models (Table 3.5) and algorithms based on the state-space model (p. 161). The differences, although small, are caused by the fact that the low-order model is used for prediction whereas all the models considered previously have the proper order of dynamics. Finally, the MPC-NPLT$_{\boldsymbol{u}(k-1)}$ algorithm based on the low-order neural multi-model is developed. Since the obtained trajectories are practically the same as in the MPC-NPLT$_{\boldsymbol{u}(k|k-1)}$ approach, they are not shown. For the MPC-NPLT$_{\boldsymbol{u}(k-1)}$ algorithm $\mathrm{SSE}_{\mathrm{sim}} = 4.0056 \times 10^8$.

Table 5.4 shows computational burden of compared nonlinear MPC-NO and MPC-NPLT$_{\boldsymbol{u}(k|k-1)}$ algorithms based on the same low-order neural multi-model for different control horizons. Depending on the control horizon, the quantitative reduction of computational burden is in the range 6.1–7.6. Because the multi-model has quite many parameters (without the submodel for $p = 1$, the total number of parameters is 542), the compared algorithms are much more computationally demanding than the ones based on the neural Wiener model which has as few as 10 parameters (Table 3.6). If one compares the discussed MPC algorithms with corresponding algorithms based on the neural state-space model (Table 4.3), the MPC-NO algorithm based on the multi-model is of lower computational burden (the prediction of all state variables and the derivatives of the predicted state trajectory are not calculated), whereas complexity of the suboptimal MPC algorithms for the state-space model and the multi-model are similar.

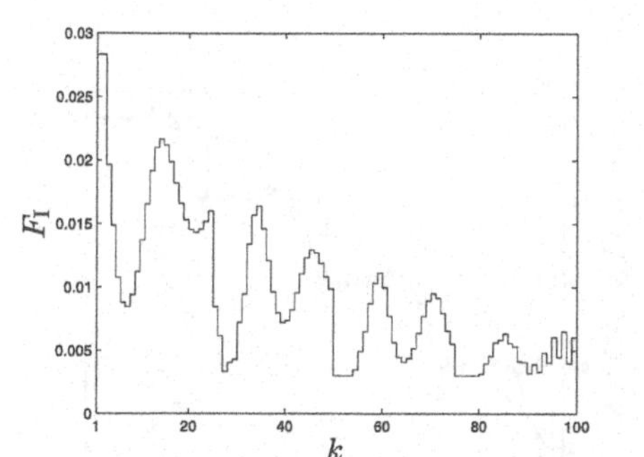

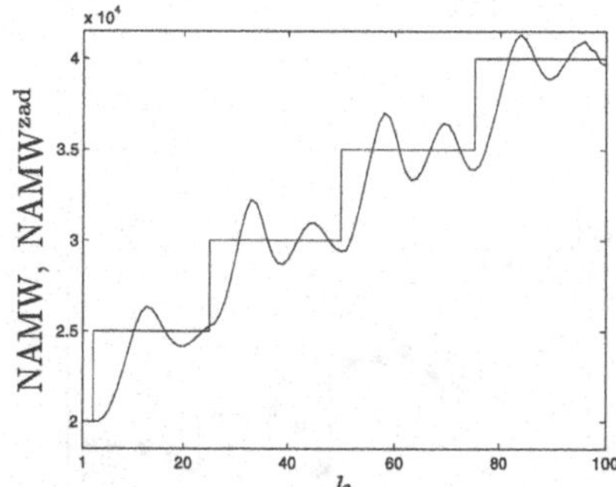

Fig. 5.6. Simulation results of the MPC-NO algorithm based on the low-order neural model in the control system of the polymerisation reactor

Table 5.4. Comparison of computational complexity (in MFLOPS) of nonlinear MPC-NO and MPC-NPLT$_{\boldsymbol{u}(k|k-1)}$ algorithms based on the low-order neural multi-model in the control system of the polymerisation reactor for different control horizons, $N = 10$

Algorithm	$N_{\mathrm{u}} = 1$	$N_{\mathrm{u}} = 2$	$N_{\mathrm{u}} = 3$	$N_{\mathrm{u}} = 4$	$N_{\mathrm{u}} = 5$	$N_{\mathrm{u}} = 10$
MPC-NO	3.98	5.50	7.27	8.91	10.57	28.13
MPC-NPLT$_{\boldsymbol{u}(k\|k-1)}$	0.66	0.82	1.03	1.28	1.58	3.70

The linear MPC algorithm based on the low-order linear multi-model is also developed. Due to significant inaccuracy of that multi-model, the resulting MPC algorithm does not work properly, the output does not converge to the set-point trajectory.

Finally, two classes of the multi-model with the proper order of dynamics (defined by $\tau = n_{\mathrm{A}} = n_{\mathrm{B}} = 2$) are also determined: linear and neural. When compared with the low-order linear multi-model (Table 5.3), the considered linear multi-model is of the same poor accuracy. When compared with the low-order neural multi-model, the errors for the considered neural multi-model is approximately two times lower. Next, the MPC-NO and suboptimal algorithms based on the considered multi-models with the proper order of dynamics are developed. The obtained results are practically the same as those when for prediction the low-order multi-model is used: the algorithm based on the linear multi-model does not work, the algorithms based on the neural multi-model give good control and suboptimal algorithms give practically the same trajectories as the MPC-NO approach. In general, control quality obtained by MPC algorithms based on the low-order neural multi-model and on the neural multi-model with the proper order of dynamics is very similar. It means that the low-order neural model is very good.

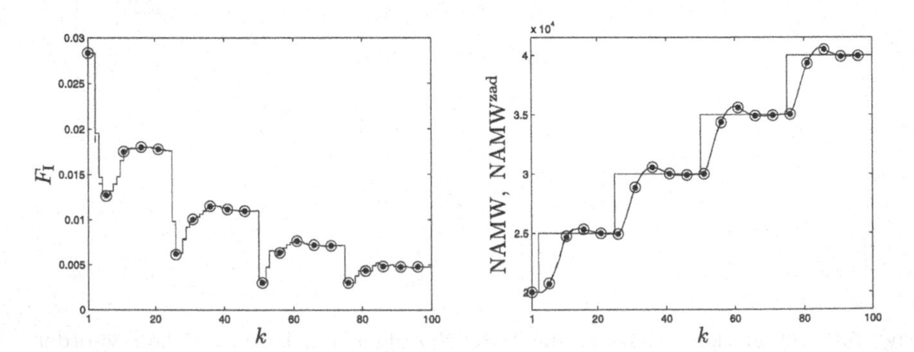

Fig. 5.7. Simulation results: the MPC-NO algorithm based on the low-order neural multi-model (*solid line with dots*) and the MPC-NPLT$_{\boldsymbol{u}(k|k-1)}$ algorithm based on the same multi-model (*dashed line with circles*) in the control system of the polymerisation reactor

5.6 Literature Review

The idea of using for prediction calculation not one, but a set of separate submodels is not new, its origins may be recognised in the adaptive MUSMAR algorithm [83]. The prediction error propagation problem, particularly important when the order of dynamics is lower than that of the process, is thoroughly discussed in [278, 279, 298]. The neural multi-models are presented in [137], although no examples which could demonstrate their superiority over a single model used recurrently are given. Conversely, the choice of the multi-model is motivated because its development (training) is simpler than in the case of a single model. Convincing examples clearly indicating advantages of the multi-model approach are given in [278, 279]. The neural multi-model and the MPC-NO algorithm based on such a model are shortly discussed in [237]. Unfortunately, no implementation details are given, the nonlinear optimisation algorithm is not specified. Computational complexity and prediction error propagation issues are not considered, but easiness of training is emphasised. The problem of rather a high number of parameters is noticed. In order to solve it, only 4 selected submodels are taken into account: for $p = 1, 2, 5, 10$, while the prediction horizon $N = 10$. The same MPC-NO algorithm based on the neural multi-model is applied in [236] to a simulated multi-input multi-output distillation process. The neural multi-model considered in this chapter, but in its single-input single-output version and without measured disturbances, is thoroughly described in [159], a suboptimal MPC algorithm is detailed, the advantages of the neural multi-model approach are demonstrated for the simulated polymerisation process. Initially, the multi-model has as many as 340 parameters, the OBD pruning algorithm [126] is used to reduce complexity of neural networks. Finally, the multi-model has only 239 parameters. The multi-model approach is universal, different structures of submodels can be used. The RBF multi-models are discussed in [167], a suboptimal MPC algorithm is detailed, simulation results of the algorithm applied to a multi-input multi-output distillation process are discussed. In addition to process control, one may find some interesting applications of the multi-model approach. For example, the multi-model may be used to predict the blood pressure [275].

The quasi-linear multi-models (1.18) are discussed in [138]. The model outputs are linear functions of the decision variables of the MPC algorithm (i.e. the future control sequence) but they depend in a nonlinear way on some process variables measured at previous sampling instants. The RBF neural networks are used in the nonlinear part of the multi-model. Unfortunately, the lack of prediction error propagation is not pointed out, but an efficient on-line training algorithm is emphasised. Thanks to a specific structure of the multi-model, the MPC cost-function is quadratic, on-line linearisation is not necessary. An explicit MPC algorithm is described, no constraints are taken into account. When compared with the neural multi-model approach, the quasi-linear structure has a limited approximation ability which is an inherent structural disadvantage.

As an alternative to the multi-model approach, one may use the so called specialised (tailored for MPC) neural model of the general form [28]

$$\begin{aligned} y(k) = f(&u(k-\tau),\ldots,u(k-n_{\mathrm{B}}-N+1),\\ &y(k-N),\ldots,y(k-n_{\mathrm{A}}-N+1)) \end{aligned}$$

where the RBF network is used for approximation. The output predictions

$$\begin{aligned} \hat{y}(k+p|k) = f(&\underbrace{u(k-\tau+p|k),\ldots,u(k|k)}_{I_{\mathrm{uf}(p)}},\\ &\underbrace{u(k-1),\ldots,u(k-n_{\mathrm{B}}-N+1+p)}_{I_{\mathrm{u}}-I_{\mathrm{uf}(p)}},\\ &\underbrace{y(k-N+p),\ldots,y(k-n_{\mathrm{A}}-N+1+p)}_{n_{\mathrm{A}}})+d(k) \end{aligned} \tag{5.12}$$

do not depend recurrently on the predictions for some previous sampling instants of the prediction horizon, $I_{\mathrm{u}} = n_{\mathrm{B}} + N - \tau$. The specialised model, similarly to the multi-model, is not used recurrently in MPC algorithms, the prediction error is not propagated, the recurrent training configuration is not necessary. The MPC-NO algorithm based on such specialised RBF models is detailed in [27], the SQP algorithm is used for nonlinear optimisation. The algorithm is next applied to an industrial process with 4 inputs and 4 outputs. Unfortunately, the necessity of on-line nonlinear optimisation and its significant computational complexity are not pointed out. The MPC-NPL algorithm based on the DLP specialised model is detailed in [164]. For the simulated polymerisation reactor advantages of the specialised model are demonstrated and the MPC-NPL and MPC-NO approaches are compared.

It is worth emphasising that in contrast to the multi-model, the specialised model has two important disadvantages. Firstly, only one model is used, the role of which is correctly calculate the predictions for the whole prediction horizon. When the horizon is long, form Eq. (5.12), it is evident that the output predictions for the initial part of the horizon depend on the output values measured quite many sampling instants ago. For the last part of the prediction horizon, the output predictions are functions of some recently measured output values. It means that if the prediction horizon is too long, it may be very difficult, or hardly possible, to find a sufficiently precise structured model. The multi-model does not have the described disadvantage, because the recently measured output signals $y(k), y(k-1), \ldots$ are arguments of all submodels. The second important disadvantage of the specialised model is the fact that the prediction horizon is the parameter of the model. Any change of the prediction horizon requires identification of a new model. The multi-model does not have that disadvantage because the prediction horizon determines the number of independent submodels. When the horizon is shortened, unnecessary submodels are simply not used. When the horizon is lengthened, only the submodels for the last part of the horizon must be found.

5.7 Chapter Summary

The structure of the multi-model is motivated by the role of the model in the MPC algorithms. The model has usually a lower order of dynamic than the real process. The models trained in the classical serial-parallel (non-recurrent) configuration are likely not to work correctly when used recurrently for prediction calculation in MPC algorithm because the prediction error is propagated. A model can be assessed as good to be used in MPC when all prediction, over the whole prediction horizon are calculated correctly. Although it is possible to use a model trained in the parallel (recurrent) configuration, but its identification is more demanding. What is more, recurrent training does not eliminate the prediction error propagation phenomenon and the obtained model may be sensitive to noise. The multi-model is a sound alternative to the classical dynamic models, because:

a) The multi-model is not used recurrently in MPC algorithms, the prediction error is not propagated.
b) The order of dynamics of the submodels which comprise the multi-model can be lower than that of the classic single model.
c) One submodel must only calculate the prediction for one sampling instant of the prediction horizon. As a result, training is carried out in the non-recurrent serial-parallel configuration, it is not necessary to train one, but usually quite complicated recurrent structure, which is expected to calculate all predictions for the whole prediction horizon.

This chapter details the structure of the neural DLP multi-model as well as the MPC-NO and MPC-NPLPT algorithms based on such a model. Using the derived formulae one may implement the MPC-NPLT algorithm and the MPC-NNPAPT one. Advantages of the neural multi-model approach are demonstrated for the polymerisation reactor. For the considered process, the suboptimal MPC algorithm is very precise and it is a few times less computationally demanding than the MPC-NO approach.

The concepts presented in this chapter may be extended in many ways. Firstly, one may use alternative multi-models, e.g. use a quasi-linear multi-model, in which the outputs are linear functions of the future control sequence and nonlinear functions of some process signals measured in the past. Such an approach makes the training process simpler. Secondly, alternative neural structures may be used in the multi-model, e.g. the RBF networks or SVM approximators. Thirdly, all suboptimal MPC algorithms may be implemented in their explicit versions. An interesting idea is to use adaptive MPC algorithms based on multi-models. The multi-model is trained in the non-recurrent configuration, which is much simpler and more reliable when used on-line than recurrent training.

For the polymerisation reactor, the total number of parameters of the low-order neural multi-model is 542 (because of the time-delay, the first submodel, i.e. for $p = 1$, can be not taken into account in MPC algorithms). If a relatively significant number of parameters is a problem in practical implementation (the available hardware platform may be unable to deal with such data sets), it is necessary to reduce the number of model parameters. The following methods can be used for this purpose:

a) The prediction horizon can be shortened. Unfortunately, too short horizons are likely to have a negative impact on control quality (or even destabilise the process).
b) Not all N submodels are used for prediction, but only some of them. In such a way the minimised MPC cost-function (1.4) takes into account some predicted control errors (for the submodels used the coefficients $\mu_{m,p} \neq 0$, for the remaining ones $\mu_{m,p} = 0$).
c) The neural networks which comprise the multi-model are pruned. For this purpose OBD [126] or OBS [93] algorithms can be used.
d) Selected submodels are used for direct prediction calculation for some sampling instants of the prediction horizon. The remaining predictions are calculated indirectly, as a result of an interpolation procedure. Unfortunately, interpolation leads to significant complication of the resulting MPC algorithms (it is necessary to develop interpolated output predictions and their derivatives with respect to the future control sequence).

Of course, any two of the above methods or all of them can be used at the same time.

6

MPC Algorithms with Neural Approximation

This chapter discusses a family of suboptimal MPC algorithms with neural approximation the characteristic feature of which is the lack of on-line linearisation. A specially designed neural network (the neural approximator) approximates on-line the step-response coefficients of the model linearised for the current operating point of the process (such an approach is used in the MPC-NPL-NA and DMC-NA algorithms which are extensions of the MPC-NPL and DMC ones). Alternatively, the neural approximator calculates on-line the derivatives of the predicted output trajectory with respect to the future control sequence (such an approach is used in the MPC-NPLT-NA algorithm which is an extension of the MPC-NPLT one). The explicit versions of MPC algorithms with neural approximation are also presented. They are very computationally efficient, because the neural approximator directly finds on-line the coefficients of the control law, successive on-line linearisation and calculations typical of the classical MPC algorithms are not necessary.

6.1 MPC-NPL Algorithm with Neural Approximation (MPC-NPL-NA)

In the MPC-NPL algorithm, the structure of which is shown in Fig. 2.3, the nonlinear model is successively linearised on-line for the current operating point of the process. The obtained linear approximation of the nonlinear model is next used for prediction. According to the suboptimal prediction equation (2.33), thanks to linearisation, the vector of predicted output signals is a linear function of the decision variables, namely $\hat{\boldsymbol{y}}(k) = \boldsymbol{G}(k)\triangle\boldsymbol{u}(k) + \boldsymbol{y}^0(k)$, where the matrix $\boldsymbol{G}(k)$ consists of the step-response coefficients of the linearised model whereas the nonlinear free trajectory $\boldsymbol{y}^0(k)$ is calculated from the full nonlinear model.

Fig. 6.1 depicts the general structure of the MPC-NPL algorithm with Neural Approximation (MPC-NPL-NA). It is inspired by the suboptimal

M. Ławryńczuk, *Computationally Efficient Model Predictive Control Algorithms*, Studies in Systems, Decision and Control 3,
DOI: 10.1007/978-3-319-04229-9_6, © Springer International Publishing Switzerland 2014

prediction equation used in the classical MPC-NPL algorithm, in which one neural dynamic model is used (for linearisation and free trajectory calculation) whereas two neural networks are used in the MPC-NPL-NA approach. The first network is the neural approximator which calculates on-line the local step-response coefficients $s_1(k), \ldots, s_N(k)$ for the current operating point (the matrices $\boldsymbol{S}_1(k), \ldots, \boldsymbol{S}_N(k)$ for multi-input multi-output processes), which comprise the dynamic matrix $\boldsymbol{G}(k)$. Successive on-line model linearisation and analytical calculation of the step-response are not necessary in the MPC-NPL-NA algorithm. The second network is the dynamic neural model, the same as in the classical MPC-NPL approach. The dynamic model is used only for free trajectory $\boldsymbol{y}^0(k)$ calculation. Because in the MPC-NPL-NA algorithms the MPC-NPL prediction equation (2.33) is used, the same quadratic optimisation problem (2.36) is solved in both MPC-NPL and MPC-NPL-NA algorithms.

For the single-input single-output process, the neural approximator consists of one neural network. The current operating point (inputs of the network) is defined by the vector of measured signals of length $\tilde{n}_{\mathrm{A}} + \tilde{n}_{\mathrm{B}} + \tilde{n}_{\mathrm{C}} + 2$

$$\widetilde{\boldsymbol{x}}(k) = [u(k-1) \ldots u(k-\tilde{n}_{\mathrm{B}})\; h(k) \ldots h(k-\tilde{n}_{\mathrm{C}})\; y(k) \ldots y(k-\tilde{n}_{\mathrm{A}})]^{\mathrm{T}} \quad (6.1)$$

In the simplest case one may assume that $\tilde{n}_{\mathrm{A}} = n_{\mathrm{A}}$, $\tilde{n}_{\mathrm{B}} = n_{\mathrm{B}}$ and $\tilde{n}_{\mathrm{C}} = n_{\mathrm{C}}$. The recent signals $h(k)$ and $y(k)$ measured at the current sampling instant and the signal $u(k-1)$ calculated for the previous iteration are used in the definition of the operating point. The network approximates the step-response coefficients of the linearised model

$$g_p(\widetilde{\boldsymbol{x}}(k)) \approx s_p(k)$$

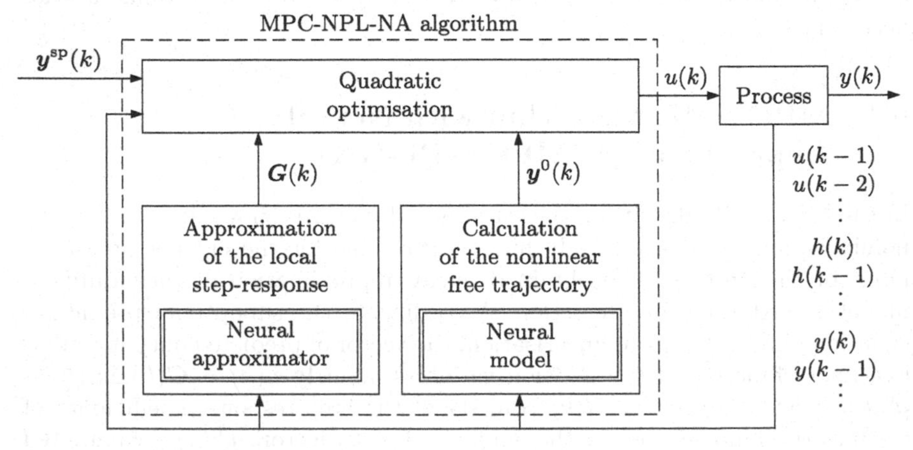

Fig. 6.1. The structure of the MPC algorithm with Neural Approximation (MPC-NPL-NA)

over the whole prediction horizon ($p = 1, \ldots, N$), $g_p \colon \mathbb{R}^{\tilde{n}_{\mathrm{A}}+\tilde{n}_{\mathrm{B}}+\tilde{n}_{\mathrm{C}}+2} \to \mathbb{R}$ are nonlinear functions which describe the approximator. One neural network can be also used as the neural approximator for the multi-input multi-output process. The current operating point in then defined by the vector of measured signals of length $n_{\mathrm{y}}\tilde{n}_{\mathrm{A}} + n_{\mathrm{u}}\tilde{n}_{\mathrm{B}} + n_{\mathrm{h}}\tilde{n}_{\mathrm{C}} + 2(n_{\mathrm{h}} + n_{\mathrm{y}})$

$$\widetilde{\boldsymbol{x}}(k) = \begin{bmatrix} u(k-1) \\ \vdots \\ u(k-\tilde{n}_{\mathrm{B}}) \\ h(k) \\ \vdots \\ h(k-\tilde{n}_{\mathrm{C}}) \\ y(k) \\ \vdots \\ y(k-\tilde{n}_{\mathrm{A}}) \end{bmatrix} \tag{6.2}$$

In the simplest case $\tilde{n}_{\mathrm{A}} = \max(n_{\mathrm{A}}^1, \ldots, n_{\mathrm{A}}^{n_{\mathrm{y}}})$, $\tilde{n}_{\mathrm{B}} = \max(n_{\mathrm{B}}^{1,1}, \ldots, n_{\mathrm{B}}^{n_{\mathrm{y}},n_{\mathrm{u}}})$ and $\tilde{n}_{\mathrm{C}} = \max(n_{\mathrm{C}}^{1,1}, \ldots, n_{\mathrm{C}}^{n_{\mathrm{y}},n_{\mathrm{u}}})$. The network approximates the step-response coefficients

$$g_p^{m,n}(\widetilde{\boldsymbol{x}}(k)) \approx s_p^{m,n}(k)$$

over the whole prediction horizon ($p = 1, \ldots, N$), for all input-output channels ($m = 1, \ldots, n_{\mathrm{y}}$, $n = 1, \ldots, n_{\mathrm{u}}$), $g_p^{m,n} \colon \mathbb{R}^{n_{\mathrm{y}}\tilde{n}_{\mathrm{A}}+n_{\mathrm{u}}\tilde{n}_{\mathrm{B}}+n_{\mathrm{h}}\tilde{n}_{\mathrm{C}}+2(n_{\mathrm{h}}+n_{\mathrm{y}})} \to \mathbb{R}$ are nonlinear functions which describe the approximator. A neural approximator comprised of independent n_{y} neural networks may turn out to be better (in terms of accuracy, the number of parameters and easiness of training). The consecutive networks approximate the step-responses of the individual multi-input single-output models. The operating point vector of the mth network is defined by the arguments of the dynamic model (2.1) as well as disturbance and output signals measured at the current sampling instant. It is of length $n_{\mathrm{A}}^m + \sum_{n=1}^{n_{\mathrm{u}}} n_{\mathrm{B}}^{m,n} + \sum_{n=1}^{n_{\mathrm{h}}}(n_{\mathrm{C}}^{m,n} + 1) + 1$ and has the following form

$$\begin{aligned} \widetilde{\boldsymbol{x}}_m(k) = [&u_1(k-1), \ldots, u_1(k-n_{\mathrm{B}}^{m,1}), \ldots, u_{n_{\mathrm{u}}}(k-1), \ldots, u_{n_{\mathrm{u}}}(k-n_{\mathrm{B}}^{m,n_{\mathrm{u}}}), \\ &h_1(k), \ldots, h_1(k-n_{\mathrm{C}}^{m,1}), \ldots, h_{n_{\mathrm{h}}}(k), \ldots, h_{n_{\mathrm{h}}}(k-n_{\mathrm{C}}^{m,n_{\mathrm{h}}}), \\ &y_m(k), \ldots, y_m(k-n_{\mathrm{A}}^m)]^{\mathrm{T}} \end{aligned} \tag{6.3}$$

The networks approximate the step-response coefficients

$$g_p^{m,n}(\widetilde{\boldsymbol{x}}_m(k)) \approx s_p^{m,n}(k)$$

where $g_p^{m,n} \colon \mathbb{R}^{n_{\mathrm{A}}^m+\sum_{n=1}^{n_{\mathrm{u}}} n_{\mathrm{B}}^{m,n}+\sum_{n=1}^{n_{\mathrm{h}}}(n_{\mathrm{C}}^{m,n}+1)+1} \to \mathbb{R}$ for all $p = 1, \ldots, N$, $m = 1, \ldots, n_{\mathrm{y}}$, $n = 1, \ldots, n_{\mathrm{u}}$ are nonlinear functions which describe the approximator. Of course, in practice one may use an alternative configuration of the neural approximator.

The neural approximator is designed for some fixed length of the prediction horizon, but it is also possible to use shorter horizons. In such a case some of the approximated coefficients are simply not used.

It is necessary to point out that the neural approximator is in fact a steady-state model. Thanks to it, its training is simple, it is carried out in the serial-parallel configuration (non-recurrent one), the prediction errors are not propagated. The data necessary for identification of the neural approximator can be obtained by means of the following methods:

1. The dynamic neural model is simulated open-loop (without any controller), as the excitation signal the data set used for training the dynamic neural model is used (or similar excitation signals are used). During simulations the model is successively linearised for the current operating points and the step-response coefficients of the linearised model are calculated.
2. The classical MPC-NPL algorithm based on the dynamic neural model is developed. Next, the algorithm is simulated (or applied to a real process) for a randomly changing set-point trajectory. During simulations the model is successively linearised for various operating conditions, the step-response coefficients of the linearised model are calculated.
3. The local step-responses of the process are recorder for different operating points. In such a case the operating point can be defined in a very simple way, by means of the most recent signals $u(k-1)$ and (or) $h(k)$ and $y(k)$. Unlike both above approaches, the step-responses of the real process, not of its model, are recorded.

6.1.1 Explicit MPC-NPL-NA Algorithm

In the explicit MPC-NPL algorithm, the structure of which is depicted in Fig. 2.8, the optimal (when no constraints are taken into account) vector of increments of the control signals for the current sampling instant is defined by Eq. (2.70), namely $\triangle u(k|k) = \boldsymbol{K}^{n_{\mathrm{u}}}(k)(\boldsymbol{y}^{\mathrm{sp}}(k) - \boldsymbol{y}^{0}(k))$, where the matrix $\boldsymbol{K}^{n_{\mathrm{u}}}(k)$ contains the first n_{u} rows of the matrix $\boldsymbol{K}(k)$, which, in turn, depends on the dynamic matrix $\boldsymbol{G}(k)$ consisting of the step-response coefficients of the locally linearised model (Eq. (2.69)). At each sampling instant of the classical MPC-NPL algorithm the neural model is linearised at the current operating point, the obtained linear approximation is used to find the step-response coefficients which comprise the matrix $\boldsymbol{G}(k)$. Next, the matrix $\boldsymbol{K}^{n_{\mathrm{u}}}(k)$ must be calculated on-line, for this purpose the computationally efficient method with the LU matrix decomposition can be used.

In the discussed explicit MPC-NPL-NA algorithm, whose structure is shown in Fig. 6.2, two neural networks are used: the neural approximator and the classical dynamic neural model, the same which is used in the numerical MPC-NPL approach. The neural approximator calculates on-line the matrix $\boldsymbol{K}^{n_{\mathrm{u}}}(k)$ for the current operating point of the process. It means that model linearisation, step-response coefficients calculation and matrix decomposition are not carried out on-line. The neural dynamic model is used to find the nonlinear free trajectory, in the same way it is done in the classical numerical or explicit MPC-NPL algorithms and in the MPC-NPL-NA approach. The current increments of the control signals are determined from Eq. (2.70), in the same way it is done in the explicit MPC-NPL algorithm. Finally, the obtained control increments are projected onto the feasible set defined by the constraints.

Thanks to the fact that the algorithm uses two neural networks, which is inspired by the suboptimal MPC approaches, the discussed explicit algorithm is computationally very uncomplicated. It is necessary to emphasise the fact that the explicit MPC-NPL-NA algorithm seems in general to be more computationally efficient than its numerical version. It is because in the numerical MPC-NPL-NA algorithm the neural approximator makes it possible to eliminate the necessity of successive on-line linearisation and the step-response calculation, but the MPC-NPL quadratic optimisation problem is still repeatedly solved on-line. In the explicit algorithm no linearisation, no step-response calculation and matrix decomposition are necessary because the neural approximator directly finds on-line, for the current operating point, the coefficients of the control law vector (2.70), i.e. the entries of the matrix $\boldsymbol{K}^{n_\mathrm{u}}(k)$ of dimensionality $n_\mathrm{u} \times n_\mathrm{y}N$. The operating point is defined in the same way it is done in the numerical MPC-NPL-NA algorithm (Eqs. (6.1), (6.2) or (6.3)). For the single-input single-output process, the approximator calculates the vector of length N

$$g_{1,p}(\widetilde{\boldsymbol{x}}(k)) \approx k_{1,p}^{n_\mathrm{u}}(k)$$

where $g_{1,p}\colon \mathbb{R}^{\tilde{n}_\mathrm{A}+\tilde{n}_\mathrm{B}+\tilde{n}_\mathrm{C}+2} \rightarrow \mathbb{R}$ are nonlinear functions which describe the approximator, $p = 1, \ldots, N$. For the multi-input multi-output process the

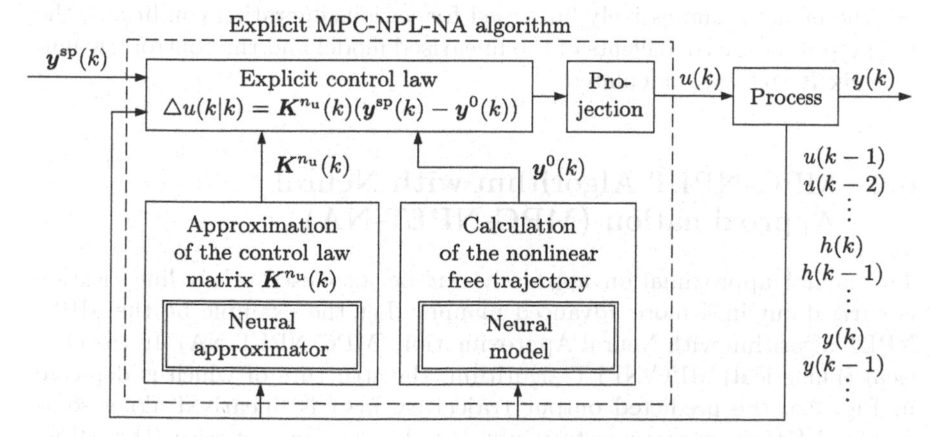

Fig. 6.2. The structure of the explicit MPC-NPL-NA algorithm

approximator finds $n_{\mathrm{u}} n_{\mathrm{y}} N$ coefficients of the control law. If only one network is used, it approximates the coefficients

$$g_{n,p}(\widetilde{\boldsymbol{x}}(k)) \approx k_{n,p}^{n_{\mathrm{u}}}(k)$$

for all $n = 1, \ldots, n_{\mathrm{u}}$, $p = 1, \ldots, N$, where the operating point $\widetilde{\boldsymbol{x}}(k)$ is defined by Eq. (6.2), $g_{n,p} \colon \mathbb{R}^{n_{\mathrm{y}}\tilde{n}_{\mathrm{A}}+n_{\mathrm{u}}\tilde{n}_{\mathrm{B}}+n_{\mathrm{h}}\tilde{n}_{\mathrm{C}}+2(n_{\mathrm{h}}+n_{\mathrm{y}})} \rightarrow \mathbb{R}$ are nonlinear functions which describe the approximator. If the approximator consists of n_{y} independent networks, they approximate the coefficients

$$g_{n,p}(\widetilde{\boldsymbol{x}}_m(k)) \approx k_{n,p}^{n_{\mathrm{u}}}(k)$$

for all $n = 1, \ldots, n_{\mathrm{u}}$, $p = 1, \ldots, N$, where the operating point $\widetilde{\boldsymbol{x}}(k)$ is defined by Eq. (6.3), $g_{n,p} \colon \mathbb{R}^{n_{\mathrm{A}}^m+\sum_{n=1}^{n_{\mathrm{u}}} n_{\mathrm{B}}^{m,n}+\sum_{n=1}^{n_{\mathrm{h}}}(n_{\mathrm{C}}^{m,n}+1)+1} \rightarrow \mathbb{R}$ are nonlinear functions which describe the approximator.

In the explicit MPC-NPL-NA algorithm the neural approximator is defined for some fixed lengths of the prediction and control horizons as well for the chosen weighting matrices $\boldsymbol{M}$ and $\boldsymbol{\Lambda}$. Its number of outputs is determined by the prediction horizon whereas the control horizon does not influence its topology and computational complexity of the control algorithm.

The data necessary for identification of the neural approximator can be obtained by means of the following methods:

1. The dynamic neural model is simulated open-loop (without any controller), as the excitation signal the data set used for training the dynamic neural model is used (or similar excitation signals are used). During simulations the model is successively linearised for the current operating points, the step-response coefficients of the linearised model and the control law matrix $\boldsymbol{K}^{n_{\mathrm{u}}}(k)$ are calculated.
2. The classical explicit MPC-NPL algorithm based on the dynamic neural model is developed. Next, the algorithm is simulated (or applied to a real process) for a randomly changing set-point trajectory. During simulations the model is successively linearised for various operating conditions, the step-response coefficients of the linearised model and the control law matrix $\boldsymbol{K}^{n_{\mathrm{u}}}(k)$ are calculated.

6.2 MPC-NPLT Algorithm with Neural Approximation (MPC-NPLT-NA)

The neural approximation approach can be also used when linearisation is carried out in a more advanced manner. Let the example be the MPC-NPLT algorithm with Neural Approximation (MPC-NPLT-NA). In the classical (numerical) MPC-NPLT algorithm, the structure of which is depicted in Fig. 2.5, the predicted output trajectory $\hat{\boldsymbol{y}}(k)$ is linearised along some assumed future input trajectory $\boldsymbol{u}^{\mathrm{traj}}(k)$. In the simplest case (the MPC-NPLT$_{\boldsymbol{u}(k-1)}$ approach) linearisation is performed along the input trajectory defined entirely by the control signals applied to the process at the previous sampling instant, which means that the predicted output trajectory corresponding to the input trajectory $\boldsymbol{u}^{\mathrm{traj}}(k)$ is in fact the nonlinear free trajectory, $\hat{\boldsymbol{y}}^{\mathrm{traj}}(k) = \boldsymbol{y}^0(k)$. Such an approach is particularly worth considering because, for $\boldsymbol{u}^{\mathrm{traj}}(k) = \boldsymbol{u}(k-1)$, the suboptimal prediction equation (2.46) reduces to

$$\hat{\boldsymbol{y}}(k) = \boldsymbol{H}(k)\boldsymbol{J}\triangle\boldsymbol{u}(k) + \boldsymbol{y}^0(k)$$

The matrix $\boldsymbol{H}(k)$, the structure of which is given by Eq. (2.45), consists of the derivatives of the predicted output trajectory with respect to the future control policy. In the classical MPC-NPLT$_{\boldsymbol{u}(k-1)}$ algorithm this matrix, similarly to the trajectory $\hat{\boldsymbol{y}}^{\mathrm{traj}}(k) = \boldsymbol{y}^0(k)$, is calculated on-line from the full dynamic model of the process. Naturally, successive on-line trajectory linearisation is much more complicated than model linearisation used on the MPC-NPL and MPC-SL algorithms. In consequence, calculation of the matrix $\boldsymbol{H}(k)$ is usually significantly more demanding than determination of the step-response matrix $\boldsymbol{G}(k)$ of the linearised model.

In the MPC-NPLT$_{\boldsymbol{u}(k-1)}$ algorithm with Neural Approximation (MPC-NPLT$_{\boldsymbol{u}(k-1)}$-NA) two neural networks are used: the first one is a neural approximator which calculates on-line the entries of the matrix $\boldsymbol{H}(k)$ for the current operating point and for the trajectory $\boldsymbol{u}^{\mathrm{traj}}(k) = \boldsymbol{u}(k-1)$, the second one is the classical neural dynamic model and it is used for calculation of the nonlinear free trajectory $\boldsymbol{y}^0(k)$. The structure of the algorithm is very similar to that of the MPC-NPL-NA one shown in Fig. 6.1, but the matrix $\boldsymbol{H}(k)$ is calculated in place of the $\boldsymbol{G}(k)$ one. Assuming that $\hat{\boldsymbol{y}}^{\mathrm{traj}}(k) = \boldsymbol{y}^0(k)$ and $\boldsymbol{u}^{\mathrm{traj}}(k) = \boldsymbol{u}(k-1)$, from the MPC-NPLT quadratic optimisation problem (2.47), one obtains the quadratic task solved at each sampling instant of the MPC-NPLT$_{\boldsymbol{u}(k-1)}$ algorithm.

For the single-input single-output process one neural network is used as the approximator. It approximates the derivatives of the predicted trajectory

$$g_{p,r}(\widetilde{\boldsymbol{x}}(k)) \approx \left.\frac{\partial \hat{y}^{\mathrm{traj}}(k+p|k)}{\partial u^{\mathrm{traj}}(k+r|k)}\right|_{\substack{\hat{y}^{\mathrm{traj}}(k+p|k)=y^0(k+p|k)\\ u^{\mathrm{traj}}(k+r|k)=u(k-1)}}$$

where $g_{p,r}\colon \mathbb{R}^{\tilde{n}_{\mathrm{A}}+\tilde{n}_{\mathrm{B}}+\tilde{n}_{\mathrm{C}}+2} \to \mathbb{R}$ are nonlinear functions which describe the approximator, $p = 1,\ldots,N$, $r = 0,\ldots,N_{\mathrm{u}}-1$. The current operating point, described by the vector $\widetilde{\boldsymbol{x}}(k)$, is defined in the same way it is done in the MPC-NPL-NA algorithm (Eq. (6.1)). One network can be also used as the approximator for the multi-input multi-output process. It approximates the derivatives

$$g_{p,r}^{m,n}(\widetilde{\boldsymbol{x}}(k)) \approx \left.\frac{\partial \hat{y}_m^{\mathrm{traj}}(k+p|k)}{\partial u_n^{\mathrm{traj}}(k+r|k)}\right|_{\substack{\hat{y}_m^{\mathrm{traj}}(k+p|k)=y_m^0(k+p|k)\\ u_n^{\mathrm{traj}}(k+r|k)=u_n(k-1)}}$$

for all $m = 1, \ldots, n_{\mathrm{y}}$, $n = 1, \ldots, n_{\mathrm{u}}$, $p = 1, \ldots, N$, $r = 0, \ldots, N_{\mathrm{u}} - 1$, where $g_{p,r}^{m,n} \colon \mathbb{R}^{n_{\mathrm{y}}\tilde{n}_{\mathrm{A}}+n_{\mathrm{u}}\tilde{n}_{\mathrm{B}}+n_{\mathrm{h}}\tilde{n}_{\mathrm{C}}+2(n_{\mathrm{h}}+n_{\mathrm{y}})} \to \mathbb{R}$ are nonlinear functions which describe the approximator. The operating point is defined by Eq. (6.2). When the approximator consists of n_{y} independent networks, the operating point of the mth network is defined by Eq. (6.3). The networks approximate the derivatives

$$g_{p,r}^{m,n}(\tilde{\boldsymbol{x}}_m(k)) \approx \left. \frac{\partial \hat{y}_m^{\mathrm{traj}}(k+p|k)}{\partial u_n^{\mathrm{traj}}(k+r|k)} \right|_{\substack{\hat{y}_m^{\mathrm{traj}}(k+p|k)=y_m^0(k+p|k) \\ u_n^{\mathrm{traj}}(k+r|k)=u_n(k-1)}}$$

where $g_{p,r}^{m,n} \colon \mathbb{R}^{n_{\mathrm{A}}^m+\sum_{n=1}^{n_{\mathrm{u}}} n_{\mathrm{B}}^{m,n}+\sum_{n=1}^{n_{\mathrm{h}}}(n_{\mathrm{C}}^{m,n}+1)+1} \to \mathbb{R}$ are nonlinear functions which describe the approximator.

Similarly to the MPC-NPL-NA algorithm, the neural approximator used in the MPC-NPLT$_{\boldsymbol{u}(k-1)}$-NA one is designed for fixed lengths of the prediction and control horizons, but it is also possible to use shorter horizons. In such a case some of the approximated entries of the matrix $\boldsymbol{H}(k)$ are simply not used.

The data necessary for identification of the neural approximator can be obtained by means of two methods very similar to those used in the case of the MPC-NPL-NA algorithm. Firstly, the matrix $\boldsymbol{H}(k)$ can be calculated as a result of open-loop simulation of the dynamic neural model for some data sets and trajectory linearisation. Alternatively, the classical MPC-NPLT$_{\boldsymbol{u}(k-1)}$ algorithm based on the dynamic neural model is developed. Next, the algorithm is simulated (or applied to a real process) for a randomly changing set-point trajectory. Of course, trajectory linearisation is performed rather than model linearisation typical of the MPC-NPL approach.

One may notice that development of the MPC-NPLT algorithm with neural approximation in which output trajectory linearisation is carried out along some other input trajectory is not so simple. For example, if the optimal input trajectory calculated at the previous sampling instant is used for linearisation, (the MPC-NPLT$_{\boldsymbol{u}(k|k-1)}$ approach), it is clear that the optimal trajectory depends on the set-point trajectory. In such a case the neural approximator should be designed for some specific set-point trajectory or the set-point trajectory must be delivered to the approximator.

6.2.1 Explicit MPC-NPLT-NA Algorithm

The MPC-NPLT-NA algorithm, similarly to the MPC-NPL-NA approach, may be also formulated in its explicit version. Due to its simplicity the following discussion is concerned with the MPC-NPLT$_{\boldsymbol{u}(k-1)}$-NA algorithm. When the predicted output trajectory is linearised along the input trajectory $\boldsymbol{u}^{\mathrm{traj}}(k) = \boldsymbol{u}(k-1)$, for which $\hat{\boldsymbol{y}}^{\mathrm{traj}}(k) = \boldsymbol{y}^0(k)$, according to Eq. (2.73), the optimal (when no constraints are taken into account) vector of increments of the control signals for the current sampling instant is

$$\triangle u(k|k) = \boldsymbol{K}^{n_{\mathrm{u}}}(k)(\boldsymbol{y}^{\mathrm{sp}}(k) - \boldsymbol{y}^{0}(k)) \tag{6.4}$$

where, from Eq. (2.72), the matrix $\boldsymbol{K}^{n_{\mathrm{u}}}(k)$ depends on the matrix $\boldsymbol{H}(k)$, defined by Eq. (2.45), which consists of the derivatives of the predicted output trajectory with respect to the future input trajectory.

Two neural networks are used in the explicit MPC-NPLT-NA algorithm: the first one is the neural approximator, which calculates on-line the entries of the control law matrix $\boldsymbol{K}^{n_{\mathrm{u}}}(k)$ for the current operating point and for the trajectory $\boldsymbol{u}^{\mathrm{traj}}(k) = \boldsymbol{u}(k-1)$ without the necessity of numerical trajectory linearisation, matrix decomposition and other algebraic operations, the second one is the classical neural dynamic model and it is used for calculation of the nonlinear free trajectory $\boldsymbol{y}^{0}(k)$. The structure of the discussed algorithm is hence very similar to that of the explicit MPC-NPL-NA algorithm shown in Fig. 6.2. The inputs and outputs of the neural approximator are defined in the same way it is done in the explicit MPC-NPL-NA approach, but the matrix $\boldsymbol{K}^{n_{\mathrm{u}}}(k)$ is different.

Similarly to the explicit MPC-NPL-NA algorithm, the neural approximator used in the explicit MPC-NPLT$_{\boldsymbol{u}(k-1)}$-NA structure is designed for some fixed lengths of the prediction and control horizons as well as for the chosen weighting matrices $\boldsymbol{M}$ and $\boldsymbol{\Lambda}$. Complexity of the approximator depends on the prediction horizon, the control horizon does not affect its complexity and computational burden of the control algorithm.

The data necessary for identification of the neural approximator can be obtained by means of two methods very similar to those used in the case of the explicit MPC-NPL-NA algorithm. Firstly, the matrix $\boldsymbol{K}^{n_{\mathrm{u}}}(k)$ can be calculated as a result of open-loop simulation of the dynamic neural model for some data sets and trajectory linearisation. Alternatively, the classical explicit the MPC-NPLT$_{\boldsymbol{u}(k-1)}$-NA algorithm based on the dynamic neural model is developed. Next, the algorithm is simulated (or applied to a real process) for a randomly changing set-point trajectory. Of course, trajectory linearisation is performed rather than model linearisation used in the MPC-NPL approach.

6.3 DMC Algorithm with Neural Approximation (DMC-NA)

Identification of all the dynamic models discussed so far, i.e. the classical DLP neural structures, neural Hammerstein and Wiener models and neural state-space ones, in particular in the multi-input multi-output cases, is usually time-consuming. It is because the order of dynamics is typically chosen experimentally, the number of hidden nodes is also selected empirically. One may recall the fact that the model used in the classical linear DMC algorithm consists of discrete-time time-responses, which describe reaction of the process to input steps. Identification of the non-parametric step-response linear model is very simple: it is only necessary to record the step-responses of the process. Easiness of model development greatly contributed to the great industrial success of the DMC algorithm.

For the single-input single-output process (with one input u and one output y) the output signal of the step-response model is [316]

$$y(k) = y(0) + \sum_{j=1}^{k} s_j \triangle u(k-j)$$

where the step-response coefficients are denoted by $s_1, \ldots, s_k$. If the process is stable and without integrated or runaway responses (i.e. for asymptotically stable processes, with self-regulating output), after a step change in the input, the output stabilises at the certain value s_∞, i.e. $\lim_{k\to\infty} s_k = s_\infty$. That is why to describe the process it is necessary to record the step-response coefficients $s_1, \ldots, s_D$, where D is the horizon of process dynamics. If the disturbance can be measured, the model is

$$y(k) = y(0) + \sum_{j=1}^{k} s_j \triangle u(k-j) + \sum_{j=1}^{k} s_j^{\mathrm{h}} \triangle h(k-j)$$

The disturbance channel is modelled in the same way the input-output channel, in order to describe the influence of the disturbance it is necessary to record the disturbance step-response coefficients $s_1^{\mathrm{h}}, \ldots, s_{D^{\mathrm{h}}}^{\mathrm{h}}$ which show reaction of the output to the disturbance step. The step-response model can be easily formulated for the multi-input multi-output process with n_{u} inputs, n_{h} measured disturbances and n_{y} outputs. It has the following form

$$\boldsymbol{y}(k) = \boldsymbol{y}(0) + \sum_{j=1}^{k} \boldsymbol{S}_j \triangle \boldsymbol{u}(k-j) + \sum_{j=1}^{k} \boldsymbol{S}_j^{\mathrm{h}} \triangle \boldsymbol{h}(k-j) \tag{6.5}$$

where the step-response matrices of the input-output and disturbance-output channels are denoted by $\boldsymbol{S}_j$ and $\boldsymbol{S}_j^{\mathrm{h}}$, respectively.

A neural step-response model is used in the discussed DMC algorithm with Neural Approximation (DMC-NA). The model is nonlinear, the step-response coefficients are not constant (as in the model (6.5)), but they depend on the current operating point of the process

$$\boldsymbol{y}(k) = \boldsymbol{y}(0) + \sum_{j=1}^{k} \boldsymbol{S}_j(k) \triangle \boldsymbol{u}(k-j) + \sum_{j=1}^{k} \boldsymbol{S}_j^{\mathrm{h}}(k) \triangle \boldsymbol{h}(k-j) \tag{6.6}$$

where the matrices

$$\boldsymbol{S}_j(k) = \begin{bmatrix} s_j^{1,1}(k) & \ldots & s_j^{1,n_\mathrm{u}}(k) \\ \vdots & \ddots & \vdots \\ s_j^{n_\mathrm{y},1}(k) & \ldots & s_j^{n_\mathrm{y},n_\mathrm{u}}(k) \end{bmatrix}, \quad \boldsymbol{S}_j^\mathrm{h}(k) = \begin{bmatrix} s_j^{\mathrm{h},1,1}(k) & \ldots & s_j^{\mathrm{h},1,n_\mathrm{h}}(k) \\ \vdots & \ddots & \vdots \\ s_j^{\mathrm{h},n_\mathrm{y},1}(k) & \ldots & s_j^{\mathrm{h},n_\mathrm{y},n_\mathrm{h}}(k) \end{bmatrix}$$

are of dimensionality $n_\mathrm{y} \times n_\mathrm{u}$ and $n_\mathrm{y} \times n_\mathrm{h}$, respectively. The current values of the step-response coefficients are calculated on-line by means of the neural approximator for the current operating point of the process, which is defined by the vector comprised of the most recent process signals $u(k-1)$ and (or) $h(k)$ and $y(k)$

$$\widetilde{\boldsymbol{x}}(k) = \begin{bmatrix} u(k-1) \\ h(k) \\ y(k) \end{bmatrix} \tag{6.7}$$

The neural approximator calculates the coefficients

$$g_j^{m,n}(\widetilde{\boldsymbol{x}}(k)) \approx s_j^{m,n}(k) \tag{6.8}$$

for all $j = 1, \ldots, D$, $m = 1, \ldots, n_\mathrm{y}$, $n = 1, \ldots, n_\mathrm{u}$ and the coefficients

$$g_j^{\mathrm{h},m,n}(\widetilde{\boldsymbol{x}}(k)) \approx s_j^{\mathrm{h},m,n}(k) \tag{6.9}$$

for all $j = 1, \ldots, D^\mathrm{h}$, $m = 1, \ldots, n_\mathrm{y}$, $n = 1, \ldots, n_\mathrm{h}$, $g_j^{m,n} \colon \mathbb{R}^{n_\mathrm{u}+n_\mathrm{h}+n_\mathrm{y}} \to \mathbb{R}$ and $g_j^{\mathrm{h},m,n} \colon \mathbb{R}^{n_\mathrm{u}+n_\mathrm{h}+n_\mathrm{y}} \to \mathbb{R}$ are nonlinear functions which describe the approximator. For the single-input single-output process only one neural network is used as the approximator or two networks are used (the first one approximates the step-response of the input-output channel, the second one – of the disturbance-output channel). For the multi-input multi-output process it is possible to use various configurations of the approximator. Only one network or two networks can be used or it may consist of n_y independent networks (each network approximates the step-response coefficients of input-output and disturbance-output channels for consecutive outputs) or as many as $2n_\mathrm{y}$ networks can be used (each network calculates separately the coefficients of the input-output or disturbance-output channels for consecutive outputs). Of course, the structure of the approximator should be designed with the particular application in mind.

In order to gather all the data necessary for identification of the neural approximator it is necessary to record the step-responses of the process for different operating points. Next, the neural network (or networks) is trained to find the relations (6.8) and (6.9) between the current operating point defined by Eq. (6.7) and the step-response coefficients. If the input and disturbance excitation steps are not unitary, they must be scaled

$$s_j^{m,n} = \frac{y_m(j) - y_m(0)}{\delta u_n}, \quad s_j^{\mathrm{h},m,n} = \frac{y_m(j) - y_m(0)}{\delta h_n}$$

where $y_m(0)$ denotes the initial value of the output signal, δu_n and δh_n are the input and disturbance steps applied to the process, respectively.

In the DMC-NA algorithm the rudimentary prediction equation (1.13) is used, where the quantities $y(k+p|k)$ are calculated from the nonlinear step-response model (6.6), the coefficients of which are determined by the neural approximator (6.8) and (6.9). The model output vector for the sampling instant $k+p$ is

$$y(k+p|k) = y(0) + \sum_{j=1}^{k+p} \boldsymbol{S}_j(k)\triangle u(k-j+p) + \sum_{j=1}^{k+p} \boldsymbol{S}_j^{\mathrm{h}}(k)\triangle h(k-j+p)$$

The prediction calculated at the current sampling instant k is given by

$$\begin{aligned}\hat{y}(k+p|k) = y(0) &+ \sum_{j=1}^{p} \boldsymbol{S}_j(k)\triangle u(k-j+p|k) + \sum_{j=p+1}^{k+p} \boldsymbol{S}_j(k)\triangle u(k-j+p) \\ &+ \sum_{j=1}^{p} \boldsymbol{S}_j^{\mathrm{h}}(k)\triangle h(k-j+p|k) + \sum_{j=p+1}^{k+p} \boldsymbol{S}_j^{\mathrm{h}}(k)\triangle h(k-j+p) + d(k)\end{aligned}$$

where, from Eqs (1.15) and (6.6), the unmeasured disturbances are estimated as

$$d(k) = y(k) - y(0) - \sum_{j=1}^{k} \boldsymbol{S}_j(k)\triangle u(k-j) - \sum_{j=1}^{k} \boldsymbol{S}_j^{\mathrm{h}}(k)\triangle h(k-j)$$

Remembering that $\boldsymbol{S}_j(k) = \boldsymbol{S}_D(k)$ for $j > D$ and $\boldsymbol{S}_j^{\mathrm{h}}(k) = \boldsymbol{S}_{D^{\mathrm{h}}}^{\mathrm{h}}(k)$ for $j > D^{\mathrm{h}}$ and assuming that the measured disturbances are constant over the whole prediction horizon, one obtains the following prediction equation

$$\hat{\boldsymbol{y}}(k) = \underbrace{\boldsymbol{G}(k)\triangle\boldsymbol{u}(k)}_{\text{future}} + \underbrace{\boldsymbol{y}(k) + \boldsymbol{G}^{\mathrm{p}}(k)\triangle\boldsymbol{u}^{\mathrm{p}}(k) + \boldsymbol{G}^{\mathrm{h}}(k)\triangle\boldsymbol{h}(k)}_{\text{past}} \tag{6.10}$$

where the dynamic matrix $\boldsymbol{G}(k)$, of dimensionality $n_{\mathrm{y}}N \times n_{\mathrm{u}}N_{\mathrm{u}}$, has the same structure as the matrix used in the MPC-NPL algorithm (Eq. (2.34)), the vector

$$\boldsymbol{y}(k) = \begin{bmatrix} y(k) \\ \vdots \\ y(k) \end{bmatrix}$$

is of length $n_{\mathrm{y}}N$, the vectors

$$\triangle\boldsymbol{u}^{\mathrm{p}}(k) = \begin{bmatrix} \triangle u(k-1) \\ \vdots \\ u(k-(\mathrm{D}-1)) \end{bmatrix}, \quad \triangle\boldsymbol{h}(k) = \begin{bmatrix} \triangle h(k) \\ \vdots \\ h(k-(\mathrm{D}^{\mathrm{h}}-1)) \end{bmatrix}$$

are of length $n_{\mathrm{y}}(\mathrm{D}-1)$ and $n_{\mathrm{h}}\mathrm{D}^{\mathrm{h}}$, respectively, the matrix

$$\boldsymbol{G}^{\mathrm{p}}(k)=\begin{bmatrix} s_2(k)-s_1(k) & s_3(k)-s_2(k) & \ldots & s_{\mathrm{D}}(k)-s_{\mathrm{D}-1}(k) \\ s_3(k)-s_1(k) & s_4(k)-s_2(k) & \ldots & s_{\mathrm{D}+1}(k)-s_{\mathrm{D}-1}(k) \\ \vdots & \vdots & \ddots & \vdots \\ s_{N+1}(k)-s_1(k) & s_{N+2}(k)-s_2(k) & \ldots & s_{N+\mathrm{D}-1}(k)-s_{\mathrm{D}-1}(k) \end{bmatrix}$$

is of dimensionality $n_{\mathrm{y}}N \times n_{\mathrm{u}}N_{\mathrm{u}}$ and the matrix

$$\boldsymbol{G}^{\mathrm{h}}(k)=\begin{bmatrix} s_1^{\mathrm{h}}(k) & s_3^{\mathrm{h}}(k)-s_2^{\mathrm{h}}(k) & \ldots & s_{\mathrm{D}^{\mathrm{h}}}^{\mathrm{h}}(k)-s_{\mathrm{D}^{\mathrm{h}}-1}^{\mathrm{h}}(k) \\ s_2^{\mathrm{h}}(k) & s_4^{\mathrm{h}}(k)-s_2^{\mathrm{h}}(k) & \ldots & s_{\mathrm{D}^{\mathrm{h}}+1}^{\mathrm{h}}(k)-s_{\mathrm{D}^{\mathrm{h}}-1}^{\mathrm{h}}(k) \\ \vdots & \vdots & \ddots & \vdots \\ s_N^{\mathrm{h}}(k) & s_{N+2}^{\mathrm{h}}(k)-s_2^{\mathrm{h}}(k) & \ldots & s_{N+\mathrm{D}^{\mathrm{h}}-1}^{\mathrm{h}}(k)-s_{\mathrm{D}^{\mathrm{h}}-1}^{\mathrm{h}}(k) \end{bmatrix}$$

is of dimensionality $n_{\mathrm{y}}N \times n_{\mathrm{h}}D^{\mathrm{h}}$. Using the obtained DMC-NA prediction equation (6.10), the general MPC optimisation problem (1.9) is transformed into the following quadratic optimisation task

$$\begin{aligned} \min_{\substack{\triangle \boldsymbol{u}(k) \\ \boldsymbol{\varepsilon}^{\min}(k),\ \boldsymbol{\varepsilon}^{\max}(k)}} \quad & \Big\{ J(k)=\big\|\boldsymbol{y}^{\mathrm{sp}}(k)-\boldsymbol{G}(k)\triangle\boldsymbol{u}(k)-\boldsymbol{y}(k) \\ & \quad -\boldsymbol{G}^{\mathrm{p}}(k)\triangle\boldsymbol{u}^{\mathrm{p}}(k)-\boldsymbol{G}^{\mathrm{h}}(k)\triangle\boldsymbol{h}(k)\big\|_{\boldsymbol{M}}^2+\|\triangle\boldsymbol{u}(k)\|_{\boldsymbol{\Lambda}}^2 \\ & \quad +\rho^{\min}\left\|\boldsymbol{\varepsilon}^{\min}(k)\right\|^2+\rho^{\max}\left\|\boldsymbol{\varepsilon}^{\max}(k)\right\|^2 \Big\} \end{aligned}$$

subject to (6.11)

$$\begin{aligned} & \boldsymbol{u}^{\min} \leq \boldsymbol{J}\triangle\boldsymbol{u}(k)+\boldsymbol{u}(k-1) \leq \boldsymbol{u}^{\max} \\ & -\triangle\boldsymbol{u}^{\max} \leq \triangle\boldsymbol{u}(k) \leq \triangle\boldsymbol{u}^{\max} \\ & \boldsymbol{y}^{\min}-\boldsymbol{\varepsilon}^{\min}(k) \leq \boldsymbol{G}(k)\triangle\boldsymbol{u}(k)+\boldsymbol{y}(k)+\boldsymbol{G}^{\mathrm{p}}(k)\triangle\boldsymbol{u}^{\mathrm{p}}(k) \\ & \qquad\qquad +\boldsymbol{G}^{\mathrm{h}}(k)\triangle\boldsymbol{h}(k) \leq \boldsymbol{y}^{\max}+\boldsymbol{\varepsilon}^{\max}(k) \\ & \boldsymbol{\varepsilon}^{\min}(k) \geq 0,\ \boldsymbol{\varepsilon}^{\max}(k) \geq 0 \end{aligned}$$

The structure of the DMC-NA algorithm is depicted in Fig. 6.3.

There is a clear similarity between the DMC-NA algorithm and the fuzzy DMC one [199, 316]. In the first approach the step-response coefficients for the current operating conditions are determined by the neural approximator whereas in the second one they are found by a fuzzy system.

The discussed DMC-NA algorithm (similarly to the classical linear DMC one) can be used for stable processes without integration. For the processes with integration it is necessary to use the incremental step-response model [316] whose coefficients are not constant but they are determined on-line by the neural approximator for the current operating point.

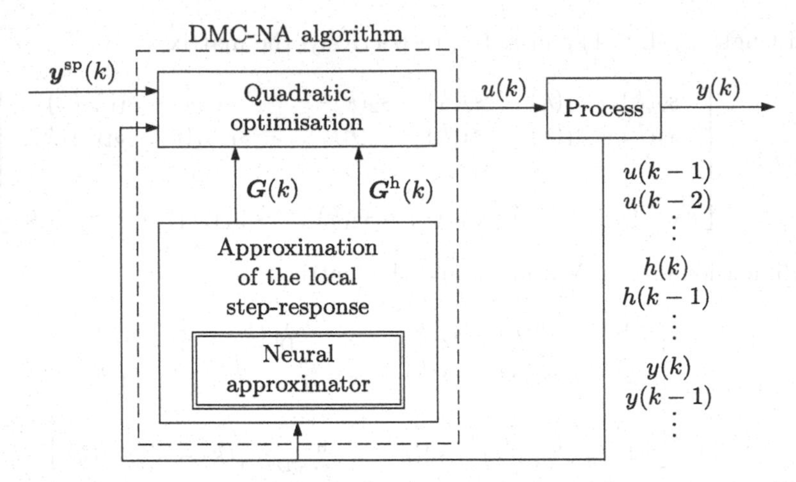

Fig. 6.3. The structure of the DMC algorithm with Neural Approximation (DMC-NA)

6.3.1 Explicit DMC-NA Algorithm

The explicit version of the DMC-NA algorithm can be also formulated. Removing the constraints, the optimisation task (6.11) reduces to minimisation of the quadratic cost-function

$$\min_{\triangle \boldsymbol{u}(k)} \Big\{ J(k) = \big\| \boldsymbol{y}^{\mathrm{sp}}(k) - \boldsymbol{G}(k)\triangle \boldsymbol{u}(k) - \boldsymbol{y}(k) \\ - \boldsymbol{G}^{\mathrm{p}}(k)\triangle \boldsymbol{u}^{\mathrm{p}}(k) - \boldsymbol{G}^{\mathrm{h}}(k)\triangle \boldsymbol{h}(k) \big\|_{\boldsymbol{M}}^{2} + \|\triangle \boldsymbol{u}(k)\|_{\boldsymbol{\Lambda}}^{2} \Big\}$$

By equating its first-order derivatives to a zeros vector of length $n_{\mathrm{u}}N_{\mathrm{u}}$, one obtains the optimal vector of future control increments in the form typical of the explicit MPC algorithms, namely $\triangle \boldsymbol{u}(k) = \boldsymbol{K}(k)(\boldsymbol{y}^{\mathrm{sp}}(k) - \boldsymbol{y}^{0}(k))$ (Eq. (2.68)), where the matrix $\boldsymbol{K}(k)$ is defined by Eq. (2.69) and the free trajectory is

$$\boldsymbol{y}^{0}(k) = \boldsymbol{y}(k) + \boldsymbol{G}^{\mathrm{p}}(k)\triangle \boldsymbol{u}^{\mathrm{p}}(k) + \boldsymbol{G}^{\mathrm{h}}(k)\triangle \boldsymbol{h}(k) \tag{6.12}$$

It is not necessary to calculate the entire matrix $\boldsymbol{K}(k)$ but only its first n_{u} rows. The control law is given by Eq. (2.70), i.e. $\triangle u(k|k) = \boldsymbol{K}^{n_{\mathrm{u}}}(k)(\boldsymbol{y}^{\mathrm{sp}}(k) - \boldsymbol{y}^{0}(k))$.

The structure of the explicit DMC-NA algorithm is depicted in Fig. 6.4. It is similar to that of the explicit MPC-NPL-NA algorithm shown in Fig. 6.2, but two neural approximators are used and the classical dynamic model necessary to find the free trajectory in the MPC-NPL approach is not used. The first neural approximator calculates on-line the control law matrix $\boldsymbol{K}^{n_{\mathrm{u}}}(k)$ for the current operating point. The second approximator is the same as in the numerical DMC-NA algorithm (Fig. 6.3), i.e. it calculates the local step-response coefficients which comprise the matrices $\boldsymbol{G}(k)$ and $\boldsymbol{G}^{\mathrm{P}}(k)$, which are next used for free trajectory calculation from Eq. (6.12).

6.4 DMC-NA and MPC-NPL-NA Algorithms with Interpolation

As many as a few dozens of the step-response coefficients are necessary to describe behaviour of a typical single-input single-output process. For multi-input multi-output processes, the total number of coefficients is the product of the number of inputs, outputs and the horizon of dynamics. The neural approximator used in the DMC-NA algorithm must calculate all the coefficients for the current operating conditions. The number of parameters of such an approximator (the number of weights) may be significant. When it is a problem, one may use the DMC-NA algorithm with interpolation. Some step-response coefficients are directly calculated by the approximator whereas the rest of them are interpolated. Thanks to interpolation, the number of model parameters is reduced, but the interpolation procedure is necessary. One may also use the interpolation approach in the MPC-NPL-NA algorithm, but it seems much more useful in the MPC-DMC-NA one, because the horizon of dynamics is usually much longer that the prediction one.

There are a few interpolation methods [259]. Of course, in the simplest case linear approximation may be used, but polynomial or spline approximation makes it possible to obtain better accuracy. Unfortunately, polynomial interpolation has some important disadvantages: i.e. its accuracy may be frequently unsatisfactory, especially at end points (Runge's phenomenon),

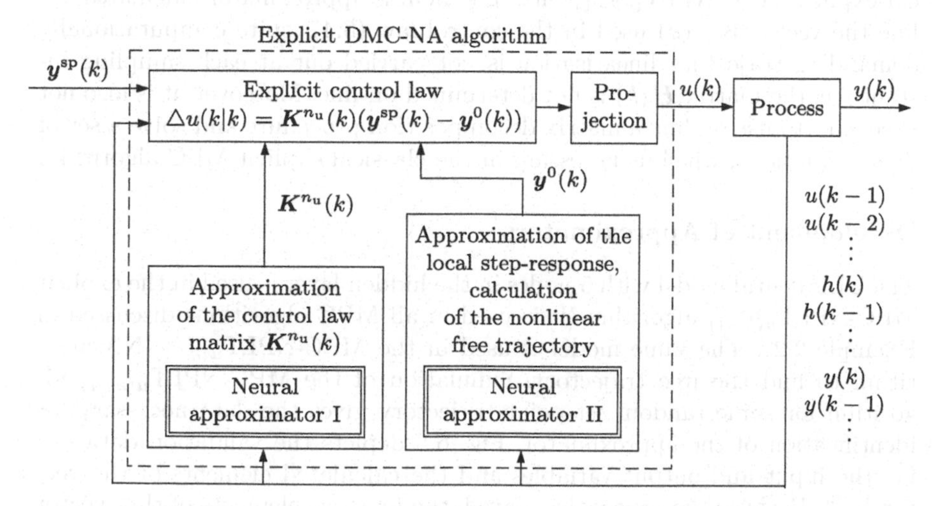

Fig. 6.4. The structure of the explicit DMC-NA algorithm

increasing the number of interpolation knots leads to increasing the polynomial order which is not a good choice from the numerical point of view. When compared to classical polynomial interpolation, splines have a few advantages. Firstly, the interpolation error can be made low even when using polynomials of a low degree. That is why third order splines (cubic splines) are frequently used. The cubic spline interpolation problem results in a linear tridiagonal set of equations which can be solved very efficiently [259]. It is necessary to remember that accuracy of the neural step-response model with interpolation must be verified for some data sets before it is used in the MPC algorithm.

6.5 Example 6.1

The considered dynamic process is the high-pressure high-purity ethylene-ethane distillation column described in Example 2.2 (p. 80). Unlike the yeast fermentation process and the polymerisation reactor, for the high-pressure distillation column the trajectories obtained in the suboptimal MPC-NPL algorithm are significantly slower than those obtained in the MPC-NO approach, which is demonstrated in Fig. 2.21. It is caused by considerably non-linear nature of the distillation process, simple model linearisation for the current operating point turns out to be insufficient for prediction calculation in MPC algorithms. That is why more advanced trajectory linearisation methods should be used. For the distillation process the MPC-NPLT$_{\boldsymbol{u}(k-1)}$ and MPC-NPLT$_{\boldsymbol{u}(k|k-1)}$ algorithms give better accuracy than the rudimentary MPC-NPL one, which is demonstrated in Fig. 2.22 and in Table 2.7.

The objective of this example is to show the advantages of the explicit MPC-NPLT$_{\boldsymbol{u}(k-1)}$-NA algorithm with neural approximation over the classical explicit MPC-NPLT$_{\boldsymbol{u}(k-1)}$ one. The neural approximator calculates on-line the vector $\boldsymbol{K}^{\mathrm{n_u}}(k)$ used in the control law (6.4), quite computationally demanding trajectory linearisation is not carried out at each sampling instant, i.e. the matrix $\boldsymbol{H}(k)$ is not determined on-line. Moreover, it is also not necessary to use on-line a matrix decomposition procedure and solve a set of linear equations, which is necessary in the classical explicit MPC algorithm.

Development of Approximators

The DLP neural model with 5 nodes in the hidden layer is used in the explicit MPC-NPLT$_{\boldsymbol{u}(k-1)}$ algorithm (it is used in all MPC algorithms discussed in Example 2.2). The same model is used in the MPC-NPLT$_{\boldsymbol{u}(k-1)}$-NA algorithm to find the free trajectory. Simulation of the MPC-NPLT$_{\boldsymbol{u}(k-1)}$ algorithm for some random set-point trajectory gives the data necessary for identification of the approximator. Fig. 6.5 depicts the validation data set: i.e. the input and output variables and the calculated elements of the vector $\boldsymbol{K}^{n_u}$. Because the process is delayed, the first two elements of that vector are always 0. Hence, the approximator has 8 outputs (for $N = 10$) and

Table 6.1. Comparison of linear and nonlinear approximators of the explicit MPC-NPLT$_{u(k-1)}$ algorithm in the control system of the high-pressure distillation column in terms of the number of parameters (NP) and accuracy (SSE)

Approximator	NP	$\text{SSE}_{\text{train}}$	SSE_{val}	SSE_{test}
Linear	24	1.3165×10^{0}	1.9882×10^{0}	–
Neural, $K = 1$	19	2.4132×10^{-1}	4.2790×10^{-1}	–
Neural, $K = 2$	30	1.9296×10^{-2}	2.5783×10^{-2}	–
Neural, $K = 3$	41	1.8875×10^{-3}	3.2891×10^{-3}	–
Neural, $K = 4$	52	6.2285×10^{-5}	2.5023×10^{-4}	1.1725×10^{-4}
Neural, $K = 5$	63	6.2911×10^{-5}	2.6146×10^{-4}	–
Neural, $K = 6$	74	8.2358×10^{-5}	2.7157×10^{-4}	–

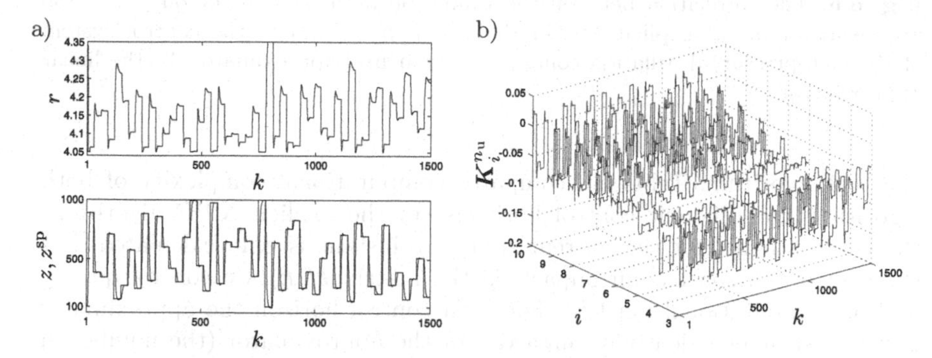

Fig. 6.5. The validation data set used during identification of the approximators of the explicit MPC-NPLT$_{u(k-1)}$ algorithm in the control system of the high-pressure distillation column: a) the trajectories of the input and output variables, b) the changes of the elements of the control law vector $\boldsymbol{K}^{n_{\text{u}}}$

2 inputs, which define the current operating point of the process: the signals $u(k-1)$ and $y(k)$ (they are selected experimentally). Table 6.1 shows the influence of the number of the hidden nodes on the number of approximator parameters and accuracy. The network with 4 hidden nodes is a straightforward choice. A linear approximator is also found for comparison. Because the process is significantly nonlinear, the control law vector must be a nonlinear function of the operating point, a linear one is insufficient. Fig. 6.6 depicts correlation between the validation data set and the outputs of both approximators.

High-Pressure Distillation Column Predictive Control

Simulation results of the classical explicit MPC-NPLT$_{u(k-1)}$ algorithm and of the explicit MPC-NPLT$_{u(k-1)}$-NA approach with neural approximation are depicted in Fig. 6.7, Table 6.2 gives the values of the SSE_{sim} index. Thanks to very good accuracy of the neural approximator the trajectories obtained in the algorithm with neural approximation are practically the same as in the

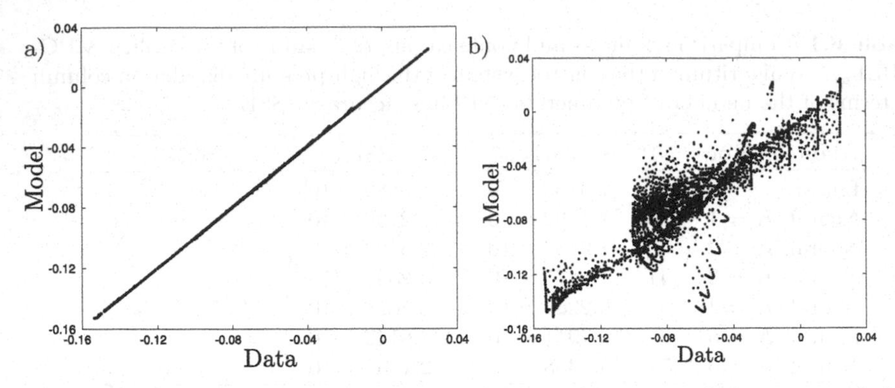

Fig. 6.6. The correlation between the validation data set and the output of the approximator of the explicit MPC-NPLT$_{\boldsymbol{u}(k-1)}$ algorithm in the control system of the high-pressure distillation column: a) the neural approximator, b) the linear approximator

classical algorithm. Table 6.3 compares computational complexity of both algorithms for different control horizons (in the explicit MPC algorithms the control signal for the current sampling instant is only calculated, but the control horizon has an impact on the matrix $\boldsymbol{H}(k)$, on control quality and on computational burden). For each control horizon the approximator is designed independently. Complexity of the approximator (the number of outputs) is determined by the prediction horizon. Since in each case the same horizon $N = 10$ is used, all approximators have the same topology (the same inputs, outputs and 4 hidden nodes). In consequence, computational burden of the explicit MPC-NPLT$_{\boldsymbol{u}(k-1)}$-NA algorithm is independent of the control horizon (it depends on the complexity of the approximator and of the dynamic model used for the free trajectory calculation). Thanks to the fact that no on-line linearisation and matrix decomposition are necessary, the explicit algorithm with neural approximation is significantly less computationally demanding than the classical explicit approach. Additionally, Table 6.3 gives computational burden of the classical numerical MPC-NPLT$_{\boldsymbol{u}(k-1)}$ algorithm with on-line quadratic optimisation. One may notice that the classical explicit algorithm is less computationally demanding than the numerical one, but the difference is not very significant. To sum up, the explicit MPC-NPLT$_{\boldsymbol{u}(k-1)}$-NA algorithm with neural approximation is very efficient.

One may notice that for the set-point trajectories 2 and 3 the explicit MPC-NPLT$_{\boldsymbol{u}(k-1)}$ algorithm is somehow faster than its numerical version (it is necessary to compare Figs. 2.22 and 6.7 or Tables 2.7 and 6.2). The optimal solution of the optimisation problem is outside the feasible set determined by the constraints. In the explicit algorithm the solution is obtained in a simplified way, i.e. an unconstrained optimisation task is solved and next its solution is projected onto the feasible set. As a result of the projection procedure, a suboptimal solution, different from the optimal one found by

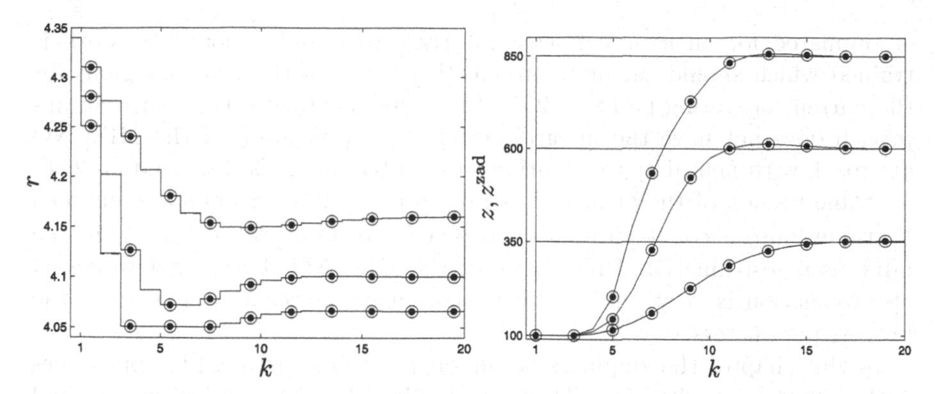

Fig. 6.7. Simulation results: the classical explicit MPC-NPLT$_{u(k-1)}$ algorithm (*solid line with dots*) and the explicit MPC-NPLT$_{u(k-1)}$-NA algorithm with neural approximation (*dashed line with circles*) in the control system of the high-pressure distillation column

Table 6.2. Comparison of control accuracy ($\mathrm{SSE_{sim}}$) of the classical explicit MPC-NPLT$_{u(k-1)}$ algorithm and of the explicit MPC-NPLT$_{u(k-1)}$-NA algorithm with neural approximation in the control system of the high-pressure distillation column

Algorithm	Trajectory 1	Trajectory 2	Trajectory 3	Sum
MPC-NPLT$_{u(k-1)}$	4.3986×10^5	1.4482×10^6	3.0113×10^6	4.8993×10^6
MPC-NPLT$_{u(k-1)}$-NA	4.3963×10^5	1.4479×10^6	3.0098×10^6	4.8973×10^6

Table 6.3. Comparison of computational complexity (in MFLOPS) of the classical numerical and explicit MPC-NPLT$_{u(k-1)}$ algorithms and of the explicit MPC-NPLT$_{u(k-1)}$-NA algorithm with neural approximation in the control system of the high-pressure distillation column for different control horizon (the summarised results are given for 3 set-points trajectories), $N = 10$

Algorithm	$N_\mathrm{u} = 1$	$N_\mathrm{u} = 2$	$N_\mathrm{u} = 3$	$N_\mathrm{u} = 4$	$N_\mathrm{u} = 5$	$N_\mathrm{u} = 10$
Num. MPC-NPLT$_{u(k-1)}$	0.14	0.22	0.33	0.45	0.61	1.91
Exp. MPC-NPLT$_{u(k-1)}$	0.13	0.20	0.27	0.34	0.42	0.92
Exp. MPC-NPLT$_{u(k-1)}$-NA	0.07	0.07	0.07	0.07	0.07	0.07

the numerical algorithm, may be determined. For the considered distillation process and for the assumed parameters, the explicit algorithms (with or without the neural approximation) give the trajectories a little bit faster. For other process the opposite may be true.

6.6 Literature Review

This chapter deliberately does not discusses the conceptually simplest MPC-NO algorithm with neural approximation. The classical MPC-NO algorithm is simulated for an assumed set-point trajectory and a neural network is trained which should calculate on-line the values of the control signals for the current operating point [3, 255]. In its rudimentary version such an approach does not have the integral action. An application of the MPC-NO approach with neural approximation to a mobile robot is discussed in [234]. A stable version of the algorithm is presented in [242]. Stability is enforced by taking into account an additional penalty term $a \left\| \hat{x}(k+N|k) \right\|_{\boldsymbol{P}}^2$ in the minimised cost-function. Unfortunately, the MPC-NO algorithm with neural approximation is very sensitive to measurement noises and changes of the process parameters.

In this chapter the emphasis is put on the suboptimal MPC approaches with neural approximation. They are inspired by the classical suboptimal MPC algorithms with on-line model or trajectory linearisation. The MPC-NPL-NA algorithm is shortly discussed in [151]. In the control system of a simulated distillation column it reduces computational burden by more than 15% in comparison with the classical MPC-NPL approach. The explicit MPC-NPL-NA and MPC-NPLT$_{\boldsymbol{u}(k-1)}$ algorithms are discussed in [142], their control quality and computational complexity are compared in the control system of the simulated high-purity ethylene-ethane distillation column. Although both algorithms with neural approximation are more computationally efficient than their classical explicit versions, approximation gives the best complexity reduction factors when the explicit algorithms MPC-NPLT$_{\boldsymbol{u}(k-1)}$ and MPC-NPLT$_{\boldsymbol{u}(k-1)}$-NA are compared. It is because computational burden of the classical explicit MPC-NPLT$_{\boldsymbol{u}(k-1)}$ algorithm is much higher than that of the explicit MPC-NPL one (trajectory linearisation repeated at each sampling instant on-line is much more demanding than simple model linearisation). The DMC-NA algorithm is discussed in [157] whereas the DMC-NA approach with interpolation is presented in [156]. Efficiency of the interpolation-based MPC approach is demonstrated for a simulated polymerisation process. Cubic splines are used for interpolation, which makes it possible to reduce the number of approximator parameters from 246 to 38.

The coefficients of the linear MPC control law are found off-line [316]. A similar approach for the Hammerstein nonlinear model is discussed in [62].

6.7 Chapter Summary

This chapter discusses three families of nonlinear MPC algorithms with neural approximation: MPC-NPL-NA, MPC-NPLT-NA, DMC-NA, numerical and explicit versions are presented. All the considered algorithms have one common feature, namely the lack of on-line model or trajectory linearisation. In the MPC-NPL-NA and DMC-NA algorithms the approximator finds on-line the step-response coefficients for the current operating conditions whereas in the MPC-NPLT-NA one the approximator calculates the derivatives of the predicted output trajectory with respect to the future control sequence. The explicit MPC algorithms with neural approximation are particularly computationally efficient, because the approximator determines directly on-line the coefficients of the time-varying control law, successive on-line model or trajectory linearisation, quadratic optimisation, matrix decomposition and linear equations solving are not necessary. For instance, replacing the classical explicit MPC-NPLT$_{\boldsymbol{u}(k-1)}$ algorithm by the explicit MPC-NPLT$_{\boldsymbol{u}(k-1)}$-NA one in the control system of the high-pressure distillation column makes it possible to reduce computational burden a few times or a dozen of times (depending on the control horizon). The discussed algorithms can be implemented on hardware platforms with limited resources and computational abilities. It is necessary to point out that the discussed algorithms are intuitive, they may be developed relatively easily. In particular, training of the approximators is carried out in the simple serial parallel (non-recurrent) configuration. In addition to the presented MPC structures, it is also possible to use alternative algorithms, e.g. the MPC-SL one with Neural Approximation (MPC-SL-NA).

From the practical point of view reduction of approximators' complexity is an interesting issue. Shortening of the prediction horizon is the simplest and naive approach, pruning of the neural networks is a straightforward option, but interpolation seems to be an interesting idea. It may be easily used in the numerical MPC-NPL-NA and DMC-NA algorithms because the neural approximator calculates only some step-response coefficients, the rest of them are interpolated. Interpolation does not lead to any modification of the algorithm itself (in contrast to the MPC algorithms based on the multi-model).

7

Stability and Robustness of MPC Algorithms

This chapter is devoted to stability and robustness issues of MPC algorithms. The most important approaches which make it possible to guarantee stability and robustness are reviewed with a view to using them in the suboptimal MPC algorithms with on-line linearisation. A modification of the dual-mode MPC strategy is thoroughly discussed which leads to the suboptimal MPC algorithm with theoretically guaranteed stability. Finally, a modification of the MPC strategy with additional state constraints is presented which leads to the suboptimal MPC algorithm with guaranteed robustness.

7.1 Stability of MPC Algorithms: A Short Review of Approaches

MPC algorithms have been successfully used in the industry since the 1970's, but the first theoretical works concerned with properties of MPC, including stability analysis, appeared a few years later [203]. The lack of theoretical analysis, however, has not hampered development and applications of new MPC solutions. Stability can be easily achieved in practice by tuning the weighting coefficients $\mu_{p,m}$ and $\lambda_{p,n}$ used in the minimised cost-function (1.4) and choosing the horizons, the prediction horizon in particular. The first tuning criteria were formulated in the 1980's [53, 74, 288], but only the explicit, unconstrained, MPC algorithms based on linear models were considered. For stability considerations the Lapunov theory [107, 112, 325] was not used, the only exception is [45].

The control law of the simplest linear MPC algorithm with no constraints is a linear function of the future control policy. For example, for the single-input single-output process, the DMC control law is [316]

$$\frac{\triangle u(k|k)}{e(k)} = \frac{k^{\mathrm{e}}}{1 + k_1^{\mathrm{u}} z^{-1} + \ldots + k_{\mathrm{D}-1}^{\mathrm{u}} z^{-(\mathrm{D}-1)}} \tag{7.1}$$

M. Ławryńczuk, *Computationally Efficient Model Predictive Control Algorithms*, Studies in Systems, Decision and Control 3,
DOI: 10.1007/978-3-319-04229-9_7, © Springer International Publishing Switzerland 2014

where $e(k) = y^{\mathrm{sp}}(k) - y(k)$ is the current control error, the coefficients k^{e} and k_i^{u} depend on the parameters of the model (i.e. on the step-response coefficients) and on the tuning parameters. Assuming no modelling errors and no disturbances, it is possible to analyse properties of the closed-loop system, in particular location of the poles and stability margins and to find the link between the tuning parameters and stability. Unfortunately, the unconstrained linear MPC algorithms have limited applicability. The possibility of taking constraints into account is an important factor which contributed to the great success of MPC algorithms. Even when a linear model is used for prediction, taking the constraints into account makes it impossible to formulate the explicit linear control law, e.g. the one given by Eq. (7.1)). In such cases stability analysis is much more difficult.

Since the 1990's, as the university communities got interested in MPC algorithms, a few stabilising modifications of MPC algorithms have been developed, not only linear but also nonlinear algorithms have been considered [203]. They are shortly discussed in the following part of this chapter. The Lapunov theory is now used for stability analysis. It is assumed that no modelling errors and no disturbances are present and that the state vector can be measured (state estimation issues fall outside the scope of this book).

7.1.1 MPC Algorithm with Additional Terminal Equality Constraint

The first stabilising modification of the MPC algorithm was to take into account an additional terminal equality constraint, the objective of which is to bring the predicted state at the end of the prediction horizon (i.e. the terminal state) to the equilibrium point. For the processes modelled by continuous-time differential equations the considered method is discussed in [45, 204], the discrete-time case is discussed in [109]. The mentioned publications refer to much earlier ones, dated back to 1970's, i.e. works [117, 123].

Let the nonlinear state-space model be

$$x(k+1) = f(x(k), u(k)) \tag{7.2}$$

It is assumed that the equilibrium point is in the origin, i.e. $f(0,0) = 0$ (if it is not the case, one must scale the state variables). In order to simplify the notation, no measured disturbances are present in the model. The optimisation problem of the MPC algorithm with the terminal equality constraint is

$$\begin{aligned}
&\min_{\boldsymbol{u}(k)} \left\{ J(k) = \sum_{p=1}^{N} L(\hat{x}(k+p|k), u(k+p-1|k)) \right\} \\
&\text{subject to} \\
&u(k+p|k) \in \mathcal{U},\ p = 0, \ldots, N-1 \\
&\hat{x}(k+p|k) \in \mathcal{X},\ p = 1, \ldots, N-1 \\
&\hat{x}(k+N|k) = 0
\end{aligned} \tag{7.3}$$

where the function $L(x,u) \geq 0$ and $L(x,u) = 0$ if and only if $x = 0$ and $u = 0$. For definition of the constraints imposed on the future control signals and on the predicted state variables the sets $\mathcal{U}$ and $\mathcal{X}$ are used, they contain the equilibrium point $(0,0)$. Of course, the box constraints are the most common, the hyper-box is defined by

$$\begin{aligned}
\mathcal{U} &= \left\{ u(k+p|k) \in \mathbb{R}^{n_\mathrm{u}} : u^{\min} \leq u(k+p|k) \leq u^{\max} \right\} \\
\mathcal{X} &= \left\{ \hat{x}(k+p|k) \in \mathbb{R}^{n_\mathrm{x}} : x^{\min} \leq \hat{x}(k+p|k) \leq x^{\max} \right\}
\end{aligned} \tag{7.4}$$

The additional equality terminal constraint $\hat{x}(k+N|k) = 0$ is only necessary to guarantee stability, it is not motivated by technological and economic issues. In the simplest case the separate control horizon is not used (the decision vector $\boldsymbol{u}(k)$ is of length $n_\mathrm{u}N$), but one may restrict the number of the decision variables assuming that $u(k+p|k) = u(k+N_\mathrm{u}-1|k)$ for $p \geq N_\mathrm{u}$.

If:

a) no modelling errors and no disturbances are present, the state vector is measured precisely,
b) at each sampling instant the feasible set of the nonlinear optimisation problem (7.3) is not empty, the equilibrium point belongs to this set,
c) at each sampling instant the global minimum of the nonlinear optimisation problem (7.3) is found,

the closed-loop system is asymptotically stable. In the proof [109], using the Lapunov stability theory, it is shown that the minimised cost-function $J(k)$ is a Lapunov function of the closed-loop system. When the function $L(x,u) \geq 0$ is continuous, all the basic conditions for the Lapunov function are satisfied. The stability proof consists in showing that the cost-function is decreasing for any $x(k) \neq 0$. For that purpose it is necessary to compare its value in two consecutive iterations of the MPC algorithm, i.e. $J^\star(k)$ and $J^\star(k+1)$, remembering, that the global minimum of the nonlinear optimisation problem is found. One has [109, 186]

$$\begin{aligned} J^\star(k+1) &= \min_{u(k)} \left\{ \sum_{p=1}^{N} L(\hat{x}(k+1+p|k+1), u(k+p|k+1)) \right\} \\ &= \min_{u(k)} \left\{ \sum_{p=1}^{N} L(\hat{x}(k+p|k+1), u(k+p-1|k+1)) \right. \\ &\quad - L(\hat{x}(k+1|k+1), u(k|k)) \\ &\quad \left. + L(\hat{x}(k+1+N|k+1), u(k+N|k)) \right\} \\ &\leq J^\star(k) - L(\hat{x}(k+1|k), u(k|k)) \\ &\quad + \min_{u(k+N|k)} \left\{ L(\hat{x}(k+1+N|k+1), u(k+N|k+1)) \right\} \end{aligned} \tag{7.5}$$

Because at the sampling instant k the terminal constraint $\hat{x}(k+N|k) = 0$ is satisfied and the equilibrium point is $(0,0)$, the condition $u(k+N|k+1) = 0$ must be true, because in such a case $\hat{x}(k+1+N|k+1) = 0$, the last part of the right side of the inequality (7.5) has the lowest possible value 0. It follows that

$$J^\star(k+1) - J^\star(k) \leq -L(\hat{x}(k+1|k), u(k|k)) \leq 0 \tag{7.6}$$

Remembering that the function $L(x,u) \geq 0$ and $L(x,u) = 0$ if and only if $x = 0$ and $u = 0$, the function $\triangle J^\star(k) = J^\star(k+1) - J^\star(k)$ is negative-definite with respect to the state $x(k)$. Taking the function $J^\star(k)$ as the Lapunov function, stability of the considered MPC algorithm with the equality constraint follows from the Lapunov theorem. The series $J^\star(k), J^\star(k+1), \ldots$ is not increasing. Remembering that the function $L(x,u) \geq 0$ and its lower value are bounded, the series is convergent to 0. For $k \to \infty$, one has $L(\hat{x}(k+1|k), u(k|k)) \to 0$ and $(\hat{x}(k+1|k), u(k|k)) \to (0,0)$. The obtained result is of course true for the most common cost-function

$$J(k) = \sum_{p=0}^{N} \left[\|\hat{x}(k+p|k)\|_{\boldsymbol{Q}}^2 + \|u(k+p|k)\|_{\boldsymbol{R}}^2 \right]$$

where the matrices $\boldsymbol{Q}$ and $\boldsymbol{R}$ are positive-definite.

The discussed MPC algorithm is very general, the model type is not limited, but two very restrictive assumptions must be satisfied. Firstly, in the general nonlinear case it is practically impossible to guarantee that the global minimum is obtained at each sampling instant. Secondly, the terminal equality constraint $\hat{x}(k+N|k) = 0$ is likely to make optimisation significantly more difficult, the feasible set is likely to be reduced. Because a linear model used for prediction leads to a quadratic cost-function with only one minimum, its global optimisation is not difficult. That is why the described stabilising approach to MPC is successfully used in MPC algorithms based on linear models. In the Constrained Receding Horizon Predictive Control (CRHPC) algorithm [52], which is an extension of the explicit GPC algorithm based on the input-output models, in place of the terminal state constraint $\hat{x}(k+N|k) = 0$ one must use the constraints

$$y(k+N+p|k) = y^{\mathrm{sp}}(k+N|k),\ p = 1, \ldots, N_{\mathrm{m}}$$

The length of the additional constraint horizon N_{m} depends on the process dynamics. When no other constraints are present, the terminal constraints are used to eliminate some of the decision variables of the optimisation problem, the optimal future increments of the manipulated variables are calculated analytically (explicitly). The numerical version of the CRHPC algorithm (with some additional constraints imposed on the manipulated and predicted output variables) is discussed in [294]. As an alternative example of the MPC algorithm with the terminal constraint and the input-output models, the Stabilizing Input Output Receding Horizon Control (SIORHC) approach may be given [23, 50, 217].

Although it is practically impossible to guarantee that for all nonlinear model types the global optimum is found at each sampling instant, using the specific structure of the Hammerstein model, one may effectively use the stabilising equality constraint to derive the explicit control law [62].

An interesting version of the considered algorithm is discussed in [291]. One may prove that for stability it is only necessary to calculate a feasible solution, which satisfies all the constraints of the optimisation problem (7.3). It means that it is not necessary to find the global or even a local minimum of the nonlinear optimisation problem at each sampling instant. Instead, the value of the cost-function must be reduced in consecutive iterations

$$J(k) - J(k-1) \leq -\mu L(x(k-1), u(k-1)) \tag{7.7}$$

where $0 < \mu \leq 1$. The suboptimal solution may lead to not very good control quality, but the lack of global nonlinear optimisation is a great advantage of the algorithm. Unfortunately, the terminal equality constraint still must be taken into account, its satisfaction may be quite difficult.

7.1.2 MPC Algorithm with Infinite Horizon

A proper choice of a sufficiently long prediction horizon is a simple, but efficient tuning method which makes it possible to stabilise the MPC algorithms. Hence, such an approach is frequently used in practice. For the finite prediction horizon N, at the sampling instant k the optimal future control sequence and a corresponding output (or state) trajectory are found for instants $k, \ldots, k+N$. At the next sampling instant $(k+1)$, the trajectories for instants $k+1, \ldots, k+N+1$ are calculated, the last part of which is not taken into account at the previous iteration. In consequence, the trajectories predicted in consecutive sampling instants may be different, even in the case of a perfect and undisturbed model. Conversely, in the MPC algorithm with infinite prediction and control horizons (assuming a constant set-point trajectory) the complete optimal trajectories are calculated at the sampling instant k. From the perspective of the optimisation procedure, no new information is available in consecutive iterations which could make it necessary to modify the previously calculated trajectories. The process follows the optimal trajectories found in the initial iteration. Hence, it is stable, stability follows from the Bellman optimality principle [186, 316].

Assuming that for prediction the nonlinear model (7.2) is used, the optimisation problem of the MPC algorithm with the infinite horizon is

$$\begin{aligned}
&\min_{\boldsymbol{u}(k)} \left\{ J(k) = \sum_{p=1}^{\infty} L(\hat{x}(k+p|k), u(k+p-1|k)) \right\} \\
&\text{subject to} \\
&u(k+p|k) \in \mathcal{U}, \ p = 0, \ldots, \infty \\
&\hat{x}(k+p|k) \in \mathcal{X}, \ p = 1, \ldots, \infty
\end{aligned} \tag{7.8}$$

Making the same assumptions as in the MPC algorithm with the terminal equality constraint $\hat{x}(k+N|k) = 0$ (i.e. no modelling errors, no disturbances, perfect measurement of the state vector) and assuming that at each sampling instant the feasible set of the nonlinear optimisation problem (7.8) is not empty, the equilibrium point belongs to this set and that at each sampling instant the global minimum is found, one may easily prove asymptotic stability of the algorithm [206]. The proof is similar to that used in the MPC algorithm with the terminal constraint, i.e. it is necessary to compare the value of the cost-function in two consecutive iterations. As a result, one may prove satisfaction of the condition (7.6). Unfortunately, such an approach cannot be directly used in practice, because an infinite number of decision variables is necessary, which is not possible. One may prove, however, that only the infinite prediction horizon is necessary for stability whereas the control horizon, which determine the number of decision variables, may be finite.

A linear model makes an efficient implementation of the MPC algorithm with the infinite horizon possible. In [220] the model

$$x(k+1) = \boldsymbol{A}x(k) + \boldsymbol{B}u(k)$$

and the cost-function

$$J(k) = \sum_{p=0}^{\infty} \left[\|\hat{x}(k+p|k)\|_{\boldsymbol{Q}}^2 + \|u(k+p|k)\|_{\boldsymbol{R}}^2 \right] \tag{7.9}$$

are use. One may notice that the condition $u(k+p|k) = 0$ must be satisfied for $p \geq N_\mathrm{u}$, because otherwise the value of the cost-function would be infinite. One obtains

$$J(k) = \sum_{p=0}^{N_\mathrm{u}-1} \left[\|\hat{x}(k+p|k)\|_{\boldsymbol{Q}}^2 + \|u(k+p|k)\|_{\boldsymbol{R}}^2 \right] + \sum_{p=N_\mathrm{u}}^{\infty} \|\hat{x}(k+p|k)\|_{\boldsymbol{Q}}^2$$

Because

$$\begin{aligned}
\hat{x}(k+N_{\mathrm{u}}+1|k) &= \boldsymbol{A}\hat{x}(k+N_{\mathrm{u}}|k)\\
\hat{x}(k+N_{\mathrm{u}}+2|k) &= \boldsymbol{A}^2\hat{x}(k+N_{\mathrm{u}}|k)\\
\hat{x}(k+N_{\mathrm{u}}+3|k) &= \boldsymbol{A}^3\hat{x}(k+N_{\mathrm{u}}|k)\\
&\vdots
\end{aligned}$$

it is possible to replace the cost-function with the infinite horizon by the one with the finite horizon and the penalty term

$$J(k) = \sum_{p=0}^{N_{\mathrm{u}}-1} \left[\|\hat{x}(k+p|k)\|^2_{\boldsymbol{Q}} + \|u(k+p|k)\|^2_{\boldsymbol{R}} \right] + \|\hat{x}(k+N_{\mathrm{u}}|k)\|^2_{\bar{\boldsymbol{Q}}} \tag{7.10}$$

For stable processes the matrix $\bar{\boldsymbol{Q}}$ is calculated from the equation

$$\bar{\boldsymbol{Q}} = \sum_{p=0}^{\infty} \left(\boldsymbol{A}^{\mathrm{T}}\right)^p \boldsymbol{Q}\boldsymbol{A}^p = \boldsymbol{Q} + \boldsymbol{A}^{\mathrm{T}}\bar{\boldsymbol{Q}}\boldsymbol{A}$$

For unstable processes it is necessary do obtain stable and unstable parts of the matrix $\boldsymbol{A}$ by the Jordan decomposition. The state prediction is calculated only for the stable part. Additionally, an additional equality constraint must be taken into account in the MPC optimisation problem to cancel the unstable modes for the sampling instant $k + N_{\mathrm{u}}$.

When the state constraints are present, the infeasibility problem may appear. To prevent such a situation from happening, the constraints' window may be used. The state constraints are defined not for the whole prediction horizon, but only for its part

$$\hat{x}(k+p|k) \in \mathcal{X}, \ p = N_1, \ldots, N_2$$

Determination of the window is discussed in [270]. A conceptually better approach is to use soft state constraints as described in [344]. The problem of existence of the feasible set is thoroughly investigated in [290].

The algorithm with the linear state-space model containing additionally the output equation $y(k) = \boldsymbol{C}x(k) + \boldsymbol{D}u(k)$ is discussed in [219]. Because the output set-point trajectory is used in the minimised cost-function, the constraints window is defined for the output constraints. An extension of the classical GPC algorithm based on the inputs-output linear models for the infinite horizon, named GPC$^{\infty}$, is discussed in [293]. Stability issues of the algorithm are investigated in [294].

To summarise, the considered MPC algorithm can be practically developed provided that the model is linear, because such an assumption makes it possible to use the finite control horizon. The necessity of finding the global solution at each sampling instant is naturally satisfied, because for the linear model one always obtains the quadratic optimisation problem.

7.1.3 Dual-Mode MPC Algorithm

The practical implementation of the MPC algorithm with the terminal equality constraint $\hat{x}(k+N|k) = 0$ is possible only when the model is linear. In such a case the minimised cost-function is quadratic, the solution of the optimisation problem (7.3) is easily calculated at each sampling instant, the equality constraint is used to eliminate some decision variables. The nonlinear case is much more difficult. Of course, one may use the suboptimal algorithm, in which it is not necessary to always find the global optimum for assuring stability (it is sufficient to find the feasible solutions which additionally satisfy the condition (7.7)). Unfortunately, from the practical point of view it is still difficult to guarantee that the nonlinear equality constraint is satisfied at each sampling instant of the algorithm. It is hence straightforward to replace the terminal equality constraint by the inequality one

$$\hat{x}(k+N|k) \in \Omega$$

the satisfaction of which is simpler from the numerical point of view. The set Ω is some neighbourhood of the equilibrium point. In the dual-mode MPC approach the MPC algorithm works only outside the set Ω. For the quadratic cost-function, the dual-mode optimisation problem is

$$\begin{aligned}
&\min_{\boldsymbol{u}(k)} \left\{ J(k) = \sum_{p=0}^{N-1} \theta(\hat{x}(k+p|k)) \left[\|\hat{x}(k+p|k)\|_{\boldsymbol{Q}}^2 + \|u(k+p|k)\|_{\boldsymbol{R}}^2 \right] \right\} \\
&\text{subject to} \\
&u(k+p|k) \in \mathcal{U},\ p = 0, \ldots, N-1 \\
&\hat{x}(k+p|k) \in \mathcal{X},\ p = 1, \ldots, N-1 \\
&\hat{x}(k+N|k) \in \Omega
\end{aligned} \tag{7.11}$$

where

$$\theta(\hat{x}(k+p|k)) = \begin{cases} 1 & \text{if } \hat{x}(k+p|k) \notin \Omega \\ 0 & \text{if } \hat{x}(k+p|k) \in \Omega \end{cases}$$

Of course, one may assume that $u(k+p|k) = u(k+N_{\mathrm{u}}-1|k)$ for $p \geq N_{\mathrm{u}}$. In order to guarantee that the state is brought to the equilibrium point, an additional controller with state feedback is used in the set Ω. The dual-mode control law is then

$$u(k) = \begin{cases} u(k|k) & \text{if } x(k) \notin \Omega \\ h(x(k)) & \text{if } x(k) \in \Omega \end{cases}$$

The linear controller $u(k) = h(x(k)) = \boldsymbol{K}x(k)$ is usually used for stabilisation of the algorithm in the neighbourhood of the equilibrium point. Alternatively, one may use an unconstrained MPC algorithm. In such a case it it best to assume that its tuning parameters are the same as those of the nonlinear

MPC algorithm which works outside the set Ω. Because the system must be stabilised, it is best to choose the CRHPC scheme or the MPC algorithm with the infinite horizon.

Originally, the dual-mode MPC algorithm was developed for processes modelled by continuous-time differential equations [209]. Unlike all MPC algorithms discussed so far, the prediction horizon is one of the decision variables of the dual-mode approach and it is adjusted on-line at each sampling instant. The problem of nonlinear constraints imposed on input and state signals is solved very efficiently. They are of course taken into account in the MPC optimisation task (7.11) when the state does not belong the set Ω. The additional stabilising controller, although formally it does not respect any constraints, is developed in such a way that all the constraints are never violated in the set Ω. The dual-mode MPC algorithm for discrete-time linear models is discussed in [49], the discrete-time nonlinear case is described in [291]. Unlike the original dual-mode strategy, the prediction horizon is constant. The discrete-time algorithm is discussed in more details in the following part of the text, because it is a basis for deriving the dual-mode MPC algorithm with on-line linearisation considered in Chapter 7.2.

If:

a) no modelling errors and no disturbances are present, the state vector is measured precisely,
b) for $x(k) \notin \Omega$ at each sampling instant the feasible set of the nonlinear optimisation problem (7.11) is not empty, the equilibrium point belongs to this set,
c) for $x(k) \notin \Omega$ the cost-function is decreased in consecutive sampling instants

$$J(k) - J(k-1) \leq -\mu \left(\|\hat{x}(k+p|k)\|_{\boldsymbol{Q}}^2 + \|u(k+p|k)\|_{\boldsymbol{R}}^2 \right) \tag{7.12}$$

where $0 < \mu \leq 1$, the closed-loop system is asymptotically stable. The stability proof is given in [291], it corresponds to the continuous-time case [209]. In short, one may show that all trajectories reach the set Ω in some finite time, which is enforced by the condition imposed on the cost-function (7.12). In the set Ω stability is guaranteed by the additional controller with state feedback.

The dual-mode MPC algorithm has two important advantages. Firstly, it is not necessary to take into account the terminal equality constraint $\hat{x}(k+N|k) = 0$, which may be difficult, it is replaced by the inequality constraint $\hat{x}(k+N|k) \in \Omega$. Secondly, the algorithm is suboptimal, because for stability it is sufficient to find a feasible solution, which satisfies the constraints of the optimisation task (7.11) and the condition (7.12). It is not necessary to calculate the global solution of the nonlinear optimisation problem at each sampling instant.

Provided that no modelling errors and no disturbances are present, for guaranteeing stability it is necessary to solve the optimisation problem only once, for the initial iteration $k = 0$. The obtained initial solution is feasible and leads to satisfaction of the condition (7.12) in the consecutive iterations. Of course, from the perspective of good control quality, it is beneficial to repeat the optimisation problem in all iterations, the solutions found in the previous iteration may be used as the initial one.

The linear approximation of the nonlinear process (7.2) in the equilibrium point is

$$x(k+1) = \boldsymbol{A}x(k) + \boldsymbol{B}u(k) \tag{7.13}$$

where

$$\boldsymbol{A} = \left.\frac{\partial f(x(k), u(k))}{\partial x(k)}\right|_{\substack{x(k)=0\\u(k)=0}}, \quad \boldsymbol{B} = \left.\frac{\partial f(x(k), u(k))}{\partial u(k)}\right|_{\substack{x(k)=0\\u(k)=0}}$$

The gain matrix $\boldsymbol{K}$ of the linear stabilising controller is chosen in such a way that the closed-loop system

$$x(k+1) = \boldsymbol{A}_{\mathrm{c}}x(k)$$

where $\boldsymbol{A}_{\mathrm{c}} = \boldsymbol{A} + \boldsymbol{B}\boldsymbol{K}$, is asymptotically stable. The convex set Ω is

$$\Omega = \left\{x(k) \in \mathbb{R}^{n_{\mathrm{x}}} : \|x(k)\|^2_{\boldsymbol{P}} \leq \alpha\right\} \tag{7.14}$$

where the matrix $\boldsymbol{P} > 0$ solves the equation

$$\boldsymbol{A}_{\mathrm{c}}^{\mathrm{T}}\boldsymbol{P}\boldsymbol{A}_{\mathrm{c}} - \boldsymbol{P} = -(\boldsymbol{Q} + \boldsymbol{K}^{\mathrm{T}}\boldsymbol{R}\boldsymbol{K}) \tag{7.15}$$

The nonlinear terminal constraint $\hat{x}(k+N|k) \in \Omega$ used in the optimisation problem (7.11) is hence

$$\|\hat{x}(k+N|k)\|^2_{\boldsymbol{P}} \leq \alpha \tag{7.16}$$

The fundamental issues are: determination of the suitable set Ω and finding the stabilising controller, which brings the state to the equilibrium point. The controller, although formally unconstrained, must be chosen in such a way that the first two constraints from the optimisation problem (7.11) are satisfied. The stabilising controller must hence satisfy the constraints

$$\begin{aligned} &x(k) \in \mathcal{X} \\ &u(k) = \boldsymbol{K}x(k) \in \mathcal{U} \end{aligned} \tag{7.17}$$

If the linear approximation (7.13) of the nonlinear process (7.2) in the equilibrium point is stabilisable, there exists the constant $\alpha \in (0, \infty)$, which define the size of the set Ω, such that:

a) $x(k) \in \mathcal{X}$, $u(k) = \boldsymbol{K}x(k) \in \mathcal{U}$ for all $x(k) \in \Omega$,
b) the set Ω is invariant for the nonlinear process (7.2) controlled by the linear controller $u(k) = \boldsymbol{K}x(k)$, i.e. for all $x(k) \in \Omega$ it is true that

$$x(k+1) = f(x(k), \boldsymbol{K}x(k)) \in \Omega \tag{7.18}$$

From the first of the above properties, one may conclude that all the constraints (7.17) are satisfied in the set Ω. The set Ω is invariant which means that all trajectories commencing at any point belonging to the set $x(k) \in \Omega$ never leave that set. A similar theorem and its proof for the continuous-time case are given in [46], the discrete-time case is discussed in [175]. For invariance investigation of the set Ω, it is necessary to consider the trajectory of the nonlinear process $x(k+1) = f(x(k), u(k))$ with the linear controller $u(k) = h(x(k)) = \boldsymbol{K}x(k)$, i.e. the trajectory of the system

$$x(k+1) = f(x(k), \boldsymbol{K}x(k)) \tag{7.19}$$

Evolution of the function $V(k) = \|x(k)\|^2_{\boldsymbol{P}}$ along the trajectory of the system (7.19) must be analysed. The following definition is used

$$\phi(k) = f(x(k), \boldsymbol{K}x(k)) - \boldsymbol{A}_{\mathrm{c}}x(k)$$

Using Eq. (7.15), one has

$$\begin{aligned} V(k+1) - V(k) &= \|\phi(k)\|^2_{\boldsymbol{P}} + 2\phi^T(k)\boldsymbol{P}\boldsymbol{A}_{\mathrm{c}}x(k) - \|x(k)\|^2_{\boldsymbol{Q}_{\mathrm{c}}} \\ &\leq \|\phi(k)\|^2_{\boldsymbol{P}} + 2\left|\phi^T(k)\boldsymbol{P}\boldsymbol{A}_{\mathrm{c}}x(k)\right| - \|x(k)\|^2_{\boldsymbol{Q}_{\mathrm{c}}} \end{aligned}$$

where $\boldsymbol{Q}_{\mathrm{c}} = \boldsymbol{A}_{\mathrm{c}}^T\boldsymbol{P}\boldsymbol{A}_{\mathrm{c}} - \boldsymbol{P} = -(\boldsymbol{Q} + \boldsymbol{K}^T\boldsymbol{R}\boldsymbol{K})$. It can be proven [175], that for all $x(k) \in \Omega$ the following inequality is true

$$\|\phi(k)\|^2_{\boldsymbol{P}} + 2\left|\phi^T(k)\boldsymbol{P}\boldsymbol{A}_{\mathrm{c}}x(k)\right| + (\kappa - 1)\|x(k)\|^2_{\boldsymbol{Q}_{\mathrm{c}}} \leq 0 \tag{7.20}$$

and

$$V(k+1) - V(k) \leq -\kappa \|x(k)\|^2_{\boldsymbol{Q}_{\mathrm{c}}}$$

where $0 < \kappa < 1$ is a tuning parameter. The function $V(k)$ can be hence the Lapunov function for the linearised system $x(k+1) = \boldsymbol{A}_{\mathrm{c}}x(k)$ and for the nonlinear one $x(k+1) = f(x(k), u(k))$ controlled by the linear law $u(k) = \boldsymbol{K}x(k)$, i.e. for the system (7.19). The condition (7.18) is satisfied, all trajectories commencing in any point belonging to the set Ω never leave that set. The set Ω is invariant. The equilibrium point is asymptotically stable for the linearised system and for the nonlinear one.

The set Ω is calculated off-line from a two-stage procedure. At first, the greatest value of the parameter $\alpha_0 \in (0, \infty)$ is found such that for all $x(k) \in \Omega$ the constraints (7.17) are satisfied. Let $d(x(k), \mathcal{X})$ be the distance from the point $x(k)$ to the set $\mathcal{X}$, let $d(\boldsymbol{K}x(k), \mathcal{U})$ be the distance from the point $\boldsymbol{K}x(k)$ to the set $\mathcal{U}$. The conditions (7.17) are satisfied for all $x(k) \in \Omega$ only if the maximal distance from any point $x(k)$ to the set $\mathcal{X}$ and the maximal distance from any point $\boldsymbol{K}x(k)$ to the set $\mathcal{U}$ are equal to 0. In different words, for the solution the value of the objective functions of the optimisation problems

$$\begin{array}{ll} \min\limits_{x(k)}(-d(x(k), \mathcal{X})) & \min\limits_{x(k)}(-d(\boldsymbol{K}x(k), \mathcal{U})) \\ \text{subject to}, & \text{subject to} \\ \|x(k)\|^2_{\boldsymbol{P}} \leq \alpha_0 & \|x(k)\|^2_{\boldsymbol{P}} \leq \alpha_0 \end{array} \tag{7.21}$$

must be 0. In the iterative procedure, the parameter α_0 is successively increased, the optimisation problems (7.21) are solved and the values of the objective functions are calculated for the optimal point.

In the set Ω, the size of which is defined by the parameter α_0, the constraints (7.17) are satisfied for all $x(k) \in \Omega$, but it may be not invariant. In order to guarantee the invariance property, it is necessary to find the greatest value $\alpha \in (0, \alpha_0]$ such that, for the chosen coefficient κ, the condition (7.20) is fulfilled for all $x(k) \in \Omega$. In different words, for the solution of the optimisation problem

$$\min_{x(k)} \left\{ (1-\kappa) \|x(k)\|^2_{\boldsymbol{Q}_c} - \|\phi(k)\|^2_{\boldsymbol{P}} - 2\left|\phi^T(k)\boldsymbol{P}\boldsymbol{A}_c x(k)\right| \right\}$$

subject to

$$\|x(k)\|^2_{\boldsymbol{P}} \leq \alpha$$

the value of the minimised objective function must be not negative. In the iterative procedure, the value of the parameter α is successively increased, the above optimisation task is solved and the value of the objective function is calculated for the optimal point. Some alternative methods for finding the invariant set may be also used, e.g. interval arithmetic, an excellent review of approaches is given in [30].

The coefficient α, which determines the size of the set Ω, is found only once, off-line. If a few equilibrium points are necessary, the linear stabilising controller and the set Ω must be found individually for each point.

7.1.4 MPC Algorithm with Quasi-Infinite Horizon

Efficient implementation of the MPC algorithm with the infinite horizon is possible when the model is linear, because in such a case it is possible to transform the cost-function (7.9) into the classical form with the finite horizon supplemented by the penalty term (7.10). Unfortunately, for nonlinear processes there are no universal methods for prediction calculation on the infinite horizon. A sound alternative is the MPC algorithm with the quasi-infinite horizon [46]. In this strategy the inequality terminal constraint $\hat{x}(k+N|k) \in \Omega$ is used, similarly to the dual-mode approach. Additionally, a penalty term is taken into account in the cost-function, its objective is to approximate the value of the neglected infinite part of the cost-function. For the quadratic cost-function defined on the infinite horizon (7.9), one has

$$\begin{aligned} J(k) = &\sum_{p=0}^{N-1} \left[\|\hat{x}(k+p|k)\|^2_{\boldsymbol{Q}} + \|u(k+p|k)\|^2_{\boldsymbol{R}} \right] \\ &+ \sum_{p=N}^{\infty} \left[\|\hat{x}(k+p|k)\|^2_{\boldsymbol{Q}} + \|u(k+p|k)\|^2_{\boldsymbol{R}} \right] \end{aligned} \tag{7.22}$$

If all trajectories for $[k+N, \infty)$ are in some neighbourhood of the equilibrium point (which is defined by the set Ω), it is possible to find the upper bound of the second sum of the above cost-function. Provided that the linear approximation (7.13) of the nonlinear process (7.2) is stabilisable in the equilibrium point, there exists the constant $\alpha \in (0, \infty)$, which defines the size of the set Ω, such that:

a) $x(k) \in \mathcal{X}$, $u(k) = \boldsymbol{K}x(k) \in \mathcal{U}$ for all $x(k) \in \Omega$,
b) the set Ω is invariant for the nonlinear process (7.2) controlled by the controller $u(k) = \boldsymbol{K}x(k)$, i.e. for all $x(k) \in \Omega$ it follows that $x(k+1) = f(x(k), \boldsymbol{K}x(k)) \in \Omega$,
c) for any $x(k+N|k) \in \Omega$ there exists the upper bound of the second sum of the cost-function (7.22) defined as

$$\sum_{p=N}^{\infty} \left[\|\hat{x}(k+p|k)\|_{\boldsymbol{Q}}^2 + \|u(k+p|k)\|_{\boldsymbol{R}}^2 \right] \leq \frac{1}{\kappa} \|\hat{x}(k+N|k)\|_{\boldsymbol{P}}^2 \tag{7.23}$$

where the positive-definite matrix $\boldsymbol{P}$ solves Eq. (7.15) whereas $0 < \kappa < 1$ is a tuning parameter.

Using the approximation (7.23) of the second sum of the cost-function (7.22), one obtains the optimisation problem of the MPC algorithm with the quasi-infinite horizon

$$\begin{aligned} &\min_{u(k)} \left\{ J(k) = \sum_{p=0}^{N-1} \left[\|\hat{x}(k+p|k)\|_{\boldsymbol{Q}}^2 + \|u(k+p|k)\|_{\boldsymbol{R}}^2 \right] + \frac{1}{\kappa} \|\hat{x}(k+N|k)\|_{\boldsymbol{P}}^2 \right\} \\ &\text{subject to} \\ &u(k+p|k) \in \mathcal{U}, \ p = 0, \ldots, N-1 \\ &\hat{x}(k+p|k) \in \mathcal{X}, \ p = 1, \ldots, N-1 \\ &\hat{x}(k+N|k) \in \Omega \end{aligned} \tag{7.24}$$

In contrast the dual-mode MPC approach, the vector of the future control signals is always calculated on-line from the above task, even in the neighbourhood of the equilibrium point the stabilising controller is not used. It is used only during development of the algorithm to satisfy the condition (7.23). The set Ω is found off-line by means of the method discussed for the dual-mode algorithm.

The proof of inequality (7.23) and the stability proof of the algorithm is given in [46] for the continuous-time case, the discrete-time case is shortly discussed in [70]. The necessity of finding the global solution of the nonlinear optimisation problem is relaxed, in a similar way it is done in the suboptimal MPC algorithm with the equality constraint and in the dual-mode MPC scheme. From the perspective of stability, it is only necessary to find a feasible solution, which satisfies the constraints of the optimisation task (7.24).

An interesting version of the algorithm is discussed in [224]. The approximation of the second sum in the cost-function (7.22) is calculated by taking into account the predictions for $p = N, \ldots, M$, where M is a sufficiently high integer number. The disadvantage of such an approach is the fact that the model must be used recurrently many times, which may be computationally demanding. An efficient procedure for finding the parameter M is used in the second version of the algorithm [189], in which separate control and prediction horizons are possible. The decision variables of the algorithm $(u(k|k), \ldots, u(k+N_\mathrm{u}-1|k))$ are calculated from a nonlinear optimisation problem whereas the future control signals $u(k+N_\mathrm{u}|k), \ldots, u(k+N-1|k)$ are determined by means of a stabilising linear controller.

In addition to the MPC algorithms with guaranteed stability discussed in this work, there are some other approaches worth mentioning. An interesting alternative is to use the MPC algorithm with the additional inequality state constraint named the contraction constraint $\|x(k+1)\| \leq \alpha \|x(k)\|$, where $0 < \alpha < 1$. For the linear models the algorithm is detailed in [253], the nonlinear case is discussed in [334]. Theoretical investigation of the algorithm is presented in [232]. A conceptually similar approach is to use the artificial Lapunov function [22]. An additional constraint is taken into account in the MPC optimisation problem, the role of which is to guarantee that some function (the Lapunov function) decreases in consecutive iterations. A different approach is described in [242], because neither the terminal equality constraint $\hat{x}(k+N|k) = 0$ nor the stabilising controller is used, but the minimised cost-function is supplemented by the penalty term $a\|\hat{x}(k+N|k)\|_{\boldsymbol{P}}^2$, stability is enforced by a proper choice of the parameter a.

7.2 Dual-Mode MPC Algorithm with On-Line Linearisation

The first two methods of enforcing stability, i.e. the terminal equality constraint and the infinite horizon, may be efficiently used only when a linear model is used for prediction. The mentioned approaches are not implementable for a nonlinear model, because in such a case it is very difficult to satisfy the nonlinear equality constraint (it may lead to numerical problems), there are no universal methods for finding nonlinear prediction over the infinite horizon. In contrast to these methods, the dual-mode MPC algorithm is computationally efficient, because:

a) it is not necessary to calculate the global solution of the optimisation problem at each sampling instant, it is only sufficient to find a feasible one, which satisfies the constraints,
b) the additional inequality constraint $\hat{x}(k+N|k) \in \Omega$ and the stabilising controller do not lead to any significant increase of computational complexity in comparison to the classical MPC algorithms with no stabilising mechanisms.

That is why the dual-mode approach is chosen to derive stable versions of the MPC algorithms with on-line linearisation discussed in the previous chapters. In comparison with the rudimentary discrete-time dual-mode MPC algorithm [291], some important modifications are introduced. They are necessary, because a nonlinear model is used for prediction in the original dual-mode algorithm whereas a linear approximation of a nonlinear model (or of a nonlinear trajectory) and quadratic optimisation are used in the prototype algorithms introduced in Chapter 2. From the perspective of stability, the dual-mode MPC algorithm does not require precise nonlinear optimisation, but the terminal nonlinear constraint $\hat{x}(k+N|k) \in \Omega$ must be satisfied. To make it possible to use on-line linearisation and quadratic optimisation, the terminal constraint is replaced by a set of linear ones. Two versions of the dual-mode MPC algorithm with on-line linearisation are discussed in the following part of the text, namely the algorithm for state-space and input-output models.

7.2.1 Implementation for State-Space Models

Taking into account separate control and prediction horizons, the box constraints (7.4) imposed on input and state variables and the set Ω given by Eq. (7.14), the rudimentary optimisation problem of the dual-mode MPC algorithm (7.11) becomes

$$\begin{aligned}
&\min_{\boldsymbol{u}(k)} \left\{ J(k) = \sum_{p=1}^{N-1} \theta(\hat{x}(k+p|k)) \left\| \hat{x}(k+p|k) \right\|_{\boldsymbol{Q}}^2 \right. \\
&\qquad\qquad \left. + \sum_{p=0}^{N_{\mathrm{u}}-1} \theta(\hat{x}(k+p|k)) \left\| u(k+p|k) \right\|_{\boldsymbol{R}}^2 \right\} \\
&\text{subject to} \qquad\qquad\qquad\qquad\qquad\qquad\qquad (7.25) \\
&u^{\min} \leq u(k+p|k) \leq u^{\max}, \ p = 0, \ldots, N_{\mathrm{u}}-1 \\
&x^{\min} \leq \hat{x}(k+p|k) \leq x^{\max}, \ p = 1, \ldots, N-1 \\
&\left\| \hat{x}(k+N|k) \right\|_{\boldsymbol{P}}^2 \leq \alpha
\end{aligned}$$

Because a nonlinear model, e.g. a neural one, is used for prediction, the predicted state variables are nonlinear functions of the future control sequence. The optimisation problem (7.25) is then nonlinear. Alternatively, the successive linearisation approach may be used. Although both model linearisation (MPC-SL and MPC-NPL algorithms) and trajectory linearisation (MPC-NPLT and MPC-NPLPT algorithms) are possible, the discussion presented next is concerned with the MPC-NPL algorithm. Using the suboptimal prediction equation (4.27) derived for the MPC-NPL algorithm based on the state-space model, the predictions for consecutive sampling instants of the prediction horizon are

$$\hat{x}(k+p|k) = \boldsymbol{P}^{(p-1)n_{\mathrm{x}}+1}(k)\Delta\boldsymbol{u}(k) + \boldsymbol{x}^{0,(p-1)n_{\mathrm{x}}+1}(k) \tag{7.26}$$

where $\boldsymbol{P}^{(p-1)n_{\mathrm{x}}+1}(k)$ is a matrix of dimensionality $n_{\mathrm{x}} \times n_{\mathrm{u}}N_{\mathrm{u}}$ comprised of n_{x} rows of the matrix $\boldsymbol{P}(k)$ (Eq. (4.28)), starting from the row $(p-1)n_{\mathrm{x}}+1$. Similarly, the vector $\boldsymbol{x}^{0,(p-1)n_{\mathrm{x}}+1}(k)$ is comprised of n_{x} elements of the vector $\boldsymbol{x}^0(k)$, starting from the element $(p-1)n_{\mathrm{x}}+1$. Because Eq. (7.26) is used for prediction, it is possible to transform the optimisation problem (7.25) to a quadratic optimisation one. Thanks to linearisation, the box state constraints $x^{\min} \leq \hat{x}(k+p|k) \leq x^{\max}$ are linear. Unfortunately, the terminal state constraint $\|\hat{x}(k+N|k)\|_{\boldsymbol{P}}^2 \leq \alpha$ is a second-order function of the future control sequence. To make it possible to use quadratic optimisation, the terminal constraint must be replaced by a set of linear ones. Fig. 7.1 depicts an example (for a process with two state variables) ellipse Ω (an ellipsoid if $n_{\mathrm{x}} > 2$) and corresponding boxes (hyperboxes in general). In the simplest case, illustrated in Fig. 7.1a, the ellipsoid is replaced by a set of linear constraints determined by vectors of the standard basis. The resulting hyperbox, which is symmetric with respect to the axes, belongs to the interior of the ellipsoid. It means that if the predicted state $\hat{x}(k+N|k)$ belongs to the hyperbox, it belongs to the ellipsoid Ω. The nonlinear constraint $\|\hat{x}(k+N|k)\|_{\boldsymbol{P}}^2 \leq \alpha$ is approximated by the linear ones

$$\tilde{x}^{\min} \leq \hat{x}(k+N|k) \leq \tilde{x}^{\max} \tag{7.27}$$

where, because of symmetry, $\tilde{x}^{\max} = -\tilde{x}^{\min}$. In order to calculate the quantities $\tilde{x}_1^{\max}, \ldots, \tilde{x}_{n_{\mathrm{x}}}^{\max}$ which define the vector $\tilde{x}^{\max}$, one may notice that the box has $2^{n_{\mathrm{x}}}$ vertices, but thanks to symmetry it is sufficient to consider only a half of them. They are described as

$$w_i = \left(\sigma_1\tilde{x}_1^{\max}, \sigma_2\tilde{x}_2^{\max}, \ldots, \sigma_{n_{\mathrm{x}}-1}\tilde{x}_{n_{\mathrm{x}}-1}^{\max}, \tilde{x}_{n_{\mathrm{x}}}^{\max}\right)$$

where $i = 1, \ldots, 2^{n_{\mathrm{x}}-1}$, $\sigma_n \in \{-1, 1\}$, $n = 1, \ldots, n_{\mathrm{x}} - 1$. The quantities $\tilde{x}_1^{\max}, \ldots, \tilde{x}_{n_{\mathrm{x}}}^{\max}$, which uniquely determine the linear constraints (7.27), are found from the following optimisation problem

$$\begin{aligned}
&\max_{\tilde{x}_1^{\max}, \ldots, \tilde{x}_{n_{\mathrm{x}}}^{\max}} \left\{ \prod_{n=1}^{n_{\mathrm{x}}} \tilde{x}_n^{\max} \right\} \\
&\text{subject to} \\
&\tilde{x}_n^{\max} > 0, \ n = 1, \ldots, n_{\mathrm{x}} \\
&\|w_i\|_{\boldsymbol{P}}^2 \leq \alpha, \ i = 1, \ldots, 2^{n_{\mathrm{x}}-1}
\end{aligned}$$

because the vertices of the box have to belong to the ellipsoid ($\|w_i\|_{\boldsymbol{P}}^2 \leq \alpha$) and the biggest possible hyperbox has to be found (that is why the volume of the hyperbox is maximised). The above optimisation problem is nonlinear, but it is solved only once off-line, during development of the algorithm (possible computational problems and significant computational complexity are hence not so important as in the case of on-line MPC optimisation). The discussed approximation method of the set Ω by a set of linear constraints may be not very efficient, because the volume of the hyperbox may be much smaller than that of the ellipsoid. A better alternative (not detailed here) is to replace the ellipsoid by the hyperbox determined by orthogonal eigenvectors of the matrix $\boldsymbol{P}$, the idea of which is shown in Fig. 7.1b.

Using for prediction a linear approximation of the nonlinear model successively calculated at the current operating point of the process, i.e. using the prediction equation (7.26), and replacing the nonlinear terminal constraint (7.16) by the linear ones (7.27), from (7.25) one obtains the optimisation problem of the dual-mode MPC algorithm with on-line linearisation

$$\min_{\Delta \boldsymbol{u}(k)} \left\{ J(k) = \sum_{p=1}^{N-1} \theta(\hat{x}(k+p|k)) \left\| \boldsymbol{P}^{(p-1)n_{\mathrm{x}}+1}(k)\Delta \boldsymbol{u}(k) + \boldsymbol{x}^{0,(p-1)n_{\mathrm{x}}+1}(k) \right\|_{\boldsymbol{Q}}^2 \right.$$
$$\left. + \sum_{p=0}^{N_{\mathrm{u}}-1} \theta(\hat{x}(k+p|k)) \left\| \sum_{i=1}^{p} \Delta u(k+p|k) + u(k-1) \right\|_{\boldsymbol{R}}^2 \right\}$$

subject to (7.28)

$$u^{\min} \leq \sum_{i=1}^{p} \Delta u(k+p|k) + u(k-1) \leq u^{\max},\ p = 0, \ldots, N_{\mathrm{u}} - 1$$
$$x^{\min} \leq \boldsymbol{P}^{(p-1)n_{\mathrm{x}}+1}(k)\Delta \boldsymbol{u}(k) + \boldsymbol{x}^{0,(p-1)n_{\mathrm{x}}+1}(k) \leq x^{\max},\ p = 1, \ldots, N-1$$
$$\tilde{x}^{\min} \leq \boldsymbol{P}^{(N-1)n_{\mathrm{x}}+1}(k)\Delta \boldsymbol{u}(k) + \boldsymbol{x}^{0,(N-1)n_{\mathrm{x}}+1}(k) \leq \tilde{x}^{\max}$$

In the obtained optimisation problem the cost-function is quadratic, all the constraints are linear. That is why it can be solved by a quadratic optimisation procedure. The decision variables are calculated taking into account the current state of the process:

1. For $k = 0$: if $x(0) \in \Omega$, the signals $u(0) = \boldsymbol{K}x(0)$ are applied to the process. Otherwise, the vector of the future control increments $\Delta \boldsymbol{u}(0)$

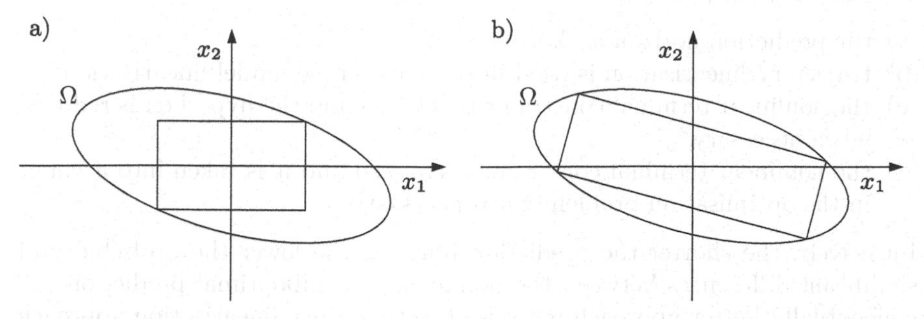

Fig. 7.1. An example ellipse (for $n_{\mathrm{x}} = 2$) and the corresponding boxes: a) the box determined by vectors of the standard basis, b) the box determined by orthogonal eigenvectors of the matrix $\boldsymbol{P}$

is calculated from the optimisation problem (7.28). The signals $u(0) = \triangle u(0|0) + u(-1)$ are applied to the process.
2. For $k \geq 1$: if $x(k) \in \Omega$, the signals $u(k) = \boldsymbol{K}x(k)$ are applied to the process. Otherwise, the vector of the future control increments $\triangle \boldsymbol{u}(k)$ is calculated from the optimisation problem (7.28). If

$$J(k) - J(k-1) \leq -\mu \left(\|x(k-1)\|^2_{\boldsymbol{Q}} + \|u(k-1)\|^2_{\boldsymbol{R}} \right) \tag{7.29}$$

where $0 < \mu \leq 1$, the signals $u(k) = \triangle u(k|k) + u(k-1)$ are applied to the process. If the condition (7.29) is not satisfied, the previously calculated feasible control signals which satisfy the condition are used, e.g. the signals calculated at the previous sampling instant.

If $x(k) \notin \Omega$, according to the general MPC-NPL algorithm shown in Fig 2.3, the model is linearised and the free trajectory is calculated at each sampling instant. For $x(k) \notin \Omega$ stability is enforced by the nonlinear terminal constraint and the condition (7.29) (which guarantees that the state converges to the set Ω within some finite time [291]) whereas for $x(k) \in \Omega$ stability is guaranteed by the stabilising controller.

Because a linear approximation of the model for the current operating point is used for prediction, one may expect some discrepancy between the nominal nonlinear predicted trajectory (calculated by means of the nonlinear model) and the suboptimal predicted trajectory (calculated from the linearised model). In different words, even provided that the model is perfect (no process-model mismatch), it is possible that the state vector predicted from the linearised model satisfies the box constraint

$$\tilde{x}^{\min} \leq \boldsymbol{P}^{(N-1)n_x+1}(k)\triangle \boldsymbol{u}(k) + \boldsymbol{x}^{0,(N-1)n_x+1}(k) \leq \tilde{x}^{\max} \tag{7.30}$$

but the true predicted state found using the nonlinear model does not satisfies the original nonlinear terminal constraint (7.16), i.e. the constraint $\|\hat{x}(k+N|k)\|^2_{\boldsymbol{P}} \leq \alpha$. When the described scenario is not taken into account, the whole algorithm may become unstable. The following methods can be used to solve the problem:

a) the prediction horizon is shortened,
b) trajectory linearisation is used in place of simple model linearisation,
c) the nonlinear terminal constraint is checked and the hyperbox is reduced when necessary,
d) the nonlinear terminal constraint is checked and it is taken into account in the optimisation problem when necessary.

Intuitively, the shorter the prediction horizon, the lower the probability of significant differences between the nonlinear and suboptimal predictions. A conceptually better approach is to use the trajectory linearisation approach (e.g. the MPC-NPLT algorithm) rather than the simple model linearisation used in the MPC-NPL algorithm.

Unfortunately, both mentioned approaches still do not guarantee that the nonlinear terminal constraint is satisfied. In the third method, at each sampling instant, satisfaction of the nonlinear terminal constraint is checked for the calculated future control policy. For this purpose the nonlinear model is used to calculate the state predictions, not the linearised one. If the nonlinear constraint is not satisfied, the volume of the hyperbox is reduced and the whole procedure (quadratic optimisation, prediction calculation and constraints' checking) is repeated. That is why the constraint (7.30) is replaced by

$$\begin{aligned}\tilde{x}^{\min} + \tilde{\varepsilon}^{\min}(k+N) \leq & \boldsymbol{P}^{(N-1)n_x+1}(k)\Delta\boldsymbol{u}(k) \\ & + \boldsymbol{x}^{0,(N-1)n_x+1}(k) \leq \tilde{x}^{\max} - \tilde{\varepsilon}^{\max}(k+N)\end{aligned} \tag{7.31}$$

in which the hyperbox is reduced using a technique very similar to the case when the original hard output or state constraints are relaxed (the soft constraints). The vectors $\tilde{\varepsilon}^{\min}(k+N)$ and $\tilde{\varepsilon}^{\max}(k+N)$ are of length n_x. Using the soft state constraints over the prediction horizon, the optimisation problem (7.28) becomes

$$\begin{aligned}\min_{\substack{\Delta\boldsymbol{u}(k)\\ \varepsilon^{\min}(k+p)\\ \varepsilon^{\max}(k+p)}} \Bigg\{ J(k) = & \sum_{p=1}^{N-1} \theta(\hat{x}(k+p|k)) \times \\ & \times \left\| \boldsymbol{P}^{(p-1)n_x+1}(k)\Delta\boldsymbol{u}(k) + \boldsymbol{x}^{0,(p-1)n_x+1}(k) \right\|_{\boldsymbol{Q}}^2 \\ & + \sum_{p=0}^{N_u-1} \theta(\hat{x}(k+p|k)) \left\| \sum_{i=1}^{p} \Delta u(k+p|k) + u(k-1) \right\|_{\boldsymbol{R}}^2 \\ & + \sum_{p=1}^{N-1} \left[\rho^{\min} \left\| \varepsilon^{\min}(k+p) \right\|^2 + \rho^{\max} \left\| \varepsilon^{\max}(k+p) \right\|^2 \right] \Bigg\}\end{aligned}$$

subject to (7.32)

$$u^{\min} \leq \sum_{i=1}^{p} \Delta u(k+p|k) + u(k-1) \leq u^{\max}, \ p = 0, \ldots, N_u - 1$$

$$\begin{aligned}x^{\min} - \varepsilon^{\min}(k+p) \leq & \boldsymbol{P}^{(p-1)n_x+1}(k)\Delta\boldsymbol{u}(k) \\ & + \boldsymbol{x}^{0,(p-1)n_x+1}(k) \leq x^{\max} + \varepsilon^{\max}(k+p), \\ & p = 1, \ldots, N-1\end{aligned}$$

$$\begin{aligned}\tilde{x}^{\min} + \tilde{\varepsilon}^{\min}(k+N) \leq & \boldsymbol{P}^{(N-1)n_x+1}(k)\Delta\boldsymbol{u}(k) \\ & + \boldsymbol{x}^{0,(N-1)n_x+1}(k) \leq \tilde{x}^{\max} - \tilde{\varepsilon}^{\max}(k+N)\end{aligned}$$

$$\varepsilon^{\min}(k+p) \geq 0, \ \varepsilon^{\max}(k+p) \geq 0, \ p = 1, \ldots, N-1$$

It is necessary to point out that the vectors $\tilde{\varepsilon}^{\min}(k+N)$ and $\tilde{\varepsilon}^{\max}(k+N)$ are not the decision variables of the above optimisation problem, they are auxiliary variables. At the beginning of calculation they are zeros vectors. After solving the optimisation problem (7.32), the state predictions are calculated using the nonlinear model for the obtained future control policy. If the condition $\|\hat{x}(k+N|k)\|_{\boldsymbol{P}}^2 \leq \alpha$ (Eq. (7.16)) is satisfied, the future control policy is accepted. Otherwise, some elements of the auxiliary vectors are increased, which leads to reduction of the volume of the hyperbox, and the optimisation procedure is repeated. Optimisation, prediction calculation and hyperbox modification are continued untill satisfaction of the linear constraints (7.31) in the quadratic optimisation task (7.32), the purpose of which is to approximate the ellipsoid, enforces satisfaction of the original nonlinear terminal constraint (7.16). In spite of a similar notation (the symbols $\varepsilon^{\min}(k+p)$, $\varepsilon^{\max}(k+p)$ and the auxiliary coefficients $\tilde{\varepsilon}^{\min}(k+N)$, $\tilde{\varepsilon}^{\max}(k+N)$), the state constraints (7.31) are not relaxed, because they guarantee stability. On the contrary, the auxiliary variables restrict the constraints which leads to reducing the hyperbox.

The described hyperbox reduction method may result in satisfaction of the original terminal constraint (7.16), but it is not guaranteed, in particular when the operating point is far from the equilibrium point. In the fourth method which may be used for solving the problem of suboptimal prediction, a hybrid approach is used: at first, the quadratic optimisation problem (7.28) is solved at each sampling instant. One may use soft state constraints, as in the optimisation task (7.32). The nonlinear state prediction is calculated for the obtained future control policy, satisfaction of the nonlinear terminal constraint (7.16) is checked. If it is satisfied, the solution is accepted. Otherwise, the MPC-NNPAPT algorithm is used at the current sampling instant, in which the terminal constraint is taken into account. The general formulation of the MPC-NNPAPT described in Chapter 2.3.5 and equations derived for the MPC-NO algorithm based on state-space models in Chapter 4.3 are used. In particular, the terminal constraint must be included in the constraints matrix (4.21) and the gradient of that constraint with respect to the decision variables must be included in the matrix (4.22).

7.2.2 Implementation for Input-Output Models

The quasi-state vector must be defined when input-output models are used for prediction [186]. Assuming for simplicity that $n_{\mathrm{A}} = \max(n_{\mathrm{A}}^m)$ for all $m = 1, \ldots, n_{\mathrm{y}}$ and $n_{\mathrm{B}} = \max(n_{\mathrm{B}}^{m,n})$ for $m = 1, \ldots, n_{\mathrm{y}}$ and $n = 1, \ldots, n_{\mathrm{u}}$, the quasi-state vector is

$$x(k) = \begin{bmatrix} x_{\mathrm{u}}(k) \\ x_{\mathrm{y}}(k) \end{bmatrix}$$

where

$$x_{\mathrm{u}}(k) = \begin{bmatrix} u(k-1) \\ \vdots \\ u(k - \max(n_{\mathrm{B}} - 1, 1)) \end{bmatrix}, \ x_{\mathrm{y}}(k) = \begin{bmatrix} y(k) \\ \vdots \\ y(k - n_{\mathrm{A}} + 1) \end{bmatrix} \tag{7.33}$$

The quasi-state has length of $n_{\mathrm{y}} n_{\mathrm{A}} + n_{\mathrm{u}} \max(n_{\mathrm{B}} - 1, 1)$. If the equilibrium point is not in the origin, it is necessary to scale the variables. In the dual-mode approach, which is typical of MPC algorithms with theoretically proven stability, the cost-function is a function of the predicted state vector and of the future control signals (as in the optimisation problem (7.25)), whereas it is straightforward in practice to minimise the predicted control errors and increments of the manipulated variables (as in the cost-function (1.4)). That is why two cost-functions are used. The first one, which takes into account the predicted control errors and the control increments, is used only for optimisation. The second one, which takes into account the predicted state (the quasi-state) and the control values, similarly as in the optimisation problem (7.25), is used to decide if the solution of the optimisation problem can be accepted or rejected (the condition (7.29)).

The following discussion is concerned with the MPC-NPL algorithm, although trajectory linearisation is also possible (MPC-NPLT and MPC-NPLPT algorithms). The optimisation problem of the prototype MPC-NPL algorithm (2.36) must take into account the nonlinear terminal constraint (7.16) present in the dual-mode MPC optimisation problem (7.25). To make it possible, the linearised model is used for prediction, which makes it possible to approximate the terminal constraint by the linear ones (7.27). Taking into account the quasi-state vector (7.33), one obtains the constraints

$$\begin{aligned} \tilde{u}^{\min} &\leq x_{\mathrm{u}}(k+N|k) \leq \tilde{u}^{\max} \\ \tilde{y}^{\min} &\leq \hat{x}_{\mathrm{y}}(k+N|k) \leq \tilde{y}^{\max} \end{aligned} \tag{7.34}$$

The first of the above constraints is linear, it can be directly included in quadratic optimisation. The vector of the predicted output signals must be expressed as a linear function of the future control policy. Using for prediction the linear approximation of the nonlinear model, similarly to Eq. (2.33), one obtains

$$\tilde{y}^{\min} \leq \widetilde{\boldsymbol{G}}(k)\triangle\boldsymbol{u}(k) + \tilde{\boldsymbol{y}}^0(k) \leq \tilde{y}^{\max} \tag{7.35}$$

where the matrix

$$\widetilde{\boldsymbol{G}}(k) = \begin{bmatrix} \boldsymbol{S}_{N-n_{\mathrm{A}}+1}(k) & \boldsymbol{S}_{N-n_{\mathrm{A}}}(k) & \cdots & \boldsymbol{0}_{n_{\mathrm{y}} \times n_{\mathrm{u}}} \\ \boldsymbol{S}_{N-n_{\mathrm{A}}+2}(k) & \boldsymbol{S}_{N-n_{\mathrm{A}}+1}(k) & \cdots & \boldsymbol{0}_{n_{\mathrm{y}} \times n_{\mathrm{u}}} \\ \vdots & \vdots & \ddots & \vdots \\ \boldsymbol{S}_{N}(k) & \boldsymbol{S}_{N-1}(k) & \cdots & \boldsymbol{S}_{N-N_{\mathrm{u}}+1}(k) \end{bmatrix}$$

is of dimensionality of $n_{\mathrm{y}} n_{\mathrm{A}} \times n_{\mathrm{u}} N_{\mathrm{u}}$ and of the structure similar to the dynamic matrix (2.34), the vector $\tilde{\boldsymbol{y}}^0(k)$ describes the free trajectory from the sampling instant $k + N - n_{\mathrm{A}}$ to $k + N$.

Even if the model is perfect, the predicted vector $\hat{x}_{\mathrm{y}}(k+N|k)$ calculated from the linearised model may satisfy the box constraint (7.35), but the true nonlinear state predicted from the nonlinear model may not satisfy the nonlinear terminal constraint (7.16). Because the same phenomenon may occur in the dual-mode MPC algorithm with on-line linearisation based on the state-space model, it is possible to use the methods mentioned on p. 228. In the third approach the auxiliary variables $\tilde{\varepsilon}^{\min}(k+N)$ and $\tilde{\varepsilon}^{\max}(k+N)$ are introduced (the vectors of length $n_{\mathrm{y}}n_{\mathrm{A}}$), which modify the constraints (7.35). One obtains

$$\tilde{y}^{\min} + \tilde{\varepsilon}^{\min}(k+N) \leq \widetilde{\boldsymbol{G}}(k)\triangle\boldsymbol{u}(k) + \tilde{\boldsymbol{y}}^{0}(k) \leq \tilde{y}^{\max} - \tilde{\varepsilon}^{\max}(k+N) \quad (7.36)$$

Taking into account the constraints (7.34) and (7.36), the MPC-NPL quadratic optimisation task (2.36) becomes

$$\min_{\substack{\triangle\boldsymbol{u}(k) \\ \boldsymbol{\varepsilon}^{\min}(k),\ \boldsymbol{\varepsilon}^{\max}(k)}} \Big\{ J(k) = \left\|\boldsymbol{y}^{\mathrm{sp}}(k) - \boldsymbol{G}(k)\triangle\boldsymbol{u}(k) - \boldsymbol{y}^{0}(k)\right\|_{\boldsymbol{M}}^{2} + \left\|\triangle\boldsymbol{u}(k)\right\|_{\boldsymbol{\Lambda}}^{2} + \rho^{\min}\left\|\boldsymbol{\varepsilon}^{\min}(k)\right\|^{2} + \rho^{\max}\left\|\boldsymbol{\varepsilon}^{\max}(k)\right\|^{2} \Big\}$$

subject to

$$\begin{aligned}
&\boldsymbol{u}^{\min} \leq \boldsymbol{J}\triangle\boldsymbol{u}(k) + \boldsymbol{u}(k-1) \leq \boldsymbol{u}^{\max} \\
&-\triangle\boldsymbol{u}^{\max} \leq \triangle\boldsymbol{u}(k) \leq \triangle\boldsymbol{u}^{\max} \\
&\boldsymbol{y}^{\min} - \boldsymbol{\varepsilon}^{\min}(k) \leq \boldsymbol{G}(k)\triangle\boldsymbol{u}(k) + \boldsymbol{y}^{0}(k) \leq \boldsymbol{y}^{\max} + \boldsymbol{\varepsilon}^{\max}(k) \\
&\boldsymbol{\varepsilon}^{\min}(k) \geq 0,\ \boldsymbol{\varepsilon}^{\max}(k) \geq 0 \\
&\tilde{u}^{\min} \leq x_{\mathrm{u}}(k+N|k) \leq \tilde{u}^{\max} \\
&\tilde{y}^{\min} + \tilde{\varepsilon}^{\min}(k+N) \leq \widetilde{\boldsymbol{G}}(k)\triangle\boldsymbol{u}(k) + \tilde{\boldsymbol{y}}^{0}(k) \leq \tilde{y}^{\max} - \tilde{\varepsilon}^{\max}(k+N)
\end{aligned}$$

At the beginning of calculations, the auxiliary variables are zeros vectors. The decision variables are calculated from the above optimisation problem and for the determined future control sequence the output predictions are found from the nonlinear model. If the condition (7.16) is satisfied, the solution is accepted. Otherwise, some elements of the auxiliary vectors are increased, which leads to reduction of the volume of the hyperbox, and the optimisation procedure is repeated. Optimisation, prediction calculation and hyperbox modification are continued until the nonlinear constraint (7.16) is satisfied.

The problem resulting from suboptimal prediction may be also solved by taking into account the nonlinear terminal constraint directly in the optimisation task, which leads to the MPC-NNPAPT algorithm. Implementation of the algorithm for input-output models presented in Chapter 2.3.5 and some equations derived for the MPC-NO algorithm in Chapter 2.2 are useful.

Of course, the set $\mathcal{U}$ must correspond with the input constraints present in the above optimisation problem, the set $\mathcal{X}$ must take into account the quasi-state vector (7.33) as well as input and output constraints. The ellipsoid may be also replaced by the box determined by orthogonal eigenvectors of the matrix $\boldsymbol{P}$, as shown in Fig. 7.1b.

Simulation results of the discussed dual-mode MPC algorithm with on-line linearisation in the control system of a high-pressure ethylene-ethane distillation column are discussed in [175]. The input-output neural model described in Example 2.2 is used.

7.3 Robustness of MPC Algorithms: A Short Review of Approaches

In all MPC algorithms with guaranteed stability discussed so far it is assumed that the model is perfect. Unfortunately, such an assumption is not realistic at all. Even in the case of very small modelling errors, the MPC algorithm may become unstable which is shown in [85]. In practice, some discrepancy between the model and properties of the process is always present. Firstly, model parameters are likely to change in time (as a result of ageing of some process components). Secondly, the model used for control is usually deliberately simplified: its order of dynamics may be lower than that of the process and its structure (e.g. a neural network) is chosen arbitrarily. Of course, in practice modelling errors are compensated by the negative feedback loop, but stability is still not guaranteed. That is why many robust MPC algorithms, in which stability is guaranteed even when the model is not perfect have been developed. A few the most important robust MPC approaches are characterised in the following text. Assessment of process-model mismatch is discussed in [210, 245, 271], it is outside the scope of this book.

7.3.1 MPC Algorithms with Min-Max Optimisation

An intuitive method of enforcing robustness of MPC algorithms is to consider all possible predicted trajectory, i.e. for all model errors. For the input-output model

$$\boldsymbol{A}(q^{-1})y(k) = \boldsymbol{B}(q^{-1})\triangle u(k-1) + \theta(k)$$

where $\theta(k)$ describes uncertainty of the model, it is possible to prove [41] that the predicted trajectory (over the prediction horizon) can be expressed as

$$\hat{\boldsymbol{y}}(k) = \boldsymbol{G}_{\mathrm{u}}\boldsymbol{u}(k) + \boldsymbol{G}_{\theta}\boldsymbol{\theta}(k) + \boldsymbol{y}^{0}(k)$$

where $\boldsymbol{G}_{\mathrm{u}}$ and $\boldsymbol{G}_{\theta}$ denote constant matrices whereas the time-varying vector $\boldsymbol{\theta}(k)$ describes model uncertainty over the prediction horizon. The vector of decision variables $\boldsymbol{u}(k)$ is found from the optimisation task

$$\min_{\boldsymbol{u}(k)} \max_{\theta(k)\in\boldsymbol{\Theta}} \left\{ J(k) = J(\boldsymbol{u}(k), \theta(k)) \right\} \tag{7.37}$$

The algorithm minimises the cost-function for the worst case, i.e. for all possible values of $\theta(k)$ which belong to the defined set of possible models $\boldsymbol{\Theta}$. Assuming the classical quadratic cost-function (1.4), one obtains a convex optimisation problem. The problem of shallow local minima does not exist, but optimisation is very computationally demanding, as a matter of fact such an approach is not practically implementable. As shown in [43], it is more useful to use the following cost-function

$$\begin{aligned} J(\boldsymbol{u}(k), \theta(k)) &= \max_{p=1,\ldots,N} \|y^{\mathrm{sp}}(k+p|k) - \hat{y}(k+p|k)\|_{\infty} \\ &= \max_{p=1,\ldots,N} \max_{m=1,\ldots,n_{\mathrm{y}}} |y_m^{\mathrm{sp}}(k+p|k) - \hat{y}_m(k+p|k)| \end{aligned}$$

In such a case the optimisation problem (7.37) becomes a linear optimisation one, in which the classical input and output linear constraints may be also included. Unfortunately, the total number of constraints may be very high. It is much better to use the cost-function

$$\begin{aligned} J(\boldsymbol{u}(k), \theta(k)) = &\sum_{p=1}^{N} \sum_{m=1}^{n_{\mathrm{y}}} \mu_{p,m} |y_m^{\mathrm{sp}}(k+p|k) - \hat{y}_m(k+p|k)| \\ &+ \sum_{p=0}^{N_{\mathrm{u}}-1} \sum_{n=1}^{n_{\mathrm{u}}} \lambda_{n,p} |\triangle u_n(k+p|k)| \end{aligned}$$

The above cost-function not only takes into account the control increments, but also, in comparison with the previously mentioned cost-functions, it makes it possible to formulate a linear optimisation MPC problem with a lower number of constrains. On the other hand, the optimisation problem, although linear, is still quite complex. That is why some methods aimed at reducing computational complexity of the discussed approach have been developed. It is possible to minimise not the cost-function, but its upper bound [265]. The min-max optimisation problem may be approximated by a quadratic optimisation task [16]. Alternatively, a neural network can be used for optimisation in place of a numerical routine [240, 266]. An application of the method introduced in [265] to robust control of a real chemical reactor is described in [87]. An application of the method detailed in [16] to robust control of the reactor is discussed in [90].

The optimisation problem (7.37) leads to the so called open-loop approach, in which only one vector of decision variables is calculated for all possible model uncertainty. The worst case is considered during optimisation, but during on-line control the negative feedback loop works, thanks to which better control quality may be obtained. Some closed-loop robust MPC algorithms are discussed in [128, 292].

A different method is detailed in [120]. The linear model

$$\begin{aligned} x(k+1) &= \boldsymbol{A}(k)x(k) + \boldsymbol{B}(k)u(k) \\ y(k) &= \boldsymbol{C}x(k) \end{aligned}$$

is used, where the matrices $\boldsymbol{A}(k)$ and $\boldsymbol{B}(k)$ are time-varying. The possible changes of model parameters are assumed to be known, i.e. $[\boldsymbol{A}(k), \boldsymbol{B}(k)] \in \mathcal{M}$. The cost-function with the infinite horizon is used

$$J(k) = \sum_{p=0}^{\infty} \left[\|\hat{x}(k+p|k)\|_{\boldsymbol{Q}}^2 + \|u(k+p|k)\|_{\boldsymbol{R}}^2 \right]$$

Similarly to the optimisation problem (7.37), the min-max optimisation task is

$$\min_{u(k)} \max_{[\boldsymbol{A}(k),\ \boldsymbol{B}(k)] \in \mathcal{M}} \left\{ J(k) \right\}$$

The above problem may be easily solved by means of the Linear Matrix Inequalities (LMI) approach. It is assumed that the future control signals are determined by the linear control law $u(k+p|k) = \boldsymbol{K}(k)x(k)$, where the time-varying matrix $\boldsymbol{K}(k)$ is calculated at each sampling instant. The discussed algorithm may be used when constraints are present or not. An improved version of the algorithm (computationally less demanding) is discussed in [121].

The most important disadvantage of the min-max approach, namely significant computational complexity, is particularly clear in the nonlinear case. For example, one may calculate the future control sequence which is determined by the future control policy [188]

$$\kappa_0(x(k)), \ldots, \kappa_{N_{\mathrm{u}}-1}(x(k+N_{\mathrm{u}}-1))$$

where, for instance, second-order polynomials may be used as the functions $\kappa_i(x(k+p))$. The dual-mode robust MPC algorithms with the min-max approach for nonlinear processes are discussed in [125], both open-loop and closed-loop cases are considered. Unfortunately, even for a very simple process, optimisation for the closed-loop algorithm turns out to be not possible, the open-loop algorithm works, but it is very computationally demanding. An interesting idea is presented in [60, 91], where the application of Volterra models leads to significant reduction of computational complexity. Unfortunately, for different other nonlinear models (e.g. neural ones), it is not possible.

7.3.2 MPC Algorithm with Additional State Constraints and Interval Arithmetic

The robust MPC algorithms with min-max optimisation are very computationally demanding in the nonlinear case. It means that they may be practically applied to relatively simple or slow processes [252]. An application of the linear model may greatly reduce computational complexity, but such an approach is valid for only some processes. Furthermore, the MPC algorithms with min-max optimisation are designed for the worst case, which is likely to result in unsatisfactory control quality.

Taking into account the fact that this book is concerned with nonlinear MPC algorithms, nonlinear robust MPC approaches are given much more attention. The MPC algorithm with some additional state constraints and interval arithmetic [134] is discussed first. The model (7.2) is named the nominal one, it is used for prediction calculation, whereas it is assumed that the real process may be described by

$$x(k+1) = f(x(k), u(k)) + w(k) \tag{7.38}$$

where the vector $w(k) \in \mathbb{R}^{n_x}$ represents additive uncertainty. It is assumed that $w(k) \in \mathcal{W}$, where the set $\mathcal{W}$ is compact and contains the origin. The dual-mode MPC approach is used in the discussed robust algorithm. Because the model is not precise, the predicted state trajectory may differ from the real process trajectory. In consequence, the state constraints taken into account in the dual-mode MPC optimisation problem (7.11) may be satisfied for the model, but not for the process. Not only the constraint $\hat{x}(k+p|k) \in \mathcal{X}$, but, first of all, also the constraint $\hat{x}(k+N|k) \in \Omega$ is crucial. In order to solve the problem resulting from uncertainty of state prediction, the state evolution sets $\mathcal{Z}_p(x(k), \boldsymbol{u}(k))$ are used (where $p = 1, 2, \ldots$), they may be also named reachable sets. The evolution sets depend on the current state of the process and on the currently calculated future control policy. They contain the state prediction of the uncertain system for the possible uncertainty scenarios, i.e. for all $w(k+p) \in \mathcal{W}$ one has

$$x^{\text{proc}}(k+p|k) \in \mathcal{Z}_p(x(k), \boldsymbol{u}(k))$$

In the general nonlinear case it is very difficult (or practically impossible) to precisely determine the state evolution sets. That is why they are replaced by the approximate evolution sets $\widetilde{\mathcal{Z}}_p(x(k), \boldsymbol{u}(k))$, which contain the evolution sets $\mathcal{Z}_p(x(k), \boldsymbol{u}(k))$ for all possible uncertainty scenarios, i.e.

$$\mathcal{Z}_p(x(k), \boldsymbol{u}(k)) \subseteq \widetilde{\mathcal{Z}}_p(x(k), \boldsymbol{u}(k))$$

In different words, the approximate sets $\widetilde{\mathcal{Z}}_p(x(k), \boldsymbol{u}(k))$ contain all possible variants of state predictions. The approximate sets are found by means of interval arithmetic [216]. If $x(k) \notin \Omega$, the robust MPC optimisation problem is

$$\begin{aligned}
&\min_{\boldsymbol{u}(k)} \left\{ J(k) = \sum_{p=0}^{N-k-1} L(\hat{x}(k+p|k), u(k+p|k)) + V(\hat{x}(k+N|k)) \right\} \\
&\text{subject to} \qquad\qquad\qquad\qquad\qquad\qquad\qquad\qquad\qquad\qquad (7.39) \\
&u(k+p|k) \in \mathcal{U},\ p = 0, \ldots, N-k-1 \\
&\widetilde{\mathcal{Z}}_p(x(k), \boldsymbol{u}(k)) \subseteq \mathcal{X},\ p = 0, \ldots, N-k \\
&\widetilde{\mathcal{Z}}_{N-k}(x(k), \boldsymbol{u}(k)) \subseteq \Omega
\end{aligned}$$

where $L(x, u) \geq 0$ and $V(x) \geq 0$. In comparison with the dual-mode MPC optimisation task (7.11), in the above problem the state constraints are replaced by the constraints imposed on the approximate state evolution sets. Thanks to interval arithmetic, computational complexity of such an approach is not significant, similar to that when the nominal model is used for prediction. One may prove [134] that the state converges to the set Ω in N iterations. The set $\Omega \subseteq \mathcal{X}$ is robustly invariant for the nonlinear process (7.38) controlled by the linear control law $u(k) = \boldsymbol{K}x(k)$, i.e. for all $x(k) \in \Omega$, $w(k) \in \mathcal{W}$ it is true that

$$x(k+1) = f(x(k), \boldsymbol{K}x(k)) + w(k) \in \Omega$$

The robustly invariant set may be found by some methods reviewed in [30].

Uncertainty of the model is described by the vector $w(k) \in \mathcal{W}$, it may not vanish to 0. The equilibrium point may be never reached precisely, as it is possible in the MPC algorithms with guaranteed stability discussed in the first part of the chapter. The objective of the robust controller is to drive the state of the process to some neighbourhood of the equilibrium point and to assure that the state remains in that neighbourhood. That is why the classical definition of asymptotic stability is not adequate. It is necessary to guarantee that the state is ultimately bounded [112].

One may prove [134], that for all initial conditions $x(0)$ the state is bounded provided that the optimisation problem (7.39) can be solved (when it is feasible). In different words, if the initial state is feasible, the set of feasible solutions is not empty for all $k \geq 0$. Thanks to the constraint $\widetilde{\mathcal{Z}}_{N-k}(x(k), \boldsymbol{u}(k)) \subseteq \Omega$, the system reaches the set Ω in N iterations. After reaching the set Ω, the linear control law $u(k) = \boldsymbol{K}x(k)$ is used, the system never leaves that set (because the set is robustly invariant). It is not necessary to calculate the optimal solution to the optimisation problem (7.39), a feasible solution is sufficient which satisfies the constraints. A more precise (and advanced) method of finding approximation of the state evolution sets is described in [35].

Feasibility and simple implementation are undoubted advantages of the described robust algorithms. Unfortunately, because the constraints in the optimisation task (7.39) are defined by the approximate evolution sets, it is not easy to use on-line linearisation.

7.3.3 MPC Algorithm with Additional State Constraints

An alternative approach to approximate state evolution sets and interval arithmetic is to use some additional state constraints which guarantee robustness. Such a robust MPC approach for linear models is discussed in [48] whereas the nonlinear case is described in [135] and shortly summarised in [187, 264]. The nominal model (7.2) is used for prediction and additive uncertainty is assumed, i.e. it is assumed that the real process may be characterised by Eq. (7.38). It is also assumed that the equilibrium point is in the origin and $f(0, 0) = 0$ and the function $f(x, u)$ is locally Lipschitz in x in the domain $\mathcal{X} \times \mathcal{U}$, i.e. there exists a Lipschitz constant $0 < L_{\mathrm{f}} < \infty$ such that for all $x_1, x_2 \in \mathcal{X}$ and all $u \in \mathcal{U}$

$$\left\| f(x_1, u) - f(x_2, u) \right\|_s \leq L_{\mathrm{f}} \left\| x_1 - x_2 \right\|_s \tag{7.40}$$

where s defines the norm, e.g. $s = 2$. Assuming that model uncertainty is bounded by the constant γ in the following way

$$\left\| w(k) \right\|_s \leq \gamma \ \forall w(k) \in \mathcal{W}$$

the difference between the nominal state prediction $\hat{x}(k+p|k)$ (calculated from the model) and the real state $x(k+p)$ is also bounded

$$\|\hat{x}(k+p|k) - x(k+p)\|_s \leq \frac{L_{\mathrm{f}}^p - 1}{L_{\mathrm{f}} - 1}\gamma$$

The above inequality may be easily proven by using the triangle inequality and the Lipschitz condition (7.40) [135]. Let $\mathcal{A}, \mathcal{B} \in \mathbb{R}^{n_{\mathrm{x}}}$ be some sets. The Pontryagin difference (it is sometimes named the Minkowski difference) of sets is defined as

$$\mathcal{A} \sim \mathcal{B} = \{x \in \mathbb{R}^{n_{\mathrm{x}}} : x + y \in \mathcal{A},\ \forall y \in \mathcal{B}\} \tag{7.41}$$

Let the set $\mathcal{X}_p$ be defined as

$$\mathcal{X}_p = \mathcal{X} \sim \mathcal{B}_p \tag{7.42}$$

where $\mathcal{B}_p$ is a hyperball

$$\mathcal{B}_p = \left\{ z \in \mathbb{R}^{n_{\mathrm{x}}} :\ \|z\|_s \leq \frac{L_{\mathrm{f}}^p - 1}{L_{\mathrm{f}} - 1}\gamma \right\} \tag{7.43}$$

The following implication is true

$$\hat{x}(k+p|k) \in \mathcal{X}_p \Rightarrow x(k+p) \in \mathcal{X},\ p = 1, \ldots, N \tag{7.44}$$

In different words, if the nominal predicted state (calculated from the model) belongs to the set $\mathcal{X}_p$, the state of the process belongs to the set $\mathcal{X}$. The implication (7.44) may be easily proven [152].

The optimisation problem of the discussed MPC algorithm with the additional state constraints is

$$\begin{aligned}
&\min_{u(k)} \left\{ J(k) = \sum_{p=0}^{N-1} L(\hat{x}(k+p|k), u(k+p|k)) + V_{\mathrm{N}}(\hat{x}(k+N|k)) \right\} \\
&\text{subject to} \\
&u(k+p|k) \in \mathcal{U},\ p = 0, \ldots, N-1 \\
&\hat{x}(k+p|k) \in \mathcal{X}_p,\ p = 1, \ldots, N-1 \\
&\hat{x}(k+N|k) \in \Omega
\end{aligned} \tag{7.45}$$

It is worth pointing out that the above problem is quite similar to the classical MPC optimisation task, in which model uncertainty is not taken into account. The min-max approach (7.37) is not necessary. The nominal model is used for prediction calculation, it is not necessary to introduce approximate state evolution sets, which are needed in the task (7.39). The cost-function $V_{\mathrm{N}}(\hat{x}(k+N|k))$, the additional state constrains $\hat{x}(k+p|k) \in \mathcal{X}_p$ over the prediction horizon (for $p = 1, \ldots, N-1$) and the terminal constraint $\hat{x}(k+N|k) \in \Omega$ guarantee robustness of the discussed algorithm.

During algorithm development the invariant set of the nominal system

$$\Omega_0 = \{x(k) \in \mathbb{R}^{n_x} : V_N(x(k)) \leq \alpha_0\}$$

where $\alpha_0 > 0$, and the linear stabilising controller $u(k) = h(x(k))$, e.g. $u(k) = \boldsymbol{K}x(k)$, are used. The terminal set is defined as

$$\Omega = \{x(k) \in \mathbb{R}^{n_x} : V_N(x(k)) \leq \alpha\}$$

where $\alpha > 0$; the terminal set is a subset of the invariant set, i.e. $\alpha \leq \alpha_0$.
If:

a) in the equilibrium point $L(0,0) = 0$ and there exists the constants $a > 0$ and $\sigma \geq 0$, for which $L(x,u) \geq a \|(x,u)\|^\sigma$, the function $L(x,u)$ satisfies the Lipschitz condition in the set $\mathcal{X} \times \mathcal{U}$ with the constant $0 < L_c < \infty$,
b) the invariant set is such that $\Omega_0 \subseteq \mathcal{X}^h = \{x(k) \in \mathcal{X}_{N-1} : h(x(k)) \in \mathcal{U}\}$, when no state constraints defined by the set $\mathcal{X}$ exist, $\mathcal{X}_{N-1} \in \mathbb{R}^{n_x}$,
c) for all $x(k) \in \Omega_0$

$$h(x(k)) \in \mathcal{U}$$
$$f(x(k), h(x(k))) \in \Omega$$
$$V_N(f(x(k), h(x(k)))) - V_N(x(k)) \leq -L(x(k), h(x(k)))$$

d) $\alpha_1(\|x\|) \leq V_N(x) \leq \alpha_2(\|x\|)$, where $\alpha_1(\cdot)$ and $\alpha_2(\cdot)$ are $\mathcal{K}_\infty$-functions,
e) the function $V_N(x)$ satisfies the Lipschitz condition in the set Ω_0 with the constant $0 < L_V < \infty$,
f) model uncertainty is bounded

$$\gamma \leq \frac{\alpha_0 - \alpha}{L_V L_f^{N-1}} \tag{7.46}$$

then the closed-loop system is asymptotically stable for any initial condition $x(0) \in \mathcal{D}$, where $\mathcal{D}$ is a set of states for which the nonlinear problem (7.45) is feasible (the set of feasible solutions exists). The stability proof is given in [135]. The input to state stability concept [303, 104] is used for this purpose.

The discussed robust MPC algorithm is suboptimal, because from the perspective of stability it is not necessary to find the global solution of the nonlinear optimisation problem (7.45) at each sampling instant, a feasible one is sufficient. For the most frequently used quadratic functions $L(x(k), u(k))$ and $V_N(x(k))$, the first, the fourth and the fifth of the above assumptions are satisfied. Finding suitable sets Ω_0, Ω and functions $V_N(x(k))$ is discussed elsewhere [30, 36, 46, 224, 109, 175, 189, 204, 208].

A new version of the algorithm is discussed in [252]. The following optimisation problem is used

$$\min_{\boldsymbol{u}(k)} \left\{ J(k) = \sum_{p=0}^{N_{\mathrm{u}}-1} L(\hat{x}(k+p|k), u(k+p|k)) \right.$$
$$\left. + \sum_{p=N_{\mathrm{u}}}^{N-1} L(\hat{x}(k+p|k), \kappa(\hat{x}(k+p|k))) + V_{\mathrm{N}}(\hat{x}(k+N|k)) \right\}$$

subject to (7.47)

$$u(k+p|k) \in \mathcal{U},\ p = 0, \ldots, N_{\mathrm{u}} - 1$$
$$\hat{x}(k+p|k) \in \mathcal{X}_p,\ p = 1, \ldots, N_{\mathrm{u}} - 1$$
$$\hat{x}(k+N_{\mathrm{u}}|k) \in \Omega$$

In comparison with the previous version (the task (7.45)), the separate control horizon is used. The additional state constraints $\hat{x}(k+p|k) \in \mathcal{X}_p$ are considered up to $p = N_{\mathrm{u}} - 1$, whereas an additional linear controller $u(k+p|k) = \kappa(\hat{x}(k+p|k))$ is used for $p = N_{\mathrm{u}}, \ldots, N-1$. The terminal constraint $\hat{x}(k+N_{\mathrm{u}}|k) \in \Omega$ takes into account the state prediction at the end of the control horizon, not the prediction one. The advantage of the new version is the lack of state constraints $\hat{x}(k+p|k) \in \mathcal{X}_p$ for $p = N_{\mathrm{u}}, \ldots, N-1$. Although the number of state constraints is lower in the new algorithm, it is not always better. Examples showing effectiveness of both algorithm versions are discussed in [264]. A further extension of the algorithm, in which no important conditions are imposed on the function $V_{\mathrm{N}}(x)$, is detailed [84].

7.4 MPC Algorithm with Additional State Constraints and On-Line Linearisation

The robust MPC algorithm with additional state constraints, in comparison with the min-max MPC approach and the algorithm with state constraints and interval arithmetic, is very computationally efficient. The nominal model is used for prediction, no approximations of the state evolution set are necessary. Furthermore, global nonlinear optimisation is not necessary, the feasible solution, which satisfies the constraints, is sufficient. That is why it is worth developing the discussed algorithm with additional state constraints and on-line linearisation [152]. Analogously to the dual-mode MPC approach, there are no equality constraints, which may lead to numerical problems. Two versions of the robust MPC algorithm with linearisation are detailed: for state-space models and for input-output ones. The following part of the text is concerned with model linearisation used in the MPC-NPL approach, but trajectory linearisation is also possible (MPC-NPLT and MPC-NPLPT strategies).

7.4.1 Implementation for State-Space Models

Using separate prediction and control horizons, the quadratic penalty term $V_{\mathrm{N}}(\hat{x}(k+N|k)) = \|\hat{x}(k+N|k)\|^2_{\boldsymbol{P}}$, the box constraints (7.4) and taking into account the set Ω defined by Eq. (7.16), the optimisation problem (7.45) of the rudimentary robust MPC algorithm with additional state constraints becomes

$$\begin{aligned}
&\min_{\boldsymbol{u}(k)} \left\{ J(k) = \sum_{p=1}^{N-1} \|\hat{x}(k+p|k)\|^2_{\boldsymbol{Q}} + \sum_{p=0}^{N_{\mathrm{u}}-1} \|u(k+p|k)\|^2_{\boldsymbol{R}} + \|\hat{x}(k+N|k)\|^2_{\boldsymbol{P}} \right\} \\
&\text{subject to} \qquad\qquad\qquad\qquad\qquad\qquad\qquad\qquad\qquad\qquad (7.48) \\
&u^{\min} \leq u(k+p|k) \leq u^{\max},\ p = 0, \ldots, N_{\mathrm{u}} - 1 \\
&\hat{x}(k+p|k) \in \mathcal{X}_p,\ 1 = 0, \ldots, N-1 \\
&\|\hat{x}(k+N|k)\|^2_{\boldsymbol{P}} \leq \alpha
\end{aligned}$$

Because the model is nonlinear, the obtained optimisation task is also nonlinear. The minimised cost-function is nonlinear, the state constraints $\hat{x}(k+p|k) \in \mathcal{X}_p$ are also nonlinear (where $\mathcal{X}_p = \mathcal{X} \sim \mathcal{B}_p$, $\mathcal{B}_p$ denote hyperballs). The terminal state constraint $\|\hat{x}(k+N|k)\|^2_{\boldsymbol{P}} \leq \alpha$ is also nonlinear, analogously to the dual-mode MPC approach. From the suboptimal prediction equations (7.26), the state predictions for the consecutive sampling instants of the prediction horizon may be expressed as the linear functions of the future control increments $\triangle \boldsymbol{u}(k)$. Thanks to suboptimal prediction, the cost-function is quadratic, the left sides of the constraints $\hat{x}(k+p|k) \in \mathcal{X}_p$ are linear.

In order to transform the nonlinear optimisation problem (7.48) into a quadratic optimisation task, the ellipsoid Ω is replaced by a set of linear constraints. For this purpose the methods similar to those introduced for the dual-mode MPC algorithm with on-line linearisation are used. For simplicity of presentation, the hyperbox determined by vectors of the standard basis is used, as depicted in Fig. 7.1a. In place of the nonlinear constraint (7.16), one obtains the linear ones (7.27). Taking into account the prediction equation (7.26), the last constraint present in the optimisation task (7.28) is obtained.

In order to solve the problem of linear approximation of the sets $\mathcal{X}_p$, the set $\mathcal{X}$ is assumed to be of the box type (7.4) and in all norms $s = 2$. Fig. 7.2a illustrates (for the process with 2 state variables, $n_{\mathrm{x}} = 2$) example balls $\mathcal{B}_1, \ldots, \mathcal{B}_{N-1}$ (hyperballs if $n_{\mathrm{x}} > 2$) and corresponding boxes $\mathcal{K}_1, \ldots, \mathcal{K}_{N-1}$ (hyperboxes in general). The hyperboxes must be circumscribed over the hyperballs, not inscribed, it results from Eqs. (7.41) and (7.42). The hyperboxes are defined for all $i = 1, \ldots, n_{\mathrm{x}}$, $p = 1, \ldots, N-1$ by

$$x_{p,i}^{\mathcal{K},\min} \leq x_{p,i} \leq x_{p,i}^{\mathcal{K},\max}$$

where, taking into account the definition of the hyperball (7.43), one obtains

$$x_{p,i}^{\mathcal{K},\min} = -\frac{L_{\mathrm{f}}^p - 1}{L_{\mathrm{f}} - 1}\gamma, \; x_{p,i}^{\mathcal{K},\max} = \frac{L_{\mathrm{f}}^p - 1}{L_{\mathrm{f}} - 1}\gamma$$

Fig. 7.2b depicts successive Pontryagin differences $\tilde{\tilde{\mathcal{X}}}_1 = \mathcal{X} - \mathcal{K}_1, \ldots, \tilde{\tilde{\mathcal{X}}}_{N-1} = \mathcal{X} - \mathcal{K}_{N-1}$. They are hyperboxes

$$\tilde{\tilde{\mathcal{X}}}_p = \left\{\hat{x}(k+p|k) \in \mathbb{R}^{n_{\mathrm{x}}} : \tilde{\tilde{x}}_p^{\min} \leq \hat{x}(k+p|k) \leq \tilde{\tilde{x}}_p^{\max}\right\}$$

for all $p = 1, \ldots, N-1$, where

$$\tilde{\tilde{x}}_{p,i}^{\min} = x_i^{\min} - x_{p,i}^{\mathcal{K},\min}, \; \tilde{\tilde{x}}_{p,i}^{\max} = x_i^{\max} - x_{p,i}^{\mathcal{K},\max}$$

if $x_i^{\min} < x_{p,i}^{\mathcal{K},\min}$ and $x_i^{\max} > x_{p,i}^{\mathcal{K},\max}$, otherwise the Pontryagin difference is an empty set. In consequence, in place of the nonlinear constraints $\hat{x}(k+p|k) \in \mathcal{X}_p$, using the prediction equation of the MPC-NPL algorithm (7.26), the following linear constraints are obtained

$$\tilde{\tilde{x}}_p^{\min} \leq \boldsymbol{P}^{(p-1)n_{\mathrm{x}}+1}(k)\triangle\boldsymbol{u}(k) + \boldsymbol{x}^{0,(p-1)n_{\mathrm{x}}+1}(k) \leq \tilde{\tilde{x}}_p^{\max} \tag{7.49}$$

for all $p = 1, \ldots, N-1$. Satisfaction of the above linear constraints guarantees that the original nonlinear constraints $\hat{x}(k+p|k) \in \mathcal{X}_p$ are also satisfied.

Assuming that the nonlinear model is successively linearised on-line for the current operating point, i.e. using the prediction equation (7.26), and replacing the nonlinear state equations by the linear ones, the optimisation

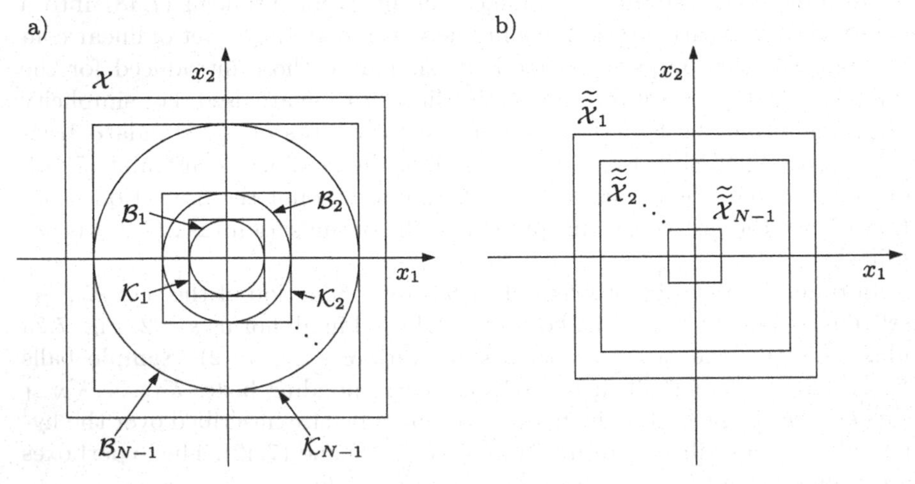

Fig. 7.2. a) Example balls $\mathcal{B}_1, \ldots, \mathcal{B}_{N-1}$ (for $n_{\mathrm{x}} = 2$) and the corresponding boxes $\mathcal{K}_1, \ldots, \mathcal{K}_{N-1}$, b) Pontryagin differences $\tilde{\tilde{\mathcal{X}}}_1 = \mathcal{X} - \mathcal{K}_1, \ldots, \tilde{\tilde{\mathcal{X}}}_{N-1} = \mathcal{X} - \mathcal{K}_{N-1}$

problem (7.48) of the discussed robust MPC algorithm with additional state constraints becomes the following quadratic optimisation task

$$\min_{\Delta\boldsymbol{u}(k)}\left\{J(k)=\sum_{p=1}^{N-1}\left\|\boldsymbol{P}^{(p-1)n_{\mathrm{x}}+1}(k)\Delta\boldsymbol{u}(k)+\boldsymbol{x}^{0,(p-1)n_{\mathrm{x}}+1}(k)\right\|_{\boldsymbol{Q}}^{2}\right.$$
$$+\sum_{p=0}^{N_{\mathrm{u}}-1}\left\|\sum_{i=1}^{p}\Delta u(k+p|k)+u(k-1)\right\|_{\boldsymbol{R}}^{2}$$
$$\left.+\left\|\boldsymbol{P}^{(N-1)n_{\mathrm{x}}+1}(k)\Delta\boldsymbol{u}(k)+\boldsymbol{x}^{0,(N-1)n_{\mathrm{x}}+1}(k)\right\|_{\boldsymbol{P}}^{2}\right\}$$

subject to (7.50)

$$u^{\min}\leq u(k+p|k)\leq u^{\max},\ p=0,\ldots,N_{\mathrm{u}}-1$$
$$\tilde{\tilde{x}}_{p}^{\min}\leq \boldsymbol{P}^{(p-1)n_{\mathrm{x}}+1}(k)\Delta\boldsymbol{u}(k)+\boldsymbol{x}^{0,(p-1)n_{\mathrm{x}}+1}(k)\leq\tilde{\tilde{x}}_{p}^{\max},\ p=1,\ldots,N-1$$
$$\tilde{x}^{\min}\leq \boldsymbol{P}^{(N-1)n_{\mathrm{x}}+1}(k)\Delta\boldsymbol{u}(k)+\boldsymbol{x}^{0,(N-1)n_{\mathrm{x}}+1}(k)\leq\tilde{x}^{\max}$$

Because a linear approximation of the nonlinear model calculated at the current operating point is used for prediction, the differences between the nominal predicted state trajectory (determined from the nonlinear model) and the suboptimal one (found from the linearised model) are possible. The predicted state vector calculated using the linearised model may satisfy the box constraints (7.30), which approximate the ellipsoid, but the real nonlinear prediction may not satisfy the original nonlinear terminal constraint (7.16). Analogously, satisfaction of the linear constraints (7.49) does not guarantee that the nonlinear constraints $\hat{x}(k+p|k)\in\tilde{\tilde{\mathcal{X}}}_p$ are fulfilled. The nature of the described problem is the same as in the dual-mode MPC algorithm with on-line linearisation. Hence, it can be solved by the methods mentioned on p. 228. In the third approach, at each sampling instant, for the calculated future control increments and using the nonlinear model, it is necessary to check whether or not the original nonlinear constraints are satisfied. If they are violated, the volume of the corresponding hyperboxes is reduced and the whole procedure (quadratic optimisation, prediction calculation and constraints' checking) is repeated. Analogously to the dual-mode MPC algorithm with on-line linearisation (in which the quadratic optimisation task (7.32) is solved), the state constraints must be modified in the following way

$$\tilde{\tilde{x}}_{p}^{\min}+\tilde{\tilde{\varepsilon}}^{\min}(k+p)\leq \boldsymbol{P}^{(p-1)n_{\mathrm{x}}+1}(k)\Delta\boldsymbol{u}(k)$$
$$+\boldsymbol{x}^{0,(p-1)n_{\mathrm{x}}+1}(k)\leq\tilde{\tilde{x}}_{p}^{\max}-\tilde{\tilde{\varepsilon}}^{\max}(k+p),$$
$$p=0,\ldots,N-1$$

$$\tilde{x}^{\min}+\tilde{\varepsilon}^{\min}(k+N)\leq \boldsymbol{P}^{(N-1)n_{\mathrm{x}}+1}(k)\Delta\boldsymbol{u}(k)$$
$$+\boldsymbol{x}^{0,(N-1)n_{\mathrm{x}}+1}(k)\leq\tilde{x}^{\max}-\tilde{\varepsilon}^{\max}(k+N)$$

At the beginning of calculations, the auxiliary variables $\tilde{\varepsilon}^{\min}(k+p)$, $\tilde{\varepsilon}^{\max}(k+p)$, $\tilde{\varepsilon}^{\min}(k+N)$ and $\tilde{\varepsilon}^{\max}(k+N)$ are zeros vectors of length n_{x}. Next, the decision variables are calculated and the nonlinear state is predicted using the nonlinear model. If the constraints $\hat{x}(k+p|k) \in \mathcal{X}_p$ and $\|\hat{x}(k+N|k)\|_{\boldsymbol{P}}^2 \leq \alpha$ are fulfilled, the obtained solution is accepted and applied to the process. Otherwise, some elements of the auxiliary vectors are increased and the optimisation problem is repeated. It is necessary to point out that the state constraints are not soft. They are not relaxed, because it may lead to instability. If the set of feasible solutions of the optimisation problem is empty, the state constraints $\hat{x}(k+p|k) \in \mathcal{X}$, which are not motivated by stability issues, must be relaxed. In such a case, the hyperboxes $\mathcal{K}_p$ and the Pontryagin differences $\tilde{\tilde{\mathcal{X}}}_p = \mathcal{X} - \mathcal{K}_p$ for all $p = 1, \ldots, N-1$, must be recalculated using the procedure illustrated in Fig. 7.2.

Of course, the problem of inaccurate suboptimal prediction possible at some sampling instants may be directly solved by taking into account the nonlinear terminal constraint (7.16) and the nonlinear constraints $\tilde{\tilde{x}}_p^{\min} \leq \hat{x}(k+p|k) \leq \tilde{\tilde{x}}_p^{\max}$ for $p = 1, \ldots, N-1$ directly into the optimisation problem, similarly to the dual-mode MPC approach with on-line linearisation. It leads to the MPC-NNPAPT algorithm.

7.4.2 Implementation for Input-Output Models

In the robust MPC algorithm with state constraints and on-line linearisation for the input-output models, similarly to the dual-mode MPC strategy with on-line linearisation, some modifications of the basic algorithm formulated for the state-space models are necessary. The quasi-state vector is defined by Eq. (7.33). Because, from the point of stability, it is sufficient to find only a feasible decision vector, the MPC-NPL optimisation problem (2.36) must take into account the constraints of the robust algorithm, which are present in the optimisation task (7.50).

In the currently discussed robust algorithm the nonlinear terminal constraint is treated in the same way it is done in the stable dual-mode MPC algorithm with on-line linearisation. In the similar way the constraints $\tilde{\tilde{x}}_p^{\min} \leq \hat{x}(k+p|k) \leq \tilde{\tilde{x}}_p^{\max}$, for $p = 1, \ldots, N-1$, are treated. Similarly to Eq. (7.34), one has

$$\begin{aligned} \tilde{\tilde{u}}^{\min} &\leq x_{\mathrm{u}}(k+p|k) \leq \tilde{\tilde{u}}^{\max} \\ \tilde{\tilde{y}}^{\min} &\leq \hat{x}_{\mathrm{y}}(k+p|k) \leq \tilde{\tilde{y}}^{\max} \end{aligned} \tag{7.51}$$

Using for prediction the successively linearised model, analogously to Eq. (7.36), one obtains

$$\tilde{\tilde{y}}^{\min} + \tilde{\varepsilon}^{\min}(k+p) \leq \tilde{\tilde{\boldsymbol{G}}}_p(k)\triangle\boldsymbol{u}(k) + \tilde{\tilde{\boldsymbol{y}}}_p^0(k) \leq \tilde{\tilde{y}}^{\max} - \tilde{\varepsilon}^{\max}(k+p) \tag{7.52}$$

where the matrices

$$\widetilde{\widetilde{G}}_p(k) = \begin{bmatrix} S_{p-n_A+1}(k) & S_{p-n_A}(k) & \dots & 0_{n_y \times n_u} \\ S_{p-n_A+2}(k) & S_{p-n_A+1}(k) & \dots & 0_{n_y \times n_u} \\ \vdots & \vdots & \ddots & \vdots \\ S_p(k) & S_{p-1}(k) & \dots & S_{p-N_u+1}(k) \end{bmatrix}$$

are of dimensionality $n_y n_A \times n_u N_u$ and they have the structure similar to that of the matrix (2.34) whereas the vector $\tilde{\tilde{y}}^0(k)$ describes the nonlinear free trajectory from the sampling instant $k+p-n_A$ to $k+p$. The auxiliary vector variables $\tilde{\tilde{\varepsilon}}^{\min}(k+p)$ and $\tilde{\tilde{\varepsilon}}^{\max}(k+p)$ of length $n_y n_A$ are used to modify the linear constraints to satisfy the original nonlinear state constraints $\tilde{\tilde{x}}_p^{\min} \leq \hat{x}(k+p|k) \leq \tilde{\tilde{x}}_p^{\max}$.

Taking into account Eqs. (7.34), (7.36), (7.51) and (7.52), the MPC-NPL quadratic optimisation problem (2.36) becomes

$$\min_{\substack{\Delta u(k) \\ \varepsilon^{\min}(k),\ \varepsilon^{\max}(k)}} \Big\{ J(k) = \left\| y^{\mathrm{sp}}(k) - G(k)\Delta u(k) - y^0(k) \right\|_M^2 + \left\| \Delta u(k) \right\|_\Lambda^2 + \rho^{\min} \left\| \varepsilon^{\min}(k) \right\|^2 + \rho^{\max} \left\| \varepsilon^{\max}(k) \right\|^2 \Big\}$$

subject to

$$\begin{aligned}
& u^{\min} \leq J\Delta u(k) + u(k-1) \leq u^{\max} \\
& -\Delta u^{\max} \leq \Delta u(k) \leq \Delta u^{\max} \\
& y^{\min} - \varepsilon^{\min}(k) \leq G(k)\Delta u(k) + y^0(k) \leq y^{\max} + \varepsilon^{\max}(k) \\
& \varepsilon^{\min}(k) \geq 0,\ \varepsilon^{\max}(k) \geq 0 \\
& \tilde{u}^{\min} \leq x_u(k+N|k) \leq \tilde{u}^{\max} \\
& \tilde{y}^{\min} + \tilde{\varepsilon}^{\min}(k+N) \leq \widetilde{G}(k)\Delta u(k) + \tilde{y}^0(k) \leq \tilde{y}^{\max} - \tilde{\varepsilon}^{\max}(k+N) \\
& \tilde{\tilde{u}}^{\min} \leq x_u(k+p|k) \leq \tilde{\tilde{u}}^{\max},\ p = 1, \dots, N-1 \\
& \tilde{\tilde{y}}^{\min} + \tilde{\tilde{\varepsilon}}^{\min}(k+p) \leq \widetilde{\widetilde{G}}_p(k)\Delta u(k) \\
& \qquad + \tilde{\tilde{y}}_p^0(k) \leq \tilde{\tilde{y}}^{\max} - \tilde{\tilde{\varepsilon}}^{\max}(k+p),\ p = 1, \dots, N-1
\end{aligned}$$

At the beginning of calculations the auxiliary variables are zeros vectors. The optimisation problem is solved and for the obtained future control policy the nonlinear output predictions are calculated using the nonlinear models. If the constraints $\|\hat{x}_y(k+N|k)\|_P^2 \leq \alpha$ and $\hat{x}(k+p|k) \in \mathcal{X}_p$, $p = 1, \dots, N-1$ are fulfilled, the obtained solution is accepted and applied to the process. Otherwise, some elements of the auxiliary vectors are increased and the optimisation problem is repeated. The set $\mathcal{U}$ must correspond with the input constraints present in the above quadratic optimisation task, the set $\mathcal{X}$ must take into account the quasi-state vector (7.33) and the constraints imposed on input and output signals.

Similarly to the robust MPC algorithm with state constraints and on-line linearisation, at some sampling instants the problem of the suboptimal prediction may be important. The methods described earlier may be used, in particular the nonlinear constraints may be directly taken into account in the optimisation problem, which leads to the MPC-NNPAPT algorithm.

The described methods of replacing the nonlinear constraints by the linear ones may be also used in the robust MPC algorithm with on-line linearisation and with the state constrains $\hat{x}(k+p|k) \in \mathcal{X}_p$ only from $p = N_{\mathrm{u}} - 1$ and the additional linear controller $u(k+p|k) = \kappa(\hat{x}(k+p|k))$ for $p = N_{\mathrm{u}}, \ldots, N-1$. To make it possible it is necessary to modify the optimisation problem (7.47), of course taking into account the nature of the model (both the state-space and input-output models may be used).

7.5 Example 7.1

Let the example dynamic process be described by the following equation [135]

$$x(k+1) = \boldsymbol{A}x(k) + \boldsymbol{B}u(k) + \boldsymbol{C}x(k)u(k) + \delta u(k)$$

where $x(k) \in \mathbb{R}^2$, $u(k) \in \mathbb{R}$, the vector $\delta = [\delta_1\ \delta_2]^{\mathrm{T}}$ describes model uncertainty and the parameters are

$$\boldsymbol{A} = \begin{bmatrix} 0.55 & 0.12 \\ 0 & 0.67 \end{bmatrix}, \quad \boldsymbol{B} = \begin{bmatrix} 0.01 \\ 0.15 \end{bmatrix}, \quad \boldsymbol{C} = \begin{bmatrix} -0.6 & 1 \\ 1 & -0.8 \end{bmatrix}$$

The following MPC algorithms are compared:

a) the robust MPC algorithm with the additional state constraints and nonlinear optimisation (the optimisation problem (7.45)),
b) the robust MPC algorithm with the additional state constraints and on-line linearisation in which the model is successively linearised on-line for the current operating point, as it is done in the MPC-NPL strategy (the quadratic optimisation problem (7.50)).

In both algorithms the same nominal model is used, in which $\delta_1 = \delta_2 = 0$. The tuning parameters of both algorithms are the same: $N = N_{\mathrm{u}} = 5$, $L(x,u) = \|x\|_{\boldsymbol{Q}}^2 + \|u\|_{\boldsymbol{R}}^2$, where $\boldsymbol{Q} = \boldsymbol{I}_{2\times 2}$, $\boldsymbol{R} = 1$. The same constraints are imposed on the input signal: $u^{\min} = -0.1$, $u^{\max} = 0.1$, and on the predicted state variable: $x^{\min} = -0.1$, $x^{\max} = 0.1$.

The Lipschitz constant for the considered model is

$$L_{\mathrm{f}} \geq \max_{u \in \mathcal{U}} \|\boldsymbol{A} + \boldsymbol{C}u\|_2 = 0.7633$$

From the solution of the Ricatti equation, one has

$$\boldsymbol{P} = \begin{bmatrix} 1.4332 & 0.1441 \\ 0.1441 & 1.8316 \end{bmatrix}$$

and the linear stabilising controller $u(k) = h(x(k)) = \boldsymbol{K}x(k)$ is described by the vector $\boldsymbol{K} = [0.0190\ \ -0.1818]$. Analogously as it is done in [264],

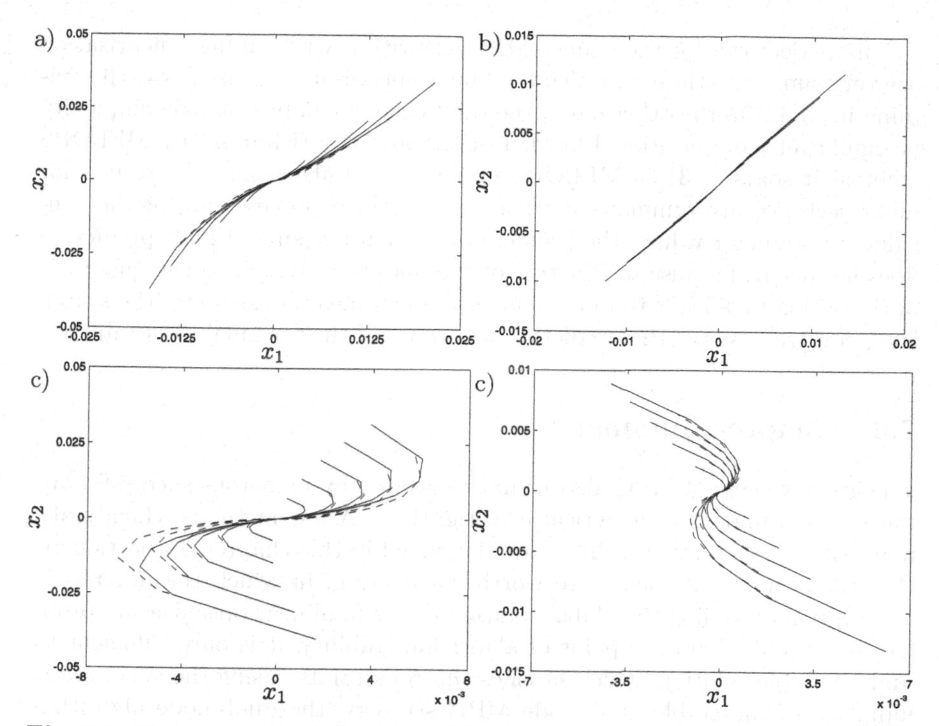

Fig. 7.3. The trajectories of the robust MPC algorithm with nonlinear optimisation (*solid line*) and of the robust MPC algorithm with on-line model linearisation and quadratic optimisation (*dashed line*) for different changes of the process parameters: a) $\delta_1 = 0.08$, $\delta_2 = 0.08$, b) $\delta_1 = 0.08$, $\delta_2 = -0.08$, c) $\delta_1 = -0.08$, $\delta_2 = 0.08$, d) $\delta_1 = -0.08$, $\delta_2 = -0.08$; all trajectories converge to the equilibrium point $(0, 0)$

the penalty term $V_{\mathrm{N}}(\hat{x}(k+N|k)) = \beta \|\hat{x}(k+N|k)\|_{\boldsymbol{P}}^2 \leq \alpha$ is used. In order to satisfy all assumptions of the robust MPC algorithm with the additional state constraints (p. 239), the following parameters are used: $\alpha_0 = 2$, $\alpha = 1$, $\beta = 1200$. It is possible to calculate the upper bound of the Lipschitz constant of the penalty term $V_{\mathrm{N}}(\hat{x}(k+N|k))$ in the following way

$$L_{\mathrm{V}} \geq 2 \max_{x \in \Omega_0} \|\boldsymbol{P}x\|_2 = 258.8954$$

From Eq. (7.46) it follows that model uncertainty is bounded by $\gamma \leq 0.0114$. assuming that $\delta_1 = \delta_2$, the maximal possible model uncertainty is $\delta_1 = \delta_2 = 0.0805$.

Fig. 7.3 depicts example trajectories of both compared robust MPC algorithms for different initial conditions. 4 scenarios of the process changes are assumed, they are described by the uncertainty parameters δ_1, δ_2. For all model changes, the simulations are carried out for the same initial conditions (imposed on the state vector) determined by the same values of the input variable at the steady-state, namely for: $u = \pm 0.01, \pm 0.02, \pm 0.03, \pm 0.04, \pm 0.05$. In all cases 20 iterations are simulated.

The trajectories of the robust MPC algorithm with on-line linearisation are very similar to those in which nonlinear optimisation is used at each sampling instant. On the other hand, the difference of computational complexity is significant: computational burden of the first algorithm is 9.07 MFLOPS whereas it soars to 31.54 MFLOPS in the second algorithm (the results for all trajectories are summarised). For the described process and for the considered parameter values, the problem of inaccurate (suboptimal) prediction does not occur, because satisfaction of the linear constraints in the quadratic optimisation task leads to fulfillment of the nonlinear constraints (the stabilising constraints over the prediction horizon and the terminal constraint).

7.6 Chapter Summary

The great success of MPC algorithms, measured by numerous successful applications, stimulates theoretical investigations. A few methods which make it possible to guarantee stability are described in this chapter. In particular, the suboptimal approaches are worth considering, in which the unrealistic assumption of finding the global solution of the nonlinear optimisation problem is relaxed. From the point of algorithm stability, it is only sufficient to find a feasible solution, which satisfies the constraints. Using the general formulation of the stable dual-mode MPC strategy, the dual-mode algorithm with successive on-line linearisation described in this chapter may be derived (its version for input-output models is described in [175]). Not only model, but also trajectory linearisation is possible. There are no limitations imposed on the model structure. Hence, the algorithm may use the neural models discussed in the previous parts of the book (of the input-output or state-space types) or some alternative model structures. A few chosen methods of enforcing robustness are discussed in the second part of this chapter. The robust MPC algorithm with the additional constraints is detailed because, analogously to the dual-mode MPC strategy, it is a suboptimal approach, in which feasibility, not optimality, is important for robustness. Next, the robust MPC algorithm with the additional constraints and on-line linearisation is discussed (its version for state-space models is described in [152]). The algorithm is universal, it does not restrict the structure of the model, which means that it may use the neural models discussed in this book or alternative structures.

It is necessary to emphasise the fact that stability and robustness investigations of MPC algorithms are currently continued in many scientific centres (robustness issues are more demanding). Although the progress made over the last two decades is huge, it seems that it is necessary to develop computationally efficient stable and robust MPC approaches. Quite frequently, because the algorithms presented in the literature are very complicated and they need on-line nonlinear optimisation, their application is practically impossible. The illustrative simulation results are usually concerned with very simple artificial academic dynamic systems. Simulation results of stable or robust MPC algorithms for real processes are rare [114, 60]. Similarly, there are very few works which describe applications of the stable or robust MPC algorithms to real processes. The rare examples include the chemical reactor [87, 90] and the solar power plant [42]. Relatively simple linear models and the robust min-max MPC algorithm are used in the majority of the cited works, the exception is the nonlinear Volterra model used in the algorithm described in [91, 60]. Thanks to a specific structure of the model, the robust MPC algorithm is not much more complicated than the rudimentary linear MPC strategy.

8

Cooperation between MPC Algorithms and Set-Point Optimisation Algorithms

This chapter is concerned with cooperation between the suboptimal MPC algorithms with on-line linearisation and set-point optimisation algorithms. At first, the classical multi-layer control system structure is discussed, the main disadvantage of which is the necessity of on-line nonlinear optimisation. Three control structures with on-line linearisation for set-point optimisation are presented next: the multi-layer structure with steady-state target optimisation, the integrated structure and the structure with predictive optimiser and constraint supervisor. Implementation details are given for three classes of neural models.

8.1 Classical Multi-Layer Control System Structure

The main objective of automatic control is usually of economic nature. Typically, the financial profit must be maximised. Of course, it is also necessary to take into account the existing product quality and safety requirements. That is why one may distinguish the following three partial control objectives:

a) safe process operation must be ensured (the frequency of emergency situations must be at the acceptable level),
b) key process outputs (e.g. product quality, pressure etc.) must be maintained within the necessary ranges,
c) effectiveness of process operation must be maximised, i.e. profit must be maximised or production cost must be minimised.

The above partial control objectives are also of the economic nature and they are motivated by the main control objective. Failures, which usually lead to abnormal situations, reduction of product quality, reduction of production level or even production breaks, are likely to badly affect the profit. Furthermore, all the constraints must be satisfied. On the one hand, when some key process variables are not within the necessary limits, the product is likely to be of low quality, which means that it is impossible to sell it for a good

M. Ławryńczuk, *Computationally Efficient Model Predictive Control Algorithms*, Studies in Systems, Decision and Control 3,
DOI: 10.1007/978-3-319-04229-9_8, © Springer International Publishing Switzerland 2014

price, which leads to serious economic losses. On the other hand, the better the product quality (e.g. purity in the distillation process), the higher the cost necessary for process operation. That is why the typical operating point is usually located close to the low (minimal) constraints of product quality, which is very efficient since such an approach gives acceptable product quality and minimises costs, but precise control is necessary. The control objectives must not neglect cooperation between different processes, e.g. in a refinery.

Taking into account the basic control objective (of economic nature) and the resulting partial control objectives, it is very straightforward to use the multi-layer (hierarchical) control system structure depicted in Fig. 8.1. Its unique feature is the fact that each layer plays a different role, corresponding to consecutive partial control objectives. Each layer has its own period of intervention. In general, decomposition of the control system significantly simplifies the design, because it makes it possible to define separately the task of each layer. Such an approach turns out to be very beneficial when the considered technological processes are complex. Furthermore, in many cases the decomposition approach is the only efficient method as the decentralised control system, although possible it theory, is likely to be very complex, which means high costs and unreliability.

The basic control layer is responsible for safe operation of the process. Unlike other layers, this layer has a direct access to input variables of the process. The basic control layer includes fault detection and isolation systems. Usually, the PID controllers, including the fuzzy PID controllers and the gain-scheduling PID structures, or the IMC controllers are used in this layer. Alternatively, some simplified MPC algorithms, i.e. based on linear models, typically explicit ones, may be applied. The period of intervention of the basic control layer is the shortest when compared to those of higher layers, it is equal to seconds or even shorter.

The second layer, i.e. the supervisory control layer (or the advanced control layer), has no direct access to the input variables of the process. The objective of this layer is to control key process outputs, its period of intervention is much longer than that of the basic layer, it is equal to dozens of seconds or, which is more typical, to minutes. Taking into account the necessity of constraints satisfaction and the fact that the majority of processes are multivariable, with strong cross-coupling and delays, the MPC technique is a natural choice for the supervisory control layer. For instance, the direct layer used in the control system of the high-pressure high-purity distillation column depicted in Fig. 2.17 is comprised of three PID controllers the objective of which is to stabilise the levels in reflux and bottom product tanks as well as the temperature on the controlled tray, respectively. The PID controllers have a direct access to the manipulated variables, which are: the bottom product stream, the reflux stream and the amount of heat delivered to the evaporator. From the perspective of the supervisory control layer, the distillation column has one manipulated variable (the reflux ratio r) and one controlled variable (impurity z of the product). The dynamic model used in

MPC algorithms should describe not only the distillation process itself, but also the direct PID controllers.

The third layer, i.e. the Local Steady-State Optimisation (LSSO) layer of the considered process, calculates on-line economically optimal set-points for the supervisory control layer in such a way that the production profit is maximised and the constraints are satisfied. For optimisation purposes the steady-state model is used, a dynamic model is used rarely. The period of intervention depends on variability of disturbances which move the optimal operating point. When the disturbances are slowly-varying, the typical period is equal to hours.

The last layer, the management layer, determines the operating conditions for the LSSO layer, i.e. the nature and the parameters of the minimised cost-function as well as the constraints. Many factors must be taken into account: the market conditions, the costs of raw materials, the possible cost of the product, the quality and safety norms. The period of intervention of this layer is quite long, equal to a few days or even longer. The models used for management are of the steady-state type.

Some modifications of the multi-layer structure are possible. Historically, only one control layer was used. Availability of advanced algorithms (typically MPC ones) and efficient hardware platforms resulted in emerging the supervisory control layer. On the other hand, the supervisory layer may be not used provided that the LSSO layer calculates the set-points for the direct control layer. Usually, it may be possible for only some process outputs, which is illustrated in Fig. 8.1. In order to simplify the discussion, in the following part of the text it is assumed that the LSSO layer is not connected to the direct control layer (such a situation is the most common in practice).

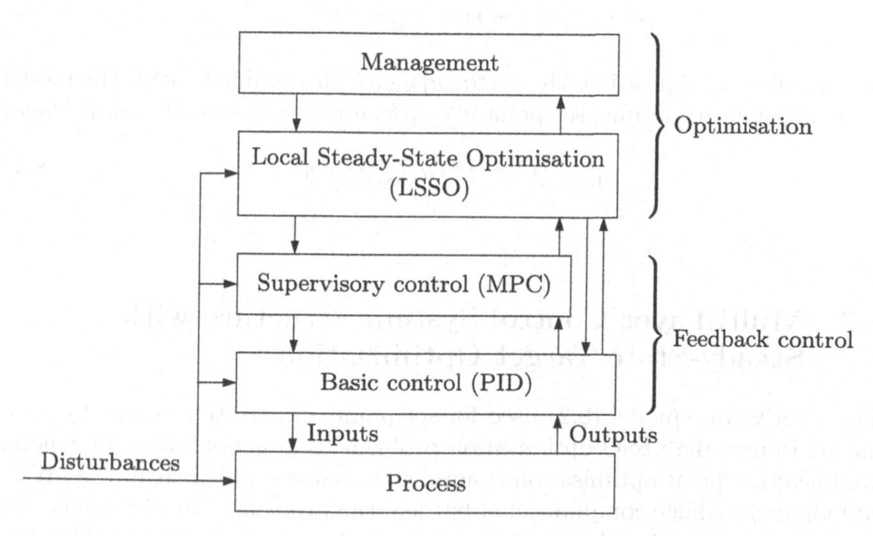

Fig. 8.1. The multi-layer control system structure

Taking into account that the production profit must be maximised and the constraints must be satisfied, the optimisation problem solved in the LSSO layer is

$$\begin{aligned}&\min_{u^{\mathrm{ss}}} \left\{ J_{\mathrm{E}} = c_{\mathrm{u}}^{\mathrm{T}} u^{\mathrm{ss}} - c_{\mathrm{y}}^{\mathrm{T}} y^{\mathrm{ss}} \right\}\\ &\text{subject to}\\ &u^{\min} \leq u^{\mathrm{ss}} \leq u^{\max}\\ &y^{\min} \leq y^{\mathrm{ss}} \leq y^{\max}\\ &y^{\mathrm{ss}} = f^{\mathrm{ss}}(u^{\mathrm{ss}}, h^{\mathrm{ss}})\end{aligned} \tag{8.1}$$

For simplification it is assumed that the constraints in the LSSO layer are the same as those in the supervisory control layer (in MPC). The vectors $c_{\mathrm{u}} = \left[c_{\mathrm{u},1} \ldots c_{\mathrm{u},n_{\mathrm{u}}}\right]^{\mathrm{T}}$, $c_{\mathrm{y}} = \left[c_{\mathrm{y},1} \ldots c_{\mathrm{y},n_{\mathrm{y}}}\right]^{\mathrm{T}}$ represent economic prices, the superscript "ss" denotes the steady state. For set-point optimisation the following steady-state model is used

$$y^{\mathrm{ss}} = f^{\mathrm{ss}}(u^{\mathrm{ss}}, h^{\mathrm{ss}}) \tag{8.2}$$

where $f^{\mathrm{ss}} \colon \mathbb{R}^{n_{\mathrm{u}}+n_{\mathrm{h}}} \to \mathbb{R}^{n_{\mathrm{y}}}$ and the current measurements (or estimations) of measurements are denoted by $h^{\mathrm{ss}} = h(k)$. Using the steady-state model, the LSSO optimisation problem (8.1) becomes

$$\begin{aligned}&\min_{u^{\mathrm{ss}}} \left\{ J_{\mathrm{E}} = c_{\mathrm{u}}^{\mathrm{T}} u^{\mathrm{ss}} - c_{\mathrm{y}}^{\mathrm{T}} (f^{\mathrm{ss}}(u^{\mathrm{ss}}, h^{\mathrm{ss}})) \right\}\\ &\text{subject to}\\ &u^{\min} \leq u^{\mathrm{ss}} \leq u^{\max}\\ &y^{\min} \leq f^{\mathrm{ss}}(u^{\mathrm{ss}}, h^{\mathrm{ss}}) \leq y^{\max}\end{aligned} \tag{8.3}$$

As a result of optimisation, the vector $u_{\mathrm{LSSO}}^{\mathrm{ss}}(k)$ is obtained. Next, the model is used to find the optimal set-point $y_{\mathrm{LSSO}}^{\mathrm{ss}}(k)$ for the supervisory control layer

$$y_{\mathrm{LSSO}}^{\mathrm{ss}}(k) = f^{\mathrm{ss}}(u_{\mathrm{LSSO}}^{\mathrm{ss}}(k), h^{\mathrm{ss}}) \tag{8.4}$$

8.2 Multi-Layer Control System Structure with Steady-State Target Optimisation

The steady-state model (8.2) used for set-point optimisation is usually nonlinear. Hence, the LSSO optimisation problem (8.3) is nonlinear. In general nonlinear set-point optimisation results in the same problem as in the MPC-NO algorithm: huge computational burden, the problem of local minima and no guarantee of finding the solution within the expected time period. The simplest approach is to use a linear MPC algorithm (which requires quadratic optimisation) and to lengthen the period of intervention of the LSSO layer. Although the set-point optimisation task is nonlinear, it is repeated quite infrequently, e.g. every few hours or a dozen of hours. Unfortunately, the classical control system structure with low frequency of intervention of the set-point optimisation layer may give satisfying results provided that disturbances are slowly-varying (when compared to the dynamics of the process). Very often disturbances (flow rates, properties of feed and energy streams etc.) vary significantly and not much slower than the dynamics of the process. In such cases operation in the classical structure with frequency of set-point optimisation much lower than that of MPC may result in economically not optimal operating points and in a significant loss of economic effectiveness.

Ideally, it would be beneficial to calculate the set-point frequently, e.g. as frequently as the supervisory control layer is activated. The simplest approach is to use for set-point optimisation a constant linear steady-state model

$$y^{\mathrm{ss}} = y^0(k+N|k) + \boldsymbol{G}^{\mathrm{ss}}(u^{\mathrm{ss}} - u(k-1)) \tag{8.5}$$

where $\boldsymbol{G}^{\mathrm{ss}}$ denotes the steady-state gain matrix derived from the linear dynamic model used in MPC. Such an approach leads to the Steady-State Target Optimisation (SSTO) control system structure depicted in Fig. 8.2. The SSTO layer is activated as frequently as the MPC algorithm. From the linear model (8.5) and from the LSSO optimisation problem (8.3) one obtains the SSTO task

$$\begin{aligned}
&\min_{u^{\mathrm{ss}}} \left\{ J_{\mathrm{E}} = c_{\mathrm{u}}^{\mathrm{T}} u^{\mathrm{ss}} - c_{\mathrm{y}}^{\mathrm{T}} (y^0(k+N|k) + \boldsymbol{G}^{\mathrm{ss}}(u^{\mathrm{ss}} - u(k-1))) \right\} \\
&\text{subject to} \\
&u^{\min} \leq u^{\mathrm{ss}} \leq u^{\max} \\
&y^{\min} \leq y^0(k+N|k) + \boldsymbol{G}^{\mathrm{ss}}(u^{\mathrm{ss}} - u(k-1)) \leq y^{\max}
\end{aligned}$$

The above problem is of a linear optimisation type (the cost-function and the constraints are linear). Of course, the LSSO layer still solves the nonlinear optimisation problem (for this purpose the nonlinear model is used), but it is done much less frequently in comparison with the SSTO layer (and MPC). In order to guarantee that the feasible set is not empty, it is better to solve the following SSTO optimisation problem with soft output constraints

$$\begin{aligned}
&\min_{\substack{u^{\mathrm{ss}} \\ \tilde{\varepsilon}^{\min},\, \tilde{\varepsilon}^{\max}}} \Big\{ J_{\mathrm{E}} = c_{\mathrm{u}}^{\mathrm{T}} u^{\mathrm{ss}} - c_{\mathrm{y}}^{\mathrm{T}} (y^0(k+N|k) + \boldsymbol{G}^{\mathrm{ss}}(u^{\mathrm{ss}} - u(k-1))) \\
&\qquad\qquad + (\tilde{\rho}^{\min})^{\mathrm{T}} \tilde{\varepsilon}^{\min} + (\tilde{\rho}^{\max})^{\mathrm{T}} \tilde{\varepsilon}^{\max} \Big\} \\
&\text{subject to} \\
&u^{\min} \leq u^{\mathrm{ss}} \leq u^{\max} \\
&y^{\min} - \tilde{\varepsilon}^{\min} \leq y^0(k+N|k) + \boldsymbol{G}^{\mathrm{ss}}(u^{\mathrm{ss}} - u(k-1)) \leq y^{\max} + \tilde{\varepsilon}^{\max} \\
&\tilde{\varepsilon}^{\min} \geq 0, \ \tilde{\varepsilon}^{\max} \geq 0
\end{aligned}$$

where the vectors $\tilde{\varepsilon}^{\min}$, $\tilde{\varepsilon}^{\max} \in \mathbb{R}^{n_y}$ determine the degree of constraints violation and the vectors $\tilde{\rho}^{\min}$, $\tilde{\rho}^{\max} > 0 \in \mathbb{R}^{n_y}$ denote the penalties.

Unfortunately, in case of nonlinear processes the application of a linear MPC algorithm and a linear model in the SSTO layer may give economically and numerically wrong operating points. For such processes a modification of the multi-layer control system structure with the SSTO layer may be used, namely the structure with the Adaptive Steady-State Target Optimisation (ASSTO), the details of which is depicted in Fig. 8.3. The full (i.e. not simplified) nonlinear steady-state model is used in the LSSO layer. Although it requires nonlinear optimisation, it is activated infrequently. At each sampling instant of the suboptimal MPC algorithm the dynamic neural model is linearised and the steady-state model is also linearised (the same steady-state model is used in the LSSO and SSTO layers). Any computationally efficient suboptimal MPC algorithm may be used, the notation in Fig. 8.3 is concerned with the MPC-NPL one. In contrast to the simple control structure with the SSTO layer (in which a nonlinear model is used in the LSSO layer whereas a linear one, corresponding the dynamic model used in MPC, is applied in the SSTO layer), in the discussed structure with the ASSTO layer the same steady-state model is used (or its linear approximation). For simplicity, some additional signals (measurements, including disturbances) are not denoted in Fig. 8.3, but they are necessary for linearisation and optimisation.

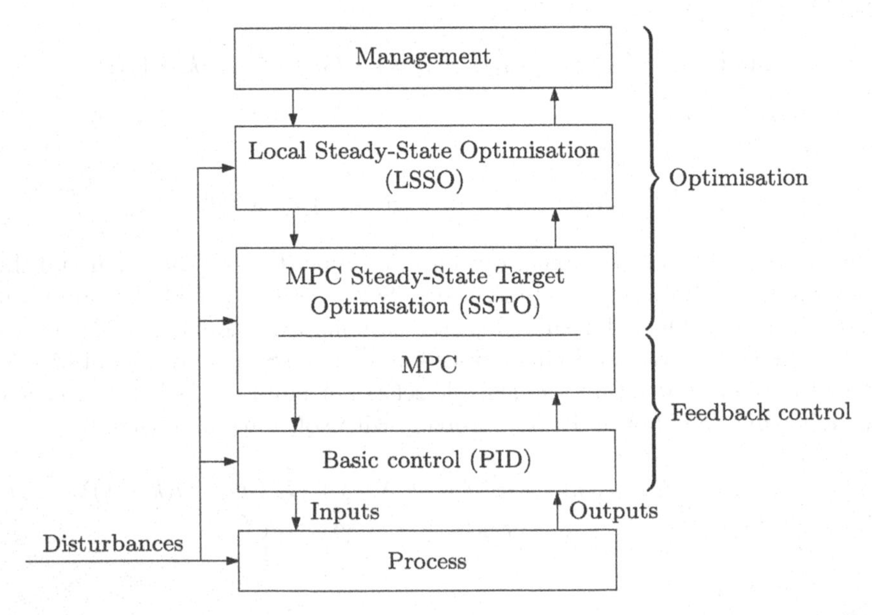

Fig. 8.2. The multi-layer control system structure with Steady-State Target Optimisation (SSTO)

Using the Taylor series expansion formula (2.25), the linear approximation of the nonlinear steady-state model (8.2) is

$$\begin{aligned} y^{\mathrm{ss}} &= f^{\mathrm{ss}}(u^{\mathrm{ss}}, h^{\mathrm{ss}})\Big|_{\substack{u^{\mathrm{ss}}=u(k-1)\\ h^{\mathrm{ss}}=h(k)}} + \boldsymbol{H}^{\mathrm{ss}}(k)(u^{\mathrm{ss}} - u(k-1)) \\ &= f^{\mathrm{ss}}(u(k-1), h(k)) + \boldsymbol{H}^{\mathrm{ss}}(k)(u^{\mathrm{ss}} - u(k-1)) \end{aligned} \tag{8.6}$$

where the current operating point is defined by the most recent (available) signals $u(k-1)$ and $h(k)$. The matrix

$$\boldsymbol{H}^{\mathrm{ss}}(k) = \frac{\mathrm{d}f^{\mathrm{ss}}(u^{\mathrm{ss}}, h^{\mathrm{ss}})}{\mathrm{d}u^{\mathrm{ss}}}\Bigg|_{\substack{u^{\mathrm{ss}}=u(k-1)\\ h^{\mathrm{ss}}=h(k)}} = \frac{\mathrm{d}f^{\mathrm{ss}}(u(k-1), h(k))}{\mathrm{d}u(k-1)}$$

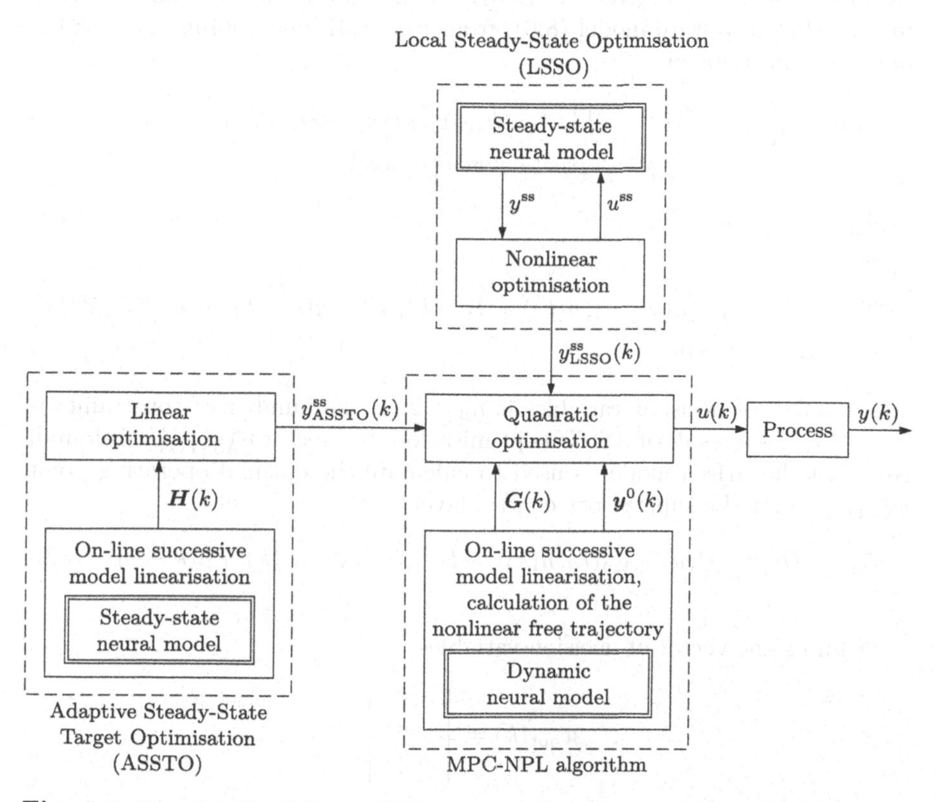

Fig. 8.3. The details of the multi-layer control system structure with Adaptive Steady-State Target Optimisation (ASSTO); the same steady-state model is used in LSSO and SSTO layers

of dimensionality $n_y \times n_u$, consists of partial derivatives of the nonlinear model function f^{ss} with respect to the inputs

$$\boldsymbol{H}^{ss}(k) = \begin{bmatrix} \frac{\partial f_1^{ss}(u(k-1),h(k))}{\partial u_1(k-1)} & \cdots & \frac{\partial f_1^{ss}(u(k-1),h(k))}{\partial u_{n_u}(k-1)} \\ \vdots & \ddots & \vdots \\ \frac{\partial f_{n_y}^{ss}(u(k-1),h(k))}{\partial u_1(k-1)} & \cdots & \frac{\partial f_{n_y}^{ss}(u(k-1),h(k))}{\partial u_{n_u}(k-1)} \end{bmatrix} \tag{8.7}$$

It is interesting to notice that both the inputs u^{ss} and the disturbances h^{ss} are the arguments of the nonlinear model (8.2) whereas its linear approximation (8.6) has only the inputs as the arguments; the current values of the disturbances are the parameters which influence the values of the parameters of the linearised model. Similarly, the linear approximation of the nonlinear dynamic model (2.1) is given by (2.26)). Using soft output constraints and the linearised steady-state model (8.6), from Eq. (8.3), one obtains the ASSTO optimisation problem

$$\begin{aligned} \min_{\substack{u^{ss} \\ \tilde{\varepsilon}^{min},\ \tilde{\varepsilon}^{max}}} \ & \Big\{ J_E = c_u^T u^{ss} - c_y^T (f^{ss}(u(k-1),h(k)) + \boldsymbol{H}^{ss}(k)(u^{ss} - u(k-1))) \\ & \quad + (\tilde{\rho}^{min})^T \tilde{\varepsilon}^{min} + (\tilde{\rho}^{max})^T \tilde{\varepsilon}^{max} \Big\} \\ \text{subject to} & \\ & u^{min} \leq u^{ss} \leq u^{max} \\ & y^{min} - \tilde{\varepsilon}^{min} \leq f^{ss}(u(k-1),h(k)) + \boldsymbol{H}^{ss}(k)(u^{ss} - u(k-1)) \leq y^{max} + \tilde{\varepsilon}^{max} \\ & \tilde{\varepsilon}^{min} \geq 0,\ \tilde{\varepsilon}^{max} \geq 0 \end{aligned} \tag{8.8}$$

The number of decision variables is $n_u + 2n_y$, the number of constraints is $2n_u + 4n_y$. As a result of ASSTO optimisation, the vector $u^{ss}_{ASSTO}(k)$ is found. Next, the linearised model is used to calculate the optimal operating point $y^{ss}_{ASSTO}(k)$ for the supervisory control layer

$$y^{ss}_{ASSTO}(k) = f^{ss}(u^{ss}_{ASSTO}(k), h(k)) + \boldsymbol{H}^{ss}(k)(u^{ss}_{ASSTO}(k) - u(k-1)) \tag{8.9}$$

Defining the vector of decision variables

$$\boldsymbol{x}_{opt}(k) = \begin{bmatrix} u^{ss} \\ \tilde{\varepsilon}^{min} \\ \tilde{\varepsilon}^{max} \end{bmatrix}$$

and the following auxiliary matrices

$$\begin{aligned} \boldsymbol{N}_1 &= \begin{bmatrix} \boldsymbol{I}_{n_u \times n_u} & \boldsymbol{0}_{n_u \times 2n_y} \end{bmatrix} \\ \boldsymbol{N}_2 &= \begin{bmatrix} \boldsymbol{0}_{n_y \times n_u} & \boldsymbol{I}_{n_y \times n_y} & \boldsymbol{0}_{n_y \times n_y} \end{bmatrix} \\ \boldsymbol{N}_3 &= \begin{bmatrix} \boldsymbol{0}_{n_y \times (n_u + n_y)} & \boldsymbol{I}_{n_y \times n_y} \end{bmatrix} \end{aligned}$$

the ASSTO optimisation problem (8.8) may be transformed to the classical linear optimisation problem [226]

$$\min_{\boldsymbol{x}^{\mathrm{opt}}(k)} \left\{ \boldsymbol{f}_{\mathrm{opt}}^{\mathrm{T}}(k) \boldsymbol{x}_{\mathrm{opt}}(k) \right\}$$
$$\text{subject to}$$
$$\boldsymbol{A}_{\mathrm{opt}}(k) \boldsymbol{x}_{\mathrm{opt}}(k) \leq \boldsymbol{b}_{\mathrm{opt}}(k)$$

which may be solved by the simplex method or by the interior point approach [226]. The minimised cost-function is defined by

$$\boldsymbol{f}_{\mathrm{opt}}(k) = c_{\mathrm{u}}^{\mathrm{T}} \boldsymbol{N}_1 - c_{\mathrm{y}}^{\mathrm{T}} \boldsymbol{H}^{\mathrm{ss}}(k) \boldsymbol{N}_1 + (\tilde{\rho}^{\min})^{\mathrm{T}} \boldsymbol{N}_2 + (\tilde{\rho}^{\max})^{\mathrm{T}} \boldsymbol{N}_3$$

whereas the constraints by

$$\boldsymbol{A}_{\mathrm{opt}}(k) = \begin{bmatrix} -\boldsymbol{N}_1 \\ \boldsymbol{N}_1 \\ -\boldsymbol{H}^{\mathrm{ss}}(k)\boldsymbol{N}_1 - \boldsymbol{N}_2 \\ \boldsymbol{H}^{\mathrm{ss}}(k)\boldsymbol{N}_1 - \boldsymbol{N}_3 \\ -\boldsymbol{N}_2 \\ -\boldsymbol{N}_3 \end{bmatrix}$$

and

$$\boldsymbol{b}_{\mathrm{opt}}(k) = \begin{bmatrix} -u^{\min} \\ u^{\max} \\ -y^{\min} - \boldsymbol{H}^{\mathrm{ss}}(k)u(k-1) + f^{\mathrm{ss}}(u(k-1), h(k)) \\ y^{\max} + \boldsymbol{H}^{\mathrm{ss}}(k)u(k-1) - f^{\mathrm{ss}}(u(k-1), h(k)) \\ 0_{n_{\mathrm{y}}} \\ 0_{n_{\mathrm{y}}} \end{bmatrix}$$

At each sampling instant (algorithm iteration k) the following steps are repeated on-line:

1. The local linear approximation (8.6) of the nonlinear (e.g. neural) steady-state model is obtained for the current operating point of the process, i.e. the matrix $\boldsymbol{H}^{\mathrm{ss}}(k)$ given by Eq. (8.7) is calculated.
2. The ASSTO linear optimisation problem (8.8) is solved to calculate the set-point $u^{\mathrm{ss}}_{\mathrm{ASSTO}}(k)$. Using the linearised steady-state model, the optimal output set-point $y^{\mathrm{ss}}_{\mathrm{ASSTO}}(k)$ is found from Eq. (8.9).
3. If the calculated set point is verified, the nonlinear LSSO optimisation problem (8.3) is solved to calculate the set-point $u^{\mathrm{ss}}_{\mathrm{LSSO}}(k)$. Using the nonlinear steady-state model, the optimal output set-point $y^{\mathrm{ss}}_{\mathrm{LSSO}}(k)$ is found from Eq. (8.4).
4. The output set-point $y^{\mathrm{ss}}_{\mathrm{ASSTO}}(k)$ or $y^{\mathrm{ss}}_{\mathrm{LSSO}}(k)$ is passed to MPC.
5. The nonlinear dynamic model is used in the suboptimal MPC algorithm (e.g. in the MPC-NPL algorithm it is linearised at the current operating point, the unmeasured disturbances are estimated and the free trajectory is determined).

6. The quadratic optimisation MPC problem is solved to calculate the future control increments vector $\triangle \boldsymbol{u}(k)$.
7. The first n_{u} elements of the determined sequence are applied to the process, i.e. $u(k) = \triangle u(k|k) + u(k-1)$.
8. The iteration of the algorithm is increased, i.e. $k\colon = k+1$, the algorithm goes to step 1.

A linear optimisation problem (in the ASSTO layer) and a quadratic optimisation one (in MPC) are solved at each sampling instant whereas nonlinear optimisation (in the LSSO layer) is used infrequently for verification. In practice, the LSSO layer may be never activated.

It is possible to use a second-order approximation of the nonlinear steady-state model in the ASSTO layer. Provided that such an approximation is used in the minimised cost-function whereas a linear approximation is used in the output constraints, one obtains a quadratic optimisation task. Unfortunately, the obvious inconsistency is likely to give the results far from optimality, because typical operating points are usually located on the border of the feasible set, determined by the constraints.

Implementation for Neural Model

Analogously as it is done in the dynamic model used in MPC, it is assumed that the steady-state neural model of the multi-input multi-output process consists of n_{y} multi-input single-output models. The structure of the model of the mth output ($m = 1, \ldots, n_{\mathrm{y}}$) is depicted in Fig. 8.4. Each network has $n_{\mathrm{u}} + n_{\mathrm{h}}$ inputs, $K^{m,\mathrm{ss}}$ hidden nodes with the nonlinear transfer function $\varphi\colon \mathbb{R} \rightarrow \mathbb{R}$ and one output y_m^{ss}. The weights of the first layer are denoted by $w_{i,j}^{1,m,\mathrm{ss}}$, where $i = 1, \ldots, K^{m,\mathrm{ss}}$, $j = 0, \ldots, n_{\mathrm{u}} + n_{\mathrm{h}}$, $m = 0, \ldots, n_{\mathrm{y}}$, the weights of the second layer are denoted by $w_i^{2,m,\mathrm{ss}}$, where $i = 0, \ldots, K^{m,\mathrm{ss}}$, $m = 0, \ldots, n_{\mathrm{y}}$. The steady-state neural model of the mth output may be described by the following equation

$$y_m^{\mathrm{ss}} = f_m^{\mathrm{ss}}(u^{\mathrm{ss}}, h^{\mathrm{ss}}) = w_0^{2,m,\mathrm{ss}} + \sum_{i=1}^{K^{m,\mathrm{ss}}} w_i^{2,m,\mathrm{ss}} \varphi(z_i^{m,\mathrm{ss}}(u^{\mathrm{ss}}, h^{\mathrm{ss}})) \tag{8.10}$$

where the sum of the input signals connected to the ith hidden ($i = 1, \ldots, K^{m,\mathrm{ss}}$) is

$$z_i^{m,\mathrm{ss}}(u^{\mathrm{ss}}, h^{\mathrm{ss}}) = w_{i,0}^{1,m,\mathrm{ss}} + \sum_{n=1}^{n_{\mathrm{u}}} w_{i,n}^{1,m,\mathrm{ss}} u_n^{\mathrm{ss}} + \sum_{n=1}^{n_{\mathrm{h}}} w_{i,n_{\mathrm{u}}+n}^{1,m,\mathrm{ss}} h_n^{\mathrm{ss}} \tag{8.11}$$

From Eqs. (8.10) and (8.11), the entries of the matrix $\boldsymbol{H}^{\mathrm{ss}}(k)$, the structure of which is given by Eq. (8.7), are calculated from

$$\frac{\partial f_m^{\mathrm{ss}}(u(k-1), h(k))}{\mathrm{d}u_n(k-1)} = \sum_{i=1}^{K^{m,\mathrm{ss}}} w_i^{2,m,\mathrm{ss}} \frac{\mathrm{d}\varphi(z_i^{m,\mathrm{ss}}(u(k-1), h(k)))}{\mathrm{d}z_i^{m,\mathrm{ss}}(u(k-1), h(k))} w_{i,n}^{1,m,\mathrm{ss}} \tag{8.12}$$

for $m = 1, \ldots, n_{\mathrm{y}}$, $n = 1, \ldots, n_{\mathrm{u}}$. Using Eqs. (8.6), (8.10), (8.11) and (8.12), the linearised model of the mth output is

$$y_m^{\mathrm{ss}} = w_0^{2,m,\mathrm{ss}} + \sum_{i=1}^{K^{m,\mathrm{ss}}} w_i^{2,m,\mathrm{ss}} \varphi(z_i^{m,\mathrm{ss}}(u(k-1), h(k-1)))$$
$$+ \sum_{n=1}^{n_{\mathrm{u}}} \sum_{i=1}^{K^{m,\mathrm{ss}}} w_i^{2,m,\mathrm{ss}} \frac{\mathrm{d}\varphi(z_i^{m,\mathrm{ss}}(u(k-1), h(k)))}{\mathrm{d}z_i^{m,\mathrm{ss}}(u(k-1), h(k))} w_{i,n}^{1,m,\mathrm{ss}}(u_n^{\mathrm{ss}} - u_n(k-1))$$

The steady-state model is used in nonlinear LSSO optimisation and in linear ASSTO optimisation, the dynamic model is only used in MPC. An interesting alternative is to use only one dynamic cascade model of Hammerstein or Wiener structure. Such a model is used in MPC. For the current operating point the dynamic model is also used to find on-line the corresponding steady-state model and its linear approximation. These steady-state models are next used in LSSO and ASSTO layers. Thanks to a very specific structure of both mentioned cascade models, derivation of the steady-state model is relatively easy.

Implementation for Neural Hammerstein Model

The Hammerstein structure with n_{v} separate steady-state parts depicted in Fig. 3.1b is considered. From Eq. (3.9), in the steady-state one has

$$y_m^{\mathrm{ss}} = \sigma \sum_{s=1}^{n_{\mathrm{v}}} \sum_{l=1}^{n_{\mathrm{B}}} b_l^{m,s} \left[w_0^{2,s} + \sum_{i=1}^{K^s} w_i^{2,s} \varphi(z_i^s(u^{\mathrm{ss}}, h^{\mathrm{ss}})) \right] \tag{8.13}$$

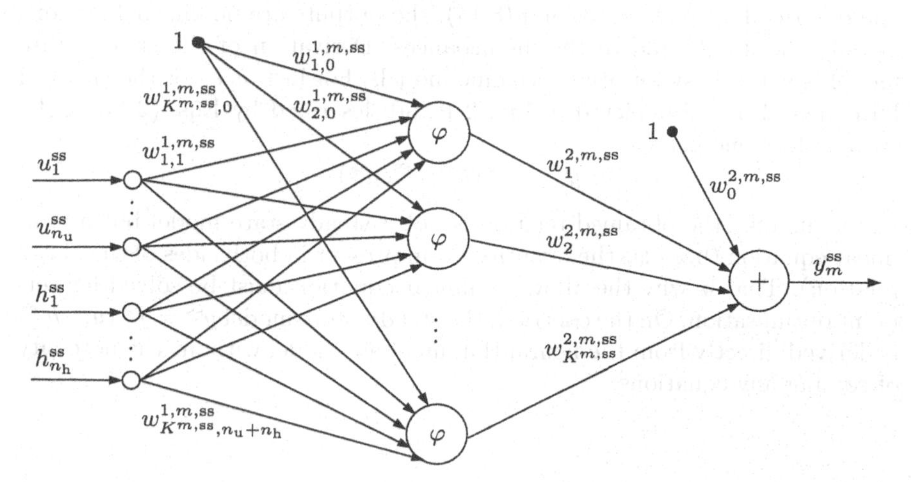

Fig. 8.4. The structure of the steady-state multi-input single-output neural model of the mth output, $m = 1, \ldots, n_{\mathrm{y}}$

where $\sigma = 1/(1+\sum_{l=1}^{n_\mathrm{A}} a_l^m)$ and

$$z_i^s(u^{\mathrm{ss}}, h^{\mathrm{ss}}) = w_{i,0}^{1,s} + \sum_{j=1}^{n_\mathrm{u}} w_{i,j}^{1,s} u_j^{\mathrm{ss}} + \sum_{j=1}^{n_\mathrm{h}} w_{i,n_\mathrm{u}+j}^{1,s} h_j^{\mathrm{ss}} \tag{8.14}$$

The nonlinear steady-state model is used in LSSO optimisation, which is activated infrequently. Conversely, at each sampling instant of the MPC algorithm, the ASSTO layer is used. For this purpose the linearised steady-state model is necessary. From Eqs. (8.13) and (8.14), the entries of the matrix $\boldsymbol{H}^{\mathrm{ss}}(k)$, the structure of which is given by Eq. (8.7), are calculated from

$$\frac{\partial f_m^{\mathrm{ss}}(u(k-1), h(k))}{\mathrm{d}u_n(k-1)} = \sigma \sum_{s=1}^{n_\mathrm{v}} \sum_{l=1}^{n_\mathrm{B}} b_l^{m,s} \sum_{i=1}^{K^s} w_i^{2,s} \frac{\mathrm{d}\varphi(z_i^s(u(k-1), h(k)))}{\mathrm{d}z_i^s(u(k-1), h(k))} \times w_{i,n}^{1,s} \tag{8.15}$$

for all $m = 1, \ldots, n_\mathrm{y}$, $n = 1, \ldots, n_\mathrm{u}$. From Eqs. (8.6), (8.13), (8.14) and (8.15), the linearised model of the mth output may be expressed in the following way

$$\begin{aligned} y_m^{\mathrm{ss}} &= \sigma \sum_{s=1}^{n_\mathrm{v}} \sum_{l=1}^{n_\mathrm{B}} b_l^{m,s} \left[w_0^{2,s} + \sum_{i=1}^{K^s} w_i^{2,s} \varphi(z_i^s(u(k-1), h(k))) \right] \\ &+ \sum_{n=1}^{n_\mathrm{u}} \sigma \sum_{s=1}^{n_\mathrm{v}} \sum_{l=1}^{n_\mathrm{B}} b_l^{m,s} \sum_{i=1}^{K^s} w_i^{2,s} \frac{\mathrm{d}\varphi(z_i^s(u(k-1), h(k)))}{\mathrm{d}z_i^s(u(k-1), h(k))} w_{i,n}^{1,s} (u_n^{\mathrm{ss}} - u_n(k-1)) \end{aligned}$$

It is quite easy to derive the steady-state description from the Hammerstein model, because in the dynamic model (3.9) and in the resulting steady-state one described by Eqs. (8.13) and (8.14), the outputs are nonlinear functions of only the inputs and of the disturbances. Derivation of the steady-state model is not so easy for other dynamic models. For instance, for the classical DLP neural model depicted in Fig. 2.1 and described by Eqs. (2.1), in the steady-state one has

$$y^{\mathrm{ss}} = f^{\mathrm{ss}}(u^{\mathrm{ss}}, h^{\mathrm{ss}}, y^{\mathrm{ss}})$$

Unfortunately, the obtained relation is not a steady-state model but a nonlinear equation (because the quantity y^{ss} in present in both sides of the above relation). That is why the above nonlinear equation must be solved for set-point optimisation. On the contrary, the steady-state model $y^{\mathrm{ss}} = f^{\mathrm{ss}}(u^{\mathrm{ss}}, h^{\mathrm{ss}})$ is derived directly from the neural Hammerstein model without the necessity of solving any equations.

Implementation for Neural Wiener Model

The Wiener structure with n_{y} separate steady-state parts depicted in Fig. 3.4b is considered. From Eq. (3.23), in the steady-state one has

$$y_m^{\mathrm{ss}} = w_0^{2,m} + \sum_{i=1}^{K^m} w_i^{2,m}\varphi\left(w_{i,0}^{1,m} + \sum_{s=1}^{n_{\mathrm{v}}} w_{i,s}^{1,m}\left[\sum_{n=1}^{n_{\mathrm{u}}}\sum_{l=1}^{n_{\mathrm{B}}} b_l^{s,n}u_n^{\mathrm{ss}} + \sum_{n=1}^{n_{\mathrm{h}}}\sum_{l=1}^{n_{\mathrm{B}}} b_l^{s,n_{\mathrm{u}}+n}h_n^{\mathrm{ss}} - \sum_{l=1}^{n_{\mathrm{A}}} a_l^s v_s^{\mathrm{ss}}\right]\right) \tag{8.16}$$

where, from Eq. (3.17), one finds the vector

$$v^{\mathrm{ss}} = \boldsymbol{G}^{\mathrm{ss}}\begin{bmatrix} u^{\mathrm{ss}} \\ h^{\mathrm{ss}} \end{bmatrix}$$

Because $\boldsymbol{G}^{\mathrm{ss}} = \boldsymbol{A}^{-1}(1)\boldsymbol{B}(1)$, it is true that

$$v_s^{\mathrm{ss}} = \sum_{n=1}^{n_{\mathrm{u}}} g_{s,n}^{\mathrm{ss}}u_n^{\mathrm{ss}} + \sum_{n=1}^{n_{\mathrm{h}}} g_{s,n_{\mathrm{u}}+n}^{\mathrm{ss}}h_n^{\mathrm{ss}} \tag{8.17}$$

From Eqs. (8.16) and (8.17), the steady-state model is

$$y_m^{\mathrm{ss}} = w_0^{2,m} + \sum_{i=1}^{K^m} w_i^{2,m}\varphi(z_i^m(u^{\mathrm{ss}}, h^{\mathrm{ss}})) \tag{8.18}$$

where

$$z_i^m(u^{\mathrm{ss}}, h^{\mathrm{ss}}) = w_{i,0}^{1,m} + \sum_{s=1}^{n_{\mathrm{v}}} w_{i,s}^{1,m}\left[\sum_{n=1}^{n_{\mathrm{u}}}\left\{\sum_{l=1}^{n_{\mathrm{B}}} b_l^{s,n} - \sum_{l=1}^{n_{\mathrm{A}}} a_l^s g_{s,n}^{\mathrm{ss}}\right\}u_n^{\mathrm{ss}} + \sum_{n=1}^{n_{\mathrm{h}}}\left\{\sum_{l=1}^{n_{\mathrm{B}}} b_l^{s,n_{\mathrm{u}}+n} - \sum_{l=1}^{n_{\mathrm{A}}} a_l^s g_{s,n_{\mathrm{u}}+n}^{\mathrm{ss}}\right\}h_n^{\mathrm{ss}}\right] \tag{8.19}$$

It is interesting to notice that in the obtained steady-state model, similarly to the one derived from the dynamic Hammerstein structure (Eqs. (8.13) and (8.14)), the consecutive outputs are nonlinear functions of only the inputs and the disturbances, the outputs and the auxiliary variables are not present in the right side of the model.

From Eqs. (8.18) and (8.19), the entries of the matrix $\boldsymbol{H}^{\mathrm{ss}}(k)$, the structure of which is given by Eq. (8.7), are calculated from

$$\frac{\partial f_m^{\mathrm{ss}}(u(k-1), h(k))}{\mathrm{d}u_n(k-1)} = \sum_{i=1}^{K^m} w_i^{2,m}\frac{\mathrm{d}\varphi(z_i^m(u(k-1), h(k)))}{\mathrm{d}z_i^m(u(k-1), h(k))} \times\left[\sum_{l=1}^{n_{\mathrm{B}}} b_l^{s,n} - \sum_{l=1}^{n_{\mathrm{A}}} a_l^s g_{s,n}^{\mathrm{ss}}\right] \tag{8.20}$$

for all $m = 1, \ldots, n_{\mathrm{y}}$, $n = 1, \ldots, n_{\mathrm{u}}$. From Eqs. (8.6), (8.18), (8.19) and (8.20), the linearised model of the mth output is

$$\begin{aligned} y_m^{\mathrm{ss}} &= w_0^{2,m} + \sum_{i=1}^{K^m} w_i^{2,m} \varphi(z_i^m(u(k-1), h(k))) \\ &+ \sum_{n=1}^{n_{\mathrm{u}}} \sum_{i=1}^{K^m} w_i^{2,m} \frac{\mathrm{d}\varphi(z_i^m(u(k-1), h(k)))}{\mathrm{d}z_i^m(u(k-1), h(k))} \left[\sum_{l=1}^{n_{\mathrm{B}}} b_l^{s,n} - \sum_{l=1}^{n_{\mathrm{A}}} a_l^s g_{s,n}^{\mathrm{ss}} \right] \\ &\times (u_n^{\mathrm{ss}} - u_n(k-1)) \end{aligned}$$

8.3 Integrated Control System Structure

Two separate optimisation problems are solved with the same period of intervention in the multi-layer structure with the ASSTO layer: a linear ASSTO one and a quadratic MPC one. It is also possible to integrate both optimisation tasks. Taking into account the MPC-NPL optimisation problem (2.36), although one may also use any other suboptimal MPC algorithm, and the ASSTO optimisation task (8.8), one obtains the quadratic optimisation problem

$$\begin{aligned} \min_{\substack{\triangle \boldsymbol{u}(k) \\ \boldsymbol{\varepsilon}^{\min}(k),\ \boldsymbol{\varepsilon}^{\max}(k) \\ u^{\mathrm{ss}},\ \tilde{\varepsilon}^{\min},\ \tilde{\varepsilon}^{\max}}} \Big\{ J(k) + \alpha J_{\mathrm{E}} = & \left\| \boldsymbol{y}^{\mathrm{ss}}(k) - \boldsymbol{G}(k)\triangle\boldsymbol{u}(k) - \boldsymbol{y}^0(k) \right\|_{\boldsymbol{M}}^2 + \left\| \triangle\boldsymbol{u}(k) \right\|_{\boldsymbol{\Lambda}}^2 \\ & + \rho^{\min} \left\| \boldsymbol{\varepsilon}^{\min}(k) \right\|^2 + \rho^{\max} \left\| \boldsymbol{\varepsilon}^{\max}(k) \right\|^2 \\ & + \alpha(c_{\mathrm{u}}^{\mathrm{T}} u^{\mathrm{ss}} - c_{\mathrm{y}}^{\mathrm{T}} (f^{\mathrm{ss}}(u(k-1), h(k)) \\ & \qquad + \boldsymbol{H}^{\mathrm{ss}}(k)(u^{\mathrm{ss}} - u(k-1)))) \\ & + (\tilde{\rho}^{\min})^{\mathrm{T}} \tilde{\varepsilon}^{\min} + (\tilde{\rho}^{\max})^{\mathrm{T}} \tilde{\varepsilon}^{\max} \Big\} \end{aligned}$$

subject to (8.21)

$$\begin{aligned} & \boldsymbol{u}^{\min} \leq \boldsymbol{J}\triangle\boldsymbol{u}(k) + \boldsymbol{u}(k-1) \leq \boldsymbol{u}^{\max} \\ & -\triangle\boldsymbol{u}^{\max} \leq \triangle\boldsymbol{u}(k) \leq \triangle\boldsymbol{u}^{\max} \\ & \boldsymbol{y}^{\min} - \boldsymbol{\varepsilon}^{\min}(k) \leq \boldsymbol{G}(k)\triangle\boldsymbol{u}(k) + \boldsymbol{y}^0(k) \leq \boldsymbol{y}^{\max} + \boldsymbol{\varepsilon}^{\max}(k) \\ & \boldsymbol{\varepsilon}^{\min}(k) \geq 0,\ \boldsymbol{\varepsilon}^{\max}(k) \geq 0 \\ & u^{\min} \leq u^{\mathrm{ss}} \leq u^{\max} \\ & y^{\min} - \tilde{\varepsilon}^{\min} \leq f^{\mathrm{ss}}(u(k-1), h(k)) + \boldsymbol{H}^{\mathrm{ss}}(k)(u^{\mathrm{ss}} - u(k-1)) \leq y^{\max} + \tilde{\varepsilon}^{\max} \\ & \tilde{\varepsilon}^{\min} \geq 0,\ \tilde{\varepsilon}^{\max} \geq 0 \end{aligned}$$

where the vector

$$\boldsymbol{y}^{\mathrm{ss}}(k) = \begin{bmatrix} y^{\mathrm{ss}}(k) \\ \vdots \\ y^{\mathrm{ss}}(k) \end{bmatrix}$$

has length $n_y N$ and $\alpha > 0$ is an auxiliary coefficient. The number of decision variables is $n_u N_u + n_u + 2n_y N + 2n_y$, the number of constraints is $4n_u N_u + 2n_u + 4n_y N + 4n_y$. If the vectors $\varepsilon^{\min}(k) \in \mathbb{R}^{n_y}$ and $\varepsilon^{\max}(k) \in \mathbb{R}^{n_y}$ are constant over the prediction horizon (as in the optimisation problem (1.11)), the number of decision variables drops to $n_u N_u + n_u + 4n_y$ and the number of constraints to $4n_u N_u + 2n_u + 8n_y$.

The integrated multi-layer control system structure is depicted in Fig. 8.5, its details are shown in Fig. 8.6. At each sampling instant (algorithm iteration k) the following steps are repeated on-line:

1. The local linear approximation (8.6) of the nonlinear (e.g. neural) steady-state model is obtained for the current operating point of the process, i.e. the matrix $\boldsymbol{H}^{\text{ss}}(k)$ given by Eq. (8.7) is calculated.
2. The nonlinear dynamic model is used in the suboptimal MPC algorithm (e.g. in the MPC-NPL algorithm it is linearised at the current operating point, the unmeasured disturbances are estimated and the free trajectory is determined).
3. The integrated set-point optimisation and MPC optimisation quadratic optimisation problem (8.21) is solved to calculate the set-point $u^{\text{ss}}_{\text{INT}}(k)$ and the future control increments $\triangle \boldsymbol{u}(k)$.
4. If the calculated set point is verified, the nonlinear LSSO optimisation problem (8.3) is solved to calculate the set-point $u^{\text{ss}}_{\text{LSSO}}(k)$. Using the nonlinear steady-state model, the optimal output set-point $y^{\text{ss}}_{\text{LSSO}}(k)$ is found from Eq. (8.4). Next, the MPC quadratic optimisation problem is solved for the current optimal set-point $y^{\text{ss}}_{\text{LSSO}}(k)$ to calculate the future control increments $\triangle \boldsymbol{u}(k)$.

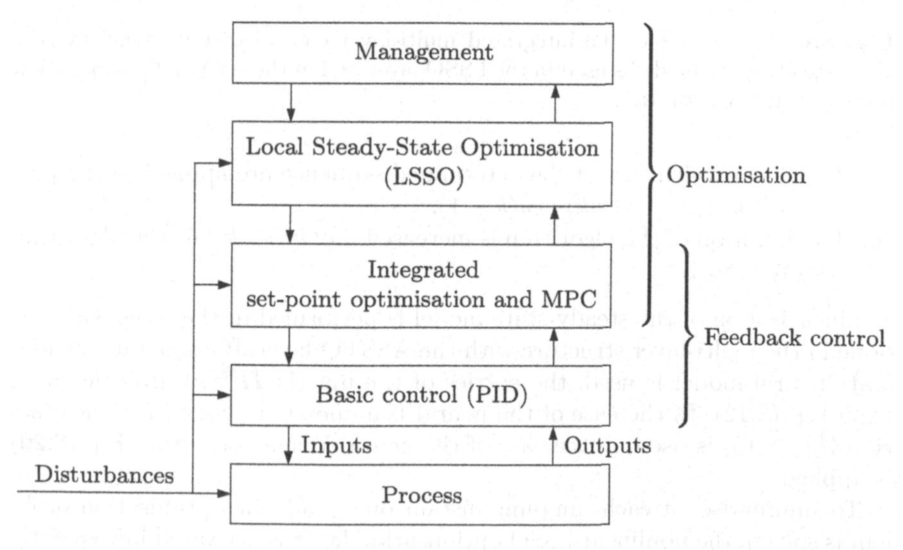

Fig. 8.5. The integrated multi-layer control system structure

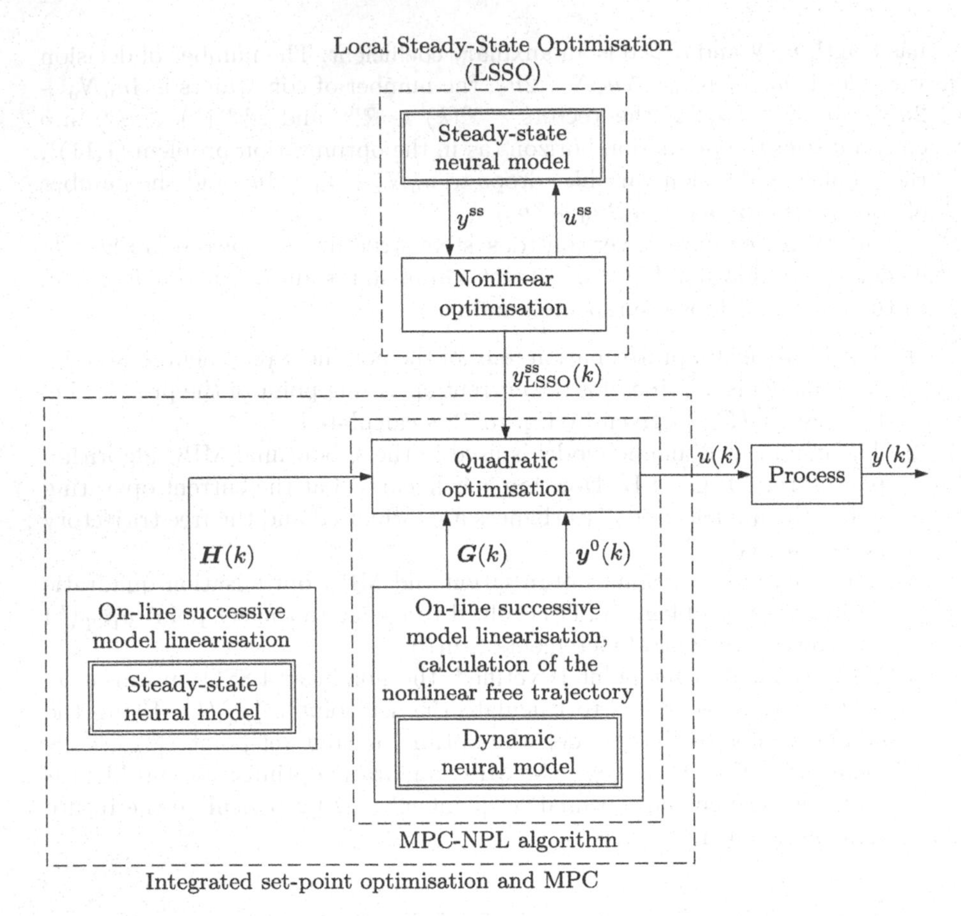

Fig. 8.6. The details of the integrated multi-layer control system structure; the same steady-state model is used in the LSSO layer and in the set-point optimisation layer integrated with MPC

5. The first n_{u} elements of the determined sequence are applied to the process, i.e. $u(k) = \triangle u(k|k) + u(k-1)$.
6. The iteration of the algorithm is increased, i.e. $k\colon = k+1$, the algorithm goes to step 1.

Linearisation of the steady-state model is performed in the same way it is done in the multi-layer structure with the ASSTO layer. If a separate steady-state neural model is used, the entries of the matrix $\boldsymbol{H}^{\mathrm{ss}}(k)$ are calculated from Eq. (8.12). In the case of the neural Hammerstein model, for linearisation Eq. (8.15) is used, in the case of the neural Wiener structure Eq. (8.20) is applied.

To summarise, at each sampling instant one quadratic optimisation problem is solved, the nonlinear LSSO optimisation layer is activated infrequently for verification. In practice, the LSSO layer may be never activated.

Significant complexity is the basic disadvantage of the integrated structure. It may work satisfactorily (i.e. it may generate similar trajectories as the multi-layer structure with the ASSTO layer) for some simple processes, numerical problem are likely for big-scale systems. The number of decision variables is high, it may turn out that it is much easier to solve two separate set-point (in the ASSTO layer) and MPC optimisation problems. Finally, the choice of the tuning parameter α may be quite difficult.

8.4 Control System Structure with Predictive Optimiser and Constraint Supervisor

The classical PID controllers and the explicit simple MPC algorithms are used for numerous processes. Unfortunately, they do not take into account constraints in a systematic manner although the optimal operating point must be frequently located on the boundary of the feasible set. The output constraints are particularly important, because they determine quality and safety of the production process, but they are not respected by PID and simple MPC algorithms. In such a case, for set-point optimisation and constraint satisfaction the control system structure with a predictive optimiser and constraint supervisor (governor) may be used. The structure uses the existing explicit controller. Although PID, IMC and simple linear MPC algorithms may be used for control, the following decision is concerned with the MPC-NPL scheme.

The details of the discussed control system structure are depicted in Fig. 8.7. It consists of two parts: the quadratic set-point optimisation block and an explicit MPC algorithm. The steady-state nonlinear model of the process, e.g. a neural one, is linearised for the current operating point, in the same way it is done in the multi-layer structure with the ASSTO layer and in the integrated structure. The obtained linear approximation is used for set-point optimisation, which leads to a quadratic optimisation problem. It is essential that the explicit MPC control law and all the necessary constraints are taken into account during optimisation. The set-point is calculated in such a way that the economic cost-function is minimised and the constraints are satisfied. During calculations the second model, i.e. the dynamic (neural) model is successively linearised for the current operating point. Fig. 8.7 deliberately does not show the optional LSSO layer, because it uses the full nonlinear model and requires nonlinear optimisation. The LSSO layer may be connected and activated (for verification) significantly less frequently than the predictive optimiser and constraint supervisor, but the discussed structure is intended to improve the existing not advanced control systems in which the LSSO layer is not present.

The quadratic optimisation problem solved in the discussed structure with predictive optimiser and constraint supervisor is

$$\min_{\substack{\boldsymbol{\varepsilon}^{\min}(k),\, \boldsymbol{\varepsilon}^{\max}(k)\\ u^{\mathrm{ss}},\, \tilde{\varepsilon}^{\min},\, \tilde{\varepsilon}^{\max}}} \Big\{ J(k) = \rho^{\min} \left\| \boldsymbol{\varepsilon}^{\min}(k) \right\|^2 + \rho^{\max} \left\| \boldsymbol{\varepsilon}^{\max}(k) \right\|^2 + c_{\mathrm{u}}^{\mathrm{T}} u^{\mathrm{ss}} - c_{\mathrm{y}}^{\mathrm{T}} (f^{\mathrm{ss}}(u(k-1), h(k)) + \boldsymbol{H}^{\mathrm{ss}}(k)(u^{\mathrm{ss}} - u(k-1))) + (\tilde{\rho}^{\min})^{\mathrm{T}} \tilde{\varepsilon}^{\min} + (\tilde{\rho}^{\max})^{\mathrm{T}} \tilde{\varepsilon}^{\max} \Big\}$$

subject to (8.22)

$$\boldsymbol{u}^{\min} \leq \boldsymbol{J}\boldsymbol{K}(k)(\boldsymbol{y}^{\mathrm{ss}}(k) - \boldsymbol{y}^0(k)) + \boldsymbol{u}(k-1) \leq \boldsymbol{u}^{\max}$$
$$-\triangle \boldsymbol{u}^{\max} \leq \boldsymbol{K}(k)(\boldsymbol{y}^{\mathrm{ss}}(k) - \boldsymbol{y}^0(k)) \leq \triangle \boldsymbol{u}^{\max}$$
$$\boldsymbol{y}^{\min} - \boldsymbol{\varepsilon}^{\min}(k) \leq \boldsymbol{G}(k)\boldsymbol{K}(k)(\boldsymbol{y}^{\mathrm{ss}}(k) - \boldsymbol{y}^0(k)) + \boldsymbol{y}^0(k) \leq \boldsymbol{y}^{\max} + \boldsymbol{\varepsilon}^{\max}(k)$$
$$\boldsymbol{\varepsilon}^{\min}(k) \geq 0,\ \boldsymbol{\varepsilon}^{\max}(k) \geq 0$$
$$u^{\min} \leq u^{\mathrm{ss}} \leq u^{\max}$$
$$y^{\min} - \tilde{\varepsilon}^{\min} \leq f^{\mathrm{ss}}(u(k-1), h(k)) + \boldsymbol{H}^{\mathrm{ss}}(k)(u^{\mathrm{ss}} - u(k-1)) \leq y^{\max} + \tilde{\varepsilon}^{\max}$$
$$\tilde{\varepsilon}^{\min} \geq 0,\ \tilde{\varepsilon}^{\max} \geq 0$$

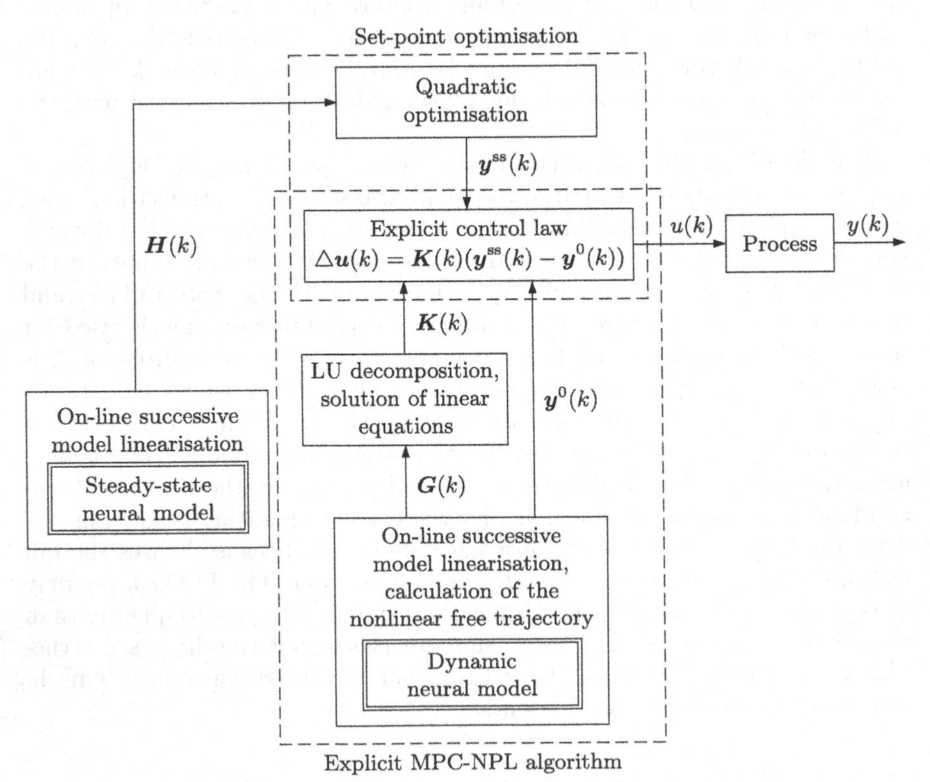

Fig. 8.7. The details of the control system structure with predictive optimiser and constraint supervisor (the optional LSSO layer is not shown)

The economic part $c_{\mathrm{u}}^{\mathrm{T}} u^{\mathrm{ss}} - c_{\mathrm{y}}^{\mathrm{T}} y^{\mathrm{ss}}$ is taken into account in the minimised cost-function, the same that in LSSO or ASSTO optimisation layers. The constraints typical of the MPC-NPL algorithms are used (as in the task (2.36)) and the set-point optimisation constraints (as in the task (8.3)). The linearised steady-state model (8.6) is used for set-point optimisation, the linearised dynamic model is used in MPC. Soft output constraints are used. The number of decision variables is $n_{\mathrm{u}} + 2n_{\mathrm{y}}N + 2n_{\mathrm{y}}$, the number of constraints is $4n_{\mathrm{u}}N_{\mathrm{u}} + 2n_{\mathrm{u}} + 4n_{\mathrm{y}}N + 4n_{\mathrm{y}}$. If the vectors $\varepsilon^{\min}(k) \in \mathbb{R}^{n_{\mathrm{y}}}$ and $\varepsilon^{\max}(k) \in \mathbb{R}^{n_{\mathrm{y}}}$ are constant over the whole prediction horizon (as in the task (1.11)), the number of decision variables drops to $n_{\mathrm{u}} + 4n_{\mathrm{y}}$ and the number of constraints to $4n_{\mathrm{u}}N_{\mathrm{u}} + 2n_{\mathrm{u}} + 8n_{\mathrm{y}}$. The unconstrained MPC explicit control law (2.68) is used during optimisation, where, knowing that $\boldsymbol{y}^{\mathrm{sp}}(k) = \boldsymbol{y}^{\mathrm{ss}}(k)$, one has

$$\triangle \boldsymbol{u}(k) = \boldsymbol{K}(k)(\boldsymbol{y}^{\mathrm{ss}}(k) - \boldsymbol{y}^{0}(k)) \tag{8.23}$$

It is necessary to emphasise the fact that the optimisation problem solved in the integrated structure has more decision variables and constraints than in the currently considered structure.

At each sampling instant (algorithm iteration k) the following steps are repeated on-line:

1. The local linear approximation (8.6) of the nonlinear (e.g. neural) steady-state model is obtained for the current operating point of the process, i.e. the matrix $\boldsymbol{H}^{\mathrm{ss}}(k)$ given by Eq. (8.7) is calculated.
2. The nonlinear dynamic model is used in the suboptimal MPC algorithm (e.g. in the MPC-NPL algorithm it is linearised at the current operating point, the unmeasured disturbances are estimated and the free trajectory is determined).
3. The matrix $\boldsymbol{K}(k)$, which defines the explicit MPC control law (2.68), is obtained from the LU decomposition task.
4. The predictive optimiser and constraint supervisor quadratic optimisation problem (8.22) is solved to calculate the set-point $u^{\mathrm{ss}}_{\mathrm{POCS}}(k)$. Using the linearised steady-state model (8.6), the optimal output set-point $y^{\mathrm{ss}}_{\mathrm{POCS}}(k)$ is found in a similar way it is done in Eq. (8.9), i.e. $y^{\mathrm{ss}}_{\mathrm{POCS}}(k) = f^{\mathrm{ss}}(u^{\mathrm{ss}}_{\mathrm{POCS}}(k), h(k)) + \boldsymbol{H}^{\mathrm{ss}}(k)(u^{\mathrm{ss}}_{\mathrm{POCS}}(k) - u(k-1))$.
5. If the calculated set point is verified, the nonlinear LSSO optimisation problem (8.3) is solved to calculate the set-point $u^{\mathrm{ss}}_{\mathrm{LSSO}}(k)$. Using the nonlinear steady-state model, the output set-point $y^{\mathrm{ss}}_{\mathrm{LSSO}}(k)$ is calculated from Eq. (8.4).
6. For the obtained set-point, the future control increments vector $\triangle \boldsymbol{u}(k)$ is determined from Eq. (8.23), with $y^{\mathrm{ss}}(k) = y^{\mathrm{ss}}_{\mathrm{POCS}}(k)$ or $y^{\mathrm{ss}}(k) = y^{\mathrm{ss}}_{\mathrm{LSSO}}(k)$.
7. The first n_{u} elements of the determined sequence are applied to the process, i.e. $u(k) = \triangle u(k|k) + u(k-1)$.
8. The iteration of the algorithm is increased, i.e. $k := k+1$, the algorithm goes to step 1.

Eq. (8.12) is used for linearisation of the separate steady-state model, whereas Eqs. (8.15) and (8.20) are used for linearisation when the steady-state description is derived from Hammerstein and Wiener dynamic models.

8.5 Example 8.1

The objectives of the example are twofold:

1. Demonstration of the advantages of the multi-layer control system structure with the ASSTO layer (in comparison with the simple structure with the SSTO layer, in which linear models are used in set-point optimisation and MPC, and in comparison with the computationally demanding classical multi-layer structure with nonlinear optimisation in the LSSO layer)
2. Demonstration of the advantages of the multi-input multi-output neural Hammerstein model (not only in comparison with the linear one, but, first of all, in comparison with the polynomial Hammerstein one).

The process under consideration is a multivariable neutralisation reactor (the pH reactor) depicted in Fig. 8.8. In the tank acid (HNO_3), base (NaOH) and buffer ($NaHCO_3$) are continuously mixed. The liquid level $\tilde{h}$ in the tank and the value of pH of the product stream q_4 are are controlled by manipulating acid and base flow rates q_1 and q_3, respectively. It means that the process has two inputs (q_1, q_3) and two outputs ($\tilde{h}$, pH). The buffer flow rate q_2 is the measured disturbance. The first-principle model of the process consists of the following nonlinear differential equations [155, 179]

$$\begin{aligned}
\frac{\mathrm{d}W_{\mathrm{a}_4}(t)}{\mathrm{d}t} &= \frac{q_1(t)(W_{\mathrm{a}_1} - W_{\mathrm{a}_4}(t))}{A\tilde{h}(t)} + \frac{q_2(t)(W_{\mathrm{a}_2} - W_{\mathrm{a}_4}(t))}{A\tilde{h}(t)} \\
&\quad + \frac{q_3(t)(W_{\mathrm{a}_3} - W_{\mathrm{a}_4}(t))}{A\tilde{h}(t)} \\
\frac{\mathrm{d}W_{\mathrm{b}_4}(t)}{\mathrm{d}t} &= \frac{q_1(t)(W_{\mathrm{b}_1} - W_{\mathrm{b}_4}(t))}{A\tilde{h}(t)} + \frac{q_2(t)(W_{\mathrm{b}_2} - W_{\mathrm{b}_4}(t))}{A\tilde{h}(t)} \\
&\quad + \frac{q_3(t)(W_{\mathrm{b}_3} - W_{\mathrm{b}_4}(t))}{A\tilde{h}(t)} \\
\frac{\mathrm{d}\tilde{h}(t)}{\mathrm{d}t} &= \frac{q_1(t) + q_2(t) + q_3(t) - C_{\mathrm{V}}\sqrt{\tilde{h}(t)}}{A}
\end{aligned}$$

and the nonlinear steady-state equation for the pH(t) variable

$$W_{\mathrm{a}_4}(t) + 10^{\mathrm{pH}(t)-14} - 10^{-\mathrm{pH}(t)} + \frac{W_{\mathrm{b}_4}(t)(1 + 2 \times 10^{\mathrm{pH}(t)+\log_{10} K_{\mathrm{a}_2}})}{1 + 10^{-\log_{10} K_{\mathrm{a}_1} - \mathrm{pH}(t)} + 10^{\mathrm{pH}(t)+\log_{10} K_{\mathrm{a}_2}}} = 0$$

Model parameters are given in Table 8.1, the nominal operating point is given in Table 8.2.

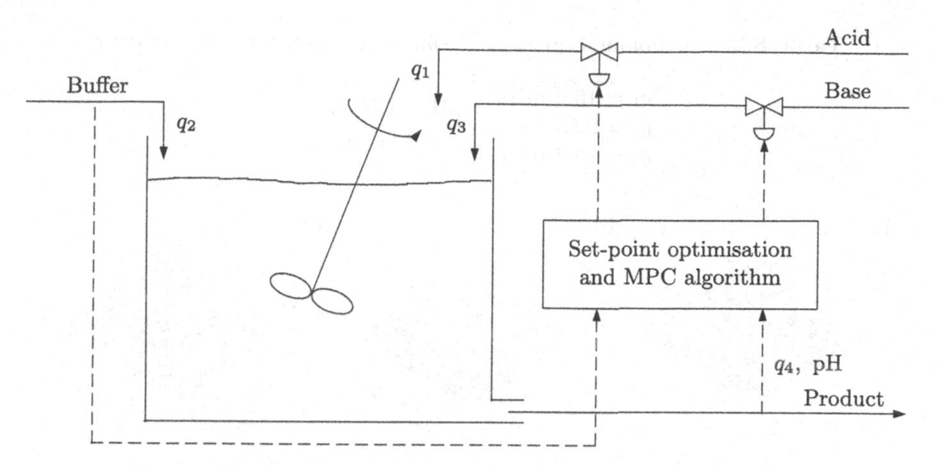

Fig. 8.8. The neutralisation reactor control system structure

Table 8.1. The parameters of the first-principle model of the neutralisation reactor

$A = 207\ \mathrm{cm}^2$	$W_{\mathrm{a}_2} = -0.03$ M
$C_{\mathrm{V}} = 8.75$ ml/(cm × s)	$W_{\mathrm{a}_3} = -0.00305$ M
$K_{\mathrm{a}_1} = 4.47 \times 10^{-7}$	$W_{\mathrm{b}_1} = 0$ M
$K_{\mathrm{a}_2} = 5.62 \times 10^{-11}$	$W_{\mathrm{b}_2} = 0.03$ M
$W_{\mathrm{a}_1} = 0.003$ M	$W_{\mathrm{b}_3} = 0.00005$ M

The neutralisation process exhibits severe nonlinear properties. Steady-state characteristics of the reactor, $\tilde{h}(q_1, q_3)$ and $\mathrm{pH}(q_1, q_3)$, are depicted in Fig. 8.9 for different values of the disturbance q_2. In particular, the relation $\mathrm{pH}(q_1, q_3)$ is nonlinear. Moreover, dynamic properties of the process are also nonlinear. As demonstrated elsewhere, e.g. in [155], for positive and negative changes in manipulated variables (q_1, q_3) the time-constants of obtained step-responses are different. Moreover, the time-constants also depend on the operating point. As a result, the classical linear MPC algorithm does not work properly for the neutralisation process [199]. Due to its nonlinear nature, the reactor is frequently used for comparing different nonlinear models and control algorithms [3, 47, 63, 96, 100, 155, 179, 247]. A simplified variant of process is usually considered, with one input q_1 and one output pH.

Neutralisation Reactor Modelling

During the research carried out the first-principle model is treated as the "real" process. The model differential equations are solved by means of the Euler method (for the sampling period 10 seconds it practically gives

Table 8.2. The nominal operating point of the neutralisation reactor

$q_1 = 16.6$ ml/s $q_2 = 0.55$ ml/s $q_3 = 15.6$ ml/s	pH = 7.0255 $\tilde{h} = 14.0090$ cm

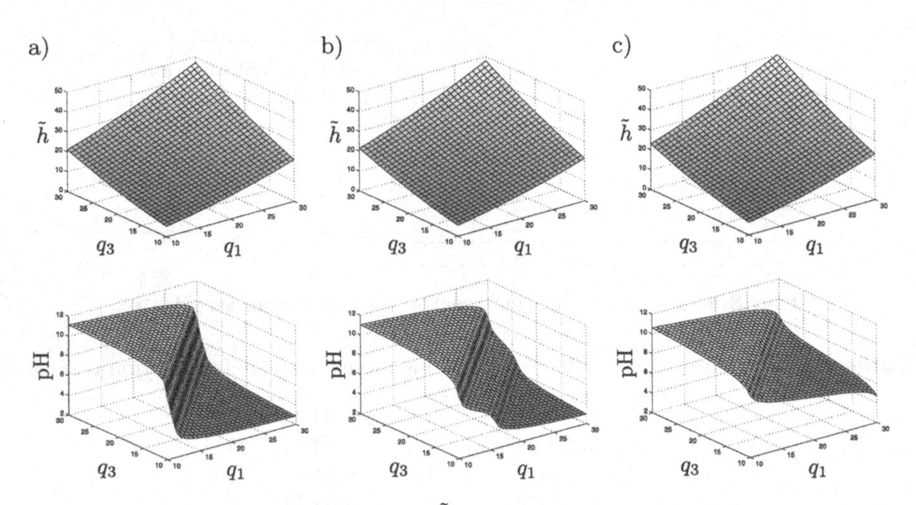

Fig. 8.9. Steady-state characteristics $\tilde{h}(q_1, q_3)$ and $\mathrm{pH}(q_1, q_3)$ of the neutralisation reactor for different values of the disturbance q_2: a) $q_2 = 0.1$ ml/s, b) $q_2 = 0.55$ ml/s, c) $q_2 = 2$ ml/s

the same results as the Runge-Kutta method of order 45). The second-order polynomial and neural Hammerstein models have the following general form

$$\begin{aligned} y_1(k) &= f_1(u_1(k-1), u_1(k-2), u_2(k-1), u_2(k-2), \\ &\qquad h(k-1), h(k-2), y_1(k-1), y_1(k-2)) \\ y_2(k) &= f_2(u_1(k-1), u_1(k-2), u_2(k-1), u_2(k-2), \\ &\qquad h(k-1), h(k-2), y_2(k-1), y_2(k-2)) \end{aligned}$$

Process signals are scaled: $u_1 = (q_1 - q_{1,\mathrm{nom}})/15$, $u_2 = (q_3 - q_{3,\mathrm{nom}})/15$, $h = (q_2 - q_{2,\mathrm{nom}})/2$, $y_1 = (\tilde{h} - \tilde{h}_{\mathrm{nom}})/35$, $y_2 = (\mathrm{pH} - \mathrm{pH}_{\mathrm{nom}})/4$, where $q_{1,\mathrm{nom}}$, $q_{2,\mathrm{nom}}$, $q_{3,\mathrm{nom}}$, $\tilde{h}_{\mathrm{nom}}$, $\mathrm{pH}_{\mathrm{nom}}$ correspond to the nominal operating point (Table 8.2). The Hammerstein models with $n_{\mathrm{v}} = 2$ separate steady-state parts are used (Fig. 3.1b). In the neural Hammerstein model two DLP neural networks, each of which has one hidden layer containing K^1 and K^2 hidden nodes, respectively, are used as the steady-state parts. In the polynomial Hammerstein structure two polynomials of order o_1 and o_2 are used, respectively. The dynamic Hammerstein model (the neural or the polynomial one) is used in the MPC algorithm. Moreover, the dynamic model is also used to derive on-line, for the current operating point, the corresponding steady-state description, which is next used for set-point optimisation (in LSSO and ASSTO layers). That is why the model must be precise in two respects: it must mimic the dynamic behaviour of the process and reflect its steady-state properties. In contrast to the identification procedure of all dynamic models considered so far in the previous chapters (which is based on a series of dynamic signals), both dynamic and steady-state data sets are used for identification of the neutralisation reactor. The fundamental model is simulated open-loop in order to obtain dynamic training, validation and test data sets. Training and test data sets contain 3000 samples, the validation data set contains 1000 samples. The sampling time is 10 s. Output signals contain small measurement noise. Next, the steady-state fundamental model is derived from the dynamic fundamental model and it is simulated to generate the steady-state training, validation and test data sets. The training and test sets have 8000 samples, the validation set has 2500 samples. The domain of interest for input and disturbance signals are: 10 ml/s $\leq q_1$, $q_3 \leq$ 30 ml/s, 0.1 ml/s $\leq q_2 \leq$ 2 ml/s. Because for model identification both dynamic and steady-state data sets are used, the following Mean of Squared Errors (MSE) quality index

$$\begin{aligned}\mathrm{MSE} = &\frac{1}{n_{\mathrm{p}}}\sum_{m=1}^{2}\sum_{k=1}^{n_{\mathrm{p}}}(y_m^{\mathrm{mod}}(k) - y_m(k))^2\\ &+\eta\frac{1}{n_{\mathrm{p}}^{\mathrm{ss}}}\sum_{m=1}^{2}\sum_{s=1}^{n_{\mathrm{p}}^{\mathrm{ss}}}(y_m^{\mathrm{mod,ss}}(s) - y_m^{\mathrm{ss}}(s))^2\end{aligned} \tag{8.24}$$

in place of the classical SSE one defined by Eq. (1.20) must be taken into account. The value of the MSE index depends on accuracy of the dynamic behaviour of the model (its first part) and on accuracy of the steady-state model derived from the dynamic one (its second part). The output of the steady-state model derived from the dynamic one is denoted by $y_m^{\mathrm{mod,ss}}(s)$, $y_m^{\mathrm{ss}}(s)$ is the real value of the recorded process output. Because dynamic and steady-state sets have a different number of samples, the MSE index rather than the SSE one is necessary, n_{p} and $n_{\mathrm{p}}^{\mathrm{ss}}$ denote the number of dynamic and steady-state samples, respectively. The coefficient η determines the relation of two parts of the MSE function. Its value is adjusted by trial and error to obtain models which have good dynamic and steady-state accuracy. During calculations $\eta = 0.1$.

As far as the steady state parts of the Hammerstein models are concerned, neural networks with $K^1 = K^2 = 1, \ldots, 10$ hidden nodes and polynomials of order $o_1 = o_2 = o_3 = 2, \ldots, 10$ are used. Table 8.3 shows properties of compared neural Hammerstein structures in terms of the number of parameters and the MSE performance index. Similarly, Table 8.4 compares properties of the polynomial Hammerstein models. In general, neural models turn out to be more precise and to have a significantly lower number of parameters. It is because polynomials have a very important disadvantage when compared with neural networks, namely they have much worse approximation abilities. Since the neutralisation process is multi-input multi-output one, the total number of model parameters soars when the polynomial order is increased. At the same time, the neural models have a significantly lower number of parameters. For example, the neural model with $K^1 = K^2 = 3$ hidden nodes (44 parameters) gives the value of the $\mathrm{MSE_{val}}$ index comparable with that obtained by the polynomial model of the 7th order (524 parameters). More complex neural models (i.e. with $K^1 = K^2 > 3$) are characterised by even lower $\mathrm{MSE_{val}}$ values. Hence, considering two compared classes of Hammerstein models, the neural Hammerstein structure is a straightforward choice. The neural Hammerstein system with $K^1 = K^2 = 5$ hidden nodes is finally chosen as a compromise, its number of parameters is 64. Further increase of the number of hidden nodes does not lead to any significant reduction of the $\mathrm{MSE_{val}}$ index. For the chosen model the test error ($\mathrm{MSE_{test}}$) is also calculated. Its value is low, which means that the model generalises well. One may also notice that all high-order polynomial models are even worse than the chosen neural one.

Table 8.3. Comparison of empirical neural Hammerstein models of the neutralisation reactor in terms of the number of parameters (NP) and accuracy (MSE)

$K^1 = K^2$	NP	$\mathrm{MSE_{train}}$	$\mathrm{MSE_{val}}$	$\mathrm{MSE_{test}}$
1	24	9.4391×10^{-2}	1.1603×10^{-1}	–
2	34	3.0922×10^{-2}	3.1471×10^{-2}	–
3	44	2.6041×10^{-2}	2.5673×10^{-2}	–
4	54	2.3771×10^{-2}	2.3204×10^{-2}	–
5	64	2.3288×10^{-2}	2.1752×10^{-2}	2.5830×10^{-2}
6	74	2.3170×10^{-2}	2.1514×10^{-2}	–
7	84	2.2671×10^{-2}	2.1385×10^{-2}	–
8	94	2.2534×10^{-2}	2.1360×10^{-2}	–
9	104	2.2463×10^{-2}	2.1275×10^{-2}	–
10	114	2.2187×10^{-2}	2.0502×10^{-2}	–

Table 8.4. Comparison of empirical polynomial Hammerstein models of the neutralisation reactor in terms of the number of parameters (NP) and accuracy (MSE)

$o_1 = o_2 = o_3$	NP	$\mathrm{MSE_{train}}$	$\mathrm{MSE_{val}}$	$\mathrm{MSE_{test}}$
2	39	5.2803×10^{-2}	5.3057×10^{-2}	–
3	76	3.8156×10^{-2}	3.8061×10^{-2}	–
4	137	3.2508×10^{-2}	3.1623×10^{-2}	–
5	228	2.9626×10^{-2}	2.8917×10^{-2}	–
6	335	2.7415×10^{-2}	2.6584×10^{-2}	–
7	524	2.5575×10^{-2}	2.5296×10^{-2}	–
8	741	2.4695×10^{-2}	2.4077×10^{-2}	–
9	1012	2.4381×10^{-2}	2.3585×10^{-2}	–
10	1343	2.4203×10^{-2}	2.3418×10^{-2}	–

For comparison, the following linear model is also found

$$\begin{aligned}y_1(k) &= b_1^{1,1}u_1(k-1) + b_2^{1,1}u_1(k-2) + b_1^{1,2}u_2(k-1) + b_2^{1,2}u_2(k-2)\\&\quad + c_1^1 h(k-1) + c_2^1 h(k-2) - a_1^1 y_1(k-1) - a_2^1 y_1(k-2)\\y_2(k) &= b_1^{2,1}u_1(k-1) + b_2^{2,1}u_1(k-2) + b_1^{2,2}u_2(k-1) + b_2^{2,2}u_2(k-2)\\&\quad + c_1^2 h(k-1) + c_2^2 h(k-2) - a_1^2 y_2(k-1) - a_2^2 y_2(k-2)\end{aligned}$$

It has the same arguments as the Hammerstain model. The dynamic data sets are only used for identification, the MSE index for the linear model takes into account only the first part of Eq. (8.24). Model errors are: $\text{MSE}_{\text{train}} = 4.3056 \times 10^{-2}$, $\text{MSE}_{\text{val}} = 4.9998 \times 10^{-2}$. Fig. 8.10 compares the validation dynamic data set and the output pH of the neural Hammerstein system and of the linear model.

From the dynamic Hammerstein model its steady-state description is derived on-line and next used for set-point optimisation. It is interesting to compare approximation accuracy of the steady-state properties of the process for both considered classes of the Hammerstein model. Fig. 8.11 depicts example steady-state characteristics $\text{pH}(q_1, q_3)$ (for $q_2 = 0.55$ ml/s) obtained from polynomial and neural Hammerstein models. The polynomials of the 5th, 7th and 10th order are compared, which means that the polynomial Hammerstein models have as many as 228, 524 and 1343 parameters, respectively, whereas the chosen neural model has only 64 parameters (Tables 8.3 and 8.4). The steady-state characteristics obtained from the neural model is not only precise (in comparison with the process characteristics depicted in Fig. 8.9b), but also it is "smooth". Unfortunately the steady-state descriptions derived from the polynomial models are not precise. The steady-state characteristics obtained from the Hammerstein model in which polynomials of the 5th order are used is not flat when necessary, in the central part it is not characteristically curved. Of course, by increasing the polynomial order it is possible

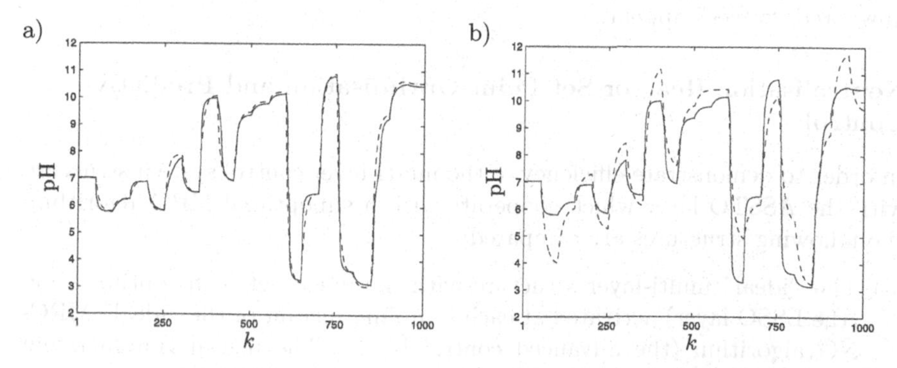

Fig. 8.10. The validation dynamic data set (*solid line*) vs. the output pH of the model of the neutralisation reactor (*dashed line*): a) the neural Hammerstein model, b) the linear model

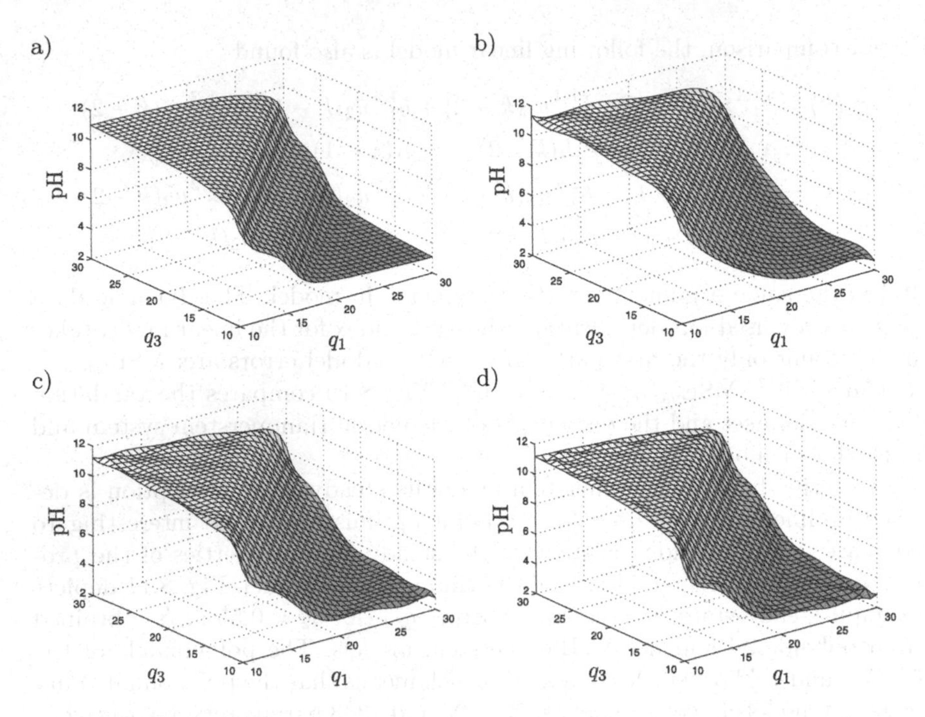

Fig. 8.11. The steady-state characteristics pH(q_1, q_3) of the neutralisation reactor for $q_2 = 0.55$ ml/s obtained from the dynamic Hammerstein models of different structures: a) the neural Hammerstein model, b) the polynomial Hammerstein model with the 5th order steady-state parts, c) the polynomial Hammerstein model with the 7th order steady-state parts, d) the polynomial Hammerstein model with the 10th order steady-state parts

to improve the shape of the steady-state characteristics of models, but some inaccuracies are still present, it is not sufficiently flat when necessary, some unwanted "waves" appear.

Neutralisation Reactor Set-Point Optimisation and Predictive Control

In order to demonstrate efficiency of the multi-layer control system structure with the ASSTO layer which cooperate with a suboptimal MPC algorithm, the following structures are compared:

a) The "ideal" multi-layer structure with nonlinear set-point optimisation (the LSSO layer) activated at each sampling instant of the "ideal" MPC-NO algorithm (the advanced control layer). The neural Hammerstein model is used for set-point LSSO optimisation and MPC.

b) The multi-layer structure with the ASSTO layer and the MPC-NPL algorithm, the LSSO layer is activated for verification 100 times less frequently than the MPC-NPL algorithm. The neural Hammerstein model is used for set-point LSSO and ASSTO optimisation and MPC.
c) The classical multi-layer structure with the LSSO layer activated 100 times less frequently than the linear MPC algorithm. The SSTO layer recalculates the set-point as frequently as the MPC algorithm is activated. The constant linear models are used for set-point SSTO optimisation and MPC, the neural Hammerstein model is used for set-point LSSO optimisation.

In the first structure two nonlinear optimisation problems (LSSO and MPC-NO tasks) are solved at each sampling instant. In spite of the fact that from the practical point of view the "ideal" structure is not realistic (it is very computationally demanding), it is treated as the reference. In the second structure nonlinear LSSO optimisation is repeated infrequently, but every iteration of the MPC algorithm is preceded by set-point ASSTO optimisation, in which a linear approximation of the model is used (the resulting linear optimisation problem is solved by means of the simplex algorithm). The MPC-NPL algorithm with quadratic optimisation is used in the advanced control layer. The same neural Hammerstein model is used in both structures, i.e. the dynamic model is used in MPC (in MPC-NO and MPC-NPL structures) whereas the steady-state model derived from the dynamic one is used in set-point optimisation (LSSO and ASSTO layers). In the last structure a nonlinear (neural) steady-state model is only used in nonlinear set-point LSSO optimisation. The linear MPC algorithm is based on a constant linear dynamic model. The steady-state gain matrix is obtained from the dynamic model and used in SSTO set-point optimisation. In contrast to the first two structures, in which the same dynamic Hammerstein model and the corresponding nonlinear steady-state description are used, different models are used in the last structure in LSSO optimisation (the nonlinear one) and in SSTO optimisation (the linear one). Such an approach, although applied in practice, is not coherent.

In order to maximise the amount of production, the following cost-function is used in set-point optimisation [148, 182]

$$J_{\mathrm{E}} = c_1 q_1^{\mathrm{ss}} + c_3 q_3^{\mathrm{ss}} - c_4 q_4^{\mathrm{ss}}$$

Taking into account that in the steady-state $q_4^{\mathrm{ss}} = q_1^{\mathrm{ss}} + q_2^{\mathrm{ss}} + q_3^{\mathrm{ss}}$ and using the scaled signals, the cost-function becomes

$$J_{\mathrm{E}} = 15(c_1 - c_4)u_1^{\mathrm{ss}} + 15(c_3 - c_4)u_2^{\mathrm{ss}} + (c_1 - c_4)q_{1,\mathrm{nom}} + (c_3 - c_4)q_{3,\mathrm{nom}} - c_4 q_2^{\mathrm{ss}}$$

The vectors of coefficients are: $c_{\mathrm{u}} = [15(c_1 - c_4)\ 15(c_3 - c_4)]^{\mathrm{T}}$, $c_{\mathrm{y}} = [0\ 0]^{\mathrm{T}}$. The price parameters are: $c_1 = 1$, $c_3 = 2$, $c_4 = 5$ [179]. The same constraints are considered in set-point optimisation and MPC: 10 ml/s $\leq q_1 \leq$ 30 ml/s,

$10 \text{ ml/s} \leq q_3 \leq 30 \text{ ml/s}$, $35 \text{ cm} \leq \tilde{h} \leq 40 \text{ cm}$, $5 \leq \text{pH} \leq 6$. The disturbance scenario is

$$q_2(k) = \begin{cases} q_{2,\text{nom}} & \text{if } k < 150 \\ q_{2,\text{nom}} + 0.45(\sin(\alpha(k-150))) \text{ ml/s} & \text{if } 150 \leq k \leq 700 \end{cases}$$

where $\alpha = 2\pi/(700-150+1)$, i.e. the disturbance changes sinusoidally from the sampling instant $k = 150$, its amplitude is 0.45. The tuning parameters of MPC are: $N = 10$, $N_{\text{u}} = 2$, $\boldsymbol{M}_p = \text{diag}(1,1)$, $\boldsymbol{\Lambda}_p = \text{diag}(1,1)$.

Fig. 8.12 depicts simulation results obtained in the first two structures: the "ideal" classical structure with nonlinear set-point LSSO optimisation and the MPC-NO algorithm as well as the structure with the ASSTO layer and the MPC-NPL algorithm are compared. Both structures use the same neural Hammerstein model. The set-point trajectories calculated by LSSO and ASSTO layers ($\tilde{h}^{\text{ss}}$, pH^{ss}) and the real (dynamic) process input trajectories (q_1, q_3) as well as output trajectories ($\tilde{h}$, pH) are shown. At the beginning of simulation (i.e. for $k = 1$), the LSSO layer calculates the optimal set-point $\tilde{h}^{\text{ss}} = 40$ cm, $\text{pH}^{\text{ss}} = 5.7387$. Because the nominal initial conditions of the process are different (Table 8.2), MPC algorithms need some 100 iterations to bring the process to the new operating point. As far as the second structure is concerned, dots and circles denote iterations in which the nonlinear LSSO layer is activated ($k = 150, 250, 350, 450, 550, 650$) whereas in all remaining iterations the set-point is calculated by the ASSTO layer based on a linearised model. The generated set-point trajectories ($\tilde{h}^{\text{ss}}$, pH^{ss}) are very similar to those calculated in the "ideal" classical structure with nonlinear set-point LSSO optimisation, which means that the structure with the ASSTO layer is very successful. Moreover, the obtained dynamic trajectories of the MPC-NPL algorithm are very close to those of the MPC-NO one. The bottom part of Fig. 8.12 shows an enlarged fragment of the output pH trajectory. This particular fragment is chosen because differences between two compared structures are relatively the biggest (but still quite small) when the set-point is changed from pH = 5 to pH = 6. For the whole simulation horizon (700 sampling instants) the economic performance index

$$J_{\text{Esim}} = \sum_{k=1}^{700} J_{\text{E}}(k) = \sum_{k=1}^{700} (c_1 q_1(k) + c_3 q_3(k) - c_4 q_4(k))$$

is calculated after simulations. Obtained values are very similar, in the first structure $J_{\text{Esim}} = -13423.30$, in the second structure $J_{\text{Esim}} = -13422.58$ (the greater the negative value the better).

Because trajectories and economic results obtained in both structures are similar, it is interesting to compare their computational complexity. The computational cost (in terms of floating point operations) of both structures is assessed. Next, the computational complexity reduction factor F is calculated (the first structure with respect to the second one). Because overall computational complexity is strongly influenced by the control horizon, for $N_u = 2$: $F = 5.40$, for $N_u = 5$: $F = 13.15$ and for $N_u = 10$: $F = 30.93$.

Finally, the classical multi-layer structure with the LSSO layer activated 100 times less frequently than the linear MPC algorithm and the SSTO layer which recalculates the set-point as frequently as MPC is activated is verified. In MPC a linear dynamic model is used, in the SSTO layer its steady-state

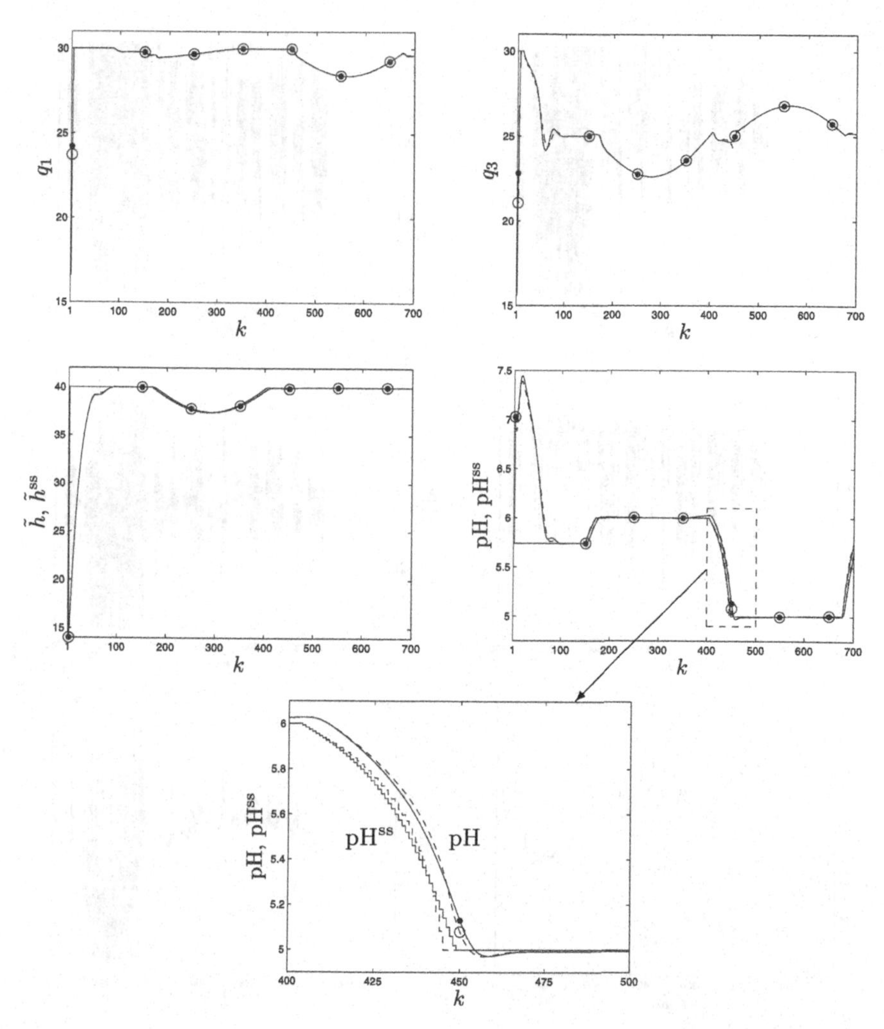

Fig. 8.12. Simulation results: the "ideal" classical structure with nonlinear set-point LSSO optimisation and the MPC-NO algorithm (*solid line with dots*) and the structure with the ASSTO layer and the MPC-NPL algorithm (*dashed line with circles*). In both structures the same same neural Hammerstein model is used. The bottom panel shows an enlarged fragment of the output trajectory.

version is used. Simulation results are depicted in Fig. 8.13. Because the process is significantly nonlinear and the linear model is very inaccurate as shown in Fig. 8.10, when it is used for MPC and set-point optimisation, one obtains numerically wrong results. System trajectories are completely different from those obtained in the "ideal" multi-layer structure and in the structure with the ASSTO layer (Fig. 8.12).

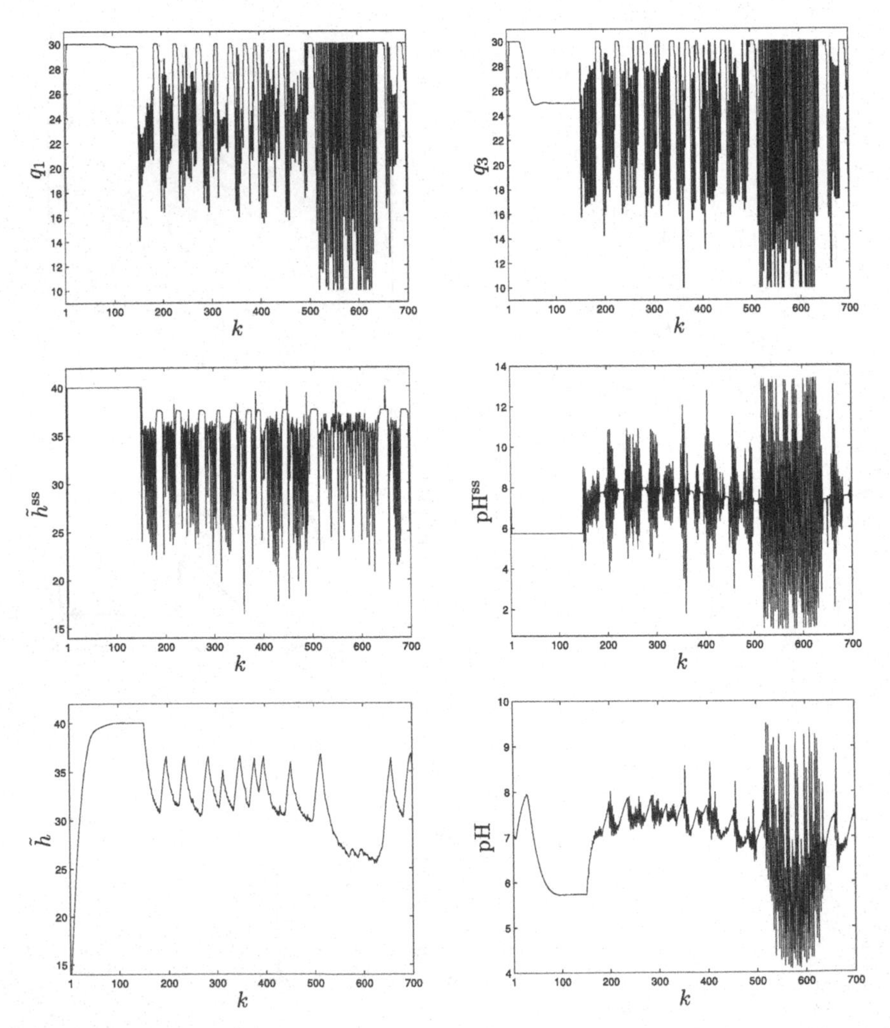

Fig. 8.13. Simulation results: the classical multi-layer structure with the SSTO layer and MPC based on linear models

8.6 Literature Review

The multi-layer control system structure appeared in the 1960s [130]. It is very universal, because each layer is responsible for taking decisions of a different type and each layer has a different period of intervention. The monographs [38, 71] are concerned with control and optimisation of complex systems whereas the problem of cooperation between MPC algorithms and set-point optimisation structures are considered in [315] and in the last chapter of the book [316]. An excellent review of possible approaches to control and optimisation of large-scale systems is given in [287]. A broader outlook on process optimisation is presented in [68], optimisation of complex systems comprised of numerous processes (plantwide control) is discussed in [301].

The majority of commercially available software packages for industrial MPC include the set-point optimisation layer [262]. Typically, the simplest approach is used, i.e. the linear MPC algorithm and the SSTO layer based on a constant linear steady-state model derived from the constant linear dynamic one used in MPC. The optimal set-point is usually on the intersection of constraints, which is typical of linear optimisation. When the modelling errors are significant or measurements are not sufficiently precise, the optimal set-point is likely to change rapidly in consecutive iterations of the ASSTO layer (as in Example 8.1). A method which accounts for uncertainty of the linear steady-state model in the set-point optimisation task is described in [5, 108]. An application of set-point optimisation of an industrial catalytic cracking process is discussed in [124] where the linear DMC algorithm is used and a linear steady-state model derived from the dynamic one is used in set-point optimisation (the SSTO approach). An application of the integrated structure to the industrial catalytic cracking process is described in [324, 341, 342], where a linear DMC algorithm is used for advanced control and a nonlinear model is used in nonlinear set-point optimisation (the LSSO layer is only used). A comparison of the multi-layer and integrated structures is presented in [324, 341]. Important aspects of practical implementation of both structures are discussed in [280], the catalytic cracking process and a distillation column are considered. A simplification in which the nonlinear steady-state model included in the cost-function necessary in set-point optimisation is replaced by a second-order approximation of its reduced gradient is detailed in [304]. Such an approach is functionally similar to linearisation.

An interesting variant of the integrated structure with on-line identification of the linear dynamic model is presented in [18]. The key issue is to generate the input and disturbance signals in such a way that the process is not affected negatively. A simplified version of the structure with predictive optimiser and constraint supervisor is discussed in [284]. A unique feature of the presented approach is the fact that the constraints are neglected. Although it greatly simplifies calculations, the determined set-points are likely to be not the economically optimal ones. Furthermore, when the controller is unconstrained, the technological requirements may be not met.

In spite of the fact that the multi-layer control structure with the SSTO layer based on a constant steady-state model is frequently used in practice, the ASSTO layer with successive model linearisation is used in more advanced cases. The latter approach is thoroughly discussed in [172, 179, 313, 315, 316]. An interesting case of that structure, in which some output variables are controlled whereas the rest of them are only constrained (range control), is considered in [180]. Thanks to such a structure it is possible to improve process economic performance in comparison with the classical approach in which all the outputs are controlled (and may be constrained in MPC). The general integrated structure is presented in [179, 182, 313, 315, 316], the general structure with predictive optimiser and constraint supervisor in [174, 183, 313, 314]. In all the mentioned works, for simplicity, a constant linear model is used in MPC. Derivation of the steady-state description from the dynamic Hammerstein model is discussed in [168], where it is also shown how such an approach may be used in the multi-layer structure with the ASSTO layer and in the integrated structure. Efficiency of the considered structures are usually demonstrated on the simulated single-input single-output neutralisation reactor (with one measured disturbance) or the reactor with two inputs and two outputs is considered, but the range of set-point changes is significantly smaller than in Example 8.1. All control system structures discussed in this chapter cooperate with nonlinear suboptimal MPC algorithms with on-line linearisation, in place of the classical linear MPC one. Implementation of the multi-layer structure with the ASSTO layer in which two separate DLP neural models (i.e. a steady-state one in LSSO and ASSTO, a dynamic one in MPC) are used, is detailed in [147] (simulations are concerned with the yeast fermentation process), implementation of the same structure for two separate RBF neural models is given in [171] (simulations for the polymerisation reactor are presented), implementation for the single DLP neural Hammerstein model is discussed in [148] (simulations for the multi-input multi-output neutralisation reactor are presented). Implementation of the integrated structure for two separate DLP neural models is presented in [176] (simulations for the polymerisation reactor are given). Implementation of the structure with predictive optimiser and constraint supervisor for two separate DLP neural models is detailed in [166] (simulations for the single-input single-output neutralisation reactor are presented).

Of course, the discussed multi-layer control structures are universal, i.e. different model types may be used. For example, implementation of the structure with predictive optimiser and constraint supervisor for a fuzzy Takagi-Sugeno step-response model is discussed in [201] (simulations are concerned with a multi-input multi-output evaporation process), implementation of the multi-layer structure with the ASSTO layer and for a Takagi-Sugeno fuzzy model is presented in [196] (simulations for the van de Vusse reactor are given). Thanks to a specific structure of the dynamic fuzzy model is it possible to derive on-line its local linear approximation in a very easy way.

Although, typically, the nonlinear steady-state model is successively linearised on-line in the structures discussed in this chapter, it is also possible to use a more advanced second-order model approximation. Details of such an approach in the structure with the ASSTO layer are presented in [172], the integrated structure case is discussed in [181]. A clear disadvantage of the idea is the fact that the second-order model approximation is only used in the minimised cost-function whereas a linear approximation of the output constraints is used, which may lead to not optimal set-points.

For numerous different processes successive linearisation of the steady-state model (in all three discussed structures) makes it possible to remove the necessity of on-line nonlinear optimisation and gives good set-points (very close to those generated by the LSSO layer with nonlinear optimisation). In some rare cases, however, determination of the optimal set-point is difficult or even impossible because of the shape of the minimised cost-function. Remembering that the LSSO approach is computationally demanding, two efficient methods of coping with the situation are possible. In the first approach the multi-layer structure with the ASSTO layer is used, but piecewise-linear approximation is performed in place of Taylor linearisation [173, 316, 318]. In the simplest case, if the solution of the set-point optimisation problem is known to belong to some set and dimensionality of the task is not significant, the optimisation problem may be practically solved by checking a mesh of possible set-points. In the second approach a neural network replaces the set-point optimisation ASSTO layer [160]. For the current operating point the network calculates (approximates) the optimal set-point. In consequence, numerical optimisation is not necessary in the ASSTO layer. The data sets necessary for training are generated by solving the set-point optimisation problem for different disturbances and constraints scenarios. The neural network may be used in place of numerical optimisation not only when model linearisation leads to numerical problems, but also when the ASSTO layer works correctly, yet significant reduction of computational burden is beneficial (e.g. when the period of intervention must be very short).

In all the discussed structures it is assumed that the steady-state model used in set-point optimisation is "sufficiently" precise and the disturbances are measured (or estimated) precisely. If it is not true, one may use the Integrated System Optimisation and Parameter Estimation algorithm (ISOPE) with on-line model adaptation [38, 316, 319].

8.7 Chapter Summary

Because the optimal set-point is frequently located in the vicinity of constraints or, more typically, on the border of the feasible set, good, precise control is necessary. Taking into account the main advantages of MPC algorithms (in particular the ability to handle multi-input multi-output processes, the possibility of taking into account the constraints and measured disturbances), one does not doubt that they are natural for the advanced control layer. Thanks to using the dynamic model for prediction, MPC algorithms are expected to give good control quality, much better than that of the classical PID controllers. This chapter discusses three computationally efficient alternatives to the classical multi-layer control system structure. Moreover, implementation details of the discussed structures for three types of neural models are given. Two separate neural model are used in the first case: a steady-state one is used in set-point optimisation and a dynamic one in MPC control. Alternatively, it is possible to use only one dynamic neural Hammerstein of Wiener model. The model is used on-line to derive a corresponding steady-state description for the current operating point. In order to obtain computational efficient control structures, the models are successively linearised on-line (the steady-state ones in set-point optimisation and the dynamic ones in MPC), which eliminates the necessity of solving on-line difficult nonlinear optimisation problems. Finally, efficiency of the multi-layer structure with the ASSTO layer is illustrated on the simulated multi-input multi-output neutralisation process, the MPC-NPL algorithm is used in the advanced control layer. The obtained trajectories are very similar to those calculated in the "ideal" multi-layer structure, in which nonlinear LSSO set-point optimisation and MPC-NO optimisation problems are solved at each sampling instants. It is also interesting to notice advantages of the neural Hammerstein model discussed in this chapter, namely good accuracy and a low number of parameters, the polynomial Hammerstein system is much worse.

The steady-state neural model turns out to be successful in the discussed set-point optimisation structures, similarly to the dynamic neural ones in the MPC algorithms. In addition to obvious advantages of neural models (i.e. good accuracy, a low number of parameters), it is necessary to point out that the steady-state neural models require rather simple calculations in set-point optimisation whereas the first-principle models are comprised of nonlinear equations, which are supposed to be repeatedly solved on-line. Although the discussion presented in this chapter is limited to the MPC-NPL algorithm, any other suboptimal MPC strategy may be used in all considered control system structures. It is also possible to use an alternative dynamic model in MPC, e.g. the multi-model.

The multi-layer control system structure with the ASSTO layer and the structure with predictive optimiser and constraint supervisor are the most important. The latter structure may be applied when the existing control algorithm (e.g. PID, IMC or a simplified version of MPC) should be accompanied by a set-point optimisation procedure and a constraint satisfaction mechanism. The integrated structure is not as universal as the previous two ones. In spite of the fact that the models are linearised on-line (in set-point optimisation and MPC), which eliminates the necessity of nonlinear optimisation, the integrated approach may lead to quite a complex optimisation problem which must be solved on-line. On the other hand, reduction of computational complexity is usually very important in practice.

9

Concluding Remarks and Further Research Directions

If a linear model is used in the MPC algorithm for prediction, one obtains a quadratic optimisation problem which may be efficiently solved in practice using the currently available software and hardware platforms. Thousands of industrial applications of the linear MPC algorithms confirm their practical usefulness [262]. The majority of those processes have many inputs and many outputs and are characterised by significant delays. For such processes, the linear MPC algorithms give good control quality, much better than that of the previously used classical PID algorithms (often single-loop ones). Furthermore, the MPC technique makes it possible to take into account all the necessary constraints in a systematic manner, the satisfaction of which is of paramount importance very frequently.

When the process is nonlinear, a nonlinear MPC algorithm is a natural option rather than a linear one. In such a case the choice of the model and the way it is used on-line are crucial. This book tries to answer both questions. Among many types of empirical models, the neural structures of the DLP type are chosen. The choice is motivated by numerous advantages of the neural models, the most important of which are: good modelling accuracy, a relatively low number of parameters, a simple, regular structure, availability of many efficient structure optimisation and training algorithms. The literature shows a huge number of examples which indicate usefulness of neural networks in modelling of various processes.

A direct application of a neural model for prediction calculation leads to a nonlinear, usually non-convex, optimisation problem, which must be solved on-line at each sampling instant of the MPC algorithm. The disadvantages of such an approach are obvious: huge computational burden and no guarantee that the global solution is found. The MPC algorithms with nonlinear optimisation, although considered in many theoretical works, have a very limited practical importance. That is why this book is concerned with the suboptimal MPC algorithms with successive on-line linearisation. The linearisation approach makes it possible to replace the MPC nonlinear optimisation problem by a quadratic optimisation task (or a series of such problems).

M. Ławryńczuk, *Computationally Efficient Model Predictive Control Algorithms*, Studies in Systems, Decision and Control 3,
DOI: 10.1007/978-3-319-04229-9_9, © Springer International Publishing Switzerland 2014

This book summarises a few years of the author's research and includes the following original contributions (only some chosen publications are cited):

a) *The computationally efficient nonlinear MPC algorithms based on the DLP neural models [146, 149, 161, 162, 163, 165, 169, 177, 178, 185, 317].* A few rudimentary classes of the nonlinear MPC structures are discussed (the prototype algorithms). A linear approximation of the neural model is calculated on-line for the current operating point in the simplest case. Predicted trajectory linearisation in place of model linearisation is performed in more advanced solutions, it is also possible to repeat prediction calculation and trajectory linearisation a few times at each sampling instant. Convergence of internal iterations is guaranteed in the most advanced suboptimal MPC algorithm. It is necessary to emphasise the fact that publications concerned with the suboptimal MPC algorithms are rare, the simplest approach with successive model linearisation is typically used. No thorough discussion concerned with the impact of the linearisation method used in MPC on the possible control quality and on computational burden is present in the literature. On the other hand, there are many works concerned with the MPC algorithm with on-line nonlinear optimisation. Unfortunately, the disadvantages of the algorithm, namely its apparent computational inefficiency and limited practical applicability, are usually overlooked.
b) *The computationally efficient nonlinear MPC algorithms based on the neural Hammerstein and Wiener models [143, 148, 150, 153, 155].* Both cascade structures turn out to be very successful in modelling of various processes. A very typical approach to MPC based on Hammerstein and Wiener models, frequently discussed in the literature, is to use on-line the inverse of the steady-state part of the model to compensate for the nonlinear nature of the process and to use a linear MPC algorithm for a modified system. It is necessary to point out that almost all of the developed algorithms (the only exception is the MPC-NPL algorithm for the Wiener model) do not need the inverse model. In addition to the suboptimal prototype MPC algorithms with on-line linearisation (model or trajectory linearisation is possible), the MPC algorithms with simplified linearisation are described, which is possible due to a special structure of the cascade Hammerstein and Wiener models.
c) *The computationally efficient nonlinear MPC algorithms based on the neural state-space models [158].* The state-space models have a much bigger potential than the rudimentary input-output ones. They are typically used in the MPC algorithms with guaranteed stability and robustness, but they usually need nonlinear on-line optimisation. Implementation details of the computationally efficient prototype MPC algorithms are given for the neural state-space models, two cases are considered: the output set-point trajectory and the state set-point trajectory.

d) *The computationally efficient nonlinear MPC algorithms based on the neural multi-models [159, 167].* When the multi-model is used for prediction, the prediction error is not propagated, because for consecutive sampling instants of the prediction horizon independent submodels are used. They may be significantly simpler than the classical model which must calculate the predictions for the whole prediction horizon, training is also simpler as the submodels are not used recurrently for prediction. Although the multi-model concept is not new, it is only possible to find in the literature the description of the MPC algorithm with nonlinear optimisation based on such models. In this book the computationally efficient MPC algorithms with on-line linearisation and quadratic optimisation are developed for the neural multi-models.
e) *The computationally efficient nonlinear MPC algorithms with neural approximation [142, 151, 156, 157].* A unique feature of the algorithms with neural approximation is the lack of on-line linearisation. The explicit versions of the algorithms are particularly interesting and computationally efficient, in which the approximator directly calculates the coefficients of the control law. As a result, the algorithms can be implemented on hardware platforms with limited resources and computational abilities. In the literature one may find the description of neural approximation in the MPC algorithm with nonlinear optimisation. In such a case the neural network replaces the whole MPC algorithm. It calculates, for the current operating point, the values of the manipulated variables. The MPC algorithms with neural approximation presented in this book are motivated by the role of the model in predictive control and the suboptimal prediction equations used in the prototype algorithms.
f) *The computationally efficient nonlinear MPC algorithms with guaranteed stability (the dual-mode MPC approach with on-line linearisation) [175] and the MPC algorithm with guaranteed robustness (the MPC algorithm with additional state constraints and on-line linearisation) [152].* It is necessary to emphasise the fact that the stable and robust MPC algorithms discussed in the literature typically need on-line nonlinear optimisation at each sampling instant, in some of them the global solution is necessary. Conversely, on-line successive (model or trajectory) linearisation and quadratic optimisation is used in the prototype developed algorithms. From the perspective of stability (robustness), it is only necessary to find a feasible solution (which satisfies the constraints), precise on-line nonlinear optimisation is not necessary. Implementation details of stable and robust MPC algorithms with on-line linearisation are given for the state-space and input-output models.
g) *Implementation of the computationally efficient control system structures cooperating with the suboptimal MPC algorithms for different classes of neural models [147, 148, 160, 166, 176, 171], whereas the structures themselves are developed by the author's research group [168, 172, 173, 174,*

179, 180, 181, 182, 183, 314, 318]. Because advanced software and hardware platforms are more and more popular in the industry, set-point optimisation is quite frequently used in contemporary control systems, which makes it possible to improve economic efficiency of processes. Co-operation of set-point optimisation and the MPC algorithms is of great practical and economic importance. Some such simplified control system structures are used in the industry, but the subject is not popular in the literature. The simplest approach is usually discussed, in which a linear steady-state model derived from the dynamic one present in the linear MPC algorithm is used in set-point optimisation, or the most computationally demanding, in which nonlinear models are directly used in set-point optimisation, which needs on-line nonlinear optimisation. That is why this book discusses three computationally efficient control structures with on-line successive linearisation, which eliminates the necessity of nonlinear optimisation. Implementation details of the structures are given for three classes of neural models.

In short, this book discusses a few suboptimal MPC algorithms with different linearisation methods and for different structures of neural models. The algorithms are computationally efficient in two respects: qualitative and quantitative. Firstly, quadratic optimisation is used on-line in place of nonlinear optimisation, which is a huge difference. Much simpler hardware platforms may be used, the problems typical of nonlinear optimisation, namely the risk of finding a local minimum and uncertain calculation time, are not present. Secondly, for the example simulated processes considered in this book, the suboptimal MPC algorithms are a few times (or a dozen of times, it depends on the tuning parameters, on the control horizon in particular) more computationally efficient. All the algorithms are derived for the general nonlinear model. Although the implementation details for a few classes of neural models are given, one may implement the algorithms for alternative model types (the only assumption is that the model is differentiable).

The processes considered in the examples discussed in the book are nonlinear, which means that they cannot be modelled precisely by linear structures. In consequence, the classical linear MPC algorithms (not to mention the PID controller) do not work properly when the set-point is changed fast and in broad range (unwanted oscillatory or slow behaviour is observed, depending on the operating point). The simulation results presented in this book indicate obvious efficiency of the discussed suboptimal MPC algorithms with on-line linearisation: they give trajectories very similar, or practically the same, to those obtained in the computationally demanding MPC scheme with on-line nonlinear optimisation. The choice of the specific suboptimal algorithm should be motivated by the qualities of the process, namely the degree of nonlinearity, the speed of dynamics and the delays should be taken into account. The simplest MPC algorithm with model linearisation at the current operating point is sufficient for the yeast fermentation process and for the polymerisation reactor, whereas the more advanced MPC scheme with trajectory linearisation is necessary for the high-pressure high-purity distillation process (the best results are obtained when prediction calculation and trajectory linearisation are repeated a few times at each sampling instant).

The application of the presented control system structures with on-line linearisation which cooperate with the suboptimal MPC algorithms results in increasing economic efficiency when compared to the classical structure in which only linear models are used. Furthermore, the recommended structures give very similar results to those obtained in the computationally inefficient approach in which nonlinear optimisation is used in MPC and set-point calculation.

In this book a lot of attention is paid to the models. Firstly, the implementation details of the discussed MPC algorithms and set-point optimisation structures are given for a few chosen types of neural models. The choice of the particular model type should be made taking into account the process characteristics. In the simplest case the black-box DLP model can be used, but the cascade Hammerstein or Wiener structures are very good for some processes. When the prediction error propagation problem is important, the multi-model should be used. Alternatively, the state-space model may turn out to be necessary in some cases. Secondly, remembering that model identification may take as much as 90% of time spent on developing the control algorithm [215], empirical modelling of example simulated processes, in particular model selection of the structure which is next used in MPC algorithms, is thoroughly discussed. It is shown that for the considered example processes it is possible to obtain neural models which, unlike the linear ones, are very precise. It is necessary to point out that the chosen models are relatively simple: each of them contains only a few hidden nodes, they have a very limited number of parameters (the only exception is the multi-model, but it is comprised of a few submodels).

In addition to the issues discussed in the book, there are a few possible interesting further research directions:

a) Implementation of the prototype MPC algorithms and the control system structures for alternative models. Different variants of the SVM model may be given as the example.
b) Development of the suboptimal MPC algorithms with on-line linearisation for other process classes, e.g. hybrid processes with not only continuous but also discrete variables and fractional-order systems may be considered.
c) The problem of dependability analysis of software implementation of the linear MPC algorithms has been researched [79]. A few methods which make it possible to improve dependability have been developed. Continuation of the work, in particular dependability analysis of the suboptimal MPC algorithms is worth considering.
d) Analysis of suboptimal MPC algorithms in abnormal situations, e.g. in the case of sensor and (or) actuator faults.

e) Alternative formulation of the MPC problem, e.g. parametrisation in order to reduce the number of decision variables.
f) Continuation of stability and robustness analysis, development of new stable and robust variants of the suboptimal MPC algorithms with on-line linearisation.
g) Because, as it was mentioned previously, model identification may take as much as 90% of time spent on developing the control algorithm, it would be beneficial to find new nonlinear model structures the identification of which can be done in a simple manner. Next, it would be necessary to develop the suboptimal MPC algorithms for such models.
h) Development of adaptive versions of the suboptimal MPC algorithms with on-line linearisation. The key problems are then: selection of the model structure and of the data sets, efficiency and reliability of the on-line identification procedure. In spite of some progress in adaptive MPC algorithms, they are not popular in practice because of unreliability of on-line identification [262].
i) Development of optimisation algorithms, in particular optimisation methods particularly adjusted to MPC problems.
j) Practical applications of the discussed suboptimal MPC algorithms and set-point optimisation structures. As the author's practical experience indicates, there are a great number of processes which should be controlled by the nonlinear MPC algorithms. Many of such processes are significantly nonlinear, multi-input multi-output, delayed. As a results, the currently used classical single-loop PID controllers and the linear MPC algorithms do not give satisfactory control quality.

It is worth emphasising the fact that the third subject and the last four subjects are also interesting from the computer science point of view, because not only the development of efficient algorithms is necessary, but also their successful software implementation is of great importance. The hardware platforms used in practice may have some limitations, e.g. memory may be limited.

References

1. Abonyi, J., Babuška, R., Ayala Botto, M., Szeifert, F., Nagy, L.: Identification and control of nonlinear systems using fuzzy Hammerstein models. Industrial and Engineering Chemistry Research 39, 4302–4314 (2000)
2. Aggelogiannaki, E., Sarimveis, H.: A simulated annealing algorithm for prioritized multiobjective optimization–implementation in an adaptive model predictive control configuration. IEEE Transactions on Systems, Man and Cybernetics–Part B: Cybernetics 37, 902–915 (2007)
3. Åkesson, B.M., Toivonen, H.T.: A neural network model predictive controller. Journal of Process Control 16, 937–946 (2006)
4. Akpan, V.A., Hassapis, G.D.: Nonlinear model identification and adaptive model predictive control using neural networks. ISA Transactions 50, 177–194 (2011)
5. Alvarez, L.A., Odloak, D.: Robust integration of real time optimization with linear model predictive control. Computers and Chemical Engineering 34, 1937–1944 (2010)
6. Al-Duwaish, H., Karim, M.N., Chandrasekar, V.: Use of multilayer feedforward neural networks in identification and control of Wiener model. Proceedings IEE, Part D, Control Theory and Applications 143, 225–258 (1996)
7. Alexandridis, A., Sarimveis, H.: Nonlinear adaptive model predictive control based on self-correcting neural network models. AIChE Journal 51, 2495–2506 (2005)
8. Allgöwer, F., Badgwell, T.A., Qin, J.S., Rawlings, J.B., Wright, S.J.: Nonlinear predictive control and moving horizon estimation–an introductory overview. In: Frank, P.M. (ed.) Advances in control–highlights of European Control Conference 1999, pp. 391–449. Springer, London (1999)
9. Al Seyab, R.K., Cao, Y.: Nonlinear model predictive control for the ALSTOM gasifier. Journal of Process Control 16, 795–808 (2006)
10. Anderson, B.D.O., Moore, J.B.: Optimal filtering. Prentice Hall, Englewood Cliffs (1979)
11. Ayoubi, M.: Comparison between the dynamic multi-layered perceptron and generalised Hammerstein model for experimental identification of the loading process in diesel engines. Control Engineering Practice 6, 271–279 (1998)
12. Arahal, M.R., Berenguel, M., Camacho, E.F.: Neural identification applied to predictive control of a solar plant. Control Engineering Practice 6, 333–344 (1998)

13. Arto, V., Hannu, P., Halme, A.: Modeling of chromatographic separation process with Wiener-MLP representation. Journal of Process Control 11, 443–458 (2001)
14. Ayala Botto, M., van den Boom, T.J.J., Krijgsman, A., da Costa, J.S.: Predictive control based on neural network models with I/O feedback linearization. International Journal of Control 72, 1538–1554 (1999)
15. Åström, K.J., Wittenmark, B.: Computer-controlled systems: theory and design. Prentice Hall, Upper Saddle River (1997)
16. Álamo, T., Ramírez, D.R., Muñoz de la Peña, D., Camacho, E.F.: Min-max MPC using a tractable QP problem. Automatica 43, 693–700 (2007)
17. Bartlett, R.A., Biegler, L.T., Backstrom, J., Gopal, V.: Quadratic programming algorithms for large-scale model predictive control. Journal of Process Control 12, 775–795 (2002)
18. Becerra, V.M., Roberts, P.D., Griffiths, G.W.: Novel developments in process optimisation using predictive control. Journal of Process Control 8, 117–138 (1998)
19. Bellemans, T., De Schutter, B., De Moor, B.: Model predictive control for ramp metering of motorway traffic: a case study. Control Engineering Practice 14, 757–767 (2006)
20. Bemporad, A., Morari, M., Dua, V., Pistikopoulos, E.N.: The explicit linear quadratic regulator for constrained systems. Automatica 38, 3–20 (2002)
21. Bemporad, A., Morari, M.: Robust model predictive control: a survey. In: Garulli, A., Tesi, A. (eds.) Robustness in Identification and Control. LNCIS, vol. 245, pp. 207–226. Springer, Heidelberg (1999)
22. Bemporad, A.: A predictive controller with artificial Lyapunov function for linear systems with input/state constraints. Automatica 34, 1255–1260 (1998)
23. Bemporad, A., Chisci, L., Mosca, E.: On the stabilizing property of SIORHC. Automatica 30, 2013–2015 (1994)
24. Bequette, B.W.: Nonlinear control of chemical processes: a review. Industrial and Engineering Chemical Research 30, 1391–1413 (1991)
25. Berenguel, M., Arahal, M.R., Camacho, E.F.: Modelling the free response of a solar plant for predictive control. Control Engineering Practice 6, 1257–1266 (1998)
26. Betts, J.T., Frank, P.D.: A sparse nonlinear optimization algorithm. Journal of Optimization Theory and Applications 82, 519–541 (1994)
27. Bhartiya, S., Whiteley, J.R.: Benefits of factorized RBF-based NMPC. Computers and Chemical Engineering 26, 1185–1199 (2002)
28. Bhartiya, S., Whiteley, J.R.: Factorized approach to nonlinear MPC using a radial basis function model. AIChE Journal 47, 358–368 (2001)
29. Biegler, L.T., Grossmann, I.E.: Retrospective on optimization. Computers and Chemical Engineering 28, 1169–1192 (2004)
30. Blanchini, F.: Set invariance in control. Automatica 35, 1747–1767 (1999)
31. Blanco, E., de Prada, C., Cristea, S., Casas, J.: Nonlinear predictive control in the LHC accelerator. Control Engineering Practice 17, 1136–1147 (2009)
32. Bloemen, H.H.J., Chou, C.T., van den Boom, T.J.J., Verdult, V., Verhaegen, M., Backx, T.C.: Wiener model identification and predictive control for dual composition control of a distillation column. Journal of Process Control 11, 601–620 (2001)

33. Bock, H.G., Diehl, M., Kühl, P., Kostina, E., Schlöder, J.P., Wirsching, L.: Numerical methods for efficient and fast nonlinear model predictive control. In: Findeisen, R., Allgöwer, F., Biegler, L.T. (eds.) Assessment and Future Directions of Nonlinear Model Predictive Control. LNCIS, vol. 358, pp. 163–179. Springer, Heidelberg (2007)
34. Bomberger, J.D., Seborg, D.E.: Determination of model order for NARX models directly from input-output data. Journal of Process Control 8, 459–468 (1998)
35. Bravo, J.M., Álamo, T., Camacho, E.F.: Robust MPC of constrained discrete-time nonlinear systems based on approximated reachable sets. Automatica 42, 1745–1751 (2006)
36. Bravo, J.M., Limón, D., Álamo, T., Camacho, E.F.: On the computation of invariant sets for constrained nonlinear systems: An interval arithmetic approach. Automatica 41, 1583–1589 (2005)
37. Brdys, M.A., Grochowski, M., Gminski, T., Konarczak, K., Drewa, M.: Hierarchical predictive control of integrated wastewater treatment systems. Control Engineering Practice 16, 751–767 (2008)
38. Brdys, M.A., Tatjewski, P.: Iterative algorithms for multilayer optimizing control. Imperial College Press/World Scientific, London (2005)
39. Byrd, R.H., Lu, P., Nocedal, J., Zhu, C.: A limited memory algorithm for bound constrained optimization. SIAM Journal on Scientific Computing 16, 1190–1208 (1995)
40. Camacho, E.F., Bordons, C.: Robustness and robust design of MPC for nonlinear discrete-time systems. In: Findeisen, R., Allgöwer, F., Biegler, L.T. (eds.) Assessment and Future Directions of Nonlinear Model Predictive Control. LNCIS, vol. 358, pp. 1–16. Springer, Heidelberg (2007)
41. Camacho, E.F., Bordons, C.: Model Predictive Control. Springer, London (1999)
42. Camacho, E.F., Berenguel, M.: Robust adaptive model predictive control of a solar plant with bounded uncertainties. International Journal of Adaptive Control and Signal Processing 11, 311–325 (1997)
43. Campo, P.J., Morari, M.: Robust model predictive control problem. In: Proceedings of the American Control Conference (ACC 1987), Minneapolis, pp. 1021–1026 (1987)
44. Cervantes, A.L., Agamennoni, O.E., Figueroa, J.L.: A nonlinear model predictive control based on Wiener piecewise linear models. Journal of Process Control 13, 655–666 (2003)
45. Chen, C.C., Shaw, L.: On receding horizon feedback control. Automatica 18, 349–352 (1982)
46. Chen, H., Allgöwer, F.: A quasi-infinite horizon nonlinear model predictive control scheme with guaranteed stability. Automatica 34, 1205–1217 (1998)
47. Chen, J., Huang, T.C.: Applying neural networks to on-line updated PID controllers for nonlinear process control. Journal of Process Control 14, 211–230 (2004)
48. Chisci, L., Rossiter, J.A., Zappa, G.: Systems with persistent disturbances: predictive control with restriceted constraints. Automatica 37, 1019–1028 (2001)
49. Chisci, L., Lombardi, A., Mosca, E.: Dual-receding horizon control of constrained discrete time systems. European Journal of Control 2, 278–285 (1996)

50. Chisci, L., Mosca, E.: Stabilizing I-O receding horizon control of CARMA plants. IEEE Transactions on Automatic Control 39, 614–618 (1994)
51. Chu, J.Z., Tsai, P.F., Tsai, W.Y., Jang, S.S., Wong, D.S.H., Shieh, S.S., Lin, P.H., Jiang, S.J.: An experimental study of model predictive control based on artificial neural networks. In: Palade, V., Howlett, R.J., Jain, L. (eds.) KES 2003. LNCS, vol. 2773, pp. 1296–1302. Springer, Heidelberg (2003)
52. Clarke, D.W., Scattolini, R.: Constrained receding-horizon predictive control. Proceedings IEE, Part D 138, 347–354 (1991)
53. Clarke, D.W., Mohtadi, C.: Properties of generalized predictive control. Automatica 25, 859–875 (1989)
54. Clarke, D.W., Mohtadi, C., Tuffs, P.S.: Generalized predictive control. Automatica 23, 137–160 (1987)
55. Colin, G., Chamaillard, Y., Bloch, G., Corde, G.: Neural control of fast nonlinear systems–application to a turbocharged SI engine with VCT. IEEE Transactions on Neural Networks 18, 1101–1114 (2007)
56. da Cruz Meleiro, L.A., José, F., Zuben, V., Filho, R.M.: Constructive learning neural network applied to identification and control of a fuel-ethanol fermentation process. Engineering Applications of Artificial Intelligence 22, 201–215 (2009)
57. Cutler, C.R., Ramaker, B.L.: Dynamic matrix control–a computer control algorithm. In: Proceedings of the AIChE National Meeting, Houston (1979)
58. Declercq, F., de Keyser, R.: Suboptimal nonlinear predictive controllers. International Journal of Applied Mathematics and Computer Science 9, 129–148 (1999)
59. Dempsey, E.J., Westwick, D.T.: Identification of Hammerstein models with cubic spline nonlinearities. IEEE Transactions on Biomedical Engineering 51, 237–245 (2004)
60. Díaz-Mendoza, R., Budman, H.: Structured singular valued based robust nonlinear model predictive controller using Volterra series models. Journal of Process Control 20, 653–663 (2010)
61. Doering, A., Galicki, M., White, H.: Structure optimisation of neural networks with the A^* algorithm. IEEE Transactions on Neural Networks 8, 1434–1445 (1997)
62. Domek, S.: Robust predictive control of nonlinear processes, vol. 593. Scientific Works of Szczecin University of Technology, Szczecin (2006) (in Polish)
63. Dougherty, D., Cooper, D.: A practical multiple model adaptive strategy for single-loop MPC. Control Engineering Practice 11, 141–159 (2003)
64. Doyle, F.J., Pearson, R.K., Ogunnaike, B.A.: Identification and control using Volterra models. Springer, London (2002)
65. Doyle, F.J., Ogunnaike, B.A., Pearson, R.K.: Nonlinear model-based control using second-order Volterra models. Automatica 31, 697–714 (1995)
66. Duch, W., Korbicz, J., Rutkowski, L., Tadeusiewicz, R. (eds.): Neural networks, Biocybernetics and biomedical engineering, vol. 6. EXIT, Warsaw (2000) (in Polish)
67. Economou, C.G., Morari, M.: Internal model control–5. Extension to nonlinear systems. Industrial Engineering Chemical Process Design and Development 25, 403–411 (1986)
68. Engell, S.: Feedback control for optimal process operation. Journal of Process Control 17, 203–219 (2007)

69. Eskinat, E., Johnson, S., Luyben, W.L.: Use of Hammerstein models in identification of nonlinear systems. AIChE Journal 37, 255–268 (1991)
70. Findeisen, R.H., Rawlings, J.B.: Suboptimal infinite horizon nonlinear model predictive control for discrete time. Automatic Control Laboratory, Swiss Federal Institute of Thechnology (ETH), Raport 97–13, Zurich (1997)
71. Findeisen, W.M., Baileym, F.N., Brdyś, M., Malinowski, K., Tatjewski, P., Woźniak, A.: Control and coordination in hierarchical systems. John Wiley & Sons, Chichester (1980)
72. Fleming, P.J., Purshouse, R.C.: Evolutionary algorithms in control systems engineering: a survey. Control Engineering Practice 10, 1223–1241 (2002)
73. Fruzzetti, K.P., Palazoğlu, A., McDonald, K.A.: Nonlinear model predictive control using Hammerstein models. Journal of Process Control 7, 31–41 (1997)
74. Garcia, C.E., Prett, D.M., Morari, M.: Model predictive control: theory and practice–a survey. Automatica 25, 335–348 (1989)
75. Garcia, C.E., Morshedi, A.M.: Quadratic programming solution of dynamic matrix control (QDMC). Chemical Engineering Communications 46, 73–87 (1986)
76. Garcia, C.E.: Quadratic dynamic matrix control of nonlinear processes: an application to a batch reactor process. In: Proceedings of the AIChE National Meeting, San Francisco (1984)
77. Garcia, C.E., Morari, M.: Internal model control–1. A unifying review and some new results. Industrial Engineering Chemical Process Design and Development 21, 308–323 (1982)
78. Gattu, G., Zafiriou, E.: Nonlinear Quadratic Dynamic Matrix Control with state estimation. Industrial and Engineering Chemistry Research 31, 1096–1104 (1992)
79. Gawkowski, P., Ławryńczuk, M., Marusak, P., Sosnowski, J., Tatjewski, P.: Fail-bounded implementations of the numerical model predictive control algorithms. Control and Cybernetics 39, 1117–1134 (2010)
80. Goethals, I., Pelckmans, K., Suykens, J.A.K., De Moor, B.: Identification of MIMO Hammerstein models using least squares support vector machines. Automatica 41, 1263–1272 (2005)
81. Golub, G.H., Van Loan, C.F.: Matrix computations. The Johns Hopkins University Press, Baltimore (1989)
82. Gómez, J.C., Jutan, A., Baeyens, E.: Wiener model identification and predictive control of a pH neutralisation process. Proceedings IEE, Part D, Control Theory and Applications 151, 329–338 (2004)
83. Greco, C., Menga, G., Mosca, E., Zappa, G.: Performance improvement of self tuning controllers by multistep horizons: the MUSMAR approach. Automatica 20, 681–700 (1984)
84. Grimm, G., Messina, M.J., Tuna, S.E., Teel, A.R.: Nominally robust model predictive control with state constraints. IEEE Transactions on Automatic Control 52, 1856–1870 (2007)
85. Grimm, G., Messina, M.J., Tuna, S.E., Teel, A.R.: Examples when nonlinear model predictive control is nonrobust. Automatica 40, 1729–1738 (2004)
86. Grossmann, I.E., Biegler, L.T.: Part II. Future perspective on optimization. Computers and Chemical Engineering 28, 1193–1218 (2004)
87. Gruber, J.K., Ramírez, D.R., Álamo, T., Camacho, E.F.: Min-max MPC based on an upper bound of the worst case cost with guaranteed stability. Application to a Pilot Plant 21, 194–204 (2011)

88. Gruber, J.K., Guzmán, J.L., Rodríguez, F., Bordons, C., Berenguel, M., Sánchez, J.A.: Nonlinear MPC based on a Volterra series model for greenhouse temperature control using natural ventilation. Control Engineering Practice 19, 354–366 (2011)
89. Gruber, J.K., Doll, M., Bordons, C.: Design and experimental validation of a constrained MPC for the air feed of a fuel cell. Control Engineering Practice 17, 874–885 (2009)
90. Gruber, J.K., Doll, M., Bordons, C.: Control of a pilot plant using QP based min-max predictive control. Control Engineering Practice 17, 1358–1366 (2009)
91. Gruber, J.K., Ramírez, D.R., Álamo, T., Bordons, C.: Nonlinear min-max model predictive control based on Volterra models. Application to a pilot plant. In: Proceedings of the European Control Conference (ECC 2009), Budapest, pp. 1112–1117 (2009)
92. Haseltine, E.L., Rawlings, J.B.: Critical evaluation of extended Kalman filtering and moving-horizon estimation. Industrial and Engineering Chemical Research 44, 2451–2460 (2005)
93. Hassibi, B., Stork, B.: Second order derivatives for network prunning: Optimal brain surgeon. In: Touretzky, D.S. (ed.) Advances of Neural Information Processing Systems, vol. 5, Morgan Kaufmann, San Mateo (1993)
94. Haykin, S.: Neural networks–a comprehensive foundation. Prentice Hall, Upper Saddle River (1993)
95. Henson, M.A.: Nonlinear model predictive control: current status and future directions. Computers and Chemical Engineering 23, 187–202 (1998)
96. Henson, M.A., Seborg, D.E.: Adaptive nonlinear control of a pH neutralization process. IEEE Transactions on Control System Technology 2, 169–182 (1994)
97. Henson, M.A., Seborg, D.E.: Theoretical analysis of unconstrained nonlinear model predictive control. International Journal of Control 58, 1053–1080 (1993)
98. Hornik, K., Stinchcombe, M., White, H.: Multilayer feedforward networks are universal approximators. Neural Networks 2, 359–366 (1989)
99. Hosen, M.A., Hussain, M.A., Mjalli, F.S.: Control of polystyrene batch reactors using neural network based model predictive control (NNMPC): an experimental investigation. Control Engineering Practice 19, 454–467 (2011)
100. Hu, Q., Saha, P., Rangaiah, G.P.: Experimental evaluation of an augmented IMC for nonlinear systems. Control Engineering Practice 8, 1167–1176 (2000)
101. Hussain, M.A.: Review of the applications of neural networks in chemical process control–simulation and online implementation. Artificial Intelligence in Engineering 13, 55–68 (1999)
102. Iplikci, S.: Support vector machines-based generalized predictive control. International Journal of Robust and Nonlinear Control 16, 843–862 (2006)
103. Janczak, A.: Identification of nonlinear systems using neural networks and polynomial models: block oriented approach. LNCIS, vol. 310. Springer, Heidelberg (2004)
104. Jiang, Z.P., Wang, Y.: Input-to-state stability for discrete-time nonlinear systems. Automatica 37, 857–869 (2001)
105. Johansen, T.A.: Approximate explicit receding horizon control of constrained nonlinear systems. Automatica 40, 293–300 (2004)
106. Juditsky, A., Hjalmarsson, H., Benveniste, A., Delyon, B., Ljung, L., Sjöoberg, J., Zhang, O.: Nonlinear black-box models in system identification: mathematical foundations. Automatica 31, 1725–1750 (1995)

107. Kalman, R.E., Bertram, J.E.: Control system analysis and design via the "second method" of Lyapunov. Transactions of ASME, Journal of Basic Engineering 82, 371–400 (1960)
108. Kassmann, D.E., Badgwell, T.A., Hawkins, R.B.: Robust steady-state target calculation for model predictive control. AIChE Journal 46, 1007–1024 (2000)
109. Keerthi, S.S., Gilbert, E.G.: Optimal infinite-horizon feedback laws for a general class of constrained discrete-time systems: stability and moving-horizon approximations. Journal of Optimization Theory and Applications 57, 265–293 (1988)
110. Keviczky, T., Balas, G.J.: Receding horizon control of an F-16 aircraft: a comparative study. Control Engineering Practice 14, 1023–1033 (2006)
111. de Keyser, R.M.C., van Cauvenberghe, A.R.: Extended predictive self adaptive control. In: Proceedings of the 7th IFAC Symposium on Identification and System Parameter Identification, York, pp. 1255–1260 (1985)
112. Khalil, H.K.: Nonlinear systems. Prentice Hall, Upper Saddle River (1996)
113. Khodadadi, H., Jazayeri-Rad, H.: Design of a computationally efficient observer-based nonlinear model predictive control for a continuous stirred tank reactor. In: Proceedings of the IEEE Conference on Intelligent Computing and Intelligent Systems, Xianmen, paper T27–001 (2011)
114. Kim, Y.H., Kwon, W.H.: An application of min-max generalized predictive control to sintering process. Control Engineering Practice 6, 999–1007 (1998)
115. Kiran, A.U.M., Jana, A.K.: Control of continuous fed-batch fermentation process using neural network based model predictive controller. Bioprocess and Biosystems Engineering 32, 801–808 (2009)
116. Kittisupakorn, P., Thitiyasook, P., Hussain, M.A., Daosud, W.: Neural network based model predictive control for a steel pickling process. Journal of Process Control 19, 579–590 (2009)
117. Kleinman, D.: An easy way to stabilize a linear constant system. IEEE Transactions on Automatic Control 15, 692
118. Korbicz, J., Kościelny, J.M. (eds.): Modeling, diagnostics and process control: implementation in the DiaSter System. Springer, Berlin (2010)
119. Korbicz, J., Obuchowicz, A., Uciśki, D.: Artificial neural networks: princiles and applications. PLJ, Warsaw (1994) (in Polish)
120. Kothare, V.M., Balakrishnan, V., Morari, M.: Robust constrained model predictive control using linear matrix inequalities. Automatica 32, 1361–1379 (1996)
121. Kouvaritakis, B., Rossiter, J.A., Schuurmans, J.: Efficient robust predictive control. IEEE Transactions on Automatic Control 45, 1545–1549 (2000)
122. Kuure-Kinsey, M., Cutright, R., Bequette, B.W.: Computationally efficient neural predictive control based on a feedforward architecture. Industrial and Engineering Chemistry Research 45, 8575–8582 (2006)
123. Kwon, W.H., Pearson, A.E.: A modified quadratic cost problem and feedback stabilization of a linear system. IEEE Transactions on Automatic Control 22, 838–842 (1977)
124. Lautenschlager, M.L.F., Odloak, D.: Constrained multivariable control of fluid catalytic cracking converters. Journal of Process Control 5, 29–39 (1995)
125. Lazar, M., Muñoz de la Peña, D., Heemels, W.P.M.H., Álamo, T.: On input-to-state stability of min-max nonlinear model predictive control. Systems & Control Letters 57, 39–48 (2008)

126. Le Cun, Y., Denker, J., Solla, S.: Optimal brain damage. In: Advances in Neural Information Processing Systems, vol. 2, pp. 598–605. Morgan Kaufmann, San Mateo (1990)
127. Lee, E.B., Markus, L.: Foundations of optimal control theory. John Wiley & Sons, New York (1967)
128. Lee, J.H., Yu, Z.: Worst-case formulations of model predictive control for systems with bounded parameters. Automatica 33, 763–781 (1997)
129. Lee, J.H., Ricker, N.L.: Extended Kalman filter based nonlinear model predictive control. Industrial and Engineering Chemistry Research 33, 1530–1541 (1994)
130. Lefkowitz, I.: Multilevel approach applied to control system design. Transactions of ASME, Journal of Basic Engineering 88, 392–398 (1966)
131. Leontaritis, I.J., Billings, S.A.: Input–output parametric models for non-linear systems–Part I: deterministic non-linear systems. International Journal of Control 41, 303–328 (1985)
132. Li, W.C., Biegler, L.T.: Multistep, Newton-type control strategies for constrained, non-linear systems. Chemical Engineering Research and Design 67, 562–577 (1989)
133. Li, W.C., Biegler, L.T.: Process control strategies for constrained nonlinear systems. Industrial and Engineering Chemistry Research 27, 1421–1433 (1988)
134. Limón, D., Bravo, J.M., Álamo, T., Camacho, E.F.: Robust MPC of constrained nonlinear systems based on interval arithmetic. IEE Proceedings, Part D: Control Theory and Applications 152, 325–332 (2005)
135. Limón Marruedo, D., Álamo, T., Camacho, E.F.: Input-to-state stable MPC for constrained discrete-time nonlinear systems with bounded additive uncertainties. In: Proceedings of the 41th IEEE Conference on Decision and Control (CDC 2002), Las Vegas, pp. 4619–4624 (2002)
136. Ling, W.M., Rivera, D.: Nonlinear black-box identification of distillation column models – design variable selection for model performance enhancement. International Jurnal of Applied Mathematics and Computer Science 8, 793–813 (1998)
137. Liu, D., Shah, S.L., Fisher, D.G.: Multiple prediction models for long range predictive control. In: Proceedings of the 14th IFAC World Congress, Beijing, pp. 175–180 (1999)
138. Liu, G.P., Kadirkamanathan, V., Billings, S.A.: Predictive control for nonlinear systems using neural networks. International Journal of Control 71, 1119–1132 (1998)
139. Liu, S., Wang, J.: A simplified dual neural network for quadratic programming with its KWTA application. IEEE Transactions on Neural Networks 17, 1500–1510 (2006)
140. Lu, C.H., Tsai, C.C.: Adaptive predictive control with recurrent neural network for industrial processes: an application to temperature control of a variable-frequency oil-cooling machine. IEEE Transactions on Industrial Electronics 55, 1366–1375 (2008)
141. Luyben, W.L.: Process modelling, simulation and control for chemical engineers. McGraw Hill, New York (1990)
142. Ławryńczuk, M.: Explicit nonlinear predictive control algorithms with neural approximation. Neurocomputing (2014) (accepted for publication)
143. Ławryńczuk, M.: Practical nonlinear predictive control algorithms for neural Wiener models. Journal of Process Control 23, 696–714 (2013)

144. Ławryńczuk, M.: Nonlinear predictive control based on Least Squares Support Vector Machines Hammerstein models. In: Tomassini, M., Antonioni, A., Daolio, F., Buesser, P. (eds.) ICANNGA 2013. LNCS, vol. 7824, pp. 246–255. Springer, Heidelberg (2013)
145. Ławryńczuk, M.: On-line trajectory-based linearisation of neural models for a computationally efficient predictive control algorithm. In: Rutkowski, L., Korytkowski, M., Scherer, R., Tadeusiewicz, R., Zadeh, L.A., Zurada, J.M. (eds.) ICAISC 2012, Part I. LNCS, vol. 7267, pp. 126–134. Springer, Heidelberg (2012)
146. Ławryńczuk, M.: On improving accuracy of computationally efficient nonlinear predictive control based on neural models. Chemical Engineering Science 66, 5253–5267 (2011)
147. Ławryńczuk, M.: Online set-point optimisation cooperating with predictive control of a yeast fermentation process: A neural network approach. Engineering Applications of Artificial Intelligence 24, 968–982 (2011)
148. Ławryńczuk, M.: On-line set-point optimisation and predictive control using neural Hammerstein models. Chemical Engineering Journal 166, 269–287 (2011)
149. Ławryńczuk, M.: Accuracy and computational efficiency of suboptimal nonlinear predictive control based on neural models. Applied Soft Computing 11, 2202–2215 (2011)
150. Ławryńczuk, M.: Nonlinear predictive control based on multivariable neural Wiener models. In: Dobnikar, A., Lotrič, U., Šter, B. (eds.) ICANNGA 2011, Part I. LNCS, vol. 6593, pp. 31–40. Springer, Heidelberg (2011)
151. Ławryńczuk, M.: Predictive control of a distillation column using a control-oriented neural model. In: Dobnikar, A., Lotrič, U., Šter, B. (eds.) ICANNGA 2011, Part I. LNCS, vol. 6593, pp. 230–239. Springer, Heidelberg (2011)
152. Ławryńczuk, M.: Robust nonlinear MPC algorithm with quadratic optimisation. In: Proceedings of the 17th National Control Conference (KKA 2011), Kielce-Cedzyna, pp. 685–696 (2011) (in Polish)
153. Ławryńczuk, M.: Computationally efficient nonlinear predictive control based on neural Wiener models. Neurocomputing 74, 401–417 (2010)
154. Ławryńczuk, M.: Training of neural models for predictive control. Neurocomputing 73, 1332–1343 (2010)
155. Ławryńczuk, M.: Suboptimal nonlinear predictive control based on multivariable neural Hammerstein models. Applied Intelligence 32, 173–192 (2010)
156. Ławryńczuk, M.: Dynamic Matrix Control algorithm based on interpolated step response neural models. In: Rutkowski, L., Scherer, R., Tadeusiewicz, R., Zadeh, L.A., Zurada, J.M. (eds.) ICAISC 2010, Part II. LNCS, vol. 6114, pp. 297–304. Springer, Heidelberg (2010)
157. Ławryńczuk, M.: Neural Dynamic Matrix Control algorithm with disturbance compensation. In: García-Pedrajas, N., Herrera, F., Fyfe, C., Benítez, J.M., Ali, M. (eds.) IEA/AIE 2010, Part III. LNCS, vol. 6098, pp. 52–61. Springer, Heidelberg (2010)
158. Ławryńczuk, M.: Computationally efficient nonlinear predictive control based on state-space neural models. In: Wyrzykowski, R., Dongarra, J., Karczewski, K., Wasniewski, J. (eds.) PPAM 2009, Part I. LNCS, vol. 6067, pp. 350–359. Springer, Heidelberg (2010)

159. Ławryńczuk, M., Tatjewski, P.: Nonlinear predictive control based on neural multi-models. International Journal of Applied Mathematics and Computer Science 20, 7–21 (2010)
160. Ławryńczuk, M., Tatjewski, P.: Approximate neural economic set-point optimisation for control systems. In: Rutkowski, L., Scherer, R., Tadeusiewicz, R., Zadeh, L.A., Zurada, J.M. (eds.) ICAISC 2010, Part II. LNCS, vol. 6114, pp. 305–312. Springer, Heidelberg (2010)
161. Ławryńczuk, M., Marusak, P., Tatjewski, P.: Efficient predictive control algorithms based on soft computing approaches: application to glucose concentration stabilization. In: Novel Algorithms and Techniques in Telecommunications, Automation and Industrial Electronics, pp. 425–430. Springer, Berlin (2010)
162. Ławryńczuk, M.: Explicit nonlinear predictive control of a distillation column based on neural models. Chemical Engineering and Technology 32, 1578–1587 (2009)
163. Ławryńczuk, M.: Efficient nonlinear predictive control of a biochemical reactor using neural models. Bioprocess and Biosystems Engineering 32, 301–312 (2009)
164. Ławryńczuk, M.: Efficient nonlinear predictive control based on structured neural models. International Journal of Applied Mathematics and Computer Science 19, 233–246 (2009)
165. Ławryńczuk, M.: Neural networks in model predictive control. In: Nguyen, N.T., Szczerbicki, E. (eds.) Intelligent Systems for Knowledge Management. SCI, vol. 252, pp. 31–63. Springer, Heidelberg (2009)
166. Ławryńczuk, M.: A predictive control economic optimiser and constraint governor based on neural models. In: Kolehmainen, M., Toivanen, P., Beliczynski, B. (eds.) ICANNGA 2009. LNCS, vol. 5495, pp. 79–88. Springer, Heidelberg (2009)
167. Ławryńczuk, M.: Computationally efficient nonlinear predictive control based on RBF neural multi-models. In: Kolehmainen, M., Toivanen, P., Beliczynski, B. (eds.) ICANNGA 2009. LNCS, vol. 5495, pp. 89–98. Springer, Heidelberg (2009)
168. Ławryńczuk, M., Marusak, P., Tatjewski, P.: On cooperation of set-point optimisation and predictive control based on Hammerstein models. In: Proceedings of the 7th Workshop on Advanced Control and Diagnosis (ACD 2009), Zielona Góra, paper 88 (2009)
169. Ławryńczuk, M.: Modelling and nonlinear predictive control of a yeast fermentation biochemical reactor using neural networks. Chemical Engineering Journal 145, 290–307 (2008)
170. Ławryńczuk, M.: Suboptimal nonlinear predictive control with MIMO neural Hammerstein models. In: Nguyen, N.T., Borzemski, L., Grzech, A., Ali, M. (eds.) IEA/AIE 2008. LNCS (LNAI), vol. 5027, pp. 225–234. Springer, Heidelberg (2008)
171. Ławryńczuk, M.: RBF neural models in steady-state target optimisation and predictive control. In: Malinowski, K., Rutkowski, L. (eds.) Proceedings of the 16th National Control Conference (KKA 2008), Szczyrk. Recent advances in control and automation, pp. 353–362. EXIT, Warsaw (2008)
172. Ławryńczuk, M., Marusak, P., Tatjewski, P.: Cooperation of model predictive control with steady-state economic optimisation. Control and Cybernetics 37, 133–158 (2008)

173. Ławryńczuk, M., Marusak, P., Tatjewski, P.: Piecewise linear steady-state target optimization for control systems with MPC: a case study. In: Proceedings of the 17th IFAC World Congress, Soeul, pp. 13169–13174 (2008)
174. Ławryńczuk, M., Marusak, P., Tatjewski, P.: Optimising predictive controllers in control systems with constraints. In: Malinowski, K., Rutkowski, L. (eds.) Proceedings of the 16th National Control Conference (KKA 2008), Szczyrk. Control and automatisation: current problems and their solutions, pp. 437–446. EXIT, Warsaw (2008) (in Polish)
175. Ławryńczuk, M., Tadej, W.: Acomputationally efficient stable dual-mode type nonlinear predictive control algorithm. Control and Cybernetics 37, 99–132 (2008)
176. Ławryńczuk, M., Tatjewski, P.: Efficient predictive control integrated with economic optimisation based on neural models. In: Rutkowski, L., Tadeusiewicz, R., Zadeh, L.A., Zurada, J.M. (eds.) ICAISC 2008. LNCS (LNAI), vol. 5097, pp. 111–122. Springer, Heidelberg (2008)
177. Ławryńczuk, M.: A family of model predictive control algorithms with artificial neural networks. International Jurnal of Applied Mathematics and Computer Science 17, 217–232 (2007)
178. Ławryńczuk, M.: Advanced predictive control of a distillation column with neural models. Archives of Control Sciences 17, 121–148 (2007)
179. Ławryńczuk, M., Marusak, P., Tatjewski, P.: Multilayer and integrated structures for predictive control and economic optimisation. In: Proceedings of the 11th IFAC/IFORS/IMACS/IFIP Symposium on Large Scale Systems: Theory and Applications (LSS 2007), Gdańsk paper 60, vol. 60 (2007)
180. Ławryńczuk, M., Marusak, P., Tatjewski, P.: Economic efficacy of multilayer constrained predictive control structures: an application to a MIMO neutralisation reactor. In: Proceedings of the 11th IFAC/IFORS/IMACS/IFIP Symposium on Large Scale Systems: Theory and Applications (LSS 2007), Gdańsk paper 93 (2007)
181. Ławryńczuk, M., Marusak, P., Tatjewski, P.: Efficient model predictive control integrated with economic optimisation. In: Proceedings of the 15th Mediterranean Conference on Control and Automation (MED 2007), Athens, paper T27–001 (2007)
182. Ławryńczuk, M., Marusak, P., Tatjewski, P.: An efficient MPC algorithm integrated with economic optimisation for MIMO systems. In: Proceedings of the 13th IEEE International Conference on Methods and Models in Automation and Robotics (MMAR 2007), Szczecin, pp. 295–302 (2007)
183. Ławryńczuk, M., Marusak, P., Tatjewski, P.: Set-point optimisation and predictive constrained control for fast feedback controlled processes. In: Proceedings of the 13th IEEE International Conference on Methods and Models in Automation and Robotics (MMAR 2007) Szczecin, Szczecin, pp. 357–362 (2007)
184. Ławryńczuk, M., Tatjewski, P.: A computationally efficient nonlinear predictive control algorithm with RBF neural models and its application. In: Kryszkiewicz, M., Peters, J.F., Rybiński, H., Skowron, A. (eds.) RSEISP 2007. LNCS (LNAI), vol. 4585, pp. 603–612. Springer, Heidelberg (2007)
185. Ławryńczuk, M., Tatjewski, P.: An efficient nonlinear predictive control algorithm with neural models and its application to a high-purity distillation process. In: Rutkowski, L., Tadeusiewicz, R., Zadeh, L.A., Żurada, J.M. (eds.) ICAISC 2006. LNCS (LNAI), vol. 4029, pp. 76–85. Springer, Heidelberg (2006)

186. Maciejowski, J.M.: Predictive control with constraints. Prentice Hall, Harlow (2002)
187. Magni, L., Scattolini, R.: Robustness and robust design of MPC for nonlinear discrete-time systems. In: Findeisen, R., Allgöwer, F., Biegler, L.T. (eds.) Assessment and Future Directions of Nonlinear Model Predictive Control. LNCIS, vol. 358, pp. 239–254. Springer, Berlin (2007)
188. Magni, L., de Nicolao, G., Scattolini, R., Allgöwer, F.: Robust model predictive control for nonlinear discrete-time systems. International Journal of Robust and Nonlinear Control 13, 229–246 (2003)
189. Magni, L., de Nicolao, G., Magnani, L., Scattolini, R.: A stabilizing model-based predictive control algorithm for nonlinear systems. Automatica 37, 1351–1362 (2001)
190. Mahfouf, M., Linkens, D.A.: Non-linear generalized predictive control (NL-GPC) applied to muscle relaxant anaesthesia. International Journal of Control 71, 239–257 (1998)
191. Maner, B.R., Doyle, F.J., Ogunnaike, B.A., Pearson, R.K.: Nonlinear model predictive control of a simulated multivariable polymerization reactor using second-order Volterra models. Automatica 32, 1285–1301 (1996)
192. Marami, B., Haeri, M.: Implementation of MPC as an AQM controller. Computer Communications 33, 227–239 (2010)
193. Marlin, T.E.: Process control. McGraw-Hill, New York (1995)
194. Marquis, P., Broustail, J.P.: SMOC, a bridge between state space and model predictive controllers: application to the automation of hydrotreating unit. In: Proceedings of the 1988 IFAC Workshop on Model Based Process Control, Oxford, pp. 37–43 (1988)
195. Martinsen, F., Biegler, L.T., Foss, B.A.: A new optimization algorithm with application to nonlinear MPC. Journal of Process Control 14, 853–865 (2004)
196. Marusak, P.M.: Efficient predictive control and set–point optimization based on a single fuzzy model. In: Dobnikar, A., Lotrič, U., Šter, B. (eds.) ICANNGA 2011, Part II. LNCS, vol. 6594, pp. 215–224. Springer, Heidelberg (2011)
197. Marusak, P.: On prediction generation in efficient MPC algorithms based on fuzzy Hammerstein models. In: Rutkowski, L., Scherer, R., Tadeusiewicz, R., Zadeh, L.A., Zurada, J.M. (eds.) ICAISC 2010, Part I. LNCS, vol. 6113, pp. 136–143. Springer, Heidelberg (2010)
198. Marusak, P.: Application of fuzzy Wiener models in efficient MPC algorithms. In: Szczuka, M., Kryszkiewicz, M., Ramanna, S., Jensen, R., Hu, Q. (eds.) RSCTC 2010. LNCS, vol. 6086, pp. 669–677. Springer, Heidelberg (2010)
199. Marusak, P.: Advantages of an easy to design fuzzy predictive algorithm in control systems of nonlinear chemical reactors. Applied Soft Computing 9, 1111–1125 (2009)
200. Marusak, P.: Efficient model predictive control algorithm with fuzzy approximations of nonlinear models. In: Kolehmainen, M., Toivanen, P., Beliczynski, B. (eds.) ICANNGA 2009. LNCS, vol. 5495, pp. 448–457. Springer, Heidelberg (2009)
201. Marusak, P.M.: Efficient fuzzy predictive economic set–point optimizer. In: Rutkowski, L., Tadeusiewicz, R., Zadeh, L.A., Zurada, J.M. (eds.) ICAISC 2008. LNCS (LNAI), vol. 5097, pp. 273–284. Springer, Heidelberg (2008)

202. Marusak, P.: Analytical predictive controllers with efficient handling of output constraints. In: Malinowski, K., Rutkowski, L. (eds.) Proceedings of the 16th National Control Conference (KKA 2008), Szczyrk. Recent advances in control and automation, pp. 130–140. EXIT, Warsaw (2008)
203. Mayne, D.Q., Rawlings, J.B., Rao, C.V., Scokaert, P.O.M.: Constrained model predictive control: stability and optimality. Automatica 36, 789–814 (2000)
204. Mayne, D.Q., Michalska, H.: Receding horizon control of nonlinear systems. IEEE Transactions on Automatic Control 35, 814–824 (1990)
205. McCulloch, W.S., Pitts, W.H.: A logical calculus of ideas immanent in nervous activity. The Bulletin of Mathematical Biophysics 5, 115–133 (1943)
206. Meadows, E.S., Rawlings, J.B.: Receding horizon control with an infinite horizon. In: Proceedings of the American Control Conference (ACC 1993), San Francisco, pp. 2926–2930 (1993)
207. Megías, D., Serrano, J., El Ghoumari, M.Y.: Extended linearised predictive control: practical control algorithms for non-linear systems. In: Proceedings of the European Control Conference (ECC 1999), Karlsruhe, paper F0883 (1999)
208. Michalska, H., Mayne, D.Q.: Moving horizon observers and observer-based control. IEEE Transactions on Automatic Control 40, 995–1006 (1995)
209. Michalska, H., Mayne, D.Q.: Robust receding horizon control of constrained nonlinear systems. IEEE Transactions on Automatic Control 38, 1623–1633 (1993)
210. Milanese, M., Novara, C.: Set membership identification of nonlinear systems. Automatica 40, 957–975 (2004)
211. Milman, R., Davison, E.J.: A fast MPC algorithm using nonfeasible active set methods. Journal of Optimization Theory and Applications 139, 591–616 (2008)
212. Minsky, M., Papert, S.: Perceptrons. MIT Press, Cambridge (1969)
213. Mjalli, F.S.: Adaptive and predictive control of liquid-liquid extractors using neural-based instantaneous linearization technique. Chemical Engineering and Technology 29, 539–549 (2006)
214. Mohanty, S.: Artificial neural network based system identification and model predictive control of a flotation column. Journal of Process Control 19, 991–999 (2009)
215. Morari, M., Lee, J.H.: Model predictive control: past, present and future. Computers and Chemical Engineering 23, 667–682 (1999)
216. Moore, E.: Introduction to interval analysis. Prentice Hall, Upper Saddle River (1996)
217. Mosca, E., Zhang, J.: Stable redesign of predictive control. Automatica 28, 1229–1233 (1992)
218. Mu, J., Rees, D., Liu, G.P.: Advanced controller design for aircraft gas turbine engines. Control Engineering Practice 13, 1001–1015 (2005)
219. Muske, K.R., Rawlings, J.B.: Model predictive control with linear models. AIChE Journal 39, 262–287 (1993)
220. Muske, K.R., Rawlings, J.B.: Implementation of a stabilizing constrained receding horizon regulator. In: Proceedings of the American Control Conference (ACC 1992), Chicago, pp. 1594–1595 (1992)
221. Nagy, Z.K.: Model based control of a yeast fermentation bioreactor using optimally designed artificial neural networks. Chemical Engineering Journal 127, 95–109 (2007)

222. Narendra, K.S., Parthasarathy, K.: Identification and control of dynamical systems using neural networks. IEEE Transactions on Neural Networks 1, 4–27 (1990)
223. Nelles, O.: Nonlinear system identification. From classical approaches to neural networks and fuzzy models. Springer, Berlin (2001)
224. de Nicolao, G., Magni, L., Scattolini, R.: Stabilizing receding-horizon control of non-linear time-varying systems. IEEE Transactions on Automatic Control 43, 1030–1036 (1998)
225. Niederliński, A., Mościński, J., Ogonowski, Z.: Adaptive control. PWN, Warsaw (1995) (in Polish)
226. Nocedal, J., Wright, S.J.: Numerical optimization. Springer, Berlin (2006)
227. Nørgaard, M., Ravn, O., Poulsen, N.K., Hansen, L.K.: Neural networks for modelling and control of dynamic systems. Springer, London (2000)
228. Nørgaard, M., Poulsen, N.K., Ravn, O.: New developments in state estimation for nonlinear systems. Automatica 36, 1627–1638 (2000)
229. Norquay, S.J., Palazoğlu, A., Romagnoli, J.A.: Application of Wiener model predictive control (WMPC) to an industrial C2 splitter. Journal of Process Control 9 (1999)
230. Norquay, S.J., Palazoğlu, A., Romagnoli, J.A.: Model predictive control based on Wiener models. Chemical Engineering Science 53, 75–84 (1998)
231. de Oliveira, N.M.C., Biegler, L.T.: An extension of Newton-type algorithms for nonlinear process control. Automatica 31, 281–286 (1995)
232. de Oliveira Kothare, S.L., Morari, M.: Contractive model predictive control for constrained nonlinear systems. IEEE Transactions on Automatic Control 45, 1053–1071 (2000)
233. Onnen, C., Babuška, R., Kaymak, U., Sousa, J.M., Verbruggen, H.B., Iserman, R.: Genetic algorithms for optimisation in predictive control. Control Engineering Practice 5, 1363–1372 (1997)
234. Ortega, J.G., Camacho, E.F.: Mobile robot navigation in a partially structured static environment, using neural predictive control. Control Engineering Practice 4, 1669–1679 (1996)
235. Osowski, S.: Neural networks for information processing. Warsaw University of Technology Press, Warsaw (2006) (in Polish)
236. Ou, J., Rhinehart, R.R.: Grouped-neural network modeling for model-predictive control. Control Engineering Practice 11, 723–732 (2003)
237. Ou, J., Rhinehart, R.R.: Grouped-neural network modeling for model predictive control. ISA Transactions 41, 195–202 (2002)
238. Pan, Y., Wang, J.: Two neural network approaches to model predictive control. In: Proceedings of the American Control Conference (ACC 2008), Washington, pp. 1685–1690 (2008)
239. Pan, Y., Wang, J.: Nonlinear model predictive control using a recurrent neural network. In: Proceedings of the International Joint Conference on Neural Networks (IJCNN 2008), Hong Kong, pp. 2296–2301 (2008)
240. Pan, Y., Wang, J.: Robust model predictive control using a discrete-time recurrent neural network. In: Sun, F., Zhang, J., Tan, Y., Cao, J., Yu, W. (eds.) ISNN 2008, Part I. LNCS, vol. 5263, pp. 883–892. Springer, Heidelberg (2008)
241. Pan, Y., Sung, S.W., Lee, J.H.: Data-based construction of feedback-corrected nonlinear prediction model using feedback neural networks. Control Engineering Practice 9, 859–864 (2001)

242. Parisini, T., Sanguineti, M., Zoppoli, R.: Nonlinear stabilization by receding-horizon neural regulators. International Journal of Control 70, 341–362 (1998)
243. Park, J., Sandberg, I.W.: Universal approximation using radial-basis-function networks. Neural Computation 3, 246–257 (1991)
244. Patan, K.: Artificial neural networks for the modelling and fault diagnosis of technical processes. LNCIS, vol. 377. Springer, Heidelberg (2008)
245. Patan, K., Witczak, M., Korbicz, J.: Towards robustness in neural network based fault diagnosis. International Journal of Applied Mathematics and Computer Science 18, 443–454 (2008)
246. Patrinos, P., Sopasakis, P., Sarimveis, H.: A global piecewise smooth Newton method for fast large-scale model predictive control. Automatica 47, 2016–2022 (2011)
247. Patwardhan, R.S., Lakshminarayanan, S., Shah, S.L.: Constrained nonlinear MPC using Hammerstein and Wiener models: PSL framework. AIChE Journal 44, 1611–1622 (1998)
248. Pearson, R.K.: Selecting nonlinear model structures for computer control. Journal of Process Control 13, 1–26 (2003)
249. Pearson, R.K., Pottmann, M.: Gray-box identification of block-oriented nonlinear models. Journal of Process Control 10, 301–315 (2000)
250. Peterka, V.: Predictor-based self-tuning control. Automatica 20, 39–50 (1984)
251. Piche, S., Sayyar-Rodsari, B., Johnson, D., Gerules, M.: Nonlinear model predictive control using neural networks. IEEE Control Systems Magazine 20, 56–62 (2000)
252. Pin, G., Raimondo, D.M., Magni, L., Parisini, T.: Robust model predictive control of nonlinear systems with bounded and state-dependent uncertainties. IEEE Transactions on Automatic Control 54, 1681–1687 (2009)
253. Polak, E., Yang, T.H.: Moving horizon control of linear systems with input saturation and plant uncertainty. International Journal of Control 58, 613–663 (1993)
254. Potočnik, P., Grabec, I.: Nonlinear model predictive control of a cutting process. Neurocomputing 43, 107–126 (2002)
255. Pottmann, M., Seborg, D.E.: Anonlinear predictive control strategy based on radial basis function models. Computers and Chemical Engineering 21, 965–980 (1997)
256. Powell, M.J.D.: Variable metric methods for constrained optimization. In: Bachem, A., Grötschel, M., Korte, B. (eds.) Mathematical Programming: the State of the Art, pp. 288–311. Springer, New York (1983)
257. Powell, M.J.D.: A fast algorithm for nonlinearly constrained optimization calculations. In: Watson, G. (ed.) Numerical Analysis. Lecture Notes in Mathematics, vol. 630, pp. 144–157. Springer, Dundee (1978)
258. Prasad, V., Bequette, B.W.: Nonlinear system identification and model reduction using artificial neural networks. Computers and Chemical Engineering 27, 1741–1754 (2003)
259. Press, W.H., Teukolsky, S.A., Vettering, W.T., Flannery, B.P.: Numerical recipes in C: the art of scientific computing. Cambridge University Press, Cambridge (1992)
260. Propoi, A.I.: Use of linear programming methods for synthesizing sampled-data automatic systems. Automation and Remote Control 24, 837–844 (1963)
261. Psichogios, D.C., Ungar, L.H.: A hybrid neural network-first principles approach to process modelling. AIChE Journal 38, 1499–1511 (1992)

262. Qin, S.J., Badgwell, T.A.: A survey of industrial model predictive control technology. Control Engineering Practice 11, 733–764 (2003)
263. Rafal, M.D., Stevens, W.F.: Discrete dynamic optimization applied to on-line optimal control. AIChE Journal 14, 85–91 (1968)
264. Raimondo, D.M., Limón, D., Álamo, T., Magni, L.: Robust model predictive control algorithms for nonlinear systems: an input-to-state stability approach. In: Zheng, T. (ed.) Model Predictive Control, pp. 87–108. Sciyo, Rijeka (2010)
265. Ramírez, D.R., Álamo, T., Camacho, E.F., Muñoz de la Peña, D.: Min-max MPC based on a computationally efficient upper bound of the worst case cost. Journal of Process Control 16, 511–519 (2006)
266. Ramírez, D.R., Arahal, M.R., Camacho, E.F.: Min-max predictive control of a heat exchanger using a neural network solver. IEEE Transactions on Control Systems Technology 12, 776–786 (2004)
267. Rao, C.V., Rawlings, J.B., Mayne, D.Q.: Constrained state estimation for nonlinear discrete-time systems: stability and moving horizon approximations. IEEE Transactions on Automatic Control 48, 246–258 (2003)
268. Rawlings, J.B., Mayne, D.Q.: Model predictive control: theory and design. Nob Hill Publishing, Madison (2009)
269. Rawlings, J.B.: Tutorial overview of model predictive control. IEEE Control Systems Magazine 20, 38–52 (2000)
270. Rawlings, J.B., Muske, K.R.: The stability of constrained receding horizon control. IEEE Transactions on Automatic Control 38, 1512–1516 (1993)
271. Reinelt, W., Garulli, A., Ljung, L.: Set membership identification of nonlinear systems. Automatica 38, 787–803 (2002)
272. Richalet, J.A., Rault, A., Testud, J.L., Papon, J.: Model predictive heuristic control: application to an industrial processes. Automatica 14, 413–428 (1978)
273. Richalet, J.A., Rault, A., Testud, J.L., Papon, J.: Algorithmic control of industrial processes. In: Proceedings of the 4th IFAC Symposium on Identification and System Parameter Estimation, Tbilisi, pp. 1119–1167 (1976)
274. Ripley, B.D.: Pattern recognition and neural networks. Cambridge University Press, Cambridge (1996)
275. Rocha, T., Paredes, S., de Carvalho, P., Henriques, J.: Prediction of acute hypotensive episodes by means of neural network multi-models. Computers in Biology and Medicine 41, 881–890 (2011)
276. Rosen, J.B.: The gradient projection method for nonlinear programming. Part I. Linear constraints. Journal of the Society for Industrail and Applied Mathematics 8, 181–217 (1960)
277. Rosenblatt, F.: Principle of neurodynamics. Spartan, New York (1962)
278. Rossiter, J.A.: Model-based predictive control. CRC Press, Boca Raton (2003)
279. Rossiter, J.A., Kouvaritakis, B.: Modelling and implicit modelling for predictive control. International Journal of Control 74, 1085–1095 (2001)
280. Rotava, O., Zanin, A.C.: Multivariable control and real-time optimization–an industrial practical view. Hydrocarbon Processing 84, 61–71 (2005)
281. Rohani, S., Haeri, M., Wood, H.C.: Modeling and control of a continuous crystallization process, Part 2. Model predictive control. Computers and Chemical Engineering 23, 279–286 (1999)
282. Rouhani, R., Mehra, R.K.: Model algorithmic control (MAC); basic theoretical properties. Automatica 18, 401–441 (1982)
283. Rutkowski, L.: Methods and techniques of artificial intelligence. PWN, Warsaw (2005)

284. Saez, D., Cipriano, A., Ordys, A.W.: Optimisation of industrial processes at supervisory level: application to control of thermal power plants. Springer, London (2002)
285. Sarabia, D., Capraro, F., Larsen, L.F.S., de Prada, C.: Hybrid NMPC of supermarket display cases. Journal of Process Control 17, 428–441 (2009)
286. Saraswati, S., Chand, S.: Online linearization-based neural predictive control of air-fuel ratio in SI engines with PID feedback correction scheme. Neural Computing and Applications 19, 919–933 (2010)
287. Scattolini, R.: Architectures for distributed and hierarchical Model Predictive Control — A review. Journal of Process Control 19, 723–731 (2009)
288. Scattolini, R., Bittanti, S.: On the choice of the horizon in long-range predictive control–some simple criteria. Automatica 26, 915–917 (1990)
289. Schölkopf, B., Smola, A.: Learning with kernels: support vector machines, regularization, optimization, and beyond. MIT, Cambridge (2001)
290. Scokaert, P.O.M., Rawlings, J.B.: Feasibility issues in linear model predictive control. AIChE Journal 45, 1649–1659 (1999)
291. Scokaert, P.O.M., Mayne, D.Q., Rawlings, J.B.: Suboptimal model predictive control (feasibility implies stability). IEEE Transactions on Automatic Control 44, 648–654 (1999)
292. Scokaert, P.O.M., Mayne, D.Q.: Min-max feedback model predictive control for constrained linear systems. IEEE Transactions on Automatic Control 43, 1136–1142 (1998)
293. Scokaert, P.O.M.: Infinite horizon generalized predictive control. International Journal of Control 66, 161–175 (1997)
294. Scokaert, P.O.M., Clarke, D.W.: Stabilising properties of constrained predictive control. Proceedings IEE, Part D 141, 295–304 (1994)
295. Schittkowski, K.: The nonlinear programming method of Wilson, Han, and Powell with an augmented Lagrangian type line search function, Part 1: convergence analysis. Numerische Mathematik 38, 83–114 (1982)
296. Shafiee, G., Arefi, M.M., Jahed-Motlagh, M.R., Jalali, A.A.: Nonlinear predictive control of a polymerization reactor based on piecewise linear Wiener model. Chemical Engineering Journal 143, 282–292 (2008)
297. Shampine, L.F., Reichelt, M.W.: The MATLAB ODE suite. SIAM Journal on Scientific Computing 18, 1–22 (1997)
298. Shook, D.S., Mohtahdi, C., Shah, S.L.: Identification for long range predictive control. IEE Proceedings, Part D: Control Theory and Applications 138, 75–84 (1991)
299. Simon, D.: Optimal state estimation: Kalman, H_∞, and nonlinear approaches. John Wiley & Sons, New Jersey (2006)
300. Sjöberg, J., Zhang, O., Ljung, L., Benveniste, A., Delyon, B., Glorennec, P., Hjalmarsson, H., Juditsky, A.: Nonlinear black-box modeling in system identification: a unified overview. Automatica 31, 1691–1724 (1995)
301. Skogestad, S.: Plantwide control: the search for the self-optimizing control structure. Journal of Process Control 10, 487–507 (2000)
302. Smith, J.G., Kamat, S., Madhavan, K.P.: Modeling of pH process using wavenet based Hammerstein model. Journal of Process Control 17, 551–561 (2007)
303. Sontag, E.D., Wang, Y.: New characterizations of the input to state stability property. IEEE Transactions on Automatic Control 41, 1283–1294 (1996)

304. De Souza, G., Odloak, D., Zanin, A.C.: Real time optimization (RTO) with model predictive control (MPC). Computers and Chemical Engineering 34, 1999–2006 (2000)
305. Sriniwas, G.R., Arkun, Y.: A global solution to the nonlinear model predictive control algorithms using polynomial ARX models. Computers and Chemical Engineering 21, 431–439 (1997)
306. Stadler, K.S., Poland, J., Gallestey, E.: Model predictive control of a rotary cement kiln. Control Engineering Practice 19, 1–9 (2011)
307. Suárez, L.A.P., Georgieva, P., de Azevedo, S.F.: Nonlinear MPC for fed-batch multiple stages sugar crystallization. Chemical Engineering Research and Design 89, 753–767 (2010)
308. Suykens, J.A.K., Van Gestel, T., De Brabanter, J., De Moor, B., Vandewalle, J.: Least squares support vector machines. World Scientific, Singapore (2002)
309. Tadeusiewicz, R.: Neural networks. Academic Publishing House, Warsaw (1993) (in Polish)
310. Takagi, T., Sugeno, M.: Fuzzy identification of systems and its application to modeling and control. IEEE Transactions on Systems, Man and Cybernetics 15, 116–132 (1985)
311. Tan, Y., van Cauvenberghe, A.R.: Nonlinear one-step-ahead control using neural networks: control strategy and stability design. Automatica 32, 1701–1706 (1996)
312. Tatjewski, P.: Disturbance modeling and state estimation for predictive control with different state-space process models. In: Proceedings of the 18th IFAC World Congress, Milano, pp. 5326–5331 (2011)
313. Tatjewski, P.: Supervisory predictive control and on-line set-point optimization. International Journal of Applied Mathematics and Computer Science 20, 483–495 (2010)
314. Tatjewski, P., Ławryńczuk, M., Marusak, P.: Integrated predictive optimiser and constraint supervisor for processes with basic feedback control algorithm. In: Proceedings of the European Control Conference (ECC 2009), Budapest, pp. 3359–3364 (2009)
315. Tatjewski, P.: Advanced control and on-line process optimization in multilayer structures. Annual Reviews in Control 32, 71–85 (2008)
316. Tatjewski, P.: Advanced control of industrial processes, structures and algorithms. Springer, London (2007)
317. Tatjewski, P., Ławryńczuk, M.: Soft computing in model-based predictive control. International Journal of Applied Mathematics and Computer Science 16, 101–120 (2006)
318. Tatjewski, P., Ławryńczuk, M., Marusak, P.: Linking nonlinear steady-state and target set-point optimisation for model predictive control. In: Proceedings of the International Control Conference Control (ICC 2006), Glasgow, paper 172 (2006)
319. Tatjewski, P., Roberts, P.D.: Newton-like algorithm for integrated system optimization and parameter estimation technique. International Journal of Control 46, 1155–1170 (1987)
320. Temeng, K.O., Schnelle, P.D., McAvoy, T.J.: Model predictive control of an industrial packed bed reactor using neural networks. Journal of Process Control 5 (1995)
321. Tompson, M.L., Kramer, M.A.: Modelling chemical processes using prior knowledge and neural networks. AIChE Journal 40, 1328–1340 (1994)

322. Tøndel, P., Johansen, T.A., Bemporad, A.: An algorithm for multi-parametric quadratic programming and explicit MPC solutions. Automatica 39, 489–497 (2003)
323. Trajanoski, Z., Wach, P.: Neural predictive control for insulin delivery using the subcutaneous route. IEEE Transactions on Biomedical Engineering 45, 1122–1134 (1998)
324. Tvrzská de Gouvêa, M., Odloak, D.: One-layer real time optimization of LPG production in the FCC unit: procedure, advantages and disadvantages. Computers and Chemical Engineering 22, S191– S198 (1998)
325. Vidysagar, M.: Nonlinear systems analysis. Prentice Hall, Englewood Cliffs (1993)
326. Vieira, W.G., Santos, V.M.L., Carvalho, F.R., Pereira, J.A.F.R., Fileti, A.M.F.: Identification and predictive control of a FCC unit using a MIMO neural model. Chemical Engineering and Processing 44, 855–868 (2005)
327. Wang, L.X., Wan, F.: Structured neural networks for constrained model predictive control. Automatica 37, 1235–1243 (2001)
328. Wang, S.W., Yu, D.L., Gomm, J.B., Page, G.F., Douglas, S.S.: Adaptive neural network model based predictive control for air-fuel ratio of SI engines. Engineering Applications of Artificial Intelligence 19, 189–200 (2006)
329. Wang, Y., Boyd, S.: fast model predictive control using online optimization. IEEE Transactions on Control System Technology 18, 267–278 (2010)
330. Wächter, A., Biegler, L.T.: On the implementation of a primal-dual interior point filter line search algorithm for large-scale nonlinear programming. Mathematical Programming 106, 25–57 (2006)
331. Weigl, Z., Oracz, P., Dunalewicz, A., Ławryńczuk, M., Poświata, A., Ilmurzyńska, J., Plesnar, M.: Development of dynamic simulator for chemical engineering processes. Report of Institute of Industrial Chemistry and Institute of Control and Computation Engineering no 106/98, Warsaw (1998) (in Polish)
332. Witczak, M.: Modelling and estimation strategies for fault diagnosis of nonlinear systems: from analytical to soft computing approaches. LNCIS, vol. 354. Springer, Heidelberg (2007)
333. Xiong, Q., Jutan, A.: Grey-box modelling and control of chemical processes. Chemical Engineering Science 57, 1027–1039 (2002)
334. Yang, T.H., Polak, E.: Moving horizon control of nonlinear systems with input saturation and plant uncertainty. International Journal of Control 58, 875–903 (1993)
335. Ydsite, B.E.: Extended horizon adaptive control. In: Proceedings of the 9th IFAC World Congress, Budapest, pp. 133–138 (1984)
336. Yu, D.L., Gomm, J.B.: Implementation of neural network predictive control to a multivariable chemical reactor. Control Engineering Practice 11, 1315–1323
337. Yu, D.W., Yu, D.L.: Multi-rate model predictive control of a chemical reactor based on three neural models. Biochemical Engineering Journal 37, 86–97 (2007)
338. Zamarreño, J.M., Vega, P., Garcia, L.D., Francisco, M.: State-space neural network for modelling prediction and control. Control Engineering Practice 8, 1063–1075 (2000)
339. Zamarreño, J.M., Vega, P.: Neural predictive control. Application to a highly non-linear system. Engineering Applications of Artificial Intelligence 12, 149–158 (1999)

340. Zamarreño, J.M., Vega, P.: State-space neural network. Properties and application. Neural Networks 11, 1099–1112 (1998)
341. Zanin, A.C., Tvrzská de Gouvêa, M., Odloak, D.: Integrating real-time optimization into model predictive controller of the FCC system. Control Engineering Practice 10, 819–831 (2002)
342. Zanin, A.C., Tvrzská de Gouvêa, M., Odloak, D.: Industrial implementation of a real-time optimization strategy for maximizing production of LPG in a FCC unit. Computers in Chemical Engineering 24, 525–531 (2000)
343. Zheng, A.: A computationally efficient nonlinear MPC algorithm. In: Proceedings of the American Control Conference (ACC 1997), Albuquerque, pp. 1623–1627 (1997)
344. Zheng, A., Morari, M.: Stability of model predictive control with mixed constraints. IEEE Transactions on Automatic Control 40, 1818–1823 (1995)
345. Zhu, G.Y., Henson, M.A., Megan, L.: Dynamic modeling and linear model predictive control of gas pipeline networks. Journal of Process Control 11, 129–148 (2001)

Index

Printed in the United States
By Bookmasters